HOME FIELD ADVANTAGE

A CENTURY OF PARTNERSHIP BETWEEN WRIGHT-PATTERSON AIR FORCE BASE AND DAYTON, OHIO, IN THE PURSUIT OF AERONAUTICAL EXCELLENCE

History Office
Aeronautical Systems Center
Air Force Materiel Command
Air Force History and Museums Program
United States Air Force
Wright-Patterson Air Force Base, Ohio

2004

DEDICATION

To the generations of citizens in communities surrounding
Wright-Patterson Air Force Base who, over the decades,
have supported Air Force personnel, their organizations,
and their missions

FOREWORD

Aviation had just got off the ground as I was growing up. I doubt that any children in the world saw more airplanes overhead than those living in the Dayton area. The government's McCook Field north of Dayton was small, and many of its aeronautical activities were managed from offices in downtown Dayton. Military housing was scant at the field, so Army Air Service families lived throughout Dayton. When I was in high school, Jimmy Doolittle used to ride the interurban from Dayton View to McCook Field. It seemed as though the entire Dayton community was involved in collecting money to purchase land that eventually became Wright Field (later Wright-Patterson Air Force Base.) Frederick Patterson, the son of John H., had shown Daytonians the wisdom of donating money to the cause of keeping Army aviation in Dayton. When Wright Field opened, there was a grand celebration and the community was elated that Dayton was and, for the foreseeable future, would be the nation's center of military aviation.

During World War II, work at Wright and Patterson fields increased dramatically as all organizations there ran three eight-hour shifts a day and no desk ever sat empty. The bases could not begin to house all the people and activities that were involved with war work, so once again military families lived in Dayton, many renting rooms in larger houses because of the severe housing shortages. As a young woman during World War II, my friends and I believed that Dayton was the best possible place to live because the field brought many more young men to the area than there were women. Back then the troops drew the USO (United Service Organizations) shows and the big bands. Dayton was a routine stop for the show business greats—I saw Glenn Miller, Benny Goodman, Dorothy Lamour, and literally hundreds of professional shows in Dayton theaters and dance halls. Theaters abounded and, to us, downtown Dayton was as exciting a place to be as Chicago or New York. And all that was against a backdrop of war, and not knowing whether the young man I was dancing with might soon die in combat. Those days were, indeed, as Dickens' said, "...the best of times and the worst of times."

The war also changed Dayton economically. It pulled the area out of the Great Depression and revived our manufacturing industry. With Wright Field as the procurer of all Army Air Forces aircraft, the town was constantly visited by the captains of industry, and we rubbed elbows with them in the Biltmore Hotel restaurant and ballroom. The Dayton Engineers Club became the prime location for the World War II-era equivalent to the "power lunch." All commercial activities benefited by the booming economy and the wartime surge in population. Churches thrived, filled with worshipers trying to make some sense of the concept of "world war."

Following World War II, Dayton and the two airfields, which merged into Wright-Patterson Air Force Base in 1948, sank into a profound Cold War. There was the Korean War and, more chilling still, the Cuban Missile Crisis. By the early 1960s, Americans were well aware of the devastation that a nuclear war would bring—and there were communist missiles sitting just offshore Florida pointed at the United States. At the time, the Strategic Air Command had a unit at Wright-Patterson. The grim joke was that we didn't need to "dunk and cover" in Dayton. The Russian's first strike would be Wright-Patterson and we would all be vaporized immediately. There was a period when aircraft from the Strategic Air Command were taking off or landing B-52s, B-58s, and B-36s every three minutes from the base runways—day and night. That was perhaps the first time I realized how much life in the United States had changed since I was a child.

Then came Vietnam and the Civil Rights movement when the fabric of American life and the character of American citizens were sorely stressed. Wright-Patterson was in the business of developing warplanes, and it became the target of citizens exercising their freedoms of speech and belief. The base became the site of Civil Rights activities because it was a large employer of minority workers. The Air Force as a service, which officially had integrated minorities into its force in 1949, did the right thing by becoming a community leader in offering full and equal job opportunities to male employees of all races. (It would take another decade before the Air Force took similar steps on women's issues.) As minorities rose into leadership positions at Wright-Patterson, they were especially good about reaching out to the community and working with minority groups to ensure they received education and encouragement and that their civil rights were fully acknowledged and protected.

Throughout the Wright-Patterson workforce, from the entry level to the most senior executives, the spirit of volunteerism was, and continues today, to be phenomenal. Every church and synagogue in the city has members from the Wright-Patterson community, and base employees are exceedingly generous every year in donating to the Dayton and Miami

Valley United Way campaigns. They go far beyond the pledge drive; there are literally hundreds of individual fundraisers on base to bring in money for what the base workers know as the Combined Federal Campaign. In recent years there even have been "airplane pulls" where competing organizations throughout the base contribute money for the privilege of dragging a plane, e.g., in 2003 an F-111, farther than anyone else. Those folks are always creative.

Now the Cold War is over and we have troops fighting in the Balkans, Afghanistan, and Iraq and keeping the peace in many other locations worldwide. The aircraft developed at Wright-Patterson (those that we know about!) boggle the imagination. They can't be seen, can't be heard, don't need human pilots, and know just where they are going and how to get there—before they take off. This is more than I can comprehend and probably more than I want to contemplate.

Dayton, indeed, has been fundamentally shaped by the presence of Wright-Patterson in our midst, and I like to think that the base has been just as fundamentally shaped by the Dayton area and our people. Many of Dayton's citizens work at the base or provide goods and services to the base, which enjoys the strong work ethic, creative spirit, and plain, old-fashioned common sense that Daytonians bring to the workplace. The Wright brothers were from Dayton, and they were both such typical Daytonians: highly inventive, dedicated to each other and their work, and embarrassed by pomp and circumstance. The boys who met heads of state also rode the interurban car.

I have always been proud to call myself a lifetime Daytonian, and *Home Field Advantage* has caused me to reflect on all the reasons for that pride. We have long, strong roots in this area—more than 200 years—and we have a tremendous respect for knowledge, creativity, and honest hard work. And we are, just like the Wright brothers, modest about our accomplishments. *Home Field Advantage* gathers many of our accomplishments into one book and will cause even the most modest of us to smile. This is a great part of the country to live in and Wright-Patterson Air Force Base is one very big reason why. Once you read this book, I know you'll agree.

Rosamond Young
Dayton Daily News
October 30, 2003

FOREWORD

SYMBOLS OF SYMBIOSIS: DAYTON, OHIO, AND THE UNITED STATES AIR FORCE

Since the beginning of cities and warfare, certain communities and certain services have become inextricably entwined in their mutual development. Sparta might be a starting point in the past, but in the United States, we can look at the comfortable, mutually supportive relationship that developed between the U.S. Army and San Antonio, the U.S. Navy and Norfolk or San Diego, the U.S. Marines and Quantico, and the U.S. Air Force and Dayton, Ohio.

In every case, the community and the service identify their common interests, equally support each other, and live in harmony. But the relationship between Dayton and the U.S. Air Force (and its predecessors, of course) is absolutely unique in the manner in which the inventive ambience of the city inspired the research and development prowess of the service.

Dayton, as the birthplace of aviation, was naturally inclined to be supportive of the Signal Corps Aviation Section and the ever increasing and more sophisticated branches that followed it. But as a center of industry, bubbling with ideas and inventions, it did more than host aviation; it inspired and furthered it. The sense of enterprise in Dayton transferred itself to the Army personnel stationed there. Nothing was beyond the realm of possibility in Dayton—if something needed to be done, someone would find a way to do it.

Perhaps most surprising, however, is the manner in which the Army and then the Air Force embraced Dayton on such a long-term basis, using it as the heart of inventive aviation activity. From the very first, an assignment to Dayton was recognized as the *sine qua non* of an upwardly mobile career path, almost without regard to one's chosen specialty. The airfields of Dayton—McCook, Wilbur Wright, Fairfield, Wright, Wright-Patterson—were home not only to research and development but also the key factor in all military operations—logistics. The inspired pairing, in one location, of research and development and logistics was a primary factor in the ability of first the Army and later (perhaps to an even more marked degree) the Air Force to make their successive transformations from peacetime, low-budget aerial competence to wartime aerial superiority.

Dayton became the wellspring of aviation ingenuity, and was characterized by the long institutional memory it had for projects. Some technical developments take years, even decades, to solve, and this was a forte of the service organizations rooted in Dayton.

The torrent of ideas that poured forth from McCook Field and its successor organizations was dazzling for its number, for its tenacity, and for the way in which it anticipated and solved future problems. One major flow of ideas involved aircraft themselves, as whole new classes of flight vehicles had to be understood, acquired, and tested. Test flying produced a crop of aviators every bit as distinguished and as capable as today's astronauts. These men, for whom some of the streets of Wright-Patterson have been named, had to test each new product with an eye not only for its suitability but also for its impact on the future.

For example, one would not have thought that in 1919, with the Great War just ended, and when airplanes routinely flew just a few thousand feet above the ground, that anyone would be concerned with cabin pressurization. But the problem was foreseen at McCook, and the talented Harold R. Harris flew a modified USD-9A, an aircraft redesigned and built at McCook, with a pressurization system that not only worked—but worked too well! After takeoff, Harris found that the cabin had pressurized to below sea level, and he had to land to let the system equalize. The Air Corps turned to Wright Field to pursue pressurization experiments and, in 1937, the Lockheed XC-35 was awarded the Aero Club of America (Collier) Trophy for being the first successful, pressurized-cabin aircraft.

There were dozens of similar disciplines pursued at McCook and its successors. Solid investigative work was done with equal success in many areas, including flight instruments, radio-controlled vehicles, armament, materials, fuel systems, engines, and literally every aspect of aviation technology. But creating inventive new equipment was only part of the McCook/Wright/Wright-Patterson story. The process of acquiring these vehicles, so expensive and available from relatively few sources, had to be carefully developed and refined. This required a sophisticated level of engineering expertise, one that was constantly developed over the years so that it could handle the rigorous requirements of intercontinental wars and the exploration of space.

With all of these technical advances, there has also been a strong and consistent understanding of the requirement for preserving history, either in archival or museum terms. The museum experience was episodic, being interrupted by World

War II. Since the end of the conflict, however, the museum component has moved forward until today, when the United States Air Force Museum ranks as one of the finest in the world on all counts, from its collections to its buildings to its research and archival facilities.

One striking aspect of the Dayton-Air Force accord is the manner in which it has responded to the demands of both war and peace. A new conflict inevitably calls for expansion, with all of the shortages in personnel and resources implicit in a wartime situation. Then, after months or years of frantic buildup, an armistice or peace brings about the reverse situation; work declines, contracts are cancelled, and workers laid-off. Such cycles have occurred as a result of World War I, World War II, Korea, Vietnam, and the Cold War. Yet Dayton and the Air Force have always managed to handle both expansion and contraction with good grace, each institution making do with less, or taking up slack, as required.

Perhaps the most telling thing about the value of Wright-Patterson Air Force Base and its forerunners is the manner in which both the Air Force and the aviation industry turns to it not only for ideas, but for procedures and methods as well. Recent wars in Afghanistan and Iraq have highlighted advanced warfare techniques including infrared targeting systems, unmanned aerial vehicles, cruise missiles, stealth, and precision-guided munitions. All of these have either originated in or been benefited by the laboratories at Wright Patterson, often decades in advance of the explicit need.

But in every instance, from Harold R. Harris and the first pressurized cabin to the industry and government teams that are developing tomorrow's systems, everything depends upon the effort of dedicated individuals who sacrifice themselves to the goal of a better, more effective, more economic Air Force. Such individuals helped put together McCook Field so many decades ago, and such individuals today still hold up the banner of peace and progress at Wright-Patterson Air Force Base.

WALTER J. BOYNE
Aviation Historian
December 8, 2003

PREFACE

Home Field Advantage: A Century of Partnership between Wright-Patterson Air Force Base and Dayton, Ohio, in the Pursuit of Aeronautical Excellence tells the story of how Dayton and Wright-Patterson Air Force Base, Ohio, became and continue to be America's "Cradle of Aviation." Why Dayton? Good question. Being home of the Wright brothers is only part of the story (if, perhaps, the most important part). There were many aspects of Dayton, Ohio, that made it the "Air City of America" by the mid-1920s. By that time, it was home to two important airfields, Wilbur Wright Field and McCook Field, that pioneered military air power logistics and research and development. By the beginning of the 1930s, these airfields were succeeded by Wright and Patterson fields that, following the establishment of the United States Air Force, became, in 1948, Wright-Patterson Air Force Base.

Home Field Advantage recounts the history of partnering and cooperation between the Dayton area and the Air Force in five chapters. Chapter One tells the story of Dayton from its settlement, in the last decade of the eighteenth century, to the beginning of World War II. Chapter Two discusses the establishment of Wilbur Wright and McCook fields during World War I. Chapter Three focuses on the little-told story of base support and the origins of Air Force logistics, primarily at Patterson Field. Chapter Four continues the story of Air Force logistics and base support, bringing the saga up to the present day. Chapter Five resumes the story of Dayton and the Air Force base community begun in Chapter One, focusing on the personal and institutional synergies that have developed between the Air Force and greater Dayton over the past six decades.

Home Field Advantage is the companion volume to *Splendid Vision, Unswerving Purpose: Developing Air Power for the United States Air Force during the First Century of Powered Flight* (2002). The two volumes constitute a complete revision and continuation of *From Huffman Prairie to the Moon: The History of Wright-Patterson Air Force Base, Ohio* (1986). Those who enjoyed *Huffman Prairie* and *Splendid Vision* will want to read *Home Field Advantage* to learn "the rest of the story ...."

Like *Huffman Prairie* and *Splendid Vision*, *Home Field Advantage* is lavishly illustrated with photographs from Air Force and public archives. If "a picture is worth a thousand words," this volume is worth its weight in gold.

We leave readers of this volume, as the two preceding, with an admonition: to remember the many men and women of courage and fortitude whose stories are recounted on these pages. Someone once said, "If we see farther, it is because we stand on the shoulders of giants." This book tells their story.

DIANA G. CORNELISSE
Chief Historian

Last flight of Ohio Air National Guard F-51 Mustangs over Dayton, June 1955 *(Colonel Dale Shafer)*

CREDITS AND ACKNOWLEDGEMENTS

A work of this size and scope could not have been completed in a timely fashion without the participation of many individuals, organizations, and institutions.

Like its predecessor and companion volume, *Splendid Vision, Unswerving Purpose* (2002), *Home Field Advantage: A Century of Partnership between Wright-Patterson Air Force Base and Dayton, Ohio, in the Pursuit of Aeronautical Excellence* is based on the seminal and pioneering history of Lois E. Walker and Shelby E. Wickam, *From Huffman Prairie to the Moon: The History of Wright-Patterson Air Force Base* (1986).

The research, writing, and editing of *Home Field Advantage* is primarily the work of the ASC History Office team, including Dr. Jim Aldridge (chapter 1), Ms. Lori Tagg, (chapters 2 and 3), Dr. Henry Narducci (chapter 4), and Ms. Helen Kavanaugh-Jones (chapter 5). Ms. Tagg rearranged and edited much of the material in chapters 2 and 3 and also in chapter 4, and Dr. Aldridge added introductory and concluding sections to chapter 5.   Contributing sidebars to these chapters were Ms. Kavanagh-Jones, Ms. Robin Smith, and Mr. Jim Ciborski. Ms. Smith was also responsible for editing the endnotes and compiling the bibliography, and Mr. Ciborski compiled and edited the appendices. Mr. Dann Andrews and Mr. Chase Simon, two history office interns, got their "baptism of fire" by contributing sidebars and research support *on deadline*. Others contributing sidebars were Master Sergeant Dave Wolf and Major Mike Menser. Master Sergeant Tony Dawson and Senior Master Sergeant Jim Sturm helped scan many photographs. Mr. Bruce Hess, ASC/HO archivist, provided transfusions of documents and tidbits of data, as needed. Captain Elizabeth Langwell "brought up the rear" by compiling the index for the whole volume. Our esteemed colleagues of the Air Force Materiel Command History Office performed, at our request, a final proof-reading and sanity check of the entire volume. Ms. Tagg reprised her virtuoso performance on the first volume by editing the entire book, including all front and back matter and "enforcing" all deadlines.

The monumental and creative task of formatting the book, from cover to cover, was once more in the capable and creative hands of Mr. Curtis Alley of the National Air and Space Intelligence Center (NAIC). His handiwork was all the more remarkable this time 'round due to constraints of time and funding. Mr. Alley's talents were again generously placed at the disposal of the history office by his supervisor, Mr. Thomas P. Richards, with the kind permission of the commander of NAIC, Colonel Mark C. Christian.

Speaking of covers . . . the cover art for the front cover of *Home Field Advantage* we owe to the talent and generosity of artist Mr. Gene Lehman, now retired from the Air Force Institute of Technology. It is entitled *Huffman Prairie Flyers, 1904.* (An 8 X 10 copy of this print is also included in the book.)

We have our sources—and how! This book would have gone nowhere without the cooperation of numerous individuals on base and in the local community who provided often critical information: verbal, written, and pictorial—kindly, generously, and expeditiously. These individuals include: Mr. Abe Aamidor; Mr. Mark D'Angelo of the Sons of Italy, John Pirelli Lodge #1633; Mr. Oscar Boonshoft, entrepreneur; Mr. Robert B. Clayton, 88th ABW JAC; Mr. James "Steve" Cloyd, AFRL Propulsion Directorate; Ms. Verlyn Conard and Mr. Dennis Gaudette, F-16 System Program Office; Mr. David Cornelisse, National Archives and Records Administration, Dayton, Ohio; Mr. Roger Cranos; Mr. Jim Custer, local historian and raconteur; Mr. Chris Doyle, the Converse Company; Mr. Robert Durrum; Dr. Janet Ferguson, Wright-Patterson Cultural Resources Program Manager; Mr. Timothy R. Gaffney, *Dayton Daily News*; Mr. Wes Henry and Mr. Brett Stolle, United States Air Force Museum; Ms. Toni Jeske, Wright State University Libraries; Ms. Barbara Howley, Civilian Personnel at Wright-Patterson; Mr. James Hull, Veterans Affairs Medical Center; Mr. Ron Kaehr, AFRL Sensors Directorate; Mr. Andrew Kididis, ASC Directorate of Engineering; Mr. J. Robert McKee and Mr. David Tritch, Ohio Task Force 1; Mr. Bill Meixner, Society of Air Racing Historians; Mr. James F. Miller, 88 ABW Mission Support Group; Mr. Al Moyers, Air Force Weather History Office; Ms. Julie Orenstein, Sinclair Community College; Mr. Randy Parker, 88th ABW Civil Engineering Directorate; Mr. Ray Rang; Mr. R. Michael Rives, 74th Medical Group; Mr. Jim Sandegren, Woodland Cemetery naturalist, historian, and poet; Ms. Kathy Schweinfurth, ASC Educational Outreach Office; Mr. Joseph M. "Jerry" Slade; Ms. Nancy Snyder, Ms. Gladys Hopkins, and Mr. Ed Keck, Moraine Farm; Chief Master Sergeant Monte E. Tahvonen and Master Sergeant Richard E. Skelly, 88th Security Forces Squadron; Major Andy Thurling, Wright Aeroplane Company; Mr. Stephen M. VanDegrift and Mr. Lynn L. Zanow, 88th ABW Disaster Preparedness Office; and Mr. John Warlick, Wright B Flyer Inc.

We owe a special debt of thanks to Ms. Ellie Bambakidis and her cohorts in the Local History Room of the Dayton Metro Library, downtown Dayton. They guided our researchers through their extensive photographic and documentary collections, performing preliminary research and providing scanning services *gratis* without which this book would have been much poorer—and chapter 1 all but impossible. Their good humor, patience, and professionalism reminded us of *how great a debt we owe to our public libraries and librarians*: they are an investment that repays itself many times over.

Other organizations and institutions that proved helpful in seeing this project through to completion were: the United States Air Force Museum; the Paul Laurence Dunbar Library; Wright State University, particularly its Current Periodicals Section and the Special Collections Archive; the University of Dayton Research Institute; the Cox Arboretum; Woodland Cemetery; and the NCR Corporation's Moraine Farm.

One person—an institution in her own right—we would like especially to thank. Ms. Rosamond M. "Roz" Young, who for years has written a column on local history and doings around the Miami Valley for the *Dayton Daily News*, graciously agreed to write one of two Forewords to this volume. We wish her good health (we know we'll get good writing!) for many years to come.

Many thanks are also due to those who, in their own ways, "greased the skids" so that this book could go from manuscript to a printed and distributed publication smoothly, if not quite effortlessly. Mr. Steve Bortz and Mr. Mike Gallagher of the United States Government Printing Office outdid themselves in providing technical support and professional advice for the publication of *Home Field Advantage* as they did last year for *Splendid Vision*. Ms. Julie Brown and Ms. Janet Kepler of ASC's Reserve Affairs Office saw to it that the ASC History Office team was augmented with—we can testify!—some of the most dedicated and talented reservists in the Air Force. Mr. Bill Meers and Ms. Patti Traylor of ASC's Public Affairs Office conducted the security review. Mr. John D. Weber, AFMC Command Historian, provided advice, encouragement, and strong advocacy for this volume as he did for its predecessor, *sotto voce* to the right persons at the right times. And last—but not least—were the "lords provider of loaves and fishes" of the AFMC and ASC financial communities (and their lawyers!): Brigadier General Frank R. Faykes, AFMC Comptroller, and Colonel Michael J. Colopy of the AFMC Judge Advocate General's Office. (We also owe a special note of thanks to Ms. Audrey Gee, Budget Analyst for the ASC Commander's Staff, who ensured that every dollar, dime, and nickel, went where it was supposed to and when.)

Finally, we are grateful for the enlightened leadership that this center has historically enjoyed—and has continued to enjoy in recent years. We are particularly fortunate that we have had as commander a man who is a warrior, thinker, writer—and student of history—Lieutenant General Richard V. Reynolds. We were also fortunate to have had as executive director a man who *embodied* over 40 years of history—and entrepreneurial leadership—at Wright Field, Dr. Vincent J. Russo. Both men demanded much of ASC's workforce—including historians—but gave even more themselves: the essence of leadership. The same can be said of Colonel Ward D. Willis, the director of the ASC Commander's Staff. Colonel Willis, who retired in October 2003, was a team player *and team builder*, who provided the History Office with the tools and freedom to undertake this book and its predecessor volume. Finally, Colonel Michael J. Belzil, commander of the 88th Air Base Wing, followed our progress with interest and support. All four fostered an environment in which big thinking and big projects—and *risk-taking*—are possible.

Both *Splendid Vision* and *Home Field Advantage* testify to the lasting influence of air power upon history—global, national, and local. On that note, we would like to close with a final "thank you" to Mr. Walter J. Boyne, whose many books and articles on aerospace topics have instructed and entertained us over the years and whose recent utterances—both public and private—in enthusiastic support of this book and its predecessor are most appreciated. He even agreed to write one of the two Forewords to this book. We are indeed honored.

DAYTON
The Center of Aviation

Birth Place of Aviation

Industrial Center

Safe from Foreign Invasion

Center of Specialized Labor

Wealthy Community

Civic Progressiveness

Fine Railroad Facilities

Precision Center

Unlimited Patriotism

Huffman Prairie at dawn (U.S. Air Force photo by Henry Narducci). Dayton was celebrated as "The Center of Aviation" on the cover of *Slipstream* in February 1923.

Scipio, Orville Wright's pet *(Library of Congress)*

TABLE OF CONTENTS

Foreword iii

Foreword: Symbols of Symbiosis: Dayton, Ohio, and the United States Air Force v

Preface vii

Credits and Acknowledgements ix

Chapter 1
Birthplace of Aviation 1

    Homecoming 1
    Dayton: City of Invention and Progress 7
    Pioneer Settlement 12
    Early Nineteenth-Century Dayton 13
    Laying the Foundations of Modern Dayton 18
    Industrial Giant 20
    Great Flood 25
    Cradle of Aviation 30
    Dayton in the 1920s and 1930s 33
    Dayton on the Eve of World War II 41

Chapter 2
Military Aviation Comes to Dayton 47

    Early Years of Signal Corps Aviation 47
    Combat-Oriented Pilot Training 51
    Establishment of Wilbur Wright Field 54
        From the Ground Up 55
        First Flying Season 59
        A "Major" Reorganization 62
    Aviation Mechanics' School 65
    Aviation Armorers' School and Gunnery Testing 67
    Spring Flying 68
    Airplane Testing at Wilbur Wright Field 69
    Over There and Back 71

Chapter 3
The Story of Air Force Logistics 79

    World War I Origins 79
    Postwar Demobilization and Reorganization 82
    Air Service Supply and Repair Depot 82
        Engineering Repair Section 84
    Fairfield Air Intermediate Depot 86
        Property, Maintenance, and Cost Compilation Section 88
        1924 Round-the-World Flight 90
            Logistical Support Plans 92
            Douglas World Cruiser 93
            World Flight Crews 93
            Success 94
        1924 International Air Races 95
        Model Airway 100

Creation of the Fairfield Air Depot Reservation 102
    1931 Air Corps Maneuvers 103
Patterson and Wright Fields 105
    The Patterson Name 105
    A Logistics Heritage 107
    Depression Years 110
    FAD Activities during the 1930s 112
World War II Expansion 115
    Command Assignments during the War 125
    Major Organizations 130
    Military Training Programs 131
    Civilian Training Programs 133
    Assisting with the Creation of New Depots 134
Patterson Field and the End of the War 135

Chapter 4
  Fulcrum of Base Support 139

  Wright-Patterson Enters the Cold War 139
    Flight Operations 145
    Base Activities 148
  Wright-Patterson in the Vietnam Era 151
    Base Activities 153
  Wright-Patterson and the End of the Cold War 157
    National and Local Emergency Responses 159
    Airfield Operations 163
    Contracting Operations 165
    Environmental Concerns 165
    Computer Technology 166
    Base Celebrations 166
  From Desert Storm to Iraqi Freedom 171
    Combat Operations 177
    Balkan Proximity Peace Talks 183
    Base Activities 187
  Associate Organizations 190
    15th Weather Squadron 190
    17th Bombardment Wing 192
    58th Air Division 193
    97th Fighter-Interceptor Squadron 193
    445th Airlift Wing 194
    3500th U.S. Air Force Recruiting Wing 198
    4950th Test Wing 198
    Air Force Institute of Technology 199
    Air Force Research Laboratory 202
    Air Force Security Assistance Center 204
    Medical Center Wright-Patterson Air Force Base 204
    National Air and Space Intelligence Center 205
    Orientation Group, U.S. Air Force 207
    United States Air Force Band of Flight 208
    United States Air Force Museum 209
  A Half-Century of Growth 211
    Housing Complexes 211
    Community Services and Recreation Facilities 213
    Area A Command Headquarters 216
    Area B Acquisition Complex and Laboratories 218
    Area C Base Operations 224
  Wright-Patterson into the Twenty-First Century 224

Chapter 5
    Wright-Patterson and the Miami Valley 227

        Yesterday, Today, and Tomorrow 227
        Dayton during World War II 227
        The Rise of Greater Dayton 228
        Urban Renewal and Downtown Revitalization 231
        Business and Industry in Transition 239
        The Wright-Patterson Connection 243
        Wright-Patterson's Economic Impact on Surrounding Communities 246
        Mutual Benefits of Research and Development 251
            Technology Transfer 251
            Edison Materials Technology Center 255
            National Composite Center 255
                Composite Airframe and Automotive Parts 258
                Composite Bridge Decks 259
        Academics 259
            Air Force Institute of Technology 262
            Advanced Degree Programs Offered on Base by Regional Universities 264
            Co-op Program 264
            Educational Outreach Program 265
            Wright Scholars Research Assistant Program/Wright STEPP 267
        Wright-Patterson Support of Centennial of Flight Events 270
            Air Power 2003 270
            Music Power 2003 272
            Events at the United States Air Force Museum 276
            Celebration Central at Deeds Point 279
            Time Flies at Huffman Prairie Flying Field 280
            2003 Vectren Dayton Air Show 280
            International Air and Space Symposium and Exposition 282
            Dayton Aviation Heritage National Historical Park 283
            Rededication of Huffman Prairie Flying Field 286
            December 17 Memorial Ceremony 287

Postscript
    The Saga Continues 291

Endnotes 293

Appendices

Appendix 1: Organizational Evolution of the Acquisition and Logistics Functions at
Wright-Patterson Air Force Base and Its Predecessors 308

Appendix 2: Dayton Industries Mobilized to Help the War Effort 310

Appendix 3: Land Acquisitions, 1917-1978 311

Appendix 4: U.S. Air Force Units Assigned at Wright-Patterson Air Force Base and
Its Predecessors 312

Appendix 5: Wright-Patterson Air Force Base On-Base Tenant Organizations, 2003 323

Appendix 6: Wright-Patterson Air Force Base Personnel Strength, 1918-2002 325

Appendix 7: Wright-Patterson Air Force Base Memorialization Program 326

Appendix 8: Huffman Prairie Rededication Ceremony—Excerpts from Opening Remarks by General Richard V. Reynolds 330

Appendix 9: Off-Base Sites Supporting Air Force Research 332

Glossary 335

Bibliography 339

Index 355

End Paper 400

Tables

Wilbur Wright Field Signal Corps Aviation School—1917 Flying Season 59
Wilbur Wright Field Commanders 64
Chronology of Fairfield Installation 81
Patterson Field Commanders 108
Organizations and Units Trained by FAD Personnel 132
Sub-Depots under Fairfield Air Depot Control—1942 135
Chronology of Wright-Patterson Air Force Base and Its Antecedents 140
Wright-Patterson Air Force Base Commanders 158
Ohio Task Force 1 Members 180
Air Force Marathon Winners 188
Wright-Patterson's Economic Impact on the Dayton Community in the Twenty-First Century 249

Maps

Great Miami River Valley Showing Locations of Five Dams 29
Huffman Prairie and Vicinity 54
Diagram of Wilbur Wright Field, 1917-1918 55
Route of the 1924 Round-the-World Flight 90
Location of Wright-Patterson Air Force Base 311
Land Acquisitions, 1917-1978 311

Aerial Photographs

P-51s flying over Dayton, Ohio viii
NCR complex, mid-1930s 8
Dayton Soldiers' Home, 1885 lithograph 10
Dayton skyline, 1930 19
Panoramic view of 1913 flood in Dayton 26-27
McCook Field, 1921 .... 30, 73
Downtown Dayton, about 1923 35
Dayton Engineering Laboratories Company (Delco) Plant, 1953 42
B-17 bomber over downtown Dayton, late 1930s 44
Wilbur Wright Field, prior to 1923 47
Construction of Wilbur Wright Field, 1917 54-55
"New" Osborn 57
Wilbur Wright Field, spring flooding, 1922 58
Aviation Armorers' School, Wilbur Wright Field 68
McCook Field 69
Wilbur Wright Field, 1923 .... 71, 76
Baby Blimp at McCook Field, 1921 73
Wright Field, 1926 77
Fairfield Aviation General Supply Depot, 1918 79

Air Service Supply and Repair Depot, Fairfield, Ohio, 1921 84
Fairfield Air Depot and Wilbur Wright Field, April 1922 89
Air Carnival, Fairfield Air Depot Reservation, 1930 104
Air Corps Maneuvers, Fairfield Air Depot Reservation, 1931 105
Patterson Field, 1931 106
Fairfield Air Depot, Patterson Field 109
Transient Camp, Patterson Field 110
Brick Quarters Complex, Patterson Field 111
Patterson Field, 1943 115
Patterson Field Runway Complex, 1945 119
Air Service Command Complex, Patterson Field 120
Wood City Complex, Patterson Field 120
Splinter City, Patterson Field 121
Skyway Park Housing Complex 121
Wright Field, 1944 128
Very Heavy Bomber Runway, Patterson Field, 1947 136
Patterson Field, 1947 136
Wright-Patterson Air Force Base, 1950 139
Area C flightline, about 1950 143
Building 10, about 1960 152
Area C flightline, early 1960s 155
Aftermath of Xenia tornado, 1974 160
Ruins of World Trade Center, New York, 2001 181
West Ramp, Area C 193
West Ramp, facilities of 4950th Test Wing 199
Base Hospital 204
Medical Center Wright-Patterson Air Force Base, Area A 205
National Air and Space Intelligence Center, Area A 206
United States Air Force Museum, 2003 210
Page Manor, 1951 211
Woodland Hills, 1971 212
Bass Lake 216
General James H. Doolittle Acquisition Complex 219
Wright-Patterson Air Force Base, 2000 225
Dayton, Ohio, 1945 231
Dayton Army Air Field, World War II 241
Dayton International Airport 241
Gentile Air Force Station 257
Wright State University 269
Wright State University's Allyn Hall, 1964 269
United States Air Force Museum, 2003 277
Wright Memorial 328

Chapter One

BIRTHPLACE OF AVIATION

HOMECOMING

The dreaded day arrived. The shrill shriek of factory whistles, the clanging of every church bell in town announced to the proprietors of the bicycle shop at 1127 West Third Street, Dayton, Ohio, the interruption of their work day.[1] They had been at it since early that morning, feverishly preparing the packing of the last components of their machine. The first crates had already been loaded on the train and sent to Washington for a product demonstration for the U.S. Army later in the month. The deadline had already been extended, from the end of June to the end of July; no more extensions could reasonably be expected from the War Department.[2] Now this! At 9:00 a.m., they emerged, dismayed and annoyed, from the inner sanctum of their workshop. For a full 10 minutes, the racket continued. When it subsided, they could hear, from the sidewalk in front of their store, the muffled drumming and tooting of several marching

bands off in the distance toward town.[3] As the music grew louder and clearer, they could make out a large, closed carriage coming their way. The carriage finally drew up in front of the store and, at 10:00 sharp, Messrs. Wilbur and Orville Wright stepped aboard. Time to go, they sighed. Go and get it over with.

The day was June 17, 1909. *Five and a half years before*, the brothers Wright had telegraphed from Kitty Hawk, North Carolina, to their father in Dayton the news that they had, that day, December 17, 1903, done what no one had ever accomplished in human history:

> SUCCESS FOUR FLIGHTS THURSDAY MORNING ALL AGAINST TWENTY-ONE-MILE WIND STARTED FROM LEVEL WITH ENGINE POWER ALONE AVERAGE SPEED THROUGH AIR THIRTY-ONE MILES LONGEST 57 [*sic*] SECONDS INFORM PRESS HOME CHRISTMAS ORVILLE.[4]

Brother Lorin took the cable from his father and dropped by the local newspaper office. The Associated Press representative at the Dayton *Journal*, without glancing up from his copy, read over the message. "Fifty-seven seconds, hey? If it had been fifty-seven minutes then it might have been a news item."[5] No headline or story appeared in the next day's *Journal*.[6]

The following year, the brothers transferred their flight experiments from the coast of North Carolina to Huffman's prairie, a 20-minute trip on the Dayton-to-Springfield line of the interurban from their shop on Dayton's west side. Over the next two years, 1904 and 1905, they gradually extended the capabilities of their machine and their experience as aeronauts. By the autumn of 1904, they were making large circles about the field.[7] The longest flight the following year lasted 38 minutes 3 seconds.[8]

Interurban passengers, if they looked up from their newspapers when they passed

The city of Dayton turns out in fine style for the June 1909 celebration honoring the Wright brothers.

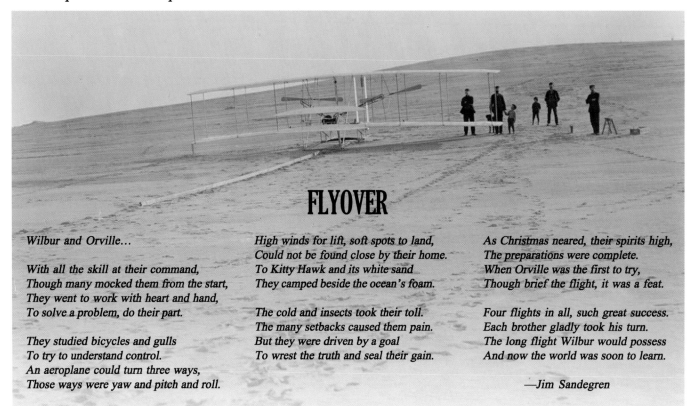

FLYOVER

Wilbur and Orville...

With all the skill at their command,
Though many mocked them from the start,
They went to work with heart and hand,
To solve a problem, do their part.

They studied bicycles and gulls
To try to understand control.
An aeroplane could turn three ways,
Those ways were yaw and pitch and roll.

High winds for lift, soft spots to land,
Could not be found close by their home.
To Kitty Hawk and its white sand
They camped beside the ocean's foam.

The cold and insects took their toll.
The many setbacks caused them pain.
But they were driven by a goal
To wrest the truth and seal their gain.

As Christmas neared, their spirits high,
The preparations were complete.
When Orville was the first to try,
Though brief the flight, it was a feat.

Four flights in all, such great success.
Each brother gladly took his turn.
The long flight Wilbur would possess
And now the world was soon to learn.

—*Jim Sandegren*

Kill Devil Hill, North Carolina, 1903. The Wright Flyer sits in front of its temporary home. The smaller building to the right served as a workshop and sleeping quarters for the two brothers. (*Library of Congress, Papers of Wilbur and Orville Wright*)

Wilbur and Orville Wright prepare their airplane for flight over Huffman Prairie. Note the American flag on the elevator at the front of the aircraft

by the prairie, might occasionally have seen what was going on. "Done anything of special interest lately?" Luther Beard, the managing editor of the Dayton *Journal* and regular rider on the interurban, politely asked Orville one day.[9] When newspapermen and journalists *did* report on the Wrights' flying machine experiments, they as often as not got the story wrong.[10] Indeed, only one amateur journalist and part-time beekeeper, Amos Root from Urbana, took the trouble to stick around the prairie and see what the Wrights were doing for himself.[11] His account, accurate and enthusiastic, appeared in his home-spun rag, *Gleanings in Bee Culture*.[12]

As sons of a bishop of the United Brethren Church, Wilbur and Orville would have been familiar with the scripture: *A prophet has no honor in his own country* (John 4:44).[13] Dismayingly, there were skeptics and naysayers in foreign parts as well. *Bluffeurs* was the common verdict in the lingua franca.[14] And all the rest of the world seemed to hail from Missouri.

Show me! So show them they did. In 1908, Wilbur and Orville took their flying machine on the road and across the sea. First near Le Mans and then Pau, Wilbur wowed the French.[15] The road led to Rome the following spring.[16] The U.S. War Department had, meanwhile, evinced interest. Under contract with the Army, Orville began a series of demonstration flights at Fort Myer, Virginia, just outside Washington, D.C.[17]

Grown men cried;[18] women swooned as the Wrights' Flyer lifted from the ground and, like some magic carpet out of the Arabian Nights,[19] swooshed through the thin air before their very eyes. Hats, caps, and *kepis* filled the sky above reviewing stands. Kerchiefs fluttered in salute to the knights of the air. Man had at last conquered the third dimension. Honors—and contracts—followed.

This was *news*—and opportunity! Back home, the management of the Dayton *Herald* conferred and began drumming up enthusiasm for a grand, local celebration of the Wrights' historic achievement.[20]

But Europe got them first. During the spring of 1909, the brothers and their sister Katharine toured the old continent, where they

Katharine Wright accompanied Orville and Wilbur on their many travels to promote their airplane.

Wright Flyer and the catapult built by the Wright brothers for launches at Huffman Prairie, shown here in Pau, France, 1909. The Wrights traveled extensively to demonstrate their airplane and its flying qualities.

SIMMS STATION

Simms Station, located about eight miles east of Dayton, was a stop on the Dayton, Springfield & Urbana Electric Railway, and was located in an area where first the "post-rider" (mounted mail carrier) and later the stagecoach ran their routes.

The 1855 Greene County Atlas listed the area as Kneisly Station and, that same year, the railroad replaced the stagecoach route. Mr. John Kneisly owned over 1,200 acres of land in the fertile valley, and a depot one mile to the west on the banks of the Mad River carried his name. Mr. W. A. Simms later purchased this land, upon which two Simms Stations came into being: one was the old Kneisly depot on the Mad River & Lake Erie Railroad; the other was a stop on the interurban Dayton, Springfield & Urbana Railway. The land passed down to Charles H.

Simms who, along with his wife Buda, founded the Dayton Country Club in 1897. They lived in a house in Dayton, in an area now occupied by the Masonic Temple. (W. A. Simms lived in a 20-room mansion on Valley Pike.) The Simms owned *The Evening News*, which James M. Cox bought in 1898 and turned into the *Dayton Daily News*.

The interurban track to Simms Station ran parallel to Dayton-Springfield Pike; Simms Station was located at the crossroads of Yellow Spring's Pike (now Dayton-Yellow Springs Road) and Dayton-Springfield Pike. The track ran from Dayton and continued on to Fairfield, old Osborn (in the approximate location of the Skyborn Drive-In Theater on State Route 235), Medway, and Springfield. The Wright brothers took the interurban from Dayton to Simms

Station to get to Huffman Prairie flying field, and groups of curious spectators soon began arriving at Simms Station in the hope of seeing a genuine flying machine. A celery farmer who lived nearby rented campstools to the crowds.

The construction of Huffman Dam in the years following the 1913 flood forced the relocation of the area's railroad and interurban lines to higher ground. The thousand-acre farm that the Simms owned, which had given the interurban station its name, eventually became part of Wright-Patterson Air Force Base.

Sources: "Mrs. Buda Simms Dies; Country Club Founder," *Dayton Daily News*, January 17, 1978, p 22; "Simms Station is Topic at Aero Chapter 536 Meeting," *Fairborn Daily Herald*, March 15, 1984, p 12; *Wright Brothers Huffman Prairie Location* folder, Aeronautical Systems Center History Office Archives.

Flight at the Wright School of Aviation near Simms Station, 1910 (*Library of Congress, Papers of Wilbur and Orville Wright*)

Large crowd of spectators gathered at the edge of a field at Simms Station, Dayton, Ohio, where a series of flights were conducted from May through July 1910, just after the opening of the Wright School of Aviation. An interurban car is visible in the background. (*Library of Congress, Papers of Wilbur and Orville Wright*)

received a series of accolades and awards. The Aero Club of the Sarthe presented them with a bronze sculpture.[21] Across the channel in England, the Aeronautical Society of Great Britain and the Aero Club of the United Kingdom wined and dined them and presented them with gold medals.[22] Back in the United States, the Aero Club of America invited them to a luncheon in New York City. The Club voted them gold medals that President William Howard Taft later formally presented at the White House.[23]

On May 13, the Wrights steamed into Dayton where the *Herald*'s boosterism had succeeded beyond expectation.[24] They were greeted at the train depot by a coach drawn by four white horses.[25] As several thousand people crowded about to shout their huzzahs, the coach—Wilbur, Orville, and Katharine aboard—accompanied by 10 other carriages crossed town to the Third Street bridge over the Great Miami River where a marching band serenaded them all the way to the Wright homestead at 7 Hawthorne Street. The house was bedecked with toy balloons, American flags, and patriotic bunting. Chinese lanterns were strung between the trees along the street front.[26] It was vintage American Midwest brouhaha of a kind later celebrated in such films as *The Music Man* and *The Wizard of Oz.*

And that was not all, folks. A month later came the *real* celebration: a two-day extravaganza when all Dayton turned out in garish display to honor her heroes.

Wilbur was not fooled or amused. In a letter to Octave Chanute in early June, he wrote glumly:

> We have been very busy on a machine for Ft. Myer and as we are interrupted very much the work goes slower than we would wish but we hope to be flying before this month ends. About a week of time will be consumed in traveling back and forth between Dayton and Washington to receive medals [from the president of the United States]. *The Dayton presentation has been made the excuse for an elaborate carnival and advertisement of the city under the guise of being an honor to us.* As it was done in spite of our known wishes, we are not as appreciative as we might be.[27] [Emphasis added.]

The boys got their revenge—sort of. When the Wrights boarded the coach on the morning of June 17, they sank far back

SUSAN KOERNER WRIGHT—MOTHER OF INGENUITY

Susan Wright in later years *(Wright State University Archives, Wright Brothers Collection)*

Orville and Wilbur Wright might not have invented the airplane if they had not inherited their mother's hankering for tinkering. Milton Wright, the boys' father, purportedly could not even hammer a nail straight; their mother Susan, however, liked to make things, and she built simple appliances to make her household chores easier and toys to entertain her children.

Susan Koerner Wright, the daughter of a German carriage-maker and a Swiss-American mother, was born in Loudon County, Virginia, on April 30, 1831. Soon after her birth, the Koerners moved to Indiana, where Susan's father continued his trade of making carriages. Susan spent a considerable amount of time with him as he practiced his craft, so she grew up in an atmosphere of design and construction.

Susan was also very interested in religion, and at age 14 she became a member of the Church of the United Brethren. She was extremely shy (a trait that Orville would inherit from her), yet she did well in school. Against the expectations for a girl of Susan's time and social class, her father sent her to Hartsville College in Huntington, Indiana, for nearly four years. While there, she became friends with Milton Wright, an ardent supporter of equality for women and a fellow student preparing for the United Brethren ministry. A year after Milton was ordained, he received a mission assignment to Oregon. He asked Susan to accompany him there as his wife, and although she agreed to marry him, she had no desire to go to Oregon. Instead, she waited two years for Milton to return and, on November 24, 1859, they married.

The couple moved many times throughout their marriage, due to Milton's vocation as a minister, and finally settled in Dayton, Ohio. Susan created a loving and supportive home environment for her husband and raised children who grew up devout, healthy, and productive. Susan had seven children—Reuchlin, Lorin, Wilbur, twins Otis and Ida (who died in infancy), Orville, and Katharine. Susan always encouraged her children's natural curiosity, and she enjoyed making things for them and with them. Katharine called her mother a genius who could make anything, and she remembered with special fondness the sleds her mother built for Reuchlin and Lorin when their father could not afford to buy them from a store.

Unfortunately, Susan was ill many times throughout her life. In 1883, she contracted tuberculosis, leaving her, three years later, permanently bedridden. At age 19, Wilbur became her primary caregiver, and Milton marveled at his son's tender devotion to his mother. Susan Wright passed away on July 4, 1889, and was buried in Woodland Cemetery. Although Susan Wright did not live to see her sons design and build the world's first airplane, her creative spirit undoubtedly inspired them.

Sources: Tom Crouch, *The Bishop's Boys: A Life of Wilbur and Orville Wright* (New York, 1989); Richard Maurer, *The Wright Sister* (Brookfield, Connecticut, 2003).

KATHARINE WRIGHT—THIRD MEMBER OF THE TEAM

Katharine Wright was born on her brother Orville's third birthday, August 19, 1874, in Dayton, Ohio. She was the Wright brothers' youngest sibling and only sister. Katharine and Orville were especially close during their childhood years; Orville insisted to his playmates that his little sister be allowed to tag along on all their excursions.

Katharine's mother died when Katharine was 15 years old, and Katharine assumed her duties as woman of the house, making a peaceful and supportive environment in which the Wright men could do their work. Katharine still found time to pursue her education and, in June 1898, she graduated from Oberlin College. Later, as one of the first women to be elected to Oberlin's board of trustees, she championed the cause of equal pay for equal work for female faculty members.

After graduation, Katharine taught Latin and English at Steele High School in Dayton, but she eventually gave up teaching to support the aviation work of her brothers. Katharine often accompanied Orville and Wilbur on trips that publicized and marketed their airplanes, and some Europeans thought Katharine was the real brain behind the invention of the airplane, since she was a college graduate and neither Wilbur nor Orville had graduated from high school, let alone attended a university.

For many years, Katharine sacrificed her personal life so that her brothers had the stable home life they needed to devote themselves without distraction to aviation matters. At age 52, however, she decided to marry former college classmate Henry J. Haskell. In response, Orville disowned her and denied her permission to be married in the Wright family home. Katharine moved to Kansas with her new husband but their marital bliss was cut tragically short when Katharine developed pneumonia. Katharine's brother Lorin convinced Orville to accompany him to Kansas to reconcile with his little sister before she died.

Katharine Wright Haskell passed away on March 3, 1929, at age 54. Orville took her body back to Dayton and buried her next to Wilbur in Woodland Cemetery. During the funeral service, three airplanes dropped roses on Katharine's grave in recognition of her contribution to aviation.

Sources: Tom Crouch, *The Bishop's Boys: A Life of Wilbur and Orville Wright* (New York, 1989); Richard Maurer, *The Wright Sister* (Brookfield, Connecticut, 2003).

Katharine Wright as a child   Katharine Wright as a young woman

Katharine canoeing in Canada,

and flying with Orville.

The June 18, 1909, celebration honored the Wright brothers and their accomplishments. Here, Orville Wright takes a moment to smile, a rare sight in photographs of him as a young man.

Wilbur and Orville Wright, dressed in top hats and tails for their 1909 homecoming celebration in Dayton

Crowds surround the Court of Honor to celebrate the homecoming of the Wright brothers, June 1909.

At the Montgomery County Fairgrounds, 2,500 schoolchildren wearing red, white, and blue formed an immense American flag, June 1909. *(Dayton Metro Library)*

in their seats while two friends and long-time associates riding with them, Ed Sines and Ed Ellis, waved to the crowds and received the plaudits of the cheering multitude along the parade route.[28] This was a pre-television age and instant recognition of celebrity was a thing of the future. It was the sort of prank that Orville found immensely entertaining.[29]

No one knew the difference and no one cared, really. Kids got out of school and doing chores, and tellers of the Winters Bank and floorwalkers of the Rike-Kumler department store had the day off to swell the cheering masses along Main Street.[30] As cash registers all over the city fell silent, cannons boomed a salute from Van Cleve Park. Dayton could offer no finer tribute.

Over the next two days, Wilbur and Orville gamely joined other members of the

Wright clan in receiving tributes from their fellow citizens. When their presence wasn't absolutely necessary, they gave the slip and returned to their shop to finalize preparations for the Army trials of the Flyer.[31]

The festivities peaked in two events. The first night, a spectacular fireworks display took place on the riverfront. At the conclusion of the show, two giant figures of the brothers wrapped in an American flag lit up the evening sky.[32] On the afternoon of the second day was the presentation of medals to the "fathers of aviation" at the Montgomery County Fairgrounds. Standing stiffly in formal attire—looking dapper in top hats and tails[33]—Wilbur and Orville each received a gold Congressional Medal from General James Allen of the Army Signal Corps, a

gold medal from the governor of Ohio, and a gaudy, diamond-studded medal from the mayor of Dayton.[34] In the background, 2,500 schoolchildren dressed in red, white, and blue formed an immense, singing American flag.[35] The celebration concluded with a parade up Main Street featuring progress in transportation, culminating in a float bearing an imaginative rendering of the Wright Flyer circling the globe.[36]

Next morning as street sweepers cleaned up the previous day's trash and confetti and workers returned to their offices and counters in the city's downtown, Dayton's two most illustrious sons boarded the 10:00 train to Washington.[37]

Neither Dayton nor the nation nor the world would ever be quite the same again.

DAYTON: CITY OF INVENTION AND PROGRESS

The city that turned out in style—albeit half a decade late—to celebrate the Wrights' invention of the airplane was, at the outset of the twentieth century, the center of invention and industry. Indeed, in 1900, Dayton, Ohio, listed more inventions than any other city in the United States. Only a community of inventors could, perhaps, for so long regard with such complete insouciance, the quiet tinkerings of a couple of hometown bicycle mechanics. Invention in Dayton was big business and the Wrights were, at the time, small fry. The city already boasted a half-dozen automobile manufacturers. In the early twentieth century, large companies also specialized in agricultural equipment, sewing machines, and bicycles.[38] However, the granddaddy of them all, a veritable

Daytonian James Ritty's great invention—the cash register (*Dayton Metro Library*)

Charles F. Kettering, inventor of the automobile self-starter and other automotive and aeronautical technologies (*Dayton Metro Library*)

patent factory, was John H. Patterson's National Cash Register Company (NCR), known to all and sundry around town simply as "The Cash."

Along with big business went big egos—and sharp elbows—as the captains of industry jockeyed with one another for pride of place. Patterson was, hands down, the leading egoist of the lot.[39] Not far behind were a number of his subordinates, whom he regularly fired as they threatened to outgrow their NCR britches.[40] One such clever country boy who got his start at The Cash—and was ultimately cashiered by John H.[41]—was Edward Deeds.[42] Another was Charles Kettering, who was admittedly less an egoist than *sui generis*, a man whom Patterson could not and would not understand—but who, while in his employ, literally electrified the cash register business.[43] Every time Patterson fired Kettering, Deeds hired him back.[44] Before Deeds and Kettering left NCR, they formed the Dayton Engineering Laboratories Company, or Delco for short, to manufacture their latest better ideas, electric ignitions and self-starters for automobiles.[45]

For Kettering, leaving NCR was the beginning of his career-long association with the automobile industry. For many years he headed up the General Motors (GM) Research Corporation in Dayton. When GM later moved the division to Detroit, Kettering commuted regularly from his home in Dayton.[46]

Deeds soon moved on. In the 'teens, he was the driving force behind the Miami

Conservancy District.[47] During World War I, he managed the nation's aircraft production and brought the Air Service's research and development operation to Dayton.[48] During the 1920s, he made a fortune reorganizing ailing corporations and managing other men's money on Wall Street.[49] In the 1930s, he returned to Dayton to steer NCR through the shoals of the Great Depression.[50] Save for John Patterson—and the Wright brothers—no other man did so much for Dayton, Ohio, in the first half of the twentieth century.

Across town, the man who brought journalism into the modern business world and who ultimately created a media empire sat perched in his office at the *Dayton Daily News* and surveyed events around him—as they were and as he would have them. James Middleton Cox[51] squinted through rimless glasses at his city, state, and country and saw boundless opportunity for progress—for his fellow citizens and for James Middleton Cox. By his late thirties, Cox had progressed from plowboy to schoolteacher to newspaperman to congressman.[52] He would soon be governor of Ohio[53] and, later, Democratic candidate for president of the United States.[54] He was not to be trifled with, as John H. Patterson and the Montgomery County Republican political machine discovered time and again.

Cox was what was called a Progressive. When a congressman, in addition to obtaining a splendid new post office for Dayton[55] and benefits for Civil War veterans at the Dayton Soldiers' Home

Edward Deeds, right, and John H. Patterson on First Street between Main and Ludlow streets in downtown Dayton, about March 1913 (*Dayton Metro Library*)

(Republican by affiliation, they thereafter voted in the majority for Cox),[56] he also supported lower tariffs, railroad regulation, and public utilities regulation.[57] He opposed the overweening power of industrial trusts and corruption in government, particularly that alleged in the Interior Department.[58] As governor, he campaigned for and signed into law much Progressive legislation.[59]

John H. Patterson was also a Progressive—if in his own peculiar way. An innovator in advertising and marketing products, Patterson also pioneered various industrial welfare schemes—paying and treating his employees exceptionally well—thereby effectively shutting out the unions. When Dayton's city council continually stymied his reformist proposals, Patterson engineered the reform of city government itself, introducing the commission-city manager form of government that soon became a model, known as the "Dayton Plan," for other cities across the nation.[60]

Dayton, then, was a progressive town. It had much to be proud of (and even more to look forward to) when it pulled out all the stops to celebrate the Wright brothers' achievement—and it could be pardoned for patting itself on the back in the process. It was a city of wide, paved streets; tall, modern buildings; excellent water purification and sewage systems; and a population of ambitious, energetic, and competitive men and women. Compared with larger towns, like Cincinnati or even Cleveland, it glistened like a rare jewel in the southwest Ohio countryside. And, in fact, it had long regarded itself as Ohio's "Gem City."[61]

NCR complex in the mid-1930s (*Dayton Metro Library*)

Dayton Daily News headquarters, Fourth and Ludlow streets (*Dayton Metro Library*)

JOHN HENRY PATTERSON—*AVE ATQUE VALE!*

John H. Patterson turned the cash register into big business through his National Cash Register Company. (*Dayton Metro Library*)

John Henry Patterson had been away from Dayton, off and on, for nearly five years. He had briefly attended Miami University down the pike in Oxford, Ohio. When the Civil War interrupted his studies, he donned Union blue. When the war concluded, he entered Dartmouth, graduating in 1867. He taught grade school briefly. It was the first thing he decided that he never wanted to do again. Armed with his bachelor of arts degree, he returned home to Dayton only to find that local businessmen had no use for a restless young fellow with a college degree—too good to work, they thought. Frustrated, he got a job as toll collector at the Dayton station of the Miami-Erie Canal. Bored by the long, languid days sitting in the tollbooth, he went into business delivering coal. Initially, the business prospered, due in large part to John Henry's fastidious and precise nature: he made sure that his coalmen delivered their loads promptly; he made sure that the coal was exactly what the customer ordered (he delivered no "dirty" coal!); and he made sure that neither coalman nor customer cheated him of the payment due. He kept precise accounts, using a mechanical device invented by Daytonian James Ritty—what Ritty called a "cash register."

But ambition got the better of John Henry. He entered into a partnership with Boston moneymen to expand his business. Over time, he lost more money than he made. Finally, he gave up and sold out.

Like many a young man of his day, he took Horace Greeley's advice and headed west to try his hand at ranching. But prospects were bleak, and he was really not cut out to ranch. Discouraged, he returned home. Then one day, he espied a "for sale" ad of a local company that made cash registers. Knowing first hand the value of the device, he bought the business—only to be laughed to scorn by all and sundry. He tried to sell the business. He even tried to *give* the business back to the former owner, but without result.

The year was 1885 and John Henry was staring in the face his 41st birthday. What to do? He knew he had a useful product. The challenge was to make the rest of the world know its value too—and to buy it. He initiated a campaign to market and improve the cash register, a campaign that ceased only with his dying breath 37 years later.

It was there and then that John Henry Patterson discovered his talent as a *teacher* that had lain dormant since he turned his back on his adolescent charges 18 years earlier. It was there and then that he discovered the value of the seemingly worthless *liberal arts* training at Miami and Dartmouth. He recalled especially Professor Bishop at Miami: how he got his students *to learn by teaching one another*. John Henry set out on a similar mission to educate the business world, starting with his own salesmen and workmen. He drilled them; he instructed them; he held conventions and pep rallies. He published manuals that employees had to memorize. He invented the "flip chart" for lecturing and made extensive use of the stereopticon, motion pictures, and other "audio-visual" devices as they became available. In the midst of the National Cash Register factory complex, he built an auditorium; he called it the "Schoolhouse." It remained an NCR and Dayton institution for nearly three quarters of a century.

In later years, he traveled extensively around the United States and Europe. Initially, his trips were to preach the gospel of marketing and salesmanship—and to inspect his burgeoning empire; later they served his peripatetic restlessness and growing hypochondriacal obsessions. As he broadened his horizons, he wanted his employees to broaden theirs, too, and sponsored excursions for deserving workers and their families to New York, Chicago, and even European points of interest. Back home in Dayton, he worked for civic improvements by badgering municipal officialdom and propagandizing from the NCR Schoolhouse. His record was mixed. He succeeded in pushing for city government reform, but he failed to "bury" the Miami-Erie Canal, which he considered both an eyesore and menace to public health—and perhaps the source of painful early memories. (When the canal, several years after his death, was finally covered over and paved, the city named the resulting highway Patterson Boulevard in his honor.)

Patterson died a millionaire—many times over. He made money, not by cutting costs and obsessing on the "bottom line," but by improving his product and aggressively expanding sales. For a wealthy man, he was relatively indifferent to money. He was the most generous employer in Dayton in wages, salaries, and worker welfare (which he pioneered). When the great flood struck Dayton in 1913, Patterson absorbed the cost of flood relief—several millions of dollars—as a routine operating expense for the year. Before he died, he turned much of his landed estate over to the city of Dayton as a recreational area.

It is said that when traveling abroad, Patterson, restless as ever, would seek evening entertainment by going from one theater and show to the next, rarely staying to see an entire performance. Sometimes he would visit several theaters in a single evening. Did he like Shakespeare? His biographer does not say. But one can imagine him slipping out of a performance of *Julius Caesar*, topcoat, walking stick, and bowler in hand, before Mark Antony had quite finished his famous oration over the great man's corpse:

Moreover, he hath left you all his walks,
His private arbors, and new-planted orchards,
On this side Tiber; he hath left them you,
And to your heirs for ever—common pleasures,
To walk abroad and recreate yourselves.
Here was a Caesar! When comes such another?

Sources: Samuel Crowther, *John H. Patterson: Pioneer in Industrial Welfare* (Garden City, New York, 1924); Ron Rollins, ed., *For the Love of Dayton: Life in the Miami Valley, 1796-1996* (Dayton, 1996).

DAYTON SOLDIERS' HOME AND VETERANS AFFAIRS MEDICAL CENTER

President Abraham Lincoln signed legislation establishing the National Home for Disabled Volunteer Soldiers on March 3, 1863. The law provided for the establishment of government-operated facilities to treat, rehabilitate and, if necessary, house the veterans of the Civil War. It represented the first time in American history that the government assumed responsibility for the welfare of a portion of its citizenry. As years passed, the soldiers' homes became a complex social service program, providing housing, community services, medical care, and lifetime support to men whose lives were forever changed by war.

The Dayton Soldiers' Home opened its doors in 1867, as the Central Branch of the National Home for Disabled Volunteer Soldiers, to provide comprehensive care to the war's veterans. Eligible veterans were all disabled soldiers honorably separated from U.S. military service, thus excluding men who had fought for the Confederacy. (Southern state governments did establish homes for Confederate veterans, but the treatment did not approach the comprehensive care provided by the National Soldiers' Home system.) The Dayton Soldiers' Home became the largest facility of its kind. The Home Hospital was completed in 1870 with 300 beds and an operating room. It was regarded as one of the best in the nation and, by 1900, several additions brought the capacity to 840 acute-care beds. During the Civil War, primary treatment for severe limb wounds was amputation with the more fortunate soldiers admitted to soldiers' homes for extensive physical and mental follow-up treatment. A barracks at the Dayton Soldiers' Home provided an additional 6,000 beds for disabled veterans in the latter category.

A military governor originally managed the Dayton Soldiers' Home. Residents, as many as 7,000 at one time, wore military uniforms and lived the regimented life of soldiers. Physical exercise and hard work were emphasized. The residents maintained the grounds and buildings, many of which were constructed by veterans, including the Home Chapel (the first permanent church built by the United States government) and the conservatories. Hundreds of thousands of people visited yearly to see relatives in residence and some just to vacation in the Home hotel, eat in the restaurant, stroll the lush gardens and grottos, and enjoy nightly entertainment in the theater. Some brought loved ones who had served in the Union army and needed treatment or visited the grave of a loved one who was buried in the impressive cemetery that, much like Arlington National Cemetery, featured long neat rows and columns of stately white grave markers standing in silent order. Many Dayton families, including that of Bishop Wright, took the trolley to the Soldiers' Home for picnics during the summer months. Visitors wandered through the grotto and a seemingly endless series of greenhouses and gardens and visited the wildlife preserve that housed alligators, deer, and bears. Concerts,

An 1885 lithograph showing an aerial view of the Dayton Soldiers' Home. The expansive grounds included the main hospital and barracks, a chapel, bandstand, store, Memorial Hall, landscaped gardens, several lakes, and a hotel. (*Dayton VA Archive*)

lectures, and theatrical performances in the Memorial Hall were well attended by the local community, as well as the residents. The Home Band entertained nightly. From the grounds, visitors could look across Gettysburg Avenue and view the Cyclorama, a life-sized, painted diorama depicting several scenes from the battle of Gettysburg.

After World War I, the National Soldiers' Home system was inundated with soldiers returning from Europe, while still caring for an aging population of Civil War veterans. Fifty new homes were built nationwide and operated by several federal agencies before the various hospitals, homes, and services combined into the Veterans Administration (VA) in 1930. At that time, the Dayton Soldiers' Home became known as the Dayton VA Medical Center (VAMC). (The Veterans Administration later became Veterans Affairs.)

The returning World War I veterans desired to recover from their wounds and quickly rejoin society. This was made possible with advances in healthcare, shifting the focus from simply relieving pain and symptoms toward repairing injured bodies, healing physical and mental diseases, and rehabilitating men for return to society and the workplace. The Dayton

VAMC remained one of the largest field stations through the 1970s with a total bed capacity of 2,153. Hospital stays became more infrequent and outpatient care more common in the 1980s. At the turn of the twenty-first century, the Dayton VAMC continued to provide a complete range of inpatient and outpatient services, including medicine, surgery, primary care, rehabilitation, mental-health programs, and geriatric and extended care. In 2003, the total capacity was 500 beds and included a nine-story patient tower that opened in 1992.

In the 1970s, the Department of Veterans Affairs (VA) and Department of Defense (DOD) Health Resources Management and Emergency Operations Act (Public Law 97-174) passed, establishing cooperative agreements and the sharing of resources between DOD and VA medical facilities. This was augmented in 1983 with a memorandum of understanding that established VA healthcare facilities as emergency hospital sites during crises. During the Persian Gulf War (Operation Desert Storm), the Dayton VAMC readied itself to serve as an emergency treatment site capable of handling thousands of potential victims of chemical and biological weapons attacks by Iraqi forces. Today,

the VAMC and Wright-Patterson Medical Center share treatment capabilities that are highly specialized but in relatively low demand. All gynecological services are provided at Wright-Patterson, while patients at the base that need inpatient psychiatric or sleep-disorder treatments are referred to the VAMC. The sharing of resources provides savings for both institutions.

Today, many of the original Victorian-era buildings remain from the days of the Dayton Soldiers' Home, and they have been listed on the National Register of Historic Places. Many, once so full of activity, now sit empty and in need of preservation. The campus of the Dayton VAMC, where medical and social medicine was pioneered in the decades following the Civil War, now stands as one more example of the progressive and innovative spirit commonly found in the Dayton area.

Sources: "Saving History," *Government Executive*, July 2003, pp 45-52; Jeff Hull, VAMC, personal communication with Diana Cornelisse, July 14, 2003 and September 15, 2003; "Monuments to Times Past," *Dayton Daily News*, August 14, 2003, p A15; "National Military Home, Dayton, Ohio: Virtual Tour and Exhibits," viewed online September 25, 2003, at http://www.dayton.med.va.gov/museum.

An alligator pond was one of many attractions that enticed Daytonians to visit the Soldiers' Home. The Home also maintained an indoor nursery for baby alligators, as well as habitats for deer, elk, bears, antelope, foxes, raccoons, a wolf, and a variety of birds. (*Dayton VA Archive*)

Veterans in residence at the Dayton Soldiers' Home relax in the grotto. The grotto included a waterfall, drinking spring, stone archway and steps, and lush foliage. (*Dayton VA Archive*)

PIONEER SETTLEMENT

It had not always been so. A little more than a century before, the site of Dayton, Ohio, was hardly more than a forested lowland area filled with tangled underbrush along a lazy bend in a wide, slow-moving stream shown on maps of the Ohio country of the Northwest Territory as the Great Miami River.

Thus, it caught the attention of a group of New Jersey land speculators in the mid-1790s. Several of their number had already reconnoitered the area a decade before and found it filled with wild game, a few white squatters and somewhat more Indians, believed friendly. They approached their friend and local congressman, a Revolutionary War hero, General Jonathan Dayton. Would he be interested in obtaining Congressional approval to purchase the land and resell it to

Jonathan Dayton, namesake of the city of Dayton. General Dayton never visited the city. (*Dayton Metro Library*)

Daniel C. Cooper, Dayton founder and philanthropist (*Dayton Metro Library*)

prospective settlers? Indeed he would! Within a year, three small bands of pioneers set out from Cincinnati, two by land, one by water, to the site of the new settlement. On April 1, 1796, the group traveling up the Great Miami River, including Benjamin Van Cleve, came ashore near the present intersection of St. Clair Street and Monument Avenue.[62] A few days later, they were joined by the overland parties led by George Newcom and Samuel Thompson.[63] The previous year, one Daniel C. Cooper—by prearrangement with the New Jersey syndicate—had already platted the land, named several streets (St. Clair, Jefferson, Ludlow), and christened the soon-to-be settlement in honor of General Dayton.[64]

Dayton, in fact, never laid eyes on Dayton. Indeed, a few years later he probably wished he had never heard of the place. As events transpired, John Cleve Symmes, one of the group who approached General Dayton, had failed to secure proper title to the land that he then proceeded to sell to his partners and unsuspecting homesteaders. He purchased his share of the land, moreover, at 67 cents an acre—well under the government's going rate of two dollars an acre. Congress got wind of what had happened and insisted on restitution of the full amount to the Treasury of the United States from those then holding the properties. Most homesteaders, then as later, were bargain-hunters and could ill afford the government rate. Many pulled up stakes and left. Others threatened to do so. Dayton, Ohio, was threatened with becoming a ghost town less than a decade after its founding.[65]

Benjamin Van Cleve, one of Dayton's earliest settlers, served as Dayton's first postmaster, librarian, and schoolmaster. (*Dayton Metro Library*)

It was the first great crisis in Dayton's history, and it was resolved, as at least one other crisis in Dayton nearly a century later, by a single individual of considerable wealth, exceptional philanthropy, and civic patriotism. That man was the selfsame Daniel Cooper. Cooper bought up all the bad shares, paid off the Treasury, and then resold the land to those willing to stay in town at prices that each was able to pay. It stopped the stampede out of Dayton and saved the day—and Cooper's own already considerable investment in the area.[66] On his death, Cooper donated land for a public park, which today bears his name[67] and houses a statue of President William McKinley. There is no statue of Cooper anywhere in Dayton, a fact that local historian Charlotte Reeve Conover thought unfortunate. But then, in the words of an eminent Englishman, "if you seek his monument, look about you."[68]

Cooper also sold Robert Patterson, newly arrived from Kentucky,[69] land south of town at the foot of the "far hills" and a stream called the Rubicon. There Patterson established a wide-ranging farmstead—after shooing Cooper's men off the land, where they continued, following the sale, to poach for trees, shrubs, and foodstuffs.[70] Patterson arrived in Dayton in 1804, a year before James Steele, Joseph Peirce, Jonathan Harshman, and Dayton's first lawyer, Joseph Crane (another immigrant from the Garden State).[71] When he came to Dayton, Patterson was already a co-founder of Lexington, Kentucky.[72] Many years later, when Patterson's grandson, John H., heard that the Kentucky locals wanted to pull down Robert's dilapidated lean-to, he moved it to Dayton and preserved it on the family homestead.[73] (John H. also saved George Newcom's rather more commodious cabin *cum* tavern from turn-of-the-century Daytonians all gung-ho for progress at any price.)[74]

Life in early Dayton was, in the words of another Englishman, "nasty, brutish, and short."[75] Daniel Cooper died in 1818 at the age of 45, the result of overstrain from transporting a large bell in a wheelbarrow to the Presbyterian church.[76] In those pioneering days, even the well-heeled literally put their shoulder to the wheel. Natural disaster, war, and disease carried off many more. Daytonians had been hearing of fire, brimstone, and the last days ever since the first itinerant preacher—a Methodist—passed through town in 1798.[77] Since that time, they had built three taverns and one (Presbyterian) church. Their comeuppance came in 1811. An earthquake shook the entire Miami Valley and aftershocks lasted a month. Then, in June, a late frost blighted many crops; a mass migration of squirrels carried off much of the rest.[78] While not quite the plagues of Egypt, it made serious men think. Then, the following year, America locked horns with the nasty and brutish British in the War of 1812.

As during the Revolution, the British allied with Indian tribes alarmed at white migration throughout the Old Northwest. (By 1805, more than 30,000 immigrants of European descent were pouring yearly into Ohio alone.[79]) Dayton was the mustering place for the First Division of the Ohio Militia—some 2,500 men under the command of General William Hull.[80] To keep idle hands occupied during the militia's sojourn in his fair city, Daniel Cooper set the soldiers to work building a levee around the town to protect it from catastrophic flooding.[81] (In 1805, Dayton had already experienced a foretaste of its later bitter cup, when in 1913 the Great Miami overflowed its banks and put much of the town under eight feet of water.[82]) Although successfully overseeing his men in this task of circumvallation, General Hull—alas, no Caesar!—led them to utter defeat in their encounter with hostiles up the Old Troy Pike. This was in August. In September, a man of sterner stuff, General William Henry Harrison raised a second army that likewise passed through Dayton and led it to victory in the Battle of Missisinewa.[83] The following year, he defeated the Indian chief, Tecumseh, at the Battle of the Thames.[84] Of these encounters, only 200 men returned to Dayton to tell the tale.[85] However, everyone loves a hero. Many years later, in 1840, when General Harrison ran for president on the Whig ticket, his Dayton campaign rally—which "Old Tippecanoe" himself attended—exceeded every local civic celebration until Dayton honored the Wright brothers.[86]

EARLY NINETEENTH-CENTURY DAYTON

War concluded (by the Treaty of Ghent in 1814), Dayton got on to the pursuits of peace. The first order of business was to find better ways to get in and out of town. Both overland and water routes were full of obstacles. Consider first the water route. When Benjamin Van Cleve's party navigated up river to found Dayton a decade and a half earlier, spring rains had swelled the Great Miami. However, as soon

Newcom's Tavern, built in 1796 on the corner of Main and Water (now Monument) streets. This home of Colonel George Newcom and his family is Dayton's oldest extant structure. It has been moved to Carillon Historical Park and preserved as part of their outdoor exhibits. (*Dayton Metro Library*)

East side of Main Street, from Third to Second streets, looking north, 1850 (*Dayton Metro Library*)

The Miami-Erie Canal reached Dayton in 1829. It was up to 40 feet wide at water level and approximately 4 feet deep. Horses and mules towed heavily laden flat-bottomed boats down the canal. In 1877, the state of Ohio abandoned the canal. Fifty years later, Dayton filled in the bed and replaced it with Patterson Boulevard, named in honor of John H. Patterson, founder of NCR. (*Dayton Metro Library*)

became apparent, the river was never as high during the remainder of the year, except for occasional flash floods. In addition to low water, narrow and irregular channels, and sand bars, man-made obstacles existed as well. Farmers dammed portions of the river along the banks to drive their gristmills, and other "factories" were dependent on waterpower, too. Indeed, the area that later became the posh suburb

Robert W. Steele, Dayton business leader and philanthropist. In 1894, Steele High School (razed in the 1950s) was named in his honor. (*Dayton Metro Library*)

Robert C. Schenck, Dayton lawyer, member of the United States House of Representatives, a general in the Union Army during the Civil War, and United States ambassador to Great Britain. Schenck is buried in Woodland Cemetery, Dayton. (*Dayton Metro Library*)

"Lower Dayton View" accommodated several mills and other light industry throughout the nineteenth century. Fishermen also established weirs along the river to catch fish driven in the downstream current. Alas, flatboat commerce during the first three decades of Dayton's history depended upon the regular viability of the river with which the mill dams and fish weirs often interfered. To regulate the placement of mills and weirs and to maintain a viable river channel at times of low water, Dayton established a Navigation Board around 1815.[87]

Overland routes were, if anything, even worse. Indeed, just getting around Dayton's streets, mired in mud and muck, could be an unpleasant chore during rainy season—and remained so throughout most of the century. Huge logs, arrayed in "corduroy" fashion, served to pave Dayton's main thoroughfares, thus limiting the depth to which man, wagon, and beast sank into the mud. (Workers laying

Dayton's first modern pavements a century later unearthed a number of these enormous tree trunks that lay buried beneath Dayton's streets like the bones of a long extinct prehistoric beast.) Roads also began to radiate out of Dayton. The Dixie Highway was one of the first.[88] The first bridge also was thrown across the Great Miami River around this time. A schedule of tolls for foot, horse, or carriage traffic kept the bridge in repair well into the 1870s.[89] In 1818, the first weekly coach service was established between Dayton and Cincinnati by way of Franklin, Middletown, and Hamilton. Even so, travel remained ponderous. It took two full days to make the trip by stage between Dayton and Cincinnati, much to the profit of Hamilton's innkeepers.[90]

Canal! In 1821, the word seemed to city fathers the solution to all of Dayton's transportation problems. Dayton got on board the "bandwagon" of canal construction early, enlisting Ohio's

Huge logs were laid over early Dayton streets to prevent wagons and carriages from sinking in the mud. Here, some of the old "corduroy" paving is uncovered, 1890s. (*Dayton Metro Library*)

governor and key legislators to push through legislation in 1825 that ran the Miami-Erie Canal right past Dayton's downtown and on to Cincinnati.[91] Much like the later I-75 interstate highway, which similarly bisected the city (and could, at times, be as slow-moving), the canal was both opportunity and eyesore. However, in its heyday, the canal maintained Dayton's commercial fortunes until the railroad finally came to town in the late 1840s.[92]

If all this sounds primitive—it was. However, it would be fallacious to conclude that lack of physical comforts, not to say amenities, meant that Dayton's citizens were uncultured or crude. Quite the contrary. From the very beginning, Dayton (for whatever reason) attracted men and women of uncommonly high caliber. Take, for instance, the multifaceted Benjamin Van Cleve. After accompanying the first party of settlers to Dayton, Van Cleve established Dayton's first post office,[93] school, and later library, all in his log house. His was also the first registered marriage in Dayton, to Miss Mary Whitten.[94] Their son, John, was, by all accounts, a child prodigy and, over time, polymath. Large of both mind and body, John Van Cleve graduated from Oxford University but returned to his hometown to serve Dayton in a variety of capacities that included city engineer and editor and publisher of the Dayton *Journal.* At 16, he was teaching Latin and Greek and translated stories and plays from French and German. Organist and choirmaster of Christ Church in town, he was also a founder of the Dayton Library Association and contributed collections of plant life, geological specimens, and fossils to Dayton's first natural history museum. He also helped found Woodland Cemetery, where he lies buried. Sensitive of his appearance—he reportedly weighed more than 300 pounds—he never married nor sat for a portrait or photograph.[95] He is remembered in a short *life* written by another remarkable nineteenth-century Daytonian, the intellectual and life-long invalid Miss Mary Steele,[96] whose father Robert, an educator and businessman *extraordinaire,* lent his name to Steele High School, built in the 1890s.[97]

Judge Joseph Crane (Dayton's first attorney, see above) recruited another Oxford grad, Robert Schenck, to Dayton's citizenry, in 1831. Schenck established a Mechanics Institute in Dayton for the self-improvement of working men before his election to the Ohio legislature and then Congress. In later years, he served as United States ambassador to Brazil and a brigadier general in the Union Army during the Civil War, and closed out his career in America's plum diplomatic posting, as ambassador to the Court of St. James.[98]

The Mechanics Institute had a library, as did the hotel built by Horatio G. Phillips at the corner of Third and Main streets in 1852.[99] Indeed, Mr. Phillips outfitted the hotel's vast reading room with gilt chandeliers, globes, tables, and bookshelves along the walls amidst Corinthian columns. The library certainly impressed Dayton physician Dr. John C. Reeve's little girl, Charlotte. Years afterwards, Charlotte recalled that she had never seen so many books in one room in her life.

The Mechanics Institute was preceded by the Dayton Lyceum Association, established in 1833 for "the diffusion of knowledge and the promotion of sociability." Its weekly lecture series wisely steered clear of politics and religion, but featured a full spectrum of other topics of interest to Dayton's citizens.[100]

Mid-nineteenth-century Dayton (*Dayton Metro Library*)

Phillips House Hotel, located on Third and Main streets in downtown Dayton. It was built in 1852 and demolished in 1926. (*Dayton Metro Library*)

Dayton's second courthouse, built on the northwest corner of Third and Main streets in 1850 for $100,000. Now called the "Old Courthouse," it once housed the Montgomery County Historical Society. It was listed on the National Register of Historic Places in 1970. (*Dayton Metro Library*)

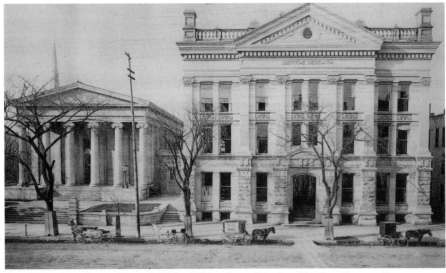

The "New Courthouse" was built in 1881 adjacent to the Old Courthouse. It was razed in 1972 to give way to "Courthouse Square." (*Dayton Metro Library*)

One topic that had begun to stir Dayton along with the rest of the nation in the 1830s was slavery and the question of its abolition. The first meeting of the Dayton Abolition Society was held in 1832. That year Federal agents arrested a fugitive slave living in Dayton called "Black Ben." Ben was well liked by several Dayton citizens, who offered to buy him from his master. When their offer was refused, Ben flung himself from a tall building, killing himself, rather than return "down river." [101]

In the 1830s, abolition was still a relatively remote issue. By the 1850s, however, two decades had been allowed to stir the pot of national debate, and the passions of men and women both north and south frequently boiled over, with occasional explosions around the country. One of the earliest outbreaks of *anti*-abolition sentiment occurred in Dayton. Dayton, located in southwest Ohio, was in very much a "border region" containing those arguing both sides of the slavery and abolition issues, right through the Civil War. Its citizens were drawn from both the New England and Mid-Atlantic states as well as southern and border states like Kentucky. A more potentially explosive combination of sectional sectarianism could hardly be imagined.

Clement Vallandigham, a Dayton lawyer, became a member of the State House of Representatives in 1845 and was elected to Congress in 1858. Vallandigham was a strong supporter of state's rights and slavery, and was imprisoned in 1863 for being a Peace Democrat. (*Dayton Metro Library*)

WOODLAND CEMETERY

Woodland Cemetery & Arboretum south of Dayton contains the graves of many men and women who advanced military and civilian aviation in the greater Miami Valley.

Woodland Cemetery opened in June 1843, on a 40-acre plot with roads sized for carriages, making it one of only five rural cemeteries in the country. The first burial there occurred on July 11, 1843, and soon thereafter 19 Revolutionary War veterans and soldiers killed in the War of 1812 were reinterred at the cemetery. By 1850, burials from a "downtown" cemetery (being encroached on by the growing city of Dayton) had been moved to either Woodland or Catholic Calvary Cemetery farther south. Woodland also contains the burials of Union soldiers from the Civil War, because the Dayton Soldiers' Home and cemetery had not yet been established. Hundreds of military members killed in action during twentieth-century wars also are buried in Woodland. The grave of Dale White, Sr., who was instrumental in the creation of the acclaimed 332nd Fighter Group, popularly called the Tuskegee Airmen of World War II, is located in the newer section of the cemetery. White also held the honor of being the first black mechanic hired at Wright-Patterson Air Force Base, where he completed a 30-year career.

Additional tombstones in Woodland trace early local aviation history. The Huffman family mausoleum stands atop one of many hills in Woodland. It was on Torrence Huffman's property east of Dayton that Wilbur and Orville Wright learned to fly in 1904 and 1905, and on which they opened their flying school in 1910. Perhaps the most famous burials in Woodland, however, are those of the ingenious brothers who "invented the airplane." Wilbur died on May 30, 1912, of typhoid; Orville on January 30, 1948, of a heart attack. The family plot near them also contains the graves of their sister Katharine; their parents, Bishop Milton Wright and his wife Susan; and the remains of their twin siblings who died as infants.

Dayton native and poet Paul Laurence Dunbar, some of whose writings Orville and Wilbur had published as printers before they became bicycle repairmen and salesmen, rests in Woodland near the Wright plot. His monument is engraved with his verse:

Lay me down beneaf de willers in de grass,
Whah de branch'll go a-singin' as it pass.
An, w'en I's a-layin' low,
I kin hyeah it as it go
Singin', "Sleep, my honey, tek yo' res' at las."

Patterson Knoll in Woodland holds the remains of Lieutenant Frank Stuart Patterson, a descendent of one of Dayton's founding families. Lieutenant Patterson died tragically at Wilbur Wright Field on June 19, 1918, while flight testing a machine-gun synchronizer on the DH-4 biplane. It is in honor of Lieutenant Patterson that the section of Wright-Patterson Air Force Base east of Huffman Dam is named.

Other notable individuals buried at Woodland Cemetery include Colonel Edward Deeds, who localized early aircraft production in Dayton and was instrumental in land purchases for Wilbur Wright Field and McCook Field; Adam Schantz, one of the men who led the fund raising to build five flood-prevention dams throughout the Miami Valley following the disastrous 1913 flood; and, on the highest hill in the cemetery, George and Mary Newcom, earliest residents of Dayton.

Woodland Arboretum contains nearly 200 species of trees, including the state's largest and oldest Birdnest Spruce and Magnolia trees. The Wright Iris Garden is filled with iris plants donated from the extensive garden of Susan and Horace Wright (Horace was a nephew of Wilbur and Orville Wright) of nearby Bellbrook, Ohio.

On Sunday, July 20, 2003, the Woodland Cemetery Foundation hosted a special memorial and wreath-laying ceremony at the Wright family plot. Participants included Wright descendents Stephen Wright and Amanda Wright Lane; astronaut Neil Armstrong, whose moonwalk occurred exactly 34 years earlier; astronaut John Glenn; and the commander of Aeronautical Systems Center, Lieutenant General Richard V. Reynolds. The ceremony was marked by a flyover of the Wright "B" Flyer.

Sources: "'Bodily remains buried, but spirit lives on," *Dayton Daily News*, July 21, 2003, p. A1; personal tour by Jim Sandegren, Woodland historian, fall 2002.

The Wright family memorial stone. Wilbur and Orville, their sister Katharine, and their parents Milton and Susan Wright are buried in the family plot.

The Patterson family knoll is marked by a large memorial stone inscribed with the names of those interred.

Poet Paul Laurence Dunbar's final resting place lies under a willow tree.

In the 1840s, an abolitionist speaking in front of Dayton's original courthouse was pelted with rotten eggs by a dissenting crowd. During the same period, anti-abolitionists attacked a church hosting an abolitionist speaker. The mob invaded the church, drove the speaker from the pulpit, hacked the pulpit, and even tore the Bible to pieces.[102] When Abraham Lincoln addressed the issue of slavery from the "new" courthouse (i.e., the present "old" courthouse) steps in 1859, Dayton's *Empire* newspaper observed that while "Mr. Lincoln is a seductive reasoner…his speech was a network of fallacies and false assumptions."[103] One of Dayton's most prominent citizens and one-time congressman, Clement Vallandigham, would have agreed. But not all his fellow citizens agreed with him. During the war, Vallandigham's home was invaded by Union soldiers, who took him into custody and deported him to the Confederacy.[104] Another anti-war Democrat was Dr. Reeve. Charlotte *Reeve* Conover recalled years later how she and her sisters listened in hushed silence from their bedroom window as local pro-Union toughs debated whether to fire the Reeve's home. They did not,[105] but others were not so lucky. In retribution for the Vallandigham affair, a mob attacked the pro-Unionist Dayton *Journal's* offices and set them ablaze.[106] Shortly afterwards, the editor of the rival *Empire* newspaper was shot dead in broad daylight. Dr. Reeve could do nothing for him.[107]

Charlotte Reeve Conover remembered many other things about the Civil War: the initial hoopla that accompanied Lincoln's call for troops following the attack on Fort Sumter; the handwritten lists of battle casualties posted at the courthouse on the corner of Third and Main streets; military funeral processions down Brown Street, flags at half staff, drums muffled, to the new city cemetery; the increasing number of older women and young ladies dressed in mourning black; the jubilation at news of General Robert E. Lee's surrender; the profound mourning and return to black crepe following Lincoln's assassination.[108]

LAYING THE FOUNDATIONS OF MODERN DAYTON

The Civil War brought down the curtain on the first half of nineteenth-century America. Dayton was no exception. The founding generation of Daytonians, moreover, had largely passed from the scene, leaving little trace save in the fading memories of their aging progeny and on tombstones in the Fifth Street cemetery. Even there they did not long rest in peace. The expanding city needed the space— now prime real estate—and the graves were dug up—with little show of ceremony or sentiment (local schoolboys watched the lugubrious procedure while perched buzzard-like atop the surrounding palisade fence)—and transferred wholesale to the new Woodland Cemetery located on a hillside overlooking town. Meanwhile, forces were set afoot following the Civil War that transformed, in a single lifetime, the way ordinary men and women lived and dressed and thought about themselves.

Survivors of the twentieth century often like to brag about all the changes that have occurred in the 100 years since 1900—the automobile, the airplane, and spaceflight. However, from the standpoint of how ordinary people live their lives, there could be no more dramatic period of change than the nineteenth century. To take but one example: Milton Wright was born in 1832 in a log cabin in Indiana.[109] During his lifetime, he moved his family to several different domiciles, the last a modest-sized, wood-frame house at 7 Hawthorne Street in Dayton.[110] When his sons, flush with success and riches from the marketing of their flying machine, planned a new home, Wilbur insisted that his bedroom have a private bathroom—and a large one at that.[111] This was thought neither extraordinary nor extravagant by the two buttoned-down bachelors. Although the Wrights were building a mansion, any reasonably commodious home in Lower Dayton View at that time had a bathroom off each of three or four bedrooms.

Indoor bathrooms meant indoor plumbing and central heating. All this had begun to invade living spaces of upper class Daytonians as early as the 1870s and 1880s. Well before Louis Pasteur and Joseph Lister's research into germ theory were published and widely disseminated, people associated bad water—or the malodorous "vapors" that wafted above it—with illness and disease.[112] Dayton had suffered recurrent bouts of typhoid, malaria, and cholera since the 1820s. Joseph Peirce and Benjamin Van Cleve succumbed to malaria in 1821.[113] Cholera epidemics were particularly devastating. In 1833, a Dayton minister buried his entire congregation of 42 souls after an outbreak of the disease.[114] The mayor and city council called for fasting and prayer—no better solution than that available to their medieval European ancestors centuries before. Perhaps at the instigation of returning Civil War veterans, who learned the hard way to secure potable water when encamped, Daytonians voted, in 1869, to engage the Holly Water Company to build water purification works to secure safe drinking water.[115]

About the same time, Daytonians decided to begin "enclosing" livestock that hitherto had been allowed to wander around town in search of fodder.[116] Up to that time, "a mother pig with her brood could nestle comfortably in a puddle on the corner of Jefferson and Fourth streets without menace from her human neighbors. Geese

Hawthorn Hill, where Orville Wright lived from 1914 until his death in 1948 (*Library of Congress, Papers of Wilbur and Orville Wright*)

Dayton skyline, about 1930. Visible are (left to right) Steele High School, Biltmore Hotel, Dayton YMCA, Rike-Kumler Department Store, Miami Hotel, and the United Brethren Building (later known as City Centre). (*Dayton Metro Library*)

squawked their way across our principal thoroughfares without molestation."[117] Cows going to and from pasture regularly got "high" on the mash left over from the local brewery. An elderly Charlotte Reeve Conover remembered Dayton in those days as "a mud-bespattered, garbage-ridden, rubbish-littered little town."[118]

But more than the stench of pig, horse, and cow manure—and swarms of flies— was wafting in the air. *Change*—rapid change—was about to descend on Dayton, as it did over towns and cities throughout the entire North American continent in the last quarter of the nineteenth century.

Daytonians had become used to communicating with their fellow countrymen near and far, *at the speed of light*, with the introduction of the telegraph. The first wire for this purpose was strung in Dayton, probably in the 1850s.[119] However, by the late 1880s, the skies overhead would darken with a veritable cat's cradle of wires, as first the telephone and then electric power penetrated the Miami Valley. This all began in 1878, when the enterprising George Phillips read about the telephone, traveled to Boston to see a demo, and returned to Dayton with the glad tidings. In less than three years, Dayton had a phone directory and long-distance

connections with West Milton, Piqua, Troy, Xenia, and Miamisburg. Electric lighting was not far behind.[120] In 1882, a demonstration, *à la* Menlo Park, illuminated Dayton's commons. By the end of the decade, the little urchins who lit Dayton's gas street lamps every evening, ladder and torch in hand, were permanently out of a job: the price of progress.[121]

While welcoming potable water, the telephone, and electric lighting—much of which was paid for by personal subscription—Daytonians were in a rut when it came to street and sidewalk improvement and the installation of sewers, and the rut was getting deeper year by year. Stone sidewalks had been installed before mid-century in the center of town, but radiated only a couple of blocks in each direction.[122] No sewers existed; backyard cesspools were common. If humanity itself had bathed more often in the nineteenth century, the stench might have been more noticeable. City streets remained, as late as the 1880s, as they had been in the 1790s, dusty or muddy, depending on the season. This had been tolerable for much of the first half of the nineteenth century, when Dayton numbered under 10,000 souls. Except for an occasional buggy or dray, there was not much traffic at high noon,

even in the center of town. However, by the 1890s, Dayton's populace had swelled to more than 60,000.[123] Civic leaders and local businessmen began to fret over Dayton's backwardness.

That backwardness embarrassed city worthies in 1884, when Dayton hosted a reunion for Civil War veterans of the Grand Army of the Republic. The highlight of the occasion was the unveiling of the soldiers' monument at the intersection of Main and Water streets (subsequently renamed Monument Avenue). As the veterans, their families, politicians, brass band, and citizens assembled and gazed upon the lofty column, the heavens opened up with a torrent of rain that, over the course of the next several hours, turned the streets into a muddy quagmire.[124] Dayton and her veterans deserved better.

However, building sewers and paving miles of city streets was an expensive proposition, one that needed the support of the entire community. Unlike bridge traffic, moreover, it could not be restricted and paid for by tolls. The standpatters held their own until 1889. In that year, William Huffman (whose son, Torrence, a dozen years later put his pasture at the disposal of pioneering aviation[125]) headed up a committee of three commissioners who

Horse and trolley car for the Southern Ohio Lunatic Asylum located on Wayne Avenue. The asylum later became the Dayton State Hospital. (*Dayton Metro Library*)

Steele High School and the soldiers' monument in downtown Dayton, early 1900s (*Library of Congress, Detroit Publishing Company Photograph Collection*)

promoted street paving and covered sewers. Subsequently, a committee of 100 concerned citizens approved a paving scheme. The following year, a system of storm and sanitary sewers was also undertaken.[126] Over the next decade, Dayton began to emerge from the medieval mire and muck that had characterized it since the first settlers clambered up the slippery banks of the Great Miami River during the spring rains of 1796.

During the last quarter of the century, Dayton also faced up to other problems of urban life. In the 1870s, the battle with demon rum was joined as the militant Temperance Movement hit Dayton and hit it hard. Ladies who otherwise subscribed to the nineteenth-century German housewives' dictum of "Kinder, Kirche, Küche" ("children, church, and kitchen") marched down to Sixth Street, invaded places of masculine refreshment there, and emptied bottles and barrels of spirits into the gutter.[127] That same decade, Dayton established its first professional police force of 35 men to maintain law and order.[128] (Lest readers infer a connection between these events, let them be assured by the chronology: the police force was established in 1873; the women marched in 1874. The efficiency—or *prudence*—of the police force was another matter.)

Twenty years later, in 1890, Dayton finally established a full-time, professional fire department. Fire fighting in Dayton had come a long way since the city organized the first volunteer force in the 1820s.[129] Volunteer crews had long contributed through their inefficiency—sometimes malicious—to urban clearance. Many a public eyesore was consumed beyond saving by the time the fire engine showed up! The first steam engine was introduced in the 1860s. In 1905, the fire

department acquired its first gasoline-powered engine,[130] and put its last two horses out to pasture in 1917.[131]

But not all improvements in urban life were of a strictly public nature or introduced by local government initiative. During the 1870s, Dayton also got its first full-time hospital when the Sisters of the Poor of St. Francis established what became St. Elizabeth's Hospital. Dr. Reeve survived his wartime politics to become the new institution's first chief of staff.[132] In 1870, Dayton formed the Young Men's Christian Association (YMCA), with a young David Sinclair as recording secretary.[133] Sometime later, Dayton's young ladies formed the Women's Christian Association, the forerunner of the Young Women's Christian Association (YWCA).[134]

In the following decade, the public library moved from its lodgings on the third floor of the city building to a home of its own, a grand Romanesque-revival structure that presided in turreted majesty over Cooper Park for the next 70 years.[135] On the third floor of the library, a natural history museum contained assorted stuffed birds and animals whose feathers and pelts would save the library staff from the wet and cold 15 years later during the Great Flood (see below).

INDUSTRIAL GIANT

Local entrepreneurs also introduced a bevy of commercial and civic improvements. The leader of the pack was John H. Patterson. By the end of the century, Patterson's National Cash Register Company dominated Dayton's cityscape south of town on the land that

his grandfather had purchased from Daniel Cooper nearly a century earlier.[136] More significantly, the manufacture and retailing of the cash register provided a *point départ* for one of the greatest innovators—and humanitarians—in civic, commercial, and industrial history. Patterson pioneered product marketing, advertising, and modern salesmanship. After building an industrial giant in Dayton's backyard, he also pioneered schemes for industrial welfare, many of which were adopted by other companies. He did not stop there. When he huffed and puffed, all Dayton took notice. He spearheaded the drive for city-manager government; managed flood relief during the 1913 deluge; and led the drive for flood control in its aftermath.[137] He died in 1922, at the outset of a campaign that he personally instigated to keep the Air Service's research and development work in his hometown.[138]

But Patterson was not the only Dayton entrepreneur with outsized plans and ego to match. Take, for instance, the leaders of the transportation industry, of which Dayton had a surfeit. As the nineteenth gave way to the twentieth century, America was on the move. The drive for the western frontier may have ended in 1890, but the *pace* of transportation picked up in a hurry. Had the 1913 flood not swept away much of Dayton's equine population, there would, in a few years, have been a mass final exit to the glue factory as the horse gave way to other, more efficient means of transportation. The bicycle and the automobile both got transportation in Dayton rolling at the turn of the century and gave birth to dozens of new manufacturing enterprises.

The advent of the modern, two-wheeled "safety" bicycle hastened the rate of change, not only in transportation, but in

IF WALLS COULD TALK

Dayton is the birthplace of aviation, but few Daytonians are aware of the aviation milestones that involved two of its oldest downtown buildings: the Reibold Building at 117 South Main Street, and the Centre City Building at 40 South Main.

On May 1, 1923, Dayton inventor Luzern Custer (who created the first, powered wheelchair) took off from the top of the Reibold Building in a "jumping" balloon and landed 22 miles away, one mile west of Middletown, setting a record for the most distance covered in an hour's time by a balloon of its type. On the following day, Custer made another memorable balloon trip from the top of the Reibold Building, when he tried to welcome two McCook Field test pilots as they flew over Dayton on the first successful, coast-to-coast, nonstop flight. Custer took off from Reibold with a "Dayton" sign streaming from his balloon. Unfortunately, Custer miscalculated the wind's direction and hit some wires and a smokestack before narrowly missing an outhouse on Fourth Street. After getting clear, he got caught in a downdraft and almost crashed into a speeding locomotive in north Dayton. Custer eventually came to earth in a cornfield, but the two history-making McCook pilots—Oakley G. Kelly and John A. Macready—never did see his sign because they flew over Dayton two hours later than anticipated.

The United Brethren Building—now known as the Centre City Building—was the site of another aviation milestone: In September 1918, McCook Field contractor Lawrence Sperry, Sr., of Long Island, New York, wanted to demonstrate his new self-contained parachute design. The typical parachute of this period was stored in the plane until the airman needed to bail out, and it often failed to open; Sperry designed his parachute to be worn by the airman and manually opened. Eager to try out his new invention, Sperry jumped out of a plane over McCook Field. He had no problem opening his parachute, but a gust of wind took control of his chute and carried him over the Great Miami River through downtown Dayton. Crowds of people looked on helplessly as his chute snagged on a corner of the United Brethren Building, 11 floors above the earth. Sperry pulled himself up onto a ledge, straightened out his chute, and then floated down into the net of the firemen below.

So the next time you're walking around downtown Dayton, give the grand old buildings at the corners of Fourth and Main a second look—and a closer listen.

Sources: Mary Ann Johnson, *A Field Guide to Flight: On the Aviation Trail in Dayton, Ohio,* revised edition (Dayton, 1996); "Reibold Building" and "United Brethren Building" viewed online September 25, 2003, at http://www.skyscrapers.com; James Custer Collection.

General Billy Mitchell called Lawrence Sperry, Sr., one of the greatest minds in the history of aviation development. In addition to designing an improved, self-contained parachute, which he tested over Dayton, Sperry developed the first amphibious flying boat, a three-way gyrostabilizer to steer bombing planes, a wing hook to enable aerial refueling, retractable landing gear, aircraft lights for night flying, the gyroscopic bank and turn indicator, radio-controlled aerial torpedoes, and the Sperry Messenger biplane. (*United States Air Force Museum*)

Intrigued spectators gather on top of the Reibold Building on May 2, 1923, as Luzern Custer's balloon is inflated with 40 cylinders of hydrogen in preparation for a failed but thrill-filled voyage to greet John A. Macready and Oakley G. Kelly as they flew over Dayton on the first nonstop, intercontinental flight in history. (*James Custer*)

The Centre City Building (formerly known as the United Brethren Building) stood 14 stories tall when Sperry's chute got snagged on the eleventh-story ledge, but the addition of a tower in 1924 made it one of the tallest concrete structures in the world. Its Room 1310 was once the office of the Wright Company Exhibition Team. (*Dayton Metro Library*)

The Reibold Building before its south and north annexes were built (*Dayton Metro Library*)

All systems are "go" for Luzern Custer, center, as he prepares to lift off from the Reibold Building. The man on the left is holding the statoscope that Custer designed to determine if his balloon was ascending or descending, especially helpful to know when passing through a cloud. (*James Custer*)

The Women's Bicycle Club was one of numerous bicycle clubs that sprang up in the Dayton area at the turn of the twentieth century. (*Dayton Metro Library*)

The Stoddard Manufacturing Company on Third and Bainbridge streets built farm machinery until 1905, when it turned to the production of automobiles. At that time, the name changed to the Dayton Motor Car Company. (*Dayton Metro Library*)

Preparing the Military Flyer for its tests at Fort Myer, Virginia, 1909

other, less obvious areas, too. Horses required constant care and feeding. For those living in town without the wherewithal to maintain a horse, there was the daily necessity of walking to the local livery to hire out an animal to ride or pull one's carriage. The bicycle could be parked anywhere near or even in one's home. Aside from occasional lubrication, a bicycle needed little maintenance. Bicycles were altogether cheaper to own and operate than horses and thus opened up to a whole segment of the population opportunities for rapid transport that hitherto had been available only to the well-off. The bicycle, moreover, emancipated women and even spawned new styles in clothing for young ladies that were less restrictive than late Victorian attire.[139] Bicycle clubs for men and women soon sprang up, Dayton's Bicycle Club being one of the oldest and most exclusive.[140] Finally, the bicycle encouraged the development of a whole industry of small-time manufacturers and proved, for one Dayton firm in particular, to be the spring-board to a higher plane of achievement.

If the bicycle provided rapid transportation for the masses, the automobile, since its advent in Germany in the 1880s and 1890s,[141] afforded the comforts of the carriage—without the inconvenience of the horse—for the well-to-do. In the first two decades of the twentieth century, Dayton sprouted a number of automobile companies. In 1905, the Stoddards, father and son, who had made their fortune in agricultural equipment, turned to making automobiles after the younger Stoddard returned from Europe, where he toured continental auto manufacturers.[142] The Stoddards competed in those years with such other companies in town as the Barney & Smith Car Company, the Speedwell Motor Car Company,[143] the Courier Car Company,[144] the Dayton Electric Car Company,[145] the Darling Motor Car Company, the Apple Automobile Company, and the Custer Specialty Company.[146] The Jitney Transportation Company began manufacturing 12-passenger gasoline buses in 1915 that gave the trolleys a run for the public's money.[147] Automobile manufacturers, in turn, spawned the start-up of other companies, such as the Dayton Steel Foundry that made small steel castings for cars,[148] and the Dayton Engineering Laboratories Company (Delco), which transformed the better ideas of Messrs. Deeds and Kettering into real mass-market money-makers.[149]

THE FLYING STINSONS

Next to the Wright brothers, the Stinsons—Katherine, Marjorie, Edward, and Jack—are among the most influential siblings in American aviation history. Although they were from the southern United States, all of them had connections to Dayton, Ohio.

A decade before Amelia Earhart began setting aviation records, Katherine Stinson was thrilling crowds with her aviatrix skills. Katherine earned her wings in 1912 at age 21, making her the fourth woman in the United States to have a pilot's license. She was one of the first pilots to skywrite in the night and to loop-the-loop. When the United States entered World War I, Katherine tried to enlist as a military pilot but was turned down because she was a woman. Determined to serve, she became an ambulance driver in war-torn France. The long days and nights of ambulance duty in miserable weather wrecked Katherine's health, and she had to give up flying. After the war, she became an architect and managed her mother's Dayton property.

Inspired by her older sister's example, Marjorie Stinson decided at age 17 to become a flier and came to Dayton to attend the Wright School of Aviation. When Marjorie soloed on August 5, 1914, she became the youngest female in the United States to qualify for a pilot's license. She and sister Katherine opened their own flying school in San Antonio in 1916, and Marjorie trained over 80 men headed for air service in World War I. Marjorie was the first woman to be accepted into the U.S. Aviation Reserve Corps, and she later became a draftsperson in the Aeronautical Division of the Navy.

Jack Stinson's extensive career in aviation began in 1916, when he became a mechanic's assistant at his sisters' flying school in San Antonio. In 1920, Katherine encouraged Jack and brother Edward to form the Stinson Aeroplane Company in Dayton. Out of a myriad of airplane parts, the Stinson brothers built a plane called the *Greyhound*, which had a mahogany-lined cockpit and an electric starter. After Edward crash-landed the plane, the business folded, and Jack worked in a variety of aviation jobs. He invented the push-pull elevator control used in early commercial airplanes. He also founded the Stinson School of Aviation in Long Island, New York, which provided hundreds of people with aviation training during World War II.

After his sisters became pilots, Edward Stinson wanted to fly, too, but they refused to teach him because of his recklessness and love of liquor. Undaunted, Eddie scraped up enough money to go to Dayton and took flying lessons at Huffman Prairie. After his sisters saw his determination to fly, they helped him get his pilot's license. Eddie became a top-notch aviator and served as a flight instructor at Kelly Field during World War I. In December 1921, he broke the world's record for airplane endurance when he flew for over 26 hours in an open cockpit in the dead of winter.

In 1920, Eddie returned to Dayton to join brother Jack in founding the Stinson Aeroplane Company. After the company closed its doors, Eddie refused to give up on his idea of Cadillac-style aircraft. In 1925, he and his friend Bill Mara went to Detroit to design and manufacture airplanes. There they built the SM-1 Detroiter, the first airplane with a heated, soundproof cabin.

Stinson airplanes also caught the attention of carmaker Errett L. Cord, who had ventured into the aviation field when he and a colleague founded the Cord Corporation, incorporated in Dayton, Ohio, with the goal of developing a tri-motor airplane. In 1929, desiring to put radial air-cooled engines into Stinson's stylish and sturdy fuselages, Cord's corporation acquired the Stinson Aircraft Corporation (the Stinson Aeroplane Company incorporated in 1926), and Stinson soon became a very prosperous division. Although Eddie died in a plane crash in 1932 and the last Stinson aircraft rolled out in 1949, many Stinson airplanes are still in use today—a fitting tribute to the Stinson family's dedication to aviation excellence.

Sources: "Stinson," *Aerofiles*, viewed online September 23, 2003, at http://www.aerofiles.com/_stin.html; John A. Bluth, *Stinson Aircraft Company* (Chicago, 2002); *The Early Birds*, viewed online September 23, 2003, at http://www.earlybirds.org/; *Genealogy of the Flying Stinsons*, viewed online September 23, 2003, at http://www.natchezbelle.org/oldtime/stinson.htm#jack; Nancy Brown Martinez, Reference Coordinator, Center for Southwest Research, email to Robin Smith, June 9, 2003; *Otero-Stinson Family Papers*, University of New Mexico, Center for Southwest Research, Box 4, Folder 5, Letters between Katherine Stinson Otero and Emma Stinson, re: Dayton, OH property, 1923-25; Jerry Slade, personal communication with Robin Smith, July 10, 2003; *Stinson Aircraft Corporation*, viewed online September 23, 2003, at http://www.centennialofflight.gov/essay/GENERAL_AVIATION/stinson/GA2.htm; *Stinson School of Aviation* brochure, 1943; Debra L. Winegarten, *Katherine Stinson: The Flying Schoolgirl* (Austin, Texas, 2002).

Katherine Stinson was dubbed "Queen of the Air" by her many fans. She inspired three younger siblings to follow her into aviation. (*The Early Birds, Carroll Gray Aeronautical Collection*)

When 17-year-old Marjorie Stinson arrived in Dayton to take flying lessons, Orville Wright had her telegraph home to confirm she had parental permission. She is shown here with Katharine Wright, left, at Huffman Prairie, waiting for her turn to fly.

A rare shot of all four Stinson siblings: Jack and Marjorie are in the car, while Edward and Katherine sit in Katherine's plane. (*Center for Southwest Research, University of New Mexico*)

Wright Company factory, about 1911. The Wright Company was incorporated in 1909 and the first factory building was completed in 1910. A second building was added the following year. Orville sold the Wright Company in 1915. (*Library of Congress, Papers of Wilbur and Orville Wright*)

Taking wing at the Wright School of Aviation, Simms Station

Students at the Wright School of Aviation on Huffman Prairie, 1910. Left to right: Duval La Chapelle, A. L. Welsh, Orville Wright, James Davis, Ralph Johnstone, and Frank Coffyn

Despite its down-to-earth industrial successes, however, Dayton's future, in fact, was, by the end of the first decade of the twentieth century, literally up in the air. After escaping the hoopla back home, Wilbur and Orville Wright detrained in Washington on June 19, 1909,[150] uncrated and assembled their flying machine at Fort Myer, and resumed the Army-sponsored flight trials that had been interrupted the previous September when the Flyer had crashed and Lieutenant Thomas Selfridge became the first military airplane fatality in history.[151] Orville, who had been severely injured in the accident, was back in the pilot seat. The final flight occurred on July 30. Orville and Lieutenant Benjamin Foulois took off around 5:30 p.m. When they landed, the Army staff and a large crowd of onlookers, including President Taft, broke out into cheers. Even Wilbur cracked a smile.[152] The Army, which had put the brothers off for nearly five years, enthusiastically shelled out the contract price of $25,000 plus a bonus of $5,000 for beating the specified speed.[153] The Army had just got into the aviation business. Dayton, too.

On returning to their hometown, Wilbur and Orville established a company to build more airplanes for the Army and all other comers. The Wright Airplane Company, while headquartered in New York City, built its manufacturing facility in Dayton.[154] The Wrights also set up a flying school on Huffman Prairie[155] and began licensing exhibition fliers to drum up business.[156] At the same time, they also set out to defend their interests against the competition. The result was a decade-long patent war with the Curtiss Aeroplane Company.[157]

Dayton was thus, in the first years of the twentieth century, an industrial and commercial center that based its wealth on state-of-the-art products, whether bicycles, automobiles, airplanes—or electric cash registers. The city counted more than 1,200 manufacturing enterprises, valued at more than $45 million, in these years.[158] Not all of its industry was heavy. The Gem City was also a center for publishing: more than 51 different periodicals rolled off Dayton's printing presses every year, reaching more than one million subscribers nationwide.[159] The dry-goods trade achieved new heights when the Rike-Kumler company built a seven-story department store in the center of town, complete with elevator, restaurant, and its own power plant.[160] This was vertical integration of a kind that would have made John H. Patterson proud. NCR also had its own power plant, laundry, library, cafeteria, recreational facilities—

and more than 95 percent of the cash register business in the nation.

The inflow of wealth that these enterprises generated, moreover, was not wholly immured in mansions and estates of top executives, board members, and stockholders—there was much of this, too, of course—but was generously diffused throughout the community by company owners who, despite the exigencies of cutthroat capitalism, had still been reared on the Beatitudes and the Golden Rule. Patterson regularly began community meetings that he was wont to summon to the NCR Schoolhouse with hymn-singing and patriotic aires. This was not hypocrisy. Patterson possessed what Europeans called *noblesse oblige*, what Americans called good-neighborliness, but what Patterson himself would probably have simply considered good business. He kept the unions out of NCR by paying some of the highest wages in town. (When Kettering started work at NCR, he tried to return his second week's $50 paycheck,

thinking that he had been engaged on a monthly basis!)[161] Patterson *listened* to his workers and their complaints and redressed their grievances with better working conditions, including side benefits like company trips out-of-town and even abroad for workers and their families at cut-rate prices. He could, on occasion, also be ruthless. He regularly fired his most successful executives. Some theorized that this was to prevent rivals in the company. Perhaps nearer the truth was Patterson's sixth sense of the workings of the Peter Principle—more than 50 years before it was formulated. In any case, NCR's loss was the national business community's gain. As NCR cast off the successful before they reached their level of incompetence, other companies swallowed them up, hoping perhaps to imbibe a modicum of Patterson's success.[162]

The diffusion of wealth among the high and low resulted in an unprecedented degree of home building and ownership

among Daytonians, making Dayton known nationwide as the "city of homes."[163] The townhouses and mansions of the wealthy sat side-by-side the more modest, but still substantial, homes of the middle class, many built in close proximity to the downtown business district. Others perched, castle-like (the Stoddard mansion comes to mind), across the river in Dayton View. Still others clustered in wooded splendor among the far hills to the south. The latter would incorporate as the city of Oakwood—the first of Dayton's suburbs and an early harbinger of the upper and middle-class exodus from urban to suburban living of the post-World War II years.

During these years, Dayton also began to loom larger on the political map of Ohio and the nation. In 1912, James M. Cox, congressman and publisher of the *Dayton Daily News*, was elected governor of Ohio.[164] Eight years later, his track record as a reformer and proven ability to garner votes landed him the top spot on the Democratic ticket for president of the United States (see below).

But there were also clouds on the horizon nearing Dayton. In February 1912, John H. Patterson, together with 29 of his top executives at NCR, was indicted by a federal grand jury for restraint of trade in the cash register business.[165] Then, at the end of May, Wilbur Wright, returning from a trip to Boston, one of numerous and exhausting legal forays in the Wrights' patent war with Curtiss, succumbed to typhoid fever and was buried, in a *private graveside ceremony*, in Woodland Cemetery.[166]

THE GREAT FLOOD

As 1912 passed to 1913, other clouds—no metaphor!—approached Dayton. The winter had been long with ice and snow still on the ground in many places in late March. As the snow slowly melted, heavy spring rains began to fall. The combination of frozen ground and abundant rainfall led to a rapid rising of the Great Miami and its tributaries.[167] Cities all along the watercourse, Piqua, Troy, but especially Dayton, were threatened with major flooding for the first time in more than a decade.

Dayton was not unused to floods. The first major inundation occurred in 1805, just nine years after Dayton's founding. Others followed in 1814, 1828, 1832, 1847, 1866, 1883, 1897, and 1898.[168] And these were

The Stoddard mansion, Lower Dayton View. Charles Stoddard built the Stoddard Manufacturing Company in the late 1880s to produce farm machinery. (*Dayton Metro Library*)

View of West Third Street looking east during the 1913 flood (*Dayton Metro Library*)

the *big* floods. There were numerous smaller inundations. The first attempt at flood control was undertaken with the building of the levee in 1812 with the assistance of Ohio's militia as it awaited its fate at the hands of the incompetent General Hull (see above). Over the years, as Dayton expanded, so too did the network of levees surrounding the city's riverside. These were not, of course, wholly water tight. The 1847 flood occurred largely because those living near the river had inadvertently weakened the levee by carrying off its dirt to fill potholes in the streets, thus undermining the structure.[169]

But the flood of 1913 was different—far worse—than anything previously experienced. The approach of danger was

Panoramic view of the 1913 flood in Dayton (*Library of Congress*)

evident on March 24 as the Great Miami surged close to the top of the city's levees. Still the rain did not stop. Early on the morning of March 25, John H. Patterson, having inspected the levee with several of his entourage, immediately called a meeting back at NCR and began firing orders: prepare food, medicine, drinking water, warm clothing, bedding—and boats, lots of boats.[170] Within the hour, the Great Miami broke past the levee at two points and water cascaded in a six to 10-foot wall through Dayton's streets, smashing railroad cars against buildings, raising frame houses off their foundations, demolishing one entire tower of Steele High School and carrying off the institution's bronze lion. Raging water coursed rapidly along the old canal bisecting the city, then submerged it entirely. Horses drowned in stables and at hitching posts; a number

could be seen wildly attempting to hold their heads above water in the swirling current. A few found refuge on porch roofs and other high points of buildings—only to be shot by their owners when the waters subsided and they were left "high and dry" but with no means of climbing down.

Meanwhile, Dayton's citizens clambered as best they could to the upper floors, attics, and roofs of houses and buildings as the floodwaters invaded interior spaces. Old Dr. Reeve and his wife were among those trapped in their downtown home. "[The water] came so fast I had to hustle to get Mother [an invalid] to the stairs," he wrote his daughter Charlotte. "Outside a raging torrent pours down Wilkinson Street, a mighty river down Third Street towards the west. No human being in sight, no sign of life—silent as the grave."[171] Afterwards he called it "a great calamity,

Aftermath of the 1913 flood in downtown Dayton. The flood caused more than $100 million in damages. (*Dayton Metro Library*)

James M. Cox served two terms as governor of Ohio, 1913 to 1915 and 1917 to 1921. He also unsuccessfully ran for president of the United States in 1920. Cox became owner and publisher of the *Dayton Daily News* in 1898. (*Dayton Metro Library*)

second, perhaps, to the Titanic, but to none other." Governor Cox, quickly learning of the extent of the disaster through his grapevine at the *Daily News*,[172] called it "the worst calamity since the Civil War." The governor declared a state of emergency and called out the National Guard for Dayton and other affected communities.[173]

The *Titanic* had gone to the bottom of the North Atlantic carrying about 1,500 of her passengers to their deaths the previous April[174] and was fresh in everyone's minds. However, the loss of life (if not property) as a result of the 1913 flood was nowhere near as extensive as in the case of the ill-starred liner. Altogether around 123 persons succumbed to the flood in Dayton, through drowning or exposure. Another 155 perished in Hamilton and Piqua.[175] Around 1,450 horses drowned.[176] But Dr. Reeve and Governor Cox can be pardoned historical hyperbole. In fact, the tragedy might have been much worse without the prompt action of Patterson's NCR, which produced more than 275 flat-bottom rowboats to carry stranded victims to warmth and safety and nourishment at The Cash.[177] It helped, too, to have a favorite son high and dry—if just barely[178]—in the governor's mansion in Columbus.

Water, moreover, was not the only cause of death and devastation. Fire broke out the night of March 25, as gas mains ruptured and exploded in several downtown locations. Soon, whole blocks of downtown real estate were ablaze, sending weary flood survivors scurrying from building to building through windows, along ledges, and over rooftops ahead of the flames. Then it began to snow. As the temperature dropped, the staff of the Dayton Public Library, ensconced in the library's third floor natural history museum, reached out for anything to keep warm, including the furred pelts of long-dead, desiccated, display specimens. Someone thought another exhibit piece—an ancient Indian canoe—might be set afloat for help. Alas, it sank—a permanent loss to the museum though no loss of life.

All the ancient elements—fire, water, earth, and air—seemed to rise up against humanity and its twentieth-century hubris. *Nemesis*, as in olden times, seemed bent on taking her full measure. Then, as suddenly, she relented. The water crested on March 25 and was at 10 feet at the Main Street Bridge on April 5.[179]

"Remember the promises you made in the attic!"[180] As Dayton's *hoi polloi* set about shoveling the thick black muck once

the flood waters subsided, the city's high and mighty furrowed their brows as they considered how to ward off a similar catastrophe in the future. Property damage had been assessed at more than $100 million (about $1 billion in today's currency).[181] The loss of 45,000 books at the Dayton Public Library[182] was deplorable; damage and destruction of banking records and business merchandise downtown was another matter altogether. The flood made national headlines due in no small part to Governor Cox's news service; Dayton's awful secret was out. If the Gem City were to continue to attract business and investment, something more would have to be done than cowering seasonally behind ever-higher dykes.

It was a turning point in Dayton's history—indeed, a watershed. And, individually and collectively, her citizens rose to the occasion. Patterson, as usual, stood in the forefront of efforts to prevent future catastrophic flooding. He presided over the Dayton Flood Prevention Committee that set about raising money for the implementation of flood-prevention measures.[183] Mawkish when required,

ARTHUR MORGAN—THAT DAM ENGINEER

Arthur E. Morgan

The Miami Valley was long plagued with flooding caused by the confluence of the Great Miami, Mad, and Stillwater rivers, and Twin and Loramie creeks. Following a disastrous flood in 1913, Dayton businessmen united to create a valley-wide flood control system. The Miami Conservancy District was established and a young engineer, Arthur Morgan, hired to devise a regional flood control system. Although Morgan lacked a formal engineering degree, he had received practical engineering training while apprenticed to his father. At age 17, he began a three-year trip throughout the American northwest, where he observed the mechanics of water flow and developed an intense interest in hydraulic engineering. He read all available literature on hydraulics and soon came to specialize in the field, eventually founding the Morgan Engineering Company. At age 35, he designed and constructed what became a remarkably simple, but highly effective, flood control system for the Miami Valley. He built five large, passive dams over the five major waterways in the area and backed them with huge, dry reservoirs that collected water during rainstorms and gradually released it into the waterways downstream. Some engineers questioned his qualifications for designing and implementing such a large project, but Morgan's fundamental understanding of hydraulic engineering principles continues to be validated yearly as spring rains flood communities along the nation's waterways, with the exception of the Miami Valley, which has not experienced serious flooding since the completion of the Miami Conservancy project.

After completion of the project, Morgan served as president of Antioch College in Yellow Springs, Ohio, where he instituted a program to develop well-rounded individuals with technical skill and a strong interest in humanitarianism. His plan for "industrial education" at Antioch included a 10-week work-study program to promote self-discipline and provide students with the kind of practical, hands-on, real life experience that Morgan himself had received from his father. During his 15 years at the "New Antioch," the college evolved from an obscure school of little note to one of the most innovative and respected liberal arts institutions in the country. The climate of creativity at Antioch drew research activities, such as the Fels Institute for the Study of Human Development, to Yellow Springs. Morgan encouraged Antioch graduates to remain in the area, establish businesses, and serve the local community.

In 1931, President Franklin D. Roosevelt selected Morgan to initiate and design the vast development project known as the Tennessee Valley Authority (TVA), which was, at the time, the largest effort at regional development of natural resources ever attempted. During his tenure with the TVA, Morgan published a biography of a man he had long respected, the nineteenth-century Utopian, Edward Bellamy. Morgan was inspired by the thought of revitalizing America's small communities into vibrant villages that were "…such live, interesting adequate places to live in that young people of quality will prefer to stay there." Morgan envisioned the TVA as an agency

of social planning, working cooperatively with local communities and the existing power producers in order to maximize resources and improve the rural living standards in the area. His attempt to fulfill this vision was doomed by politics. Morgan's idealistic notion of using government to promote social welfare was opposed by those who saw the TVA as an economic engine. After six increasingly contentious years, President Roosevelt dismissed Morgan from his post.

Morgan returned to Yellow Springs, his faith in the viability of a Utopian society still strong. He founded Community Service, Inc., to promote recognition and development of small communities and devoted the remainder of his life to social change. Arthur Morgan remained in Yellow Springs until his death in 1975, at age 97. In many ways, the village of Yellow Springs is still influenced by Arthur Morgan's ideal of a Utopian community.

Today, the Miami Conservancy District's photographic archive is located at the Wright State University Special Collections Division and the Arthur Morgan papers are part of the Antioch College Antiochiana Collection.

Sources: Diana G. Cornelisse, *The Foulois House, Its Place in the History of the Miami Valley and American Aviation* (Aeronautical Systems Division History Office, Wright-Patterson Air Force Base, 1991); Arthur E. Morgan, "The Christian Register, May, 1945;" "Arthur Morgan," viewed online July 8, 2003, at http://www.phd.antioch.edu/Pages/APhDWeb_Prospects/arthurmorgan.

Construction of one of five earthen flood dams in Arthur Morgan's flood control system, 1919. Cost of the construction amounted to more than $30 million.

Patterson urged his fellow citizens "for the love of Dayton" to donate generously. To record the accumulation of treasure, he erected a gigantic cash register at the corner of Third and Main streets, beside the old courthouse, that displayed the rising total as the dollars rolled in.[184] When the initial haul fell short of the goal by nearly half a million, Patterson summoned his fellow citizens, high and low, to a patriotic pep rally at The Cash, followed by a parade up Main Street replete with fireworks and marching band to the center of town. It worked.[185]

Tipped off by a local Democratic politico that the Dayton city council was about to appoint its own engineering firm, Patterson dispatched Edward Deeds to Memphis, Tennessee, where he contacted Arthur Morgan, an expert in dam construction and flood control.[186] Morgan came, saw, and consulted. His recommendation was a region-wide series of dry dams up and down the Miami Valley and its tributaries. Such a program was unprecedented in its scope. To fully implement it required moving hundreds of farmsteads and even a whole town—house by house. More dirt would be moved during its excavation than stone required for the Great Pyramid. Pharaoh at NCR loved it. Once more, he sent forth his grand vizier to drum up public support and cooperation. Deeds did his best, despite often raucous town meetings and even an occasional death threat from a soon-to-be uprooted farmer.[187]

One hurdle had still to be cleared: the plan, because of its length and breadth trespassing so many different municipalities, county lines, and other jurisdictions, required special enabling legislation. Governor Cox guided a bill to this effect through the Ohio Legislature[188] and, on February 18, 1914, signed the Conservancy Act of Ohio into law. Its constitutionality was almost immediately challenged by those threatened with dispossession and other inconvenience by its implementation. An additional year of court fights followed before these challenges were beaten back and the court established the Conservancy District in June 1915,[189] and still another year before the court approved the district's conservancy plan in November 1916.[190] Ground was finally broken on January 27, 1918,[191] and the last hose full of earth fill piped into position in 1922.[192] Nothing in government happens overnight—even with a push and shove from the private sector!

Miami Conservancy District's steamboat, *Dorothy Jean*, named for John H. Patterson's daughter, 1919. The steamboat pushed barges used for dredging the Miami River during construction of the flood control system.

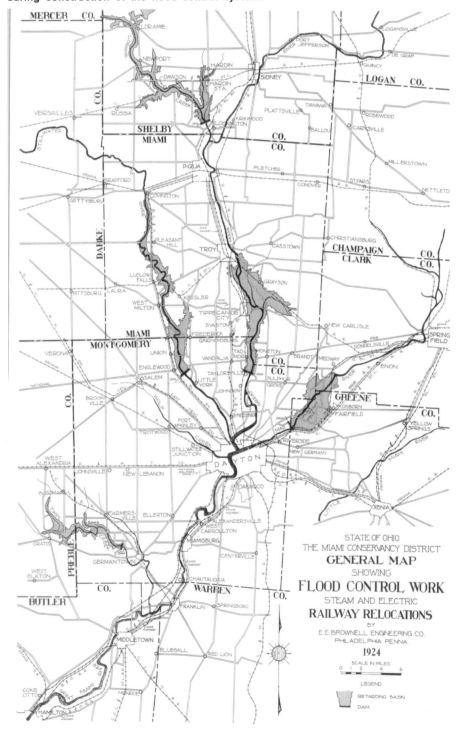

CRADLE OF AVIATION

Fortunately, Mother Nature smiled benevolently on the enterprise throughout and stayed the worst of her storm clouds and thunderbolts. Not so the gods of war. Before Arthur Morgan had excavated one cubic foot of dirt, the crowned heads of Europe shook the very earth with repercussions felt all the way, in time, to Dayton and every city, town, and farmstead in America.

War broke out in Europe in August 1914. The "guns of August"[193] were not stilled until the 11th hour of the 11th day of the 11th month, 1918. The Great War, as it was called at the time—the First World War, since the end of the *Second* World War (1939-1945)—ended nearly a century of armed peace in Europe and baptized, in blood, the most violent century in human history.

President Woodrow Wilson "kept us out of war" for nearly three years, but when Germany resumed the indiscriminate sinking of neutral shipping early in 1917, the United States had had enough. On April 2, the president called upon Congress for a declaration of war on the *Reich.*[194] Four days later, Congress obliged. The United States at last entered the Great War and thereby stepped ever so reluctantly into the limelight of the world stage.

Saul hath slain his thousands, and David his tens of thousands (1 Samuel 29: 5) would have been an appropriate text for a sermonizing pacifist at the time, for the Molech of world war had already claimed his *tens of hundreds of thousands,* nay, *millions.* America's new allies, the French and British, preparing for the next massive German onslaught, came hat in hand requesting men and materiel, including thousands of the latest weapon of war— the airplane.[195]

Of the latter item, however, the arsenal of democracy was nearly bare. When America entered the war, the U.S. Army had 55 airplanes, none of combat calibre.[196] While Europe had honed the aerial weapon during the previous three years of warfare, the United States government had frittered away precious time in intricate legal and moral posturing that left it nearly paralyzed when the moment for action arrived. Since 1912, two years before the outbreak of hostilities, the Aero Club of America had been calling for the establishment of a national *aerodynamical laboratory* to benefit both the government and industry.[197] The most Congress would do, in 1915, was to establish a National *Advisory Committee* for Aeronautics (NACA)—and this on a last minute rider to the year's naval appropriation.[198]

Back in Dayton, sharp observers of the world scene viewed the situation with alarm and dismay. Charlotte Reeve Conover was one of these. It is not surprising that this local chronicler should have so wide ranging a vision. After marrying and raising a family, she decided to complete her

Research facilities at McCook Field, Dayton's first military, aeronautical research and development site, 1922

Aerial view of McCook Field with Dayton in the background, 1921

Edward Deeds did much for his beloved hometown. Once employees of Patterson's at NCR, Deeds and Charles Kettering left to form Delco. Later Deeds returned to NCR to steer it through the Depression. He also played a large role in establishing and keeping military aeronautical research and development in Dayton. (*Moraine Farm*)

COLONEL EDWARD A. DEEDS

Edward A. Deeds (1874-1960) was a key participant in the industrial heritage of the Dayton area. Along with Charles F. Kettering, he co-founded the Dayton Engineering Laboratories Company (known as Delco), the Dayton Metal Products Company, and the Domestic Engineering Company (later Delco-Light). Over the years, he also served as president of the National Cash Register Company (NCR), formed in 1884 by John H. Patterson, and was one of the founders of the Miami Conservancy District, created after the disastrous flood of the Miami River in 1913.

Deeds was also a prominent member of the Aircraft Production Board, a body created in April 1917 to coordinate all activities of the nation's aircraft manufacturers during the massive buildup for World War I. He served on this board until August 2, 1917, when he was appointed acting chief of the newly created Equipment Division of the Signal Corps. Three weeks later, he was commissioned as a colonel in the Signal Corps Reserve and was officially promoted to the position of chief of the Equipment Division, responsible for development and supply of all Army aircraft.

Deeds was a primary influence in the establishment of military aviation in the Dayton area. Not only did he suggest the site for Wilbur Wright Field, he also pushed for the construction of a temporary facility for the centralization of aviation research and development in the Dayton area. The latter facility became McCook Field, the direct predecessor to the research and development function at Wright Field, now part of Wright-Patterson Air Force Base.

Deeds' legacy to his community persists in many ways. His estate, Moraine Farm, which had one of the first private flying fields in the United States, currently serves as a retreat and guesthouse for distinguished visitors to NCR. In 1941, Deeds and his wife, Edith, presented a carillon tower to the city of Dayton. The tower became the focal point of the Carillon Historical Park, which opened to the public in 1950 and serves to educate the public on Dayton's heritage of inventiveness and pioneering spirit.

Source: Isaac F. Marcosson, *Colonel Deeds, Industrial Builder* (New York, 1947).

Edward Deeds, college age (*Moraine Farm*)

Edith Walton Deeds, college age (*Moraine Farm*)

The Deeds family relaxes on the porch of their home at Moraine Farm. After 1921, Deeds transformed his small farmhouse into the stately mansion located on Stroop Road. NCR now uses the home as temporary quarters for distinguished visitors. (*Moraine Farm*)

SELF-STARTER—THE BETTER IDEAS OF CHARLES F. KETTERING

Charles F. Kettering in later years

Before getting into the aviation business, Charles F. Kettering (1876-1958), born in Loudonville, Ohio, made significant contributions to the auto industry from Dayton. Kettering was one member of the "Barn Gang," a group of innovative industrialists who often met in Edward Deeds' barn on Central Avenue to discuss and tinker with technology. Kettering had perfected the electric cash register in 1904, while working at the National Cash Register Company. In 1909, he developed the self-starter for automobiles, which eliminated the need for the hand crank. He immediately received a contract to build more than 8,000 electric starters for the 1909 model Cadillac. From his company, the Dayton Engineering Laboratories Company (Delco), which he co-founded with Deeds, Kettering developed a number of other automotive inventions, including lighting and ignition systems, lacquer finishes, and leaded gasoline.

In addition to his role in the development of automobiles, Kettering can claim a spot in the development of aircraft. In 1912, his interest in aviation was fueled by a flight in a Wright brothers' airplane. Five years later, Kettering joined with several Dayton industrialists to organize the Dayton Wright Airplane Company to tap into the economic potential of the military aircraft industry sure to blossom with United States' entry into World War I. The Dayton Wright Airplane Company went on to manufacture more than 3,100 DH-4 biplanes with Liberty engines for the war effort, as well as 400 Standard SJ-1 trainers.

During World War I, Kettering also designed and built the world's first guided missile, the Kettering Aerial Torpedo, nicknamed the "Bug." The $400 torpedo was a 300-pound, papier-mâché airplane with 12-foot cardboard wings and a 40-horsepower engine. It could carry 300 pounds of explosives at 50 mph. First tested on October 2, 1918, the flying bomb had impressive range and accuracy. The Army ordered a number of Bugs, but the bombs were not used in combat because officials worried about their reliability, especially when carrying explosives over Allied troops. Lessons learned from the Bug later contributed to radio-controlled drones.

Charles Kettering continued to have a leading role in the automotive and aircraft industry in his later years. General Motors Corporation purchased many of his businesses, including the Dayton Wright Airplane Company, and formed the General Motors Research Corporation. Kettering was appointed president of the corporation. With Tom Midgley, he developed the "anti-knock" Tetraethyl Lead gasoline for automobiles, which, when used as an aviation fuel, increased engine horsepower, safety, reliability, and speed. For his contributions to the field of aviation, Charles F. Kettering was enshrined in the National Aviation Hall of Fame in 1979.

Sources: "Charles Kettering," viewed online May 5, 2003, at http://www.nationalaviation.org/museum_enshrinee; "A Starting Point: Dayton's GM Story," viewed online May 5, 2003, at http://www.thinktv.org/program/archive/gmstory/gmstory.html; Stuart W. Leslie, "The Bug: 'Boss' Kettering's Cruise Missile," *Timeline* August/September 1991, pp 42-51.

Kettering "Bugs" lined up on their launch rail (*Montgomery County Historical Society*)

education—at the University of Geneva in Switzerland.[199] This was not unusual for an offspring of Dr. Reeve. Before coming to Dayton in the 1850s, young Dr. Reeve had studied medicine in England and Germany and raised his daughter with an eye to the world beyond Third and Main. Writing of Dayton's honoring the Wright brothers in 1909, she regretted in *Dayton, Ohio: An Intimate History*:

Then and there Dayton should have established those facilities which every inventor needs to carry on his work. Had this been done, in due time we should have developed into the manufacturing center of the world's airplane production. It was with a gasp of incredulity that we suddenly perceived other nations outdistancing us. The French were making extraordinary strides; England was already giving more credit to some of her own workers than to the Wrights.[200]

Not to worry. Who should be in Washington, D.C., when the United States entered the war than the ubiquitous and indispensable Edward Deeds. And how did Deeds happen to be in the nation's capital—just the right place at just the right time?

Deeds had gone to Washington early in 1917 at the request of Howard E. Coffin, the chairman of the Council of National Defense. The council was organizing a Munitions Standards Board and Coffin, who was also vice president of the Hudson Motor Company, had done business with Delco that Deeds and Kettering had established. Coffin had a high opinion of Deeds' engineering ability and business sense.[201] When the Munitions Standards Board was disestablished, shortly after the United States entered the war, Coffin called upon Deeds to join the Aircraft Production Board.[202]

Deeds agreed. He had been very much interested in the airplane and aviation, at least since making personal acquaintance with Wilbur and Orville Wright in 1909.[203] Indeed, when he acquired Moraine Farm from the proceeds of his Delco investment, Deeds set up an airfield—one of the first privately owned airfields in the country— building both a hangar there and a laboratory for aeronautical research and development. That was in 1916.[204] Early the following year, he and Kettering invited Orville Wright to survey a tract of land just north of downtown Dayton, near Triangle Park, for a second airfield. Orville approved

the site and Deeds and Kettering purchased the land with the hope of establishing a public airport near the locus of Dayton's business activity.[205]

In August 1917, Deeds was appointed acting chief and then chief of the Signal Corps' Equipment Division. Along the way, he donned a uniform and was commissioned a full colonel in the Army.[206] Thereafter, he was always "Colonel" Deeds. As chief of the Equipment Division, Deeds was in charge of all aircraft procurement for the Army. Congress had recently thrown more than a half billion dollars at the procurement problem and the public was beginning to clamor for "clouds" of warplanes to fill the sky and overwhelm the Hun.[207]

In fact, the procurement problem nearly overwhelmed the Army and its newly appointed colonel. On entering the war, the United States had virtually no infrastructure for developing and mass producing advanced warplanes. The airplane industry was small and undercapitalized and still in thrall to the Wright-Curtiss patent litigation.[208] The Army had only a handful of engineers in Washington,[209] who adjusted their spectacles and scratched their heads over contractor airplane designs before test articles were taken up to see if they could really fly. The Army's aeronautical research and development occurred mostly in connection with repairing procured aircraft that didn't fly very well, in a few scattered depots, the principal being at North Island, San Diego, California.[210]

What to do? First, the Army needed more airfields and at least one center for conducting research and development on new model aircraft. *As it happened*, two of the sites were in Dayton, Ohio. Deeds, as president of the Miami Conservancy District, put, at the disposal of the Army, conservancy land to the northeast of the projected Huffman dry dam and adjacent to Huffman Prairie, where the Wright brothers had perfected their flying machine within sight of the Dayton-Springfield interurban line. This became Wilbur Wright Field in the summer of 1917.[211] Wilbur Wright would in time become a logistics hub for the Air Service and later, as Patterson Field, for first the Air Corps and then the U.S. Air Force (see Chapter 2: Military Aviation Comes to Dayton, and Chapter 3: The Story of Air Force Logistics).

The R&D center was located in Dayton after brief consideration of Indianapolis.[212] In the event, in October 1917, the Army determined to establish an "experimental engineering field" in Dayton on the North

Field site that Deeds, Kettering, and Orville Wright had scouted out the previous year and in which Deeds and Kettering had previously divested all personal ownership. Dedicated to the memory of the "Fighting McCooks" of Civil War fame, the new center was called "McCook Field" and opened for business in December 1917 (see Chapter 2: Military Aviation Comes to Dayton).

Upon opening its gates, McCook attracted to Dayton some of the top aeronautical scientists and engineers from around the nation and even the world.[213] This represented the second largest influx of outside talent to come to Dayton during the war. The largest was the immigration of African Americans, who came looking for work in northern wartime industry and remained to swell the numbers of their brethren who had helped build the Gem City during the previous century.

Both groups, as it turned out, settled heavily in the north and west of town. West Dayton had long been a black enclave. Dayton View and North Dayton were within bicycling distance of McCook Field. Within the year, the City Register recorded names that soon would be familiar, not only around town, but in some cases, around the country, even around the world.

DAYTON IN THE 1920s AND 1930s

Demographically, economically, socially, and intellectually, World War I inaugurated a new Dayton in a new century. It was this Dayton that cheered the returning doughboys as they trooped up Main Street one last time before doffing their uniforms, having made the world safe for democracy. On either side of their route of march were buildings that had looked down upon similar parades in previous years. Within a decade many of these structures were no longer left standing, having given way to larger, more modern buildings. Of the demolitions, the most historic was the Phillips House Hotel at Third and Main, which fell to the wrecking ball in 1926. It was the library of the Phillips House that had overawed little Charlotte Reeve in the 1850s. It was the Phillips House that offered hospitality to Mr. and Mrs. Lincoln in 1859,[214] and whose grand ballroom was the center of Dayton's social scene from the age of pantaloons and hoop skirts almost to the flapper era. "So is the Phillips House and everything in it gone into the past,"

Spectators watch the International Air Races held at Wilbur Wright Field in 1924. The field commander's residence, now known as the Foulois House, is visible in the background.

Orville Wright, left, and Colonel Edward Deeds enjoy the 1924 International Air Races at Wilbur Wright Field. (*Moraine Farm*)

sighed an elderly Charlotte Reeve Conover, "to make way for a modern Dayton which we fear will neither remember nor care about those lovely, leisurely, fastidious, courteous and friendly days."[215]

However, it was not sentiment but *normalcy* that America craved once the war was over and the calendar turned to 1920. The man who introduced that term into the national lexicon was the editor of the *Marion Star* then serving in the United States Senate. Warren Gamaliel Harding *looked* like a president, said his handlers,[216] and a deadlocked Republican national convention wearily agreed. To oppose him, the Democratic Party turned to the two-time governor of Ohio, leading national Progressive, and campaigning editor of the *Dayton Daily News,* James Middleton Cox. For his running mate, Cox chose the assistant secretary of the Navy in the Wilson administration, the young, handsome, and aristocratic Franklin Delano Roosevelt of New York.[217] At the Montgomery County Fairgrounds on August 7, 1920, before a crowd of more than 100,000 Daytonians and assorted out-of-towners, Cox accepted his party's nomination.[218] The battle was joined and, for a brief moment, Dayton was in the national political spotlight.

Briefly. Cox and Roosevelt fought the good fight, campaigning vigorously for Wilson's League of Nations and completing the unfinished business of the New Freedom.[219] Harding spoke occasionally from the front porch of his home in Marion, Ohio, his speeches, in the inimitable quip of William Gibbs McAdoo, "an army of pompous phrases moving across the landscape in search of an idea."[220] Meanwhile, his vice presidential running mate, Calvin Coolidge of Massachusetts, evinced his trademark taciturnity. When the votes were counted, Harding and Coolidge won—by a landslide. When Harding chose as secretary of commerce the internationally respected Herbert Clark Hoover, Republican ascendancy was assured—and the course of the nation set—for the next dozen years.

But Dayton, in the meantime, fixed its sights on loftier goals. Billed as the "City of Factories" around the turn of the century, Dayton now styled itself "Air City of America."[221] Daytonians really needed no reminding. Nearly every day the skies over downtown and points north were filled with a veritable aerial circus. McCook's engineers and test pilots were always up to something. In 1922, shoppers downtown were treated to the antics of the "radio dog," a driverless vehicle that roamed the downtown streets, dodging traffic and observing (just installed) traffic lights[222] while remotely controlled from a circling biplane overhead.[223] That same year, McCook took its act northwards and demonstrated the possibility of "crop dusting" catalpa groves near Troy, Ohio.[224] McCook's main hangar bore the sign in big bold letters THIS FIELD IS SMALL, USE IT ALL. Sometimes it was too small and an airplane overshot the airfield and had to be fished out of the Great Miami River. Sometimes a plane crash-landed in a north-end backyard.[225] When McCook's pilots weren't buzzing the Dayton skyline, their exploits elsewhere crowded the headlines of the Dayton papers.

The apex of all this activity came in 1924. That year Dayton was chosen to host the International Air Races. Air racing had grown to be not only a sport, but also a means of showing off and pushing to the limit the nation's and the world's latest aeronautical products and technologies. The site of the meet was Wilbur Wright Field,[226] which had a longer runway than McCook and had regularly served as a flight test field for McCook's Engineering Division. (In 1923, the Barling bomber, the world's largest airplane, took off from Wilbur Wright.[227] It was at Wilbur Wright, moreover, that a young Lieutenant Frank Stuart Patterson had been killed in a flight

test experiment in 1918. Thirteen years later, he would be memorialized when the field was renamed in his honor.)

Dayton had not only the air races to cheer it in 1924. Earlier in the year, the War Department had sealed a deal to make Dayton the permanent home for its aeronautical research and development work.[228] Here is what happened.

McCook Field, from the very beginning, was never intended to be anything but a temporary set-up for the Army's aeronautical engineering activities.[229] The year before entering the war, Congress had authorized funding to build a massive aeronautical research and development complex at Hampton, Virginia.[230] Alas, the work advanced too slowly to suit the Army so that, in 1917, the service built its own installation in Dayton.[231] However, McCook was built on land leased from the Dayton Metal Products Company[232]—and every year the firm raised the rent. McCook's buildings, moreover, were built largely of wood—the prefab of the day—and presented a constant fire hazard.[233] The takeoff and landing field was, as already noted, SMALL, and increasingly inadequate for larger and faster modern aircraft.

Meanwhile, work on the first wind tunnel at the NACA's new Langley Memorial Laboratory at Hampton was completed in

1920,[234] and the experimental program—open to the Army and Navy as well as industry and academe—began with a flourish. In 1922, moreover, the Navy inaugurated its Naval Research Laboratory (NRL) to carry on R&D, in part, complementary to the Naval Aircraft Factory. Not to be outdone, the Army began to scout sites to expand its own experimental and engineering activities.

Frederick Beck Patterson, left, with his father, John H. Patterson (*Mrs. Howell Jackson*)

Downtown Dayton, about 1923

The aged but indomitable John H. Patterson got wind of the Army's intentions and decided to make a bid for keeping the Air Service's R&D work—and payroll—in his hometown.[235] Early in 1922, Patterson conferred with his son, Frederick—to whom he had turned over daily operation of NCR in 1921[236]—and other close associates.[237] His actions were cut short, however, when, train bound for Atlantic City on May 7, he abruptly expired.[238] After Frederick and the city of Dayton paused to bury Dayton's savior and first citizen, the planning and lobbying resumed. Frederick formed the Dayton Air Service Committee that raised more than $400,000 to purchase land for a new airfield.[239] The site chosen was in the floodplain below Huffman Dam, about five miles east of town.[240] Dayton turned this land—more than 4,500 acres—over to the government in August 1924.[241] Dayton's future as America's "Air City" was assured for the remainder of the century.

But what to call the new airfield? Two names immediately sprang to mind—at least to Daytonians with an opinion on the matter: Patterson and Wright. The Pattersons, father and son, had led the fight to keep the Air Service's R&D work in Dayton. Besides this were the many contributions that John H. had made over the years to his hometown. The Pattersons, moreover, were of the first generation of pioneers to found Dayton. Were it not for Wilbur and Orville Wright, on the other hand, the city of Dayton might very well not have been in the aviation game at all. One airfield, to be sure, already bore the name of the elder brother—but hadn't the two worked as a team? Dayton newspapers ran a contest to poll public support for one or the other name. In the end, the War Department decided to name the field in honor of the Wrights.[242] Wright Field was dedicated on Columbus Day, 1927, with Orville Wright as guest of honor.[243]

The "new" Wright Field included the acreage of Wilbur Wright Field. Four years later, in 1931, the partisans of the Patterson *nom* were finally quieted when the portion of Wright Field northeast of Huffman Dam (i.e., the old Wilbur Wright Field) was separated from Wright Field and named in honor of John H.'s brother Frank's son Frank Stuart—the unlucky lieutenant who lost his life at Wilbur Wright Field during the war.[244]

During the balance of 1927 and into 1928, caravans of trucks and rail cars carted all that was portable from McCook Field across and out of town to Wright Field:

wind tunnels, propeller test rigs, dynamometers—even manhole covers bearing the designation U.S. Army Signal Corps. Nothing went to waste.[245] When this had all been accomplished by the end of 1928, McCook, like ancient Carthage, was demolished and laid level with the dust. Indeed, the ground was regraded so that nearly nothing remained of America's first "Cradle of Aviation." Time to move on!

Dayton was also moving on and transmogrifying during the 1920s. Year by year, the face of the city changed—and changed dramatically—beneath the circling biplanes and occasional visiting blimps of McCook Field. In 1918, Messrs. Deeds and Kettering built a new, permanent home for the Engineers Club they had founded several years before. The new club, located on Monument Avenue across from Van Cleve Park, was a restrained, but dignified building of late antique Roman style that might have housed the Byzantine Exarch in old Ravenna. Instead, it afforded a magnificent view of airplanes flying above McCook Field to Orville Wright, who helped dedicate the building and who frequently lunched in modest obscurity in its ground floor dining room. (There, Theodore von Karman found him while visiting Dayton in 1926 and listened, fascinated, to how two local bicycle mechanics had mastered the laws of aerodynamics and solved the mystery of human flight without ever a lecture from Professor Ludwig Prandtl, Karman's *doktorvater*.)[246] Next door to the club,

Deeds built the headquarters of the Miami Conservancy District. Dayton, then as now, was big enough to be interesting while small enough to be cozy for its movers and shakers!

During the 1920s, the Hawes and Stoddard properties on the crest of Lower Dayton View gave way to two outstanding monuments of civic architecture: the Dayton Art Institute and the Masonic Temple. (When the Stoddard mansion—a tribute to the opulence of high Victorian domestic architecture—fell to the wrecking

Frederick B. Patterson, 1920s president of the National Cash Register Company, also served as president of the Dayton Air Service Committee after the death of his father. (*Mrs. Howell Jackson*)

The Dayton Chamber of Commerce placed this sign near construction of the new Wright Field, illustrating the pride of Dayton citizens in acquiring "the largest aviation field in the world."

ENGINEERS CLUB OF DAYTON

The Engineers Club of Dayton, located on Monument Street, was built between 1914 and 1918. NCR executives, Charles F. Kettering and Edward Deeds, established the private club as an organization where professional businessmen and engineers from the Dayton community could come together and share ideas on scientific and social topics. Membership included leaders from the Dayton community, as well as civilian and military personnel from nearby McCook (and later Wright) Field. The elegant brick facility was far more luxurious than those to which McCook Field personnel were accustomed.

The stately building contained a 350-person auditorium, where Kettering taught science to school-aged children; a full dining room; barbershop; a game room with pool tables for gentlemen; and an elegant ladies lounge decorated in the design and color of fine, English Wedgwood china. Orville Wright (fourth president of the club) usually reserved a two-person table near the kitchen that was partitioned from the larger dining room for his privacy.

Over the years, many of the Miami Valley community's most innovative and successful people swelled the ranks of the Engineers Club and, more recently, membership eligibility expanded to include "…anyone of integrity, from any profession." Today, photographs and artifacts line the walls to commemorate the inventive spirit and entrepreneurial savvy of past members.

Busts of Kettering and Deeds grace the lobby. These were sculpted by Gutzon Borglum, noted sculptor of Mount Rushmore and disappointed aircraft designer who set into motion a congressional court martial of Colonel Deeds by accusing him of wartime profiteering during World War I.

Source: Fact Sheet: Engineers Club Tour, ver. 10-15-2002, Dayton Engineers Club.

This historical marker honoring the Wright brothers for the invention of the first practical airplane is located in front of the Dayton Engineers Club.

Engineers Club located on Monument Avenue, built between 1914 and 1918

ball, Charlotte Reeve Conover, no stranger to the Stoddard home, could not suppress a sigh: "*Sic transit gloria Stoddard!* Which may be bad Latin but allowable as sentiment."[247]) The new Temple rivaled that of Solomon (nay, Herod!) in its dimensions but more nearly recalled, in its neoclassical facade, the Temple of Jupiter Optimus Maximus in ancient Rome. The Art Institute, a gift to the city from Mrs. Julia Shaw Patterson Carnell, was modeled on an Italian Renaissance palace of the quatrocento. Sitting majestically astride its

lofty perch above the Great Miami, it bespoke the Gem City's pretensions to a higher cultural plateau.

Another specimen of Italianate quatrocento magnificence was a great pile of a building that arose on Monument Avenue in the late 1920s to house Dayton's YMCA. The YMCA, which had always enjoyed commodious accommodations, abandoned its former headquarters at Third and Ludlow (when built, in 1909, it was the second largest YMCA building in the world) ultimately to the city of Dayton,

which thereafter made it serve as City Hall.[248]

In the late 1920s and early 1930s, other buildings also arose that defined the Dayton skyline till the 1970s and 1980s witnessed another spate of high-rise building activity. Most conspicuous were the Bell Telephone Building, a massively beautiful Art Deco structure that towered among the mansard roofs of the many Victorian homes still then extant on the west end of Dayton's downtown; and the Hulman Building, whose vertical lines—if

The Dayton Art Institute, founded in 1919 as the Dayton Museum of Fine Arts, was originally located in a downtown mansion. Once its collections outgrew the space, Mrs. Julia Shaw Patterson Carnell funded the construction of a new museum, which was completed in 1930. (*Dayton Metro Library*)

The Dayton Freemasons originally had a remodeled church building in downtown Dayton. The new Masonic Temple (now called the Masonic Center) was finished in 1928 at an estimated cost of $2.5 million. (*Dayton Metro Library*)

Julia Shaw Patterson Carnell underwrote the construction and subsequent operation of the Dayton Art Institute until her death in 1944. She was the mother of Frank Stuart Patterson, the test pilot who lost his life at Wilbur Wright Field and for whom Patterson Field was named. (*Dayton Metro Library*)

not altitude—emulated New York City's Empire State Building.

Binding the city together—east and west of the Great Miami—were great concrete bridges, most built around the turn of the century, whose classical balustrades and graceful arches more nearly resembled those of some European capital of the eighteenth century—a London, Dresden, or Paris—than an American industrial metropolis. This city of lofty architecture—Dayton old and new—thus formed a vision of American urban civilization at its most urbane. That vision was hauntingly captured in many of the photographs of Dayton's Jane Reece[249]—and none too soon! For in the 1960s, the vision was rudely dispelled by the building of the interstate highway that intruded upon, overrode, and bisected Dayton's downtown. But, as President Calvin Coolidge once remarked matter-of-factly, "The business of America is business." Keep on truckin'!

Rike-Kumler Department Store on the corner of Second and Main streets. The Miami Hotel is located to the left. Both of these buildings have been torn down. (*Dayton Metro Library*)

Main entrance to Wright Field, late 1930s. Building 12, which originally served as the Army Air Corps museum, is visible on the left.

But the circling aircraft of McCook Field and subsequently those of Wright Field also beheld developments, more transient though no less dramatic, than the changing Dayton skyline. In the 1920s, Dayton revisited the controversies, prejudices, and latent hatreds of 70 years before with a march of the white-sheeted Ku Klux Klan down Main Street to the Montgomery County Fairgrounds. It was the high point of Klan activity in Dayton, though a low point in interracial community relations among Dayton's citizens of European and African descent.[250]

Fortunately the rise of the Klan in the American heartland was short-lived. The 1920s was a period of growing prosperity for nearly all sectors of the economy—with the exception of the liquor industry. The 18th Amendment to the Constitution of the United States forbade the manufacture, selling, and consumption of alcoholic beverages. The successors of the ladies who marched for Temperance in the 1870s finally won the day and held their own for little more than a decade. The nation went dry and breweries went out of business.[251] Dayton's largest brewer, Adam Schantz, tried to market "lily water,"[252] but it was a poor substitute for what any enterprising gentleman—or lady!—could obtain under the counter, behind closed doors, and out of sight of "the Feds."

The aircraft industry also took a nosedive in the early 1920s. Colonel Deeds' Equipment Division may have had difficulty producing enough airplanes— but not airplane engines. The Liberty engine was, in fact, one of the great success stories of the war, one for which Deeds could take full credit. Alas, when the war ended, the Army had more engines—planes too!—than it knew what to do with. In addition to abruptly cancelling production contracts following the Armistice, the service also decided to dump on the market its excess aeronautical inventory. The result was a severe depression in the aircraft industry that lasted till the "Lindbergh boom" in the late 1920s.[253]

Charles Lindbergh flew into McCook Field on August 5, 1927,[254] on his triumphal tour of the Midwest following his epoch-making flight from New York to Paris in May.[255] It was perhaps the last great event in the event-filled history of the field, America's "Cradle of Aviation." A few days later, he flew over Rockport, Illinois, where young Fred Ascani looked up, saw the silvery *Spirit of St. Louis* soar across the sky, and decided that he, too, wanted to become a pilot.[256] A dozen years later, as a senior West Point cadet, Ascani visited

Wright Field for the first time. Despite the tragic crash of a demonstration airplane on that occasion, Ascani persisted in his determination.[257] In the late 1940s, he was second-in-command of Wright Field's Flight Test Division[258] and, in the 1950s, as Director of Laboratories, supervised the most far-reaching reorganization of Wright Field's aeronautical research and development activity in history.[259]

The late 1920s and early 1930s was a revolutionary period in world aviation, one in which the nation's airplane manufacturers and the Army's Materiel Division at Wright Field played a pivotal role. McCook's Alfred Verville designed a cantilevered, low-wing monoplane racer with retractable landing gear and closed cockpit in 1922.[260] It was ahead of its time by nearly 10 years. In the late 1920s, Wright Field's Carl Greene directed the development of a 55-foot, all-metal wing. The wing never flew. Instead, Wright Field's structural engineers subjected it to every kind of stress test imaginable. It passed every one with flying colors. Confident of the new, stressed-metal construction, Wright Field's Materiel Division pressured industry into thinking metal—and thinking *big*.[261]

Over the next decade, Dayton's citizens were treated to many a strange sight in the skies over Third and Main. Since the Wright brothers adapted Octave Chanute's "two-surface machine" in 1900,[262] everyone assumed that any reliable airplane needed more than one wing with supporting struts and guy wires. However, starting in the early 1930s, this familiar perception was slowly effaced as a veritable pageant of strange aircraft cast shadowy planforms along Springfield Pike on the way to Wright Field: first the B-9 (1931),[263] then the B-10 (1932),[264] and ultimately, the B-17, which flew in from Boeing's plant in Seattle— nonstop—on August 20, 1935.[265]

Indeed, the revolution was sustained largely by the military since, by the early 1930s, the nation's commercial sector was once more in the grips of economic hard times.

The term "depression" is an economists' term that describes an abnormally low level of economic activity over a lengthy period of time. The Great Depression was the economic catastrophe that afflicted the industrialized world during much of the 1930s. Indeed, the word depression—in its less severe form, "recession"—has largely replaced the term "panic" that had been used to describe earlier severe economic downturns, the Panic of 1893[266]—which John H. Patterson predicted

and congratulated himself surviving— being perhaps the most memorable.[267] Like earlier economic crises, the Depression began with a panic on the Wall Street stock and financial markets in October 1929. As the crisis spread and deepened, however, the word depression came to signify for a generation of Americans the whole awful experience of financial destitution.

The Depression came late to Dayton. While New York stockbrokers and moneymen were in freefall—sometimes literally—from the heights of the 1920s' bull market, Daytonians calmly looked up from the corner pavement of First and Main to watch the last course of bricks being laid for the new Biltmore Hotel. That was in November 1929.[268] One year later, the Biltmore was in receivership and Daytonians were beginning to apprehend the full dimensions of the economic catastrophe engulfing the rest of the nation and world.[269]

Slowly, the home-building boom that filled out Dayton's neighborhoods in the 1920s and early 1930s, particularly to the north and south, atrophied, although it never ceased altogether. The line of farthest advance can be seen, even today, where the pattern of traditional, two-story brick and frame houses suddenly gives way to smaller, more cramped, two-story and "ranch" style homes of the post-World War II era.

Banks foreclosed on many a householder no longer able to meet mortgage payments. Then, the banks themselves closed. The president of the Winters Bank took his own life in 1934. The bank itself was saved only when Charles Kettering melodramatically walked into the first floor lobby and staved off a "run" on deposits by announcing that he would underwrite the institution, if need be.[270] The homeless crowded into makeshift tent cities and lean-tos. One of these "Hoovervilles" metastasized on the banks of the Great Miami.

In the early 1930s, NCR recalled Edwards Deeds, who now became president and general manager of John H.'s tottering empire.[271] Deeds returned to his old office, which he found just as he left it more than 15 years earlier.[272] The terms of his earlier, abrupt departure were also "regularized." All forgiven if, perhaps, not quite forgotten.

In 1933, the nation retired Herbert Hoover and elected Governor Cox's running mate of 1920, Franklin D. Roosevelt, president. Roosevelt gave the banks a holiday and got many of the unemployed back to work again. "All we have to fear," he declared, "is fear itself!"[273] It was not, of course, that easy, and the Depression ground on to the advent of World War II. It was, however, a New Deal, which the nation, with few cards left to play, hopefully seized upon.

At Wright and Patterson fields, the Army laid off few personnel although it did reduce the number of workdays—and thinned paychecks—to save money.[274] Depressed conditions in industry, moreover, stanched the rapid turnover of personnel. More significantly, as the decade advanced, the Army began, for the first time since the mid-1920s, to hire more engineers. One of the newcomers was Captain Harry G. Armstrong. A medical doctor by training, Armstrong pioneered the study of human physiology in flight. In 1935, he teamed up with Carl Greene's Aircraft Branch engineers to design and test a pressurized-cabin aircraft. The result was the XC-35, which won the Collier Trophy in 1937— and contributed directly to development of the B-29 bomber.

In 1939, another young engineer full of brains and enterprise came to Wright Field. Frank Wattendorf arrived in Dayton to consult on the building of a giant new wind tunnel complex for the Materiel Division's laboratories. General Henry H. "Hap"

Arnold had broached the subject of upgrading the Air Corps' experimental program with Theodore von Karman earlier in the year, and the proposed wind tunnel complex was one result.[275] Wattendorf was one of Karman's most trusted and gifted protégés, and had just returned to the United States from China, where he had overseen the design and construction of that nation's first, major wind tunnel facilities.[276]

Wattendorf took the measure not only of Wright Field but also real estate in Dayton. He purchased a number of run-down properties in McPhersontown and Lower Dayton View, which he proceeded to repair with the object of renting out. The winds of war were blowing again, and Dayton would be the center of much aircraft research and procurement. That meant an influx of personnel not unlike the previous war, 20 years earlier. The newcomers would need some place to live not too far from the field. Wattendorf made one of the properties his domicile while living in Dayton throughout the war. Indeed, Dayton became one of the "home bases" (one other was Washington, D.C.) for the much-traveled Wattendorf clan over the next four decades.

DAYTON ON THE EVE OF WORLD WAR II

War broke out in Europe in September 1939, when Germany invaded Poland. Both Britain and France subsequently declared war on Germany. In the spring of 1940, Germany invaded France, this time overrunning the French and British defensive lines, forcing French capitulation

Charlotte Reeve Conover, doyenne of Dayton historians *(Dayton Metro Library)*

Congestion at the gate to Wright Field during World War II. The combined personnel strength of Wright and Patterson fields during the war increased to 50,000.

ON THE HOME FRONT IN THE MIAMI VALLEY

During World War II, all Americans made sacrifices, and few lives were left untouched. Women worked in defense factories or did volunteer work as civil defense and air raid wardens. They made hospital robes and gowns for patients in military hospitals, and knitted mittens, mufflers, socks, and sweaters for the fighting men overseas. They assembled Red Cross packages, containing sewing supplies, stationery, playing cards, cigarettes, soap, shoelaces, and razor blades, for the men at the front. They gave up their nylon stockings, which were made into parachutes, and their silk stockings were made into gunpowder bags. Recruiting posters at Rike's Department Store in downtown Dayton beckoned women to join the military, and many did. Several Daytonians even volunteered their dogs for military service, including Kay Carroll, wife of Brigadier General Carroll at Wright Field, who donated her beloved "Ruddy" (officially "Wright Rudder") to the cause of freedom.

Drives were held to collect all surplus tin, metal, and rubber. Rubber bands and erasers were no longer sold. Sugar, gasoline, coffee, fruit, vegetables, and meat were all rationed. Daytonians grew their own vegetables in "Victory Gardens" in their backyards. Women gave up their cooking fat to be made into gunpowder. The *Dayton Daily News* advised its readers to supply their own meat by raising rabbits.

Many Japanese-Americans, released from detention camps after proving their loyalty to the United States, resettled in Dayton through the War Relocation Authority (WRA) in Cincinnati. Dayton churches assisted them with housing and Dayton newspapers such as *The Journal Herald* championed their cause, stating that the Nisei (second-generation Japanese) had unjustly suffered persecution and hardship due to their race. Labor union members at the McCall (Printing) Corporation unanimously voted to allow the Nisei to work at their company. Other Nisei worked at the Red-Bar Battery Corporation, the Stroop Agricultural Company, and the Federal Housing Authority.

Thousands of Dayton children volunteered through the Office of Civilian Defense (located at the corner of Third and Ludlow) to hand-carry messages in case communication lines went down in the city. Children were trained to watch the skies for enemy aircraft and report anything unusual to Patterson and Wright fields. The Civil Air Patrol Cadets, founded in October 1942, provided aviation training to teenagers to prepare them for future aviation-related duties including search and rescue. Local Boy Scout troops helped with farm projects; Girl Scouts prepared food and served it to hospital patients; 4-H girls and boys grew vegetables and canned preserves. Children sold war bonds, bought war stamps, and collected much-needed rubber and scrap metal.

Literally the biggest contribution the citizens of Dayton made to the war effort was the USS *Dayton (CL-105)*, a $31 million, 600-foot long, light cruiser; its construction was underwritten with war bonds bought by Daytonians. The *Dayton* served with Admiral Halsey's Task Force 38 and received a battle star for screening the fast carrier task groups and conducting shore bombardments against Japan in the summer of 1945.

Dayton Engineering Laboratories Company (Delco) plant in Dayton, Ohio, 1953. Delco supplied the military with artillery, aircraft engine parts, and other war materiel during World War II.

Commercial industries also mobilized to supply equipment and parts to the military. Companies felt duty-bound to provide the military with necessary materiel, but they also reaped the benefits of having a guaranteed buyer, low-interest loans, and tax credits. By the end of the war, thousands of companies had been contracted to produce war materiel. While the government tried to contract to companies with specific expertise, often it was forced to turn to smaller companies with industrial backgrounds unrelated to the products they were to manufacture. Companies in the Miami Valley were not exceptions. With Wright and Patterson fields as neighbors, the industries of Dayton and surrounding cities adjusted their regular assembly lines into military supply lines. Well-known corporations such as the National Cash Register Company (NCR) and Dayton Engineering Laboratories Company (Delco) answered the call for products to strengthen the military and its equipment (see Appendix 2 for comprehensive list of companies in the Dayton area that switched to production of war materiel.)

National Cash Register (NCR) undertook another role in the defense program. Electrical engineer Joseph Desch was tasked by the United States Navy to build a computer that could take on Germany's "Enigma," a secret message encoder used by German submarine crews for planning attack missions on Allied ships. Desch and his engineers created a new "Bombe" (the British had invented an earlier version that soon became obsolete) that decrypted the Nazi submarine messages. On April 20, 1943, female Navy personnel called WAVES (Women Appointed for Voluntary Emergency Service) arrived in Dayton to assemble the bombes. The WAVES lived at NCR's Sugar Camp, built by John H. Patterson many years before for his sales force, and worked at NCR's Building 26 on Main Street in downtown Dayton. (Some of the WAVES also received assignments to Wright Field.) After they built the bombes, the WAVES went to Washington, D.C., to put them into service. Thanks to "The Bombe," invented in Dayton, Ohio, thousands of American and British lives were saved, and the war was shortened by one to two years.

Sources: Philip Shiman, *Forging the Sword: Defense Production During the Cold War* (U.S. Army Corps of Engineers Research Laboratory, 1997), p 10; "World War II Industrial Facilities: Authorized Federal Funding," Heritage Research Center, Ltd., viewed online June 30, 2003, at http://www.heritageresearch.com/War%20Facilities5.html; Curt Dalton, *Home Sweet Home Front: Dayton During World War II*, (Dayton, 2000); Deborah Anderson, email to Robin Smith, dated July 28, 2003; *Dictionary of American Naval Fighting Ships (DANFS)*, viewed online August 22, 2003, at http://www.hazegray.org/danfs/cruisers/cl105.txt.

Working in factories with bombs and gunpowder was a dangerous occupation. In comparison to the 292,000 American soldiers killed in World War II and the 671,000 wounded, nearly 300,000 American factory workers were killed while performing their job duties and another million became disabled. Women accounted for 37,000 of those who died on the job and over 200,000 of the disabled. (*National Archives and Records Administration*)

NCR during World War II. A machine paints the exterior and interior of 37-mm high-explosive shells for the war effort. (*Library of Congress, Farm Security Administration, Office of War Information Photograph Collection*)

NCR employees were encouraged to come up with catchy slogans that expressed the company's determination to "do the best job in industry in backing up the Armed Forces." (*National Archives and Records Administration*)

and British retreat back across the channel. In the fall of 1940, the German *Luftwaffe* conducted intensive bombing raids on British military, industrial, and population centers, particularly London, in the so-called "Battle of Britain." The British fought the Germans to a draw and then looked on in relief and near disbelief as the Germans, in June of 1941, turned on their erstwhile ally, the Soviet Union, and began a broad sweep across the Russian steppe toward Moscow.

As during the opening salvoes of World War I, the United States remained a neutral observer during the first several years of hostilities. However, despite the strong pacifist and neutralist protests of the America Firsters—who included many eminent men and women, among them Charles Lindbergh—the nation began preparing for the worst. The most dramatic indication of United States' resolve was President Roosevelt's call upon the Materiel Division and aircraft industry for 50,000 warplanes, with 50,000 per year subsequently. That was in the summer of 1940.[277]

A year later, in the summer of 1941, the Materiel Division began to pour concrete runways at Wright and Patterson fields. The old sod airfield would not be able to support the new class of long-range heavy bombers that the Materiel Division's engineers had begun to solicit from industry. The B-29, with a range of 5,250 nautical miles, would enter service in 1943.[278] A modified version was later prepared at Wright Field for carrying the atomic bomb. The B-36 was another airborne colossus that got its start when Wright Field issued performance specifications to industry in April 1941.[279] A whole new facility—itself colossal—was designed just to perform the static testing on the B-36.[280]

All these activities had their effect on Dayton. The wave of new recruits began to detrain in the Gem City in 1940 and 1941, booking rooms in hotels and boarding houses and renting apartments. Soon there would be a mini-building boom of mostly small, prefab houses for the city's swelling military and industrial workforce. Indeed, rumors of war foreshortened the Depression in Dayton by nearly a year, as NCR and the city's General Motors plants also began to retool for war work.[281]

Old Dayton was giving way to the new, in more ways than one. In 1940, Charlotte Reeve Conover died.[282] With her died memories stretching back nearly 85 years. When she was born, in 1855, Dayton was little more than a village and a few of Dayton's founding generation still among the quick. She grew up hearing stories of Indian raids, the mustering of troops on the Dayton commons in 1812, the building of the canal, and the tragic fate of "Black Ben." She herself witnessed the tumult of civil war on the very streets of her hometown and remembered being shunned by schoolmates as the daughter of the notorious Dr. Reeve. She recalled her debutante year in the 1870s, when young ladies giggled discretely at older, chaperoning women outfitted in out-of-fashion hoop skirts at dances in the Philips House ballroom. She remembered winter sleigh races on First Street and ice-skating on the Great Miami. She remembered—and published!—gossipy stories of the "First Families of Dayton." Remarkably, she missed the 1913 flood—she was in New York City at the time—but got a day-by-day written account from her aged father, who barely survived the deluge himself in his Wilkinson Street home.[283] Her invalid

B-17 bomber flies over downtown Dayton, late 1930s

mother died a month after the flood, possibly from the shock of it all.[284] Her father—one of the most eminent physicians of his day—lived on until 1920.[285] Already the author of numerous books and articles about Dayton and other things that amused her lively intellect, she pressed on to compile a massive, four-volume history of Dayton that she published in 1932.[286] Shortly thereafter, she lost her sight, but not her energy and love for her hometown—and of life itself. "Honeysuckle smells as sweet," she remarked a few years before her death, "when you're too old to jump on the running board of a car."[287]

Even a last whiff of honeysuckle was gainsaid many draft-age young Daytonians. As fall became winter 1941, the citizens of Dayton anxiously followed the headlines of the *Journal-Herald* and the *Dayton Daily News*. During November and into early December, the German Army inched its way toward Moscow through the bitter Russian winter. Who could doubt that the Nazi swastika would soon replace the Soviet star on the onion domes of the Kremlin! Then suddenly—unexpectedly— came a blow from the other side of the world. As Daytonians sat down to Sunday dinner on the afternoon of December 7, news crackled across radios on Cox's WHIO of a Japanese surprise attack on the main naval base of the American Pacific Fleet at Pearl Harbor, Hawaii.[288] In his war message to Congress the following day, President Roosevelt declared it "a date which will live in infamy." Hushed silence. Thunderous ovation. America was at war.

In little more than four months, a flight of B-25 bombers lumbered off the decks of the USS *Hornet*, destination Tokyo, Japan. In command was a former McCook Field test pilot, whose derring-do had always been done with the cold calculation of a professional engineer. Lieutenant Colonel James H. "Jimmy" Doolittle's raid on the heart of the Japanese Empire gave the emperor and his generals a foretaste of their ultimate, inevitable defeat.[289] It gave Americans renewed hope, and to the men and women of Wright and Patterson fields and citizens of Dayton, Ohio, justified pride and renewed determination to see the task-at-hand through to its completion.

Over the next three years, Wright and Patterson fields would nearly quadruple the number of their buildings, and the size of their workforce would increase by nearly an order of magnitude. Almost every page of the Dayton phone directory would bear one or more names of someone, somehow associated with the enterprise of flight at the twin fields and in countless local factories and businesses of the greater Dayton area. During World War II, the foundations of American air power—and beginnings of space power— were laid at Wright and Patterson fields, placing Dayton, Ohio, in the epicenter of aerospace science, technology, procurement, and logistics for the remainder of the twentieth century.

* * *

"If you were dropped down in Dayton you would hardly know it. Great improvement is going on. The streets are all busy. ...Property is selling very high."[290] Frank Wattendorf? No, John Van Cleve. The year, 1829. *Plus ça change...!*

Main Street bridge, Dayton, before 1950 *(Dayton Metro Library)*

Chapter Two

MILITARY AVIATION COMES TO DAYTON

EARLY YEARS OF SIGNAL CORPS AVIATION

On December 17, 1903, the Wright brothers made the first free, controlled, and sustained flights in a power-driven, heavier-than-air machine. Five years later, on February 10, 1908, the newly established Aeronautical Division of the Signal Corps accepted the Wrights' bid to provide the Army with its first airplane. The Wright brothers' aeronautical achievements, the Wright Exhibition Company's aerial demonstrations, and the Wright Company's manufacturing and training

Wilbur Wright Field and the Fairfield Air Intermediate Depot (FAID) prior to 1923. The Unit One hangars are located along the left side of the photograph. FAID's Building 1 is at right center, and the village of Fairfield is visible at top center.

operations of these early years imprinted on the community of Dayton, Ohio, a permanent interest in aviation, both civilian and military. One year after the Wright School of Aviation closed in February 1916, Daytonians witnessed the beginning of a new tradition that would become an enduring part of their community and heritage: the Signal Corps established two aviation fields in the area—Wilbur Wright Field for training pilots and McCook Field for developing and testing airplanes.

On August 1, 1907, the Aeronautical Division was established in the Office of the Chief Signal Officer of the Army. The division had responsibility for "all matters pertaining to military ballooning, air machines, and all kindred subjects on hand." At this time, the Signal Corps' air fleet consisted solely of two free, spherical,

hydrogen balloons. In August 1908, the Aeronautical Division also purchased a "dirigible balloon," known as Signal Corps Dirigible Number One, from Thomas Scott Baldwin for $6,750.[1] Nearly one year later, on August 2, 1909, the Aeronautical Division acquired Signal Corps Airplane Number One, designed and manufactured by Wilbur and Orville Wright, for $25,000. Congress, however, provided no further funds to continue military aeronautics. "Economy was the watchword in Washington and vision was lacking in those who held the purse-strings."[2]

Thus, during 1910 and into the early part of the following year, the Aeronautical Division's flight operations were limited to two free balloons, one dirigible, and one airplane. Lieutenant Benjamin D. Foulois was the Army's only active pilot, flying at

Signal Corps Balloon No. 10 being prepared for launch in Washington, D.C., June 1904. Until the first airplane was purchased in 1909, hydrogen balloons made up the Signal Corps' fleet. (*McCoy Album*)

instruction sufficiently comprehensive in its scope to justify its graduates being rated military aviators."[4] By mid-year, the West Coast school had seven airplanes, three instructors (including the commandant), 14 officer students, and 48 enlisted men assigned to service and support jobs.

In addition, military aviation was formally recognized for its potential as part of a field force when the Army prescribed a model organizational structure for a provisional aero squadron in 1913. The tables of organization, equipment, and allowances called for 20 officers, 90 enlisted men, eight airplanes, and six motorcycles for each squadron.[5] Available resources grew and, by the end of the year, Army aviation exceeded the strength of one squadron, with 15 airplanes and 114 commissioned and enlisted personnel, including 11 qualified pilots and nine other officers in pilot training.[6]

Much progress had been made since 1908, but the price had often been measured in terms of sacrifice. From 1908 through 1913, military airplane crashes cost the lives of 12 officers, including one noncommissioned officer. Most of these fatalities had involved pusher-type airplanes. Three of the seven fatal accidents occurring in 1913 were Wright Model C pusher airplanes, destroying pilots' confidence in that type of aircraft. On February 16, 1914, the Army grounded pushers, pending further investigation. In effect, this action eliminated the Wright fleet, for a board of pilots recommended student training at North Island be limited to the tractor-type airplanes in the fleet, including one Curtiss and four Burgess airplanes.

Military aviation continued to advance over the next year. According to Brigadier General Frank P. Lahm, who was one of the Army's first pilots:

Our early military airmen were true pioneers, men who had the courage of their convictions and who persevered in their efforts, undaunted by the conservatism of some higher authorities, by the lack of financial and moral support and by the high casualty rate demanded in the development of this new weapon.[7]

Fort Sam Houston, Texas, in Signal Corps Airplane Number One. He augmented the $150 that Congress allotted him for aviation gasoline and repairs from his meager service salary. The Aeronautical Division's total personnel strength stood at 27 officers and enlisted men.

The situation was different in Europe, where leaders proved far more visionary in recognizing the military potential of the airplane. In early 1911, the French government appropriated $1 million for its infant air arm, and other nations soon followed in establishing aviation programs as pressures built toward World War I.

Insulated from those pressures, the United States awakened more slowly to the potential of military aviation. In the opinion of a World War I historian, it was a "red-letter day in American aviation history" when Congress enacted the Appropriation Act for Fiscal Year 1912 on March 3, 1911, allocating $125,000 for American military aviation.[3] Of that amount, the Signal Corps

used $25,000 to purchase five new airplanes, including two Wright Model Bs, two Glenn Curtiss "military models," and a Burgess-Wright model. Consequently, by the end of 1911, the Signal Corps' aviation fleet consisted of five airplanes (Signal Corps Airplane Number One had retired to the Smithsonian Institution during the year), two spherical balloons, and Dirigible Number One. The 1911 personnel strength stood at 23 and included six airplane pilots. Aircraft and personnel were located at the Signal Corps Aviation School, which had summer quarters in College Park, Maryland, and winter quarters in Augusta, Georgia.

Meager appropriations over the next two years severely restricted the Aeronautical Division's activities and growth; still, two significant steps were taken. In January 1913, the flying school relocated to North Island, San Diego, California, where "year-round" flying conditions existed. The new site marked the Army's "first permanent aviation school, providing a course of

Lahm rejoiced on July 18, 1914, when "official sanction was given to the Army's wings" by Congress in the creation of an Aviation Section within the Signal Corps. The new Aviation Section inherited the general mission of the earlier Aeronautical Division and, in addition, received the specific responsibility of "training officers and enlisted men in matters pertaining to military aviation."[8]

Despite this recognition, American military aviation was still in its formative stages, particularly when compared to major European powers, at the outbreak of World War I on July 28, 1914. Germany's prewar military aviation budget was $45 million; Russia's totaled $22.5 million; France's, $12 million; Austria-Hungary's, $3 million; Great Britain's, $1 million; and Italy's, $800,000. Germany had 2,600 men in uniform flying and/or supporting a fleet of 260 airplanes. France had 3,000 military personnel and 156 airplanes. Great Britain had 154 airplanes. In stark contrast, the United States Army Aviation Section's fiscal year 1915 appropriation was $250,000; assigned personnel totaled 208; and the aircraft inventory was 23 machines.

In March 1916, fueled by the embarrassing experience of the 1st Aero Squadron during the Punitive Expedition against Pancho Villa on the Mexican border, Congress provided another $500,000 for military aeronautics. The National Defense

Captain Charles deForest Chandler, commanding officer of the Signal Corps Aviation School in College Park, in a Wright B airplane, 1911. Chandler became the first chief of the Signal Corps Aeronautical Division when it was established August 1, 1907. He qualified as pilot of balloons in 1907, of dirigibles in 1909, and of airplanes in 1911. Chandler retired in 1920 with a rank of colonel.

The Aeronautical Board that approved the first military airplane at Fort Myer in 1909 included: (left to right) Lieutenant Frank P. Lahm, Lieutenant George C. Sweet (U.S. Navy), Major Charles McK. Saltzman, Major George O. Squier, Captain Charles deF. Chandler, Lieutenant Benjamin D. Foulois, and Lieutenant Frederic E. Humphreys.

Hangars of the U.S. Army's first aviation school at College Park, Maryland, summer of 1911

The Signal Corps flying school moved from its College Park, Maryland, and Augusta, Georgia, locations to North Island, San Diego, California, in January 1913. North Island, shown here, became the Army's first, permanent aviation school. (*United States Air Force Museum*)

Wright Company Model B (shown here at Fort Sam Houston, Texas) and Model C pusher-type aircraft were used in instructing pilots at early Signal Corps aviation schools. The Wright airplanes were eliminated in 1914 due to a high number of fatal accidents with the pusher-type airplanes. Because of its "hidebound determination to stick to the biplane pusher type," the Wright Company was unable to compete to any degree in the military aviation field.

Brigadier General Benjamin D. Foulois, the Army's third airplane pilot, was also the first Army dirigible pilot. During World War I, he served as chief of the Army Air Service in France. In 1929, he became chief of the Materiel Division, Wright Field. As a major general, he was appointed chief of the Army Air Corps in 1931, the position from which he retired in 1935.

Frank P. Lahm (shown when a brigadier general), 1927. A native of Mansfield, Ohio, Lahm was qualified by Wilbur Wright on October 26, 1909, as one of the first American military airplane pilots. Lahm also qualified as a pilot of free balloons (1905) and dirigibles (1908). In 1926, at the rank of brigadier general, he was appointed assistant chief of the Army Air Corps. He retired from active duty in 1941.

Act of June 3, 1916, increased active-duty officer strength in the Aviation Section to 148 and established a Signal Corps Reserve of 297 officers and 2,000 enlisted men for the Aviation Section.

At about the same time, Lieutenant Colonel George O. Squier was appointed officer in charge of the Aviation Section. Having just returned from four years as a military attaché to the U.S. ambassador to Great Britain, Colonel Squier's observations concerning the flying advances abroad carried considerable weight with congressmen. On the strength of his testimony and that of other experts, on August 29, 1916, Congress appropriated $13,881,666 for military aviation and made a supplemental appropriation of $600,000 for the purchase of land to use as flying fields and depots. A massive nationwide search for suitable sites ensued.

When the United States declared war on Germany on April 6, 1917, the Aviation Section had 132 airplanes deployed among four Signal Corps aviation schools and the 1st, 2nd, and 3rd Aero Squadrons.[9] Total strength included 131 officers (nearly all pilots or student pilots) and 1,087 enlisted men. Unfortunately, none of the airplanes was combat-ready. Drastic measures would certainly be needed for America to meet

French premier Alexandre Ribot's urgent request for 4,500 airplanes, 5,000 pilots, and 50,000 airplane mechanics for duty on the Western Front by the first part of 1918.[10]

A wave of patriotic fervor engulfed America. Editorials in the nation's most influential newspapers called for an almost immediate air fleet. Major General Squier, now chief signal officer, supported a proposed $600 million appropriation bill for the Aviation Section and talked eloquently about "an army in the air, regiments and brigades of winged cavalry mounted on gas-driven flying horses."[11] Equally optimistic was the prophecy of Dr. James S. Ames, an eminent scientist, that Germany would be defeated "within a few months of the completion of the 22,625 planes called for in the $639,000,000 programme [*sic*], which it was estimated could be turned out at the rate of 3,500 [airplanes] a month."[12]

To accomplish this "aeronautical miracle," President Woodrow Wilson signed an aviation act on July 24, 1917, providing the Signal Corps with $640 million. It was the largest, single military appropriation in the nation's history. Although the end results were not as spectacular as envisioned, military aeronautics achieved a permanent and major role in national defense.

COMBAT-ORIENTED PILOT TRAINING

Two major problems confronted the Aviation Section with United States entry into World War I: acquiring combat-ready aircraft and training pilots to fly them. On February 3, 1917, when the United States severed diplomatic relations with Germany, none of the 132 planes on hand, or the 293 on order, was designed for combat. According to one historian, "the Aviation Section had no accurate knowledge of the equipment of a military airplane." No American aircraft, for example, had ever been mounted with a machine gun. Aviation personnel had "practically no knowledge of radiotelegraphy and telephony, photography, bombing equipment, lights for night flying, aviators' clothing, compasses used in flying, or other aviation instruments" that were well known to European military pilots.[13]

Until 1916, the Army's flight training school at San Diego, California, had graduated a number of pilots sufficient for the small fleet of the prewar period. As the

Young cadets, such as these at North Island, graduated from Signal Corps aviation schools as reserve military aviators and received commissions as second lieutenants. (*United States Air Force Museum*)

DAYTON WRIGHT AIRPLANE COMPANY

With war clouds on the nation's horizon during 1914-1916, Dayton's business leaders became attuned to the economic potential of a greatly expanded and more powerful military aviation program. When war was declared in 1917, a group of Miami Valley industrial leaders, endowed with foresight and armed with clout in political and financial circles, took action on Dayton's behalf.

Less than a week after the United States entered World War I, the Dayton Wright Airplane Company was organized. Board members included Harold E. Talbott, Sr., and Harold E. Talbott, Jr. (who later served as secretary of the Air Force from February 4, 1953, to August 13, 1955), Charles F. Kettering, Thomas P. Gaddis, George Mead, Carl Sherer, and G. M. Williams. Orville Wright served as a director and as a consulting engineer. The impetus for the new company had been generated in a meeting between Secretary of War Newton D. Baker and Ohio Governor James M. Cox on the subject of the national need for increased airplane production. Governor Cox, a Dayton newspaper publisher and financier, was well acquainted with another talented Dayton citizen, Edward A. Deeds, who had achieved national renown as an industrialist of good reputation, integrity, and patriotism. Currently serving as a member of the U.S. Munitions Standards Board in Washington, Deeds could have no part in the actual ownership or management of the Dayton Wright Airplane Company. He could and did, however, lend his expertise in the form of advice and counsel in setting the enterprise on a sound footing. Deeds was subsequently commissioned as a colonel in the Signal Corps Reserve and served as a member of the Aircraft Production Board (created in 1917 to coordinate the activities of the nation's airplane manufacturers) and as chief of the Signal Corps Equipment Division during World War I.

The new company was located in Moraine City, south of Dayton. In the fall of 1917, it began manufacturing the British DeHavilland DH-4 two-place biplane, an all-wood airplane designed for observation and light bombing. Under license from the Curtiss Aeroplane & Motor Company, Dayton Wright also produced two-place, Standard SJ-1 trainers. By the end of 1918, the Dayton Wright Airplane Company had manufactured about 3,100 of the total 4,500 DH-4s built in the United States and 400 SJ-1s.

After the war, the Dayton Wright Airplane Company continued to manufacture airplanes, including the XPS-1, the first Army airplane with retractable landing gear, and a number of commercial aircraft. In 1919, General Motors Corporation (GMC) purchased the company, but by 1923, deciding against remaining in the airplane business, GMC closed the Dayton Wright branch.

Sources: "Aircraft Production in Dayton," *NCR World*, September-October 1970, pp 21-24; Charlotte Reeve Conover, *Dayton, Ohio: An Intimate History* (Dayton, 2000), p 299; "The U.S. Aircraft Industry During World War I," viewed online July 1, 2003, at http://www.centennialofflight.gov/essay/Aerospace/Wwi/Aero5.htm.

Headquarters and principal plant of the Dayton Wright Airplane Company located in what is now the city of Moraine, south of Dayton, Ohio, as it appeared in June 1918 (*Wright State University Archives, Wright Brothers Collection*)

Fuselages of 100 DeHavilland DH-4 airplanes await their wings at the Dayton Wright Airplane Company, August 24, 1918. (*Dale Posrinan*)

Women employees work on Liberty V-12 airplane engines in the motor department of Plant 1, Dayton Wright Airplane Company, July 1918. (*United States Air Force Museum*)

number of airplanes in the inventory increased and with the likelihood of the nation entering the conflict in Europe, the Aviation Section had expanded the flight-training program. By the end of February 1917, four pilot training schools, in addition to North Island, opened: Hazelhurst Field at Minneola, Long Island, New York; Ashburn Field at Chicago, Illinois; Park Field at Memphis, Tennessee; and Gerstner Field at Essington, Pennsylvania. By June, three more schools opened, including one at Wilbur Wright Field near Fairfield, Ohio. Furthermore, in November 1916, the Aviation Section began an ambitious program to train civilian pilots in the Signal Enlisted Reserve Corps (SERC). Training was conducted at the Curtiss Company's schools of aviation in Newport News, Virginia, and in Miami, Florida. By the end of the program in June 1917, Curtiss had trained 131 enlisted reservists.

The Aviation Section adopted the British-Canadian system of training flying cadets. Ground or pre-flight schools were established at six leading American universities, including Ohio State University. The 400-hour, intensified curriculum prepared 150 men every eight weeks for admission into primary flying schools. The curriculum included classroom instruction and static demonstration in the theory and principles of flight, aerial photography and reconnaissance, communications codes, meteorology, aircraft engines and airplane structures, aircraft instruments and compasses, and aerial combat tactics. These classes, according to a World War I pilot, were of "greatest value in acclimating the men in aviation and in supplying the all-important theoretical knowledge before actual flying began."[14]

After graduating from ground school, cadets moved to one of the Signal Corps' aviation schools for flying instruction. This primary instruction lasted a period of six to eight weeks. The actual time between the initial familiarization flight with an instructor and graduation as a reserve military aviator (RMA) with the commission of second lieutenant depended on the student's progress. The curriculum covered three stages of flying and nearly 300 hours of classroom instruction in diversified subjects such as aircraft engines and structures, aerial machine guns, photographic interpretation, close-order drill, etc. The three phases of actual flying involved dual work with an instructor (four to nine hours), solo flying (24 flying hours total), and cross-country flying. From the very beginning, "the training . . . was prescribed with the utmost care, leaving just as little to chance as humanly possible. ...Step-by-step the cadet went on, always held back until he was doubly skilled in the present phase and doubly eager for the next."[15] The two-part final examination was a solo, 60-mile, cross-country flight and an altitude test of 10,000 feet.

After primary flying training, students transferred to one of three installations for advanced work, during which the pilot specialized in pursuit, bomber, or "army-corps" flying—reconnaissance flying in which the pilot "traveled about with the aerial observers in search of information and photographs—only occasionally in battle."[16] A final stage of training, labeled post-advanced or pre-combat, was conducted in England, France, and Italy, where combat-type airplanes and battle-experienced tutors were available.

Despite the dangers, between July 1917 and June 1918, more than 38,000 of the "finest of America's youth" volunteered for flying training with the Aviation Section. "Constant reports of deaths of famous aviators abroad were far outbalanced by the romance of the [air] service and the opportunities for individuality."[17] Patriotism and enthusiasm were laudable motivators, but more than a willing spirit was required. Strict physical and psychological standards eliminated 18,004 of the 38,770 candidates. Not the slightest defect was permitted in the structure and function of a candidate's cardio-respiratory system, eyes, ears, nose, throat, and other organs. Moreover, according to War Department criteria:

The candidate should be naturally athletic and have a reputation for reliability, punctuality and honesty. He should have a cool head in emergencies, good eye for distance, keen ear for familiar sounds, steady hand and solid body with plenty of reserve. He should be quick-witted,

Huffman Dam was one of five earthen dams constructed by the Miami Conservancy District after the March 1913 flood of the Miami, Mad, and Stillwater rivers. Construction of the Huffman Dam was completed in 1922. (*Miami Conservancy District*)

highly intelligent and tractable. Immature, high strung, overconfident, impatient candidates not desired.[18]

Because the curriculum was exacting in technical studies, mathematics, and the sciences, the majority of flying cadets were either college graduates or undergraduates with majors in fields transferable to aviation.

ESTABLISHMENT OF WILBUR WRIGHT FIELD

Colonel Edward Deeds, a prominent Daytonian who worked to localize aircraft production in Dayton, was a primary influence in establishing military aviation in the Dayton area (see Chapter 1: Birthplace of Aviation). He played a key role in selecting a site near Huffman Prairie, where the Wright brothers had conducted their flying experiments, for use as a Signal Corps aviation school and flying field. The War Department named this installation Wilbur Wright Field.

Through personal contacts in the Aviation Section, Deeds was aware of contingency plans to establish new aviation schools when funds became available. The August 29, 1916, congressional appropriation of $13 million for military aviation provided funds specifically to acquire land either through purchase or by lease. Deeds knew of an optimal site large enough to accommodate four training squadrons. It was located in the Mad River floodplain near the village of Fairfield in Greene County, Ohio, and was under the jurisdiction of the Miami Conservancy District, of which Deeds was president. The district was a political subdivision organized in 1915 and chartered by the state legislature for "building and maintaining flood control works in the Miami Valley." The district's mission was to prevent, by constructing five retarding dams, a recurrence of the disastrous flood of March 1913, which killed more than 400 people in Dayton and the

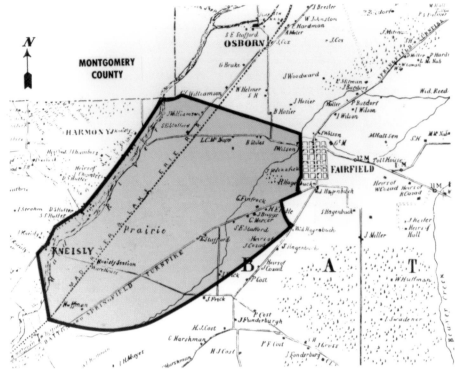

Map of Huffman Prairie and vicinity, Greene County, Ohio. The shaded area became Wilbur Wright Field in 1917.

Miami Valley and caused more than $100 million in damages.[19] The Mad River was a tributary of the Miami, and the site of one of the proposed dams. Huffman Prairie lay on its floodplain. When Deeds suggested to General Squier, chief signal officer, that the area might be suitable for an aviation school, he pointed out that the Wright brothers had trained several dozen pilots at their Simms Station school in the same locale.

On April 30, 1917, Major Benjamin D. Foulois of the Aviation Section and Lieutenant Colonel C. G. Edgar, commanding officer of the Signal Corps Construction Department, arrived in Dayton for a tour. They were escorted by Deeds and Orville Wright. Favorably impressed with the Huffman Prairie locale, Major Foulois described it as "admirably suited for aviation purposes."[20] In addition, he reported that the acreage could be acquired "at a very low cost." He had been advised that the Miami Conservancy District owned all the land, thus allowing the Signal Corps to negotiate with a sole owner for either lease or purchase of the whole parcel.

Cost aside, the selection of sites for new training schools was a difficult and "most delicate [political] matter, for . . . much

pressure was exercised in favor of various localities, and great difficulty was experienced in making unbiased decisions."[21] A board of officers, therefore, proposed selections for approval by the chief signal officer and the secretary of war. In the case of the Dayton site, on May 15, 1917, General Squier recommended that Secretary of War Newton D. Baker approve the rental with option to purchase about 2,500 acres, including Huffman Prairie.

The Miami Conservancy District did not, in fact, have title to the land. The district technically held options to purchase and could exercise the right of eminent domain as a last resort. It was generally easier and quicker to negotiate amicable settlements with property owners than to bring legal action against them and, in the situation at hand, time was very short. Deeds sent urgent messages to Ezra M. Kuhns, secretary-treasurer of the Conservancy District, to expedite the district's purchase of the farms. Two concessions were offered to the property owners, a number of whom were farmers: an inflated purchase price and an agreement to allow the dispossessed families to remain in their homes until the end of the growing season or whenever construction required their removal.

Construction of Wilbur Wright Field begins, June 9, 1917 (*Library of Congress, Pixley-Messick Company*)

"MAKING READY" U.S. AVIAT

Purchase prices reached a maximum of $40 per acre for wheat fields, $35 per acre for cornfields, $25 for oats and alfalfa fields, and $5 per acre for pastures.[22]

On May 22, 1917, Lieutenant Colonel Edgar, as the Signal Corps' agent, signed a short-term lease with the Miami Conservancy District for 2,075 acres of land between what is now Huffman Dam and the city of Fairborn. The rental agreement cost $20,000 for the initial period ending June 30, 1917, and an additional $73,000 paid to the farmers for their crops. The agreement also contained several options for continued use: renewing the lease for one year beginning July 1, 1917, to cover 2,245.20 acres (including the original 2,075 acres) at a cost of $17,600; renewing the lease for three years at a cost of $20,000 per annum; or purchasing all the acreage for $350,000.[23]

On June 6, 1917, the Office of the Chief Signal Officer issued a memorandum stating that the Signal Corps' "recently authorized aviation schools" would be located at Selfridge Field in Mt. Clemens, Michigan; Chanute Field in Rantoul, Illinois; and Wilbur Wright Field near Fairfield, Ohio.[24] Just as it had patterned flying instruction after Canadian and British flying schools, the Signal Corps Aviation Section modeled its aviation training fields according to Canadian design. During a crash, 10-day program, a civilian architectural firm in Detroit, Michigan, drew up the standard specifications for all aviation school sites from Canadian blueprints and rushed them to the various Construction Department on-site supervising officers. Wilbur Wright Field was programmed to be one of the four largest, United States aviation schools, supporting four school squadrons, 24 hangars, 1,700 personnel (including 300 flying cadets), and up to 144 airplanes.

From the Ground Up

On the strength of the May 22 interim lease signed by Colonel Edgar of the Construction Department, the job of converting a small civilian airfield on Huffman Prairie into a major military installation began. Wilbur Wright Field was

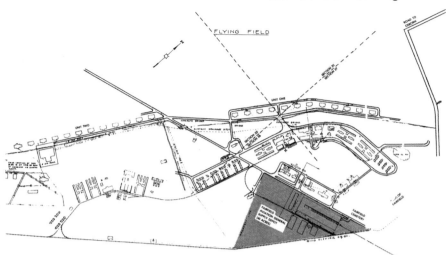

Diagram of the hangar line and major buildings constructed at Wilbur Wright Field during 1917 and 1918. The shaded area shows the location of Fairfield Aviation General Supply Depot, created to support Wilbur Wright Field and other flying installations.

intended as a two-unit, four-squadron flying field. Unit One was the Signal Corps Aviation School, to be in operation by July 15, 1917. Unit Two was to function later as an aviation armorers' school.

Captain Charles T. Waring arrived on May 25 from Ashburn Field, near Chicago, Illinois, to assume charge of the construction project. A contractors' workforce of about 3,100 laborers, together with mules, horses, wheelbarrows, steam shovels, and other machinery, awaited him. Under the press of wartime conditions, this force labored 24 hours a day, every day of the week, to have the field ready for its first contingent of flying cadets scheduled to arrive from ground-training classrooms at Ohio State University on July 15. That the airfield was ready on schedule was almost a miracle according to a detailed report published one year later by Major Arthur E. Wilbourn, who assumed command of the aviation school and Wilbur Wright Field on December 30, 1917.[25]

Approximately 1,600 of Wilbur Wright Field's 2,075 acres consisted of low-lying lands along the Mad River. The flat area on the valley floor was three-quarters of a mile wide and stretched along two miles of the river shore. Elevation above the water level in the river varied from zero to two and one-half feet. Three-quarters of a mile from the river the land sloped gently upward, with the maximum elevation about 30 feet above the river level.[26]

Most of the field's buildings were constructed on the elevated portion of the

site, while flying was done from the level land near the river. Twelve wooden hangars, each measuring 120 by 66 feet, were assigned to each of the two units and were, of necessity, located on the flightline. Each hangar could be configured for class instruction or for maintenance or experimentation on up to six airplanes. The hangars bordered a large, open, drainage ditch that crossed the installation in a northeasterly direction. Thus, flying operations were bounded on the south by the main drainage ditch and on the north by the Mad River.

During the first hectic period of construction, some attempts were made to grade the land for better drainage. Major Wilbourn, in retrospect, considered these efforts "highly unsatisfactory," pointing out that some places on the flying field were still "approximately on a level with the river."[27] He believed the drainage problem could have been resolved had all the low places been graded so as to "drain into one or two centrally located reservoirs from which water could have been pumped into the river."[28] Apparently time constraints did not permit such measures and "earth was simply hauled from the high points on the field and dumped into the low places."[29] Holes and pits made by woodchucks, chipmunks, and other animals aggravated the situation. Consequently, the field was precariously uneven for student pilots, and during each rain, the countless small depressions in the swampy turf filled with water. The drainage

IELD - FAIRFIELD, O. JUNE 9TH, 1917.

TALE OF TWO VILLAGES—FAIRFIELD AND OSBORN, OHIO

The first settler in what came to be known as the village of Fairfield arrived in 1799. The village was officially founded and platted in 1816. Located on the Old Cumberland Trail at the crossroads of four major pikes, the little village became a popular stop on the stagecoach route. Farming was the primary occupation of the village's 400 residents throughout its early history, and the town also had three hotels, three churches, several stores, two blacksmith shops, and a grist mill. That represented the apex of early Fairfield's growth, however, for earlier in the century, the town had refused to allow the Mad River and Lake Erie Railroad to pass through the farmers' fields. When the railroad replaced stagecoach travel in the late 1800s, Fairfield's prosperity waned.

Fairfield's forfeiture was another village's gain. The Mad River line was built about two miles to the north, and in 1851, the town of Osborn was founded along the line. Named for E. F. Osborn, a local railway superintendent, the town prospered. By 1874, Osborn had approximately 700 residents, making it the largest town in Bath Township. By 1900, two railroads and the electric interurban line serviced Osborn, and businesses included three flour mills, a bedspring factory, a buggy-whip factory, an egg-case factory, two banks, and four churches. Osborn also had its own water and electric plants.

Then came the great flood in 1913 (see Chapter 1: Birthplace of Aviation). While the flood itself did not reach Osborn, the little village soon suffered indirect consequences. The Miami Conservancy District, exercising the right of eminent domain, purchased the land in the floodplains above the proposed flood-control dams. Osborn was located in the floodplain of the proposed Huffman Dam. Showing the initiative that led to the founding of the village 70 years earlier, Osborn residents formed the Osborn Removal Company. The company bought all the buildings in Osborn back from the Conservancy District and purchased a new site adjacent to the village of Fairfield. Between 1921 and 1922, the removal company moved the entire town to its new location.

As Osborn started over, Fairfield also began to experience a resurgence of prosperity. In 1917, the Army built Wilbur Wright Field and the Fairfield Aviation General Supply Depot nearby. Soon, both towns reaped the benefits from their mutual cooperation and support with the military. Village residents supplied muscle, horses, and equipment to help construct the new facilities. Military personnel residing off base and civil-service employees living in Fairfield and Osborn played active roles in community affairs. Children of military families attended local schools. The two facilities also provided employment for the villages' residents. Local newspaper stories detailed happenings at the installations and covered social activities of both base and community citizens.

By the end of World War II, the individual identities of the adjacent towns of Fairfield and Osborn had largely disappeared. On January 1, 1950, the towns of Fairfield and Osborn officially merged to form the city of Fairborn—"a 'fair' city 'born' of two historic villages."

Business district of "old" Osborn, about 1920 (*Bob and Dottie Gheen*)

As an interesting side note, Ernest Morgan, the oldest son of Arthur E. Morgan, who designed the Miami Conservancy's flood-control system, related the following story to the editors of the *Dayton Daily News* in 1989:

The flood control reservoirs above Dayton are designed to control floods double the size of the great flood of 1913. Since Osborn was located in the upper portion of one of these reservoirs, the chances of floodwaters ever backing up that far were so remote as to be almost nonexistent. Arthur Morgan did not consider it necessary to move the town.

At the time the matter came to a decision, however, [Morgan] was in the hospital with a serious case of blood poisoning. Realizing that his life was in danger, he had resigned as chief engineer to minimize any complications which might arise in the event of his death….[A physician] managed to pull [Morgan] through, after which he was re-employed as chief engineer.

The decision to move (or to remove) Osborn was made by the Board of the Miami Conservancy in [Morgan's] absence. Had it not been for a bad case of blood poisoning in the foot, the city of Fairborn would probably not be in existence today.

Sources: Carl V. Roberts, "Birth of a City," *Camerica, Dayton Daily News,* December 18, 1949, pp 4-11; Dottie Grey Gheen, "The Villages of Fairfield and Osborn, Ohio," in Lois E. Walker and Shelby E. Wickam, *From Huffman Prairie to the Moon: The History of Wright-Patterson Air Force Base* (Wright-Patterson Air Force Base, Ohio, 1986); "Bath Township—Greene County, Ohio," Green County Public Library, viewed online September 26, 2003, at http://www.gcpl.lib.oh.us/bicentennial.asp; Ernest Morgan, letter to *Dayton Daily News,* July 10, 1989, on file at Aeronautical Systems Center History Office.

The Osborn Removal Company was formed to move the old town of Osborn to its new location adjacent to the village of Fairfield. Approximately 200 houses and another 200 outbuildings were moved between 1921 and 1925. Here, the railroad depot, having just been moved in sections to New Osborn, is being set up for operation, August 1921. (*Miami Conservancy District*)

Main Street of Fairfield, Ohio, early 1900s (*Dottie Gheen*)

Aerial view of "new" Osborn (*Greene County Historical Society*)

Wilbur Wright Field's aerodrome awash in spring flooding, 1922. The 12 Unit One hangars stood along what is today Skeel Avenue (Area C, Wright-Patterson Air Force Base).

problem was especially severe around the Unit Two hangars (known as the South Unit), which were located at a particularly low spot between the flying field and the drainage ditch. By mid-1918, about $50,000 had been spent in attempts to smooth the surface of the flying field to prevent landing accidents. Drainage was also improved in the area of the South Unit hangars, though little could be done about the overall problem.

The Construction Department laid all roads, composed of a mixture of sand and gravel blended with tar, within the new reservation. During the 1917-1918 winter, however, the roads proved to be "absolutely inefficient," and gravel sidewalks disappeared altogether.[30] In his annual report, Major Wilbourn noted efforts to rebuild the roads and sidewalks using screened gravel with a top covering of sand, resurfaced and rolled twice. While the Army improved roadways inside the reservation, it was powerless to do much about off-base roads, which Major Wilbourn described as being in "horrible condition."[31] For many months, Wilbur Wright Field had to rely almost entirely on its own motor transport for the hauling of all food, clothing, and supplies from

Dayton and Springfield. Post transport at the time consisted of a fleet of 40 trucks, eight touring cars (including two Cadillacs), four ambulances, one fire truck, and 20 motorcycles with sidecars.

As bad as they were, the roads were less trouble than the railroad. The nearest railroad depot was three miles away in Osborn. An electric interurban line ran between the railroad and Wilbur Wright Field. However, it was deemed "very inefficient," since the tracks, overhead wires, and pole systems caused fires and were in "poor condition to meet any increased demands."[32] In addition, the company refused to carry less-than-carload shipments. Major Wilbourn noted, "since these shipments were urgently needed it was necessary to use post transportation to haul them."[33] State and county officials were not indifferent to the problem, however, and began work in 1918 to ensure that good roads would be available to the installation at all times.

After the rapid pace of the first month, building slowed but did not stop. Between July 1917 and March 1918, the cost of constructing the aviation school and Wilbur Wright Field rose to $2,851,694, and a continuing series of control problems

became evident. The building program had been both massive and hurried; it was not surprising that Major Wilbourn found the quality of both materials and workmanship to be inferior. He based this judgment on the "excessive amount of time, labor and materials spent in [repairing] and maintaining the buildings from July 1, 1917, to May 31, 1918."[34] He could find no record that the Army had ever accepted the erected buildings. Indeed, he commented that "it is not believed that any board could have accepted these buildings, had any inspection been made of them in December, 1917."[35] But six months later, the Signal Corps Supply Division (as the Construction Department had been renamed) still continued the policy of putting up structures "in most cases without reference to the Commanding Officer" other than to ask him where he wanted them.[36] The fact that the civilian supervisor of the Supply Division was "carrying on construction work with civilian labor at Government expense" without any reference—or deference—to the post commander was a very sensitive issue.[37] This independent action, Major Wilbourn noted, "led to great confusion and duplication of effort."[38]

WILBUR WRIGHT FIELD SIGNAL CORPS AVIATION SCHOOL—1917 FLYING SEASON

	JULY	AUGUST	SEPTEMBER	OCTOBER	NOVEMBER	DECEMBER
Instructors (Military and Civilian)	10	20	22	26	28	27
Total Flying Cadets Graduated RMA/	19	169	165	168	148	127
Commissioned	0	9	16	28	20	9
Transferred	0	0	12	18	21	116
Discharged	0	1	0	0	2	2
Total Flying Time	58 hours 17 minutes	628 hours 22 minutes	1,496 hours 3 minutes	1,386 hours 5 minutes	1,466 hours 59 minutes	262 hours 41 minutes
Total Airplanes in Commission on Average	N/A	17 JN-4D 16 SJ-1	10 JN-4D 7 SJ-1	11 JN-4D 6 SJ-1	12 JN-4D 11 SJ-1	26 JN-4D 9 SJ-1
Total Accidents	1	2	0	6	3	2
Fatalities	0	1*	0	0	0	0
Destroyed Airplanes	1	2	8	6	0	0

*One enlisted man was killed when struck by a propeller.
(Numbers reflect available data)

First Flying Season

Despite Wilbur Wright Field's shortcomings, the aviation school began operations June 28, 1917, as a function of the Technical Section of the War Department's Division of Military Aeronautics. Major Arthur R. Christie arrived from Ashburn Field, Chicago, Illinois, on July 6 to assume command of the new facility. Since Ashburn Field was being inactivated, several officers and Standard SJ-1 biplane trainers also were transferred to Wilbur Wright Field.

Two days later, Wilbur Wright Field came to life with the arrival of two provisional aero squadrons of enlisted mechanics from Camp Kelly, Texas (often referred to in correspondence as Kelly Field). Captain Leo G. Heffernan, in command of the 12th Aero Squadron, led his two lieutenants and 150 enlisted men on a three-mile march down the dusty country road from the railroad depot in nearby Osborn. Captain Maxwell Kirk, in command of the two lieutenants and 150 enlisted men of the 13th Aero Squadron, repeated the trek later in the day. On October 31, 1917, the 12th and the 13th Aero Squadrons were reassigned to Garden City, Long Island, New York, and subsequently moved to France. Before the year ended, the 149th, 151st, 162nd, 166th, 172nd, and 211th Aero Squadrons had been assigned temporarily to Wilbur Wright Field

before moving to Europe. "Permanent party" squadrons included the 42nd, 43rd, 44th, and 47th.

On July 8, 1917, Lieutenant William G. Merrill and three enlisted members of the Medical Reserve Corps (MRC) arrived on post. On July 31, the 19th and 20th Aero Squadrons arrived, adding 148 and 150 men, respectively, to the number of military personnel assigned to the field. The end-of-July strength stood at approximately 800. Military strength increased dramatically in August with the addition of Cadet Squadron A and the three officers and 137 enlisted men of Company K, 3rd Ohio National Guard infantry, which had been called into active federal service. By the end of August, total strength of Wilbur Wright Field rose to 38 officers and 1,579 enlisted men. Within a year, the total personnel exceeded 3,000.

Despite the influx of personnel, preparation for the field's main function continued apace. In early July, several SJ-1 biplane trainers were shipped to the site in freight cars, reassembled, and readied for use. On July 17, Major Christie made the installation's first test flight, officially launching the military aviation history that continues today at Wright-Patterson Air Force Base.

During the first six months of operations, an average of 160 students per month were enrolled in the aviation school. Eighty-two of these graduated with the rating of

reserve military aviator (RMA) and received commissions as second lieutenants in the Reserve Corps. Surviving records indicate only five discharges, and almost 200 students continued on to other stages of training at Wilbur Wright or other installations. This record stands as an eloquent testimonial to the dedication, perseverance, and professionalism of the Wilbur Wright Field population, both military and civilian, the latter of whom were employed as flying instructors, aeronautical engineers, airplane mechanics, and "housekeeping" craftsmen. In fact, the overall effort of the "flying instruction department" was one of the few functions praised in Major Wilbourn's annual report.

Flight instruction was carried out in two basic aircraft at Wilbur Wright Field and other Signal Corps aviation schools. These were the Curtiss Aeroplane Company's JN-4D Jenny, powered by the 90-horsepower, Curtiss OX-5 engine, and the Standard Aircraft Corporation's SJ-1, powered by 100-horsepower, Hall-Scott A-7 and A-7A engines. Both were single-engine biplane trainers, with two open cockpits mounted in tandem. The Jenny evolved through the cross-breeding of an English aircraft designed by B. D. Thomas, known as Model J, with a Glenn Curtiss American design, Model N. The offspring was naturally christened "JN." The most common model of the series was the JN-4D.

Standard SJ-1 at Wilbur Wright Field, 1918. About 32 of these $6,000 airplanes were operated at the field's Signal Corps aviation school. (*Don Ream Collection*)

Curtiss JN-4D biplane trainers became the most popular trainers of the war.

Wire and metal workshop, Wilbur Wright Field, January 18, 1918. Fuselages are Curtiss JN-4D Jenny two-place biplane trainers flown by Signal Corps aviation school instructor pilots and flying cadets.

The Jenny was much easier to fly than the SJ-1, and therefore saw heavier use (6,000 JN-4D's were built by the end of 1918, as opposed to 1,601 SJ-1s). According to one historian, about 90 percent of all World War I American pilots earned their wings in the Jenny. After the war ended, hundreds of these airplanes became the mainstay "barnstormer" of the 1920s. Dozens were still being flown in the 1930s from pastures, fairgrounds, racetracks—anything that served as a makeshift airfield. Many a World War II pilot got his first taste of flying as a youngster with a five-minute flight for one dollar in a Jenny that circled a local pasture at minimum altitude.

Although he praised the instructional aspects of the flying program, Major Wilbourn found that the condition of official record keeping varied from haphazard to disastrous. Two weeks after the flying season ended for the year, he was unable to determine how many airplanes had been assigned on station since July or how much logistical support had been provided. The records of the Supply Department were "in a most chaotic condition" and the accounts of the Engineering Department, which he described as providing the "whole fabric of maintenance and operation of the field," were incomplete.[39] Fortunately, enough records existed to indicate the amount of flight instruction that had been given and supported during that first season. In total, the school at Wilbur Wright Field graduated 82 RMAs and logged 5,298 hours and 27 minutes of flight time with the loss of only 17 aircraft. At least 85 JN-4D's and 32 SJ-1s arrived on station, and 46 Jennys and 32 Standards were subsequently shipped to other primary fields. Aviation gasoline consumption was 88,036 gallons; oil usage was 1,900 gallons. Operating expenses totaled $310,000, including local purchases of aviation gasoline and oil, machinery, tools, airplane spare parts, office supplies and equipment, and other items.

On December 1, 1917, the Signal Corps Aviation Section transferred flight instruction activities from Wilbur Wright Field to more "Southern stations" for the duration of the 1917-1918 winter.[40] Five of these primary training fields were located in Texas (Barron, Carruthers, Kelly, Love, and Taliaferro), and others were scattered throughout Arkansas, Tennessee, Georgia, and Alabama.[41] By mid-month, all flying activity at Wilbur Wright Field ceased. In place of pilot training, the installation hosted a temporary school for mechanics and a permanent school for armorers. This

AN AVIATOR'S BEST FRIEND

Since the dawn of the age of flight, aviators have enjoyed the company of canine friends, as these photos attest.

Source: Ivonette Wright Miller, compiler, *Wright Reminiscences* (Air Force Museum Foundation, Wright-Patterson Air Force Base, Ohio, 1978).

Wilbur Wright rescued this famished, flea-bitten dog while visiting France in 1908 and named him "Flyer." (*Wright State University Archives, Wright Brothers Collection*)

Orville Wright and his dog, Scipio. Although Orville outlived Scipio by a quarter of a century, he always kept a photo of the beloved St. Bernard in his wallet. (*Library of Congress*)

World War I test pilot Lieutenant Frank Stuart Patterson, for whom Patterson Field is named, takes a break with a playful pup.

Lieutenant Richard E. Cole, who served as Jimmy Doolittle's copilot on the Tokyo Raid in World War II, relaxes at home in Dayton with his dog, Jinx.

After a 48-hour continuous run on a test stand at Wilbur Wright Field in 1918, a 90-horsepower Curtiss OX-5 engine is removed for subsequent examination of all its components. This four-cylinder engine powered the Curtiss JN-4D Jenny primary trainer. (*Don Ream Collection*)

Motor machine shop, Wilbur Wright Field, January 29, 1918. Building 252 on Allbrook Drive (Area C) now stands on the site of this shop.

respite in flying also provided the aviation school's commandant with an opportunity to strengthen a rather shaky organizational structure.

A "Major" Reorganization

During the first six months of operation, Wilbur Wright Field had five different commanders. Major Arthur E. Wilbourn, who served as commander from December 1917 to June 1918, was not perceived in a favorable light by his peers, one of who was Major Heffernan. As a captain, Heffernan had led the first enlisted men onto the post. As a major, he had preceded Wilbourn as commander of the post for one week in December. He had therefore spent more time at the young installation than almost any other officer, and he kept a diary of his experiences. Heffernan noted that Major Wilbourn was "a very officious type of officer and cordially disliked by all who knew him in the Air Service. He didn't last long in the game, chiefly because he could not be taken up in a plane."[42]

Though he apparently had no desire to fly—even as a passenger—Wilbourn had a good head for management and a strong hand for organization. He left an excellent, detailed, annual report dated May 31, 1918, which covered operations from the installation's inception. Though highly critical of the poor state of affairs he found upon assuming command, he detailed both the situations and the remedial actions he directed. They were sometimes drastic and, therefore, would have been considered unpopular, but they were effective. By May 31, 1918, the field was much improved.

Major Wilbourn gave praise sparingly but did give it where it was due, such as to the "flying instruction department" and the medical corps. He praised the organization and conduct of these functions and attributed the bulk of their difficulties to outside forces. Other agencies did not fare so well. Two departments in particular required various degrees of remedial action: Engineering and Supply. Wilbourn declared that the Engineering Department was "the heart of Wilbur Wright Field. Upon [it] depends the whole fabric of maintenance and operation of the field."[43] Yet, he was clearly dismayed by the conduct of the department. Engineering had charge of a large span of operations in 1917-1918 that included civil engineering, airplane field maintenance, organizational maintenance, and an aircraft engine training school. Civil engineering functions encompassed many of the same responsibilities handled by today's civil engineers, including surveying, cartography, drafting, utilities (lighting, water, and sewage), steam and emergency electrical power, and maintenance of roads and grounds, ranging from streets to flower beds.

The Engineering Department's primary problems revolved around the repair and maintenance of hastily constructed buildings and the procurement and support of aircraft used for training at the aviation school. Wilbourn believed that closer supervision and better quality control by the Engineering Department would have prevented the shoddy workmanship and inferior materials used for the post's first facilities.

About aircraft support responsibilities, Wilbourn wrote:

> It is required to secure sufficient ships; to assemble them and keep them in proper repair; supply them with fuel and lubricants; record the performance of all motors, propellers, planes etc.; to care for and replace broken parts; secure and return wrecks. [44]

To fulfill these responsibilities, the Engineering Department had to train its own airplane engine mechanics, who had not kept maintenance records of individual airplanes during the first six-month flying season. The Engineering Department could not determine the quality or quantity of work performed, and the flying instruction staff later had to reconstruct the "in-commission" rates of aircraft from incomplete monthly records.

Continuing maintenance problems, such as excessive failure of propeller blades, were compounded by equipment shortages, which nearly forced cancellation of all flying in September until new parts arrived. Many of these problems remained unsolved when the second flying season opened on April 15, 1918. At the time Major Wilbourn finished his report in May, airplane mechanics were struggling with leaking exhaust valves in the OX-5 power plants in the JN-4D's. Major Wilbourn felt the problem indicated that "the assemblers of the motors [in the factory] . . . failed to seat the valves properly."[45] Manufacturing defects were also noted in JN-4D airplanes arriving from the Curtiss Aeroplane & Motor Company's Canadian plants. The quality of dope and methods of application used on these airplanes were "inferior to a degree" that required shipping 13 of the JN-4D's to the Aviation Repair Depot at

Engine testing rig, Wilbur Wright Field, 1918 (*Don Ream Collection*)

Indianapolis, Indiana, where the entire fabric of the wings, fuselages, and tails were re-doped. At the end of May, the Engineering Department fully expected "upwards of 20" more airplanes to arrive from the factory in similar condition.[46]

Major Wilbourn was also highly displeased with "the most chaotic condition" of the Supply Department. All 1917 records, especially those pertaining to property accountability, were in "very bad shape."[47] Vouchers were missing; dozens of invoices sent out by the Supply Department were never returned; and the department had made no vigorous efforts to prepare Reports of Survey for missing property. Although the supply officer attempted to correct the deficiencies and discrepancies, Major Wilbourn relieved him from duty and appointed a board of disinterested officers to survey all Signal Corps property on Wilbur Wright Field.

The Quartermaster Department, established in early June 1917, had its own share of problems in each of its five elements (administration, finance, supply, transportation, and reclamation). The administrative division operated for nearly a year without Quartermaster Corps-qualified, administrative, enlisted men, and it was not until April 1918 that 32 qualified enlisted men arrived on station. The finance and accounting branch had, up to this point, been concerned solely with paying minor accounts in the local area, as officer and enlisted payrolls had been

handled directly by the Office of the Department Quartermaster, Central Department, in Chicago, Illinois. In December, however, the Central Department directed the local branch to pay all military personnel at both Wilbur Wright Field and the experimental testing facility at McCook Field near downtown Dayton. In addition, it was to pay the "expense accounts" of all Signal Corps civilian employees in Dayton. Monthly disbursements jumped from $1,146 in October to $168,652 the following May. Total disbursements during that period reached $622,024.

In the area of supplies, Major Wilbourn was convinced that the Quartermaster Department had made "no attempt . . . to provide an adequate supply" of food, clothing, and incidentals during 1917.[48] For example, from June through December, individual squadrons had maintained their own messes (dining halls), and organizations had purchased fresh produce and meat from local markets, making the daily runs in their own trucks. In January 1918, Major Wilbourn directed the establishment of a sales commissary on post. The Wilbur Wright Field quartermaster negotiated with local contractors for all fresh meat and vegetables. Major Wilbourn was satisfied that the quality of the meat furnished was excellent and the prices were good. In addition, on March 25, "field oven number one" began operation, freeing the installation of dependence on local,

WILBUR WRIGHT FIELD COMMANDERS—1917-1925

Captain Arthur R. Christie	July 6-September 26, 1917
Lieutenant Colonel George N. Bomford	September 26-December 19, 1917
Major Leo G. Heffernan	December 19-24, 1917
Major Walter R. Weaver	December 24-29, 1917
Major Arthur E. Wilbourn	December 29, 1917-June 28, 1918
Lieutenant Colonel Thomas E. Duncan	June 28-September 16, 1918
Lieutenant Colonel L. W. McIntosh	September 16-October 21, 1918
Major Maurice Connolly	October 21-November 2, 1918
Major Byron Q. Jones	November 2, 1918-January 10, 1919
Major Charles T. Waring	December 24, 1918-May 1, 1919*
Major Prince A. Oliver	May 1-August 2, 1919
Lieutenant Colonel George E. A. Reinburg	August 2, 1919-August 27, 1921
Major Augustine Warner Robins	August 27, 1921-August 21, 1925

* Major Waring became commander of the Fairfield Aviation General Supply Depot on December 24, 1918, and served simultaneously as the commander of Wilbur Wright Field beginning January 10, 1919, when the latter merged administratively with the depot. Between that time and the discontinuation of the designation "Wilbur Wright Field" in August 1925, the depot's commander served concurrently as the commander of Wilbur Wright Field.

Major Arthur R. Christie, the first commanding officer of Wilbur Wright Field. On July 17, 1917, Major Christie made the first test flight from Wilbur Wright Field. Later, as a lieutenant colonel, he served as chief of Air Service, V Corps, in the St. Mihiel offensive during World War I.

Major Leo G. Heffernan was commissioned as a second lieutenant in the cavalry before transferring to the Air Service. He completed pilot training in 1916 and then commanded the 12th Aero Squadron, arriving at Wilbur Wright Field on July 8, 1917. After commanding Wilbur Wright Field, Heffernan served as an Air Service officer with the American Expeditionary Forces (AEF) in France. He retired on disability in 1933.

After Major Arthur E. Wilbourn graduated from the U.S. Military Academy in 1908, he received a commission in the cavalry and was placed on active duty with the Signal Corps to command various flying fields, including Wilbur Wright Field, during World War I. He retired as a colonel in 1944.

The Unit One hospital at Wilbur Wright Field was constructed in 1917 and expanded in 1918. Though built as temporary structures, the hospital and many other buildings throughout the installation remained in use for many years.

Nurses at Wilbur Wright Field, 1918

commercial bakers. With a capacity of 216 pounds of bread per run, the oven could fill Wilbur Wright Field's needs, as well as those of McCook Field and other Army installations in the Dayton area. This reorganization in particular helped put the commissary into the black. Between January 1 and May 31, commissary sales to all Army organizations in the Dayton area totaled $57,034.

In the area of clothing supplies, records indicated that 19,904 items of clothing, including overcoats, cotton and wool coats, cotton and wool breeches, slickers (raincoats), hats, shirts, shoes, and "leggings," were issued at Wilbur Wright Field from October through May. A Reclamation Division organized in March saved resources and earned money by cleaning, repairing, or salvaging clothing and shoes. Through these changes, by the end of May 1918, the Quartermaster Department was at least functional in each

of its areas of responsibility.

The Medical Department, another essential area of post activity, was given good marks by Major Wilbourn, but he was concerned about the unsanitary environment that existed at the field. Large numbers of civilian construction employees and their families lived in primitive and squalid conditions in "squatter camps" on the field. The open-pit privies dug for the laborers were not screened, were seldom cleaned, and contaminated the wells that provided water for the post, so that large amounts of chlorine had to be added before use. Several hundred teams of horses and mules used in heavy construction work generated huge piles of manure, from which clouds of flies swarmed in all buildings throughout the field. By mid-1918, however, the domestic service "utility" functions of the field were nearing satisfactory levels of operation. Three driven wells produced 340,000 gallons of

pure water daily, and the sewage system was nearing satisfactory operation. After taking these necessary steps to establish a healthy environment on Wilbur Wright Field, Major Wilbourn considered the post to be in excellent sanitary condition.

Despite the early sanitation problems, the death rate from disease was exceptionally low. Major Wilbourn credited this to the efforts of the Medical Department, headed by Major Alfred G. Farmer, who assumed command of a staff of four on July 16, 1917. Although the first hospital building, one of the last on post to be completed, was not ready until August 21, a tent was used as an isolation center on July 18, when the first illness requiring hospitalization proved to be the highly communicable scarlet fever. By March 1918, an addition to the hospital building was completed, bringing the total beds available to 89. By mid-1918, the medical staff included 14 physicians and dentists, 11 commissioned nurses, and 63 enlisted men.

By mid-1918, the Dayton Power and Light Company was meeting the field's monthly power requirement for 660 kilowatts of electricity, and a new telephone system was completed, giving 229 instruments to Wilbur Wright Field and adjacent Fairfield Aviation General Supply Depot (see Chapter 3: The Story of Air Force Logistics). Calls averaged 5,200 a day. Fortunately for the cross-country flying program, the surrounding rural districts were fairly well covered with telephone nets, according to Major Wilbourn, and thus provided for quick notifications of all landings, forced or otherwise. A fairly common occurrence was a call from a chagrined pilot who had gotten lost and ran out of gas on his first cross-country flight.

AVIATION MECHANICS' SCHOOL

While Major Wilbourn addressed his internal problems, he also attended to his role as host for two essential Signal Corps activities that began at Wilbur Wright Field in the non-flying winter months from mid-December until April. The first of these to be organized at the field was an aviation mechanics' school. According to Allied manning experience in Europe, each combat-ready airplane required the support of 47 ground-force personnel, including officers and enlisted men in engineering, supply, administration, and maintenance.

Instructor and students examine a JN-4A training airplane at Wilbur Wright Field.

Aviation Mechanics' School students work on the fabric wings of a Curtiss JN-4D trainer.

Instructors and students at the Wilbur Wright Field Aviation Mechanics' School pose behind the hangar they used as a classroom. During 1918, the school graduated 1,181 enlisted men. (*United States Air Force Museum*)

The largest single category of these supporters was enlisted "aviation mechanicians." Experienced aviation mechanics were in short supply after the United States declared war on Germany in April 1917. The few that existed outside of the military were already working for unprecedented wages in the civilian war effort. Experienced motor-vehicle mechanics were generally in greater supply, but by fall, most of them were also occupied by the war effort, either in the Army or in civilian support functions.

By November 1917, the Aviation Section was in critical need of both types of mechanics. As a result, on November 1, the War Department directed that 5,000 mechanics be transferred immediately from the U.S. Army to the Signal Corps Aviation Section.[49] The problem of transferring

these mechanics' skills to the new area of aviation was handled through both short- and long-range training programs.

Two programs, one of which involved private industry, provided short-term results. In industrial communities such as Dayton, selected airplane and engine factories and garages were asked to open their facilities to groups of 25 soldier-students for on-the-job training. Nearly 20 companies willingly provided training for about 2,000 men.

The second program established short-term mechanics' training schools at five northern installations where flying instruction had to cease during the winter months. Such schools operated at Wilbur Wright, Chanute, Scott, Selfridge, and Hazelhurst fields. Instructors for these temporary schools came from both outside

and inside the Army. Private companies were canvassed for experienced foremen who could add their technical competence and supervisory skills to the ranks of Signal Corps instructors. A special evaluation board selected 60 foremen for such service, and after three weeks of training at Selfridge Field in December, they were distributed to the new schools to teach such specialties as "woods, propellers, wing repair, fabrics, wire work, soldering, tires, alignment, fuselage, motors, and motor transport."[50] Within the Aviation Section, the more highly skilled mechanics served as instructors in the interim period between the opening of the schools and the arrival of the newly oriented instructors from Selfridge Field.

The situation at Wilbur Wright Field was typical. The aviation mechanics' school opened December 17 in the 12 Unit One hangars that had been used for flying instruction. Major W. R. Weaver commanded the school, assisted by R. E. Dunn, who also served as chief of motor transport instruction. The 42nd and 44th Aero Squadrons, permanently assigned to Wilbur Wright Field, provided basic operating services to the school. Students arrived from various Midwest airfields on temporary duty from 20 aero squadrons: 42nd, 44th, 47th, 149th, 151st, 159th, 162nd, 163rd, 166th, 167th, 172nd, 211th, 255th, 256th, 257th, 258th, 259th, 260th, 265th, and the 827th.

Classes in the subject areas of airplane, airplane motor, and motor transport were scheduled to begin on December 17, but the instructors from Selfridge Field had not arrived by that date. In the interim, a faculty of 70 was drawn from the sharpest enlisted mechanics in the 42nd and 44th Aero Squadrons. Early in January 1918, the faculty was augmented by the arrival of two second lieutenants and 18 recent graduates from Selfridge Field. For the duration of the school's operation, the instruction staff continued to draw heavily from the ranks of the two permanent squadrons.

The school faced continuing challenges, ranging from equipment shortages to weather interruptions. Perhaps the most serious handicap faced by instructors related to the skill levels of incoming students. In theory, all recruits were screened for trade skills at the time of induction into the Army, and only those with significant mechanical aptitude were qualified for assignment in mechanical fields. However, the staff frequently encountered students with no former experience and no knowledge of the occupations or skills in which they were classified. Furthermore, a significant number of draftees were illiterate and limited even in their ability to speak English. Many had arrived in this country in the arms of immigrant parents at the beginning of the new century and had used their parents' native tongue almost exclusively while growing up. Instructing them in technical matters was difficult. Nonetheless, by its April 7, 1918, closure, the mechanics' training school at Wilbur Wright Field had 1,181 graduates: 182 in airplane motor, 386 in airplane, and 613 in motor transport courses of instruction.

In addition to its graduates, the school had also produced 85 instructors to complement the 18 instructors sent from Selfridge. When the school closed, 103 instructors transferred to other training installations. Major Wilbourn termed them "the very best . . . airplane and motor repair men" on post. Their loss was felt markedly in the ongoing operation of the flying school while their recently graduated replacements gained experience and competence. By the end of May, however, Major Wilbourn acknowledged that it was "to the very best interests of the service to cripple the flying school temporarily in order that the quality of our enlisted mechanics in general may be improved."[51]

The training school for aviation mechanics at Wilbur Wright Field contributed significantly to the overall effort of the Air Service. Its 1,181 graduates joined 1,482 more from the other four northern flying fields. Together they formed the vanguard of the 10,000 aviation mechanics trained by May 1, 1918, and provided a valuable baseline experience for the instructional competency of the Air Service.

AVIATION ARMORERS' SCHOOL AND GUNNERY TESTING

A second subject area that became increasingly important as the war progressed in Europe was armament. At the outbreak of the war, airplanes were unarmed, although pilots occasionally traded pistol shots. The German Fokker revolutionized offensive tactics with a machine gun mounted and synchronized to fire 500 rounds a minute between the blades of the propeller. Bombs were soon added to offensive aerial strategy. These new concepts in warfare precipitated new requirements in equipment and in the training of pilots.

As the Allied response developed, each aero squadron needed an armament officer and a score of men to inspect, test, and tune aerial armament before and after every flight. It was a critical responsibility for, according to a contemporary World War I source, "[s]cores of good aviators [were] killed by reason of guns jamming just at the critical moment."[52]

Two facets of the armament function were assigned to Wilbur Wright Field: (1) testing all machine guns issued to the Aviation Section to ensure that they were properly adjusted and in good firing condition and (2) training new armament officers and their enlisted assistants. Both functions geared up for operation in March 1918 using Unit Two hangars.

In preparation, a central school for both officer and enlisted instructors opened on February 4, 1918, at Ellington Field in Houston, Texas. The curriculum concentrated on mechanism and construction rather than on actual use of bombs and machine guns. Aspects of stripping, care, cleaning, causes of stoppage, loading, and testing were emphasized as important elements of the new career field.

At the conclusion of their training at Ellington, 200 of the armorers transferred to Wilbur Wright Field as the 851st Aero Repair Squadron. On March 18, the armorers' school opened for final indoctrination of the officers and enlisted men who formed the school's faculty and

A portion of the Aviation Armorers' School hangars at Wilbur Wright Field (*Wright State University Archives*)

Students at the Aviation Armorers' School learn to adjust the timing on a machine gun, Wilbur Wright Field, 1918. (*United States Air Force Museum*)

Lieutenants Mathis, Keenan, Rubin, and Skinner were instructors in the Aviation Armorers' School, Wilbur Wright Field, which operated from March 1918 to February 1919.

Soldiers from Squadron B, Armorers' School, Wilbur Wright Field, 1918 (*Darlene Gerhardt Collection*)

staff. The course of instruction was fixed at six weeks and covered a complete study of machine guns, their sights and synchronization mechanisms, and the storage and mounting of bombs.

Meanwhile, small detachments filtered in, fresh from factory training at the Marlin-Rockwell Company in New Haven, Connecticut, and the Savage Arms Corporation in Utica, New York. Together with the 96 officers and 560 enlisted men who reported as students on April 13 and 20, respectively, they formed the 874th Aero Repair Squadron. Completing the armament network was the 231st Aero Repair Squadron, which reported on April 22 from Ellington Field.

The armorers' school was organized under authority of the Signal Corps Air Division Gunnery Section and operated under the command of Major A. H. Hobley from March 18 until the conversion of effort at war's end. In April and May alone, the school hosted 95 officer and 789 enlisted students. The first class graduated June 6, with all 95 officers graduating. The enlisted program graduated 485 out of 560, reflecting the same language and trade proficiency problems that had surfaced in the mechanics' school.

The gunnery-testing function began May 1, 1918. Initially, 100 Lewis and 100 Marlin machine guns were inspected and tested each day. As operations hit stride, the capacity increased to 100 Lewis and 200 Marlin guns per day.

SPRING FLYING

With the advent of spring, airplanes, cadets, and instructors migrated northward from their southern "winter quarters." As the flying instruction program reorganized at Wilbur Wright Field, officials discovered that only six of the 18 reserve military aviators sent to serve as instructors had themselves flown more than 50 hours. The remaining 12 needed special, accelerated, instructors' training before joining the staff. In the meantime, pilot training resumed on April 15, 1918, using the six already qualified instructors.

Students and instructors alike faced hazardous field conditions. In late March, a contractor had begun smoothing and seeding the turf of the flying field. There were no hard-surfaced runways. This work was not finished by the time flight training resumed, and large numbers of laborers with assorted equipment were constantly on the field, providing daily hazards for students. To keep the number of cadets on the flying field at a manageable level, the cadets were divided into two groups. While one group spent the morning in class, the other was on the flightline, with reversed schedules in the afternoon. At first the flying field was divided into two

Curtiss JN-4D Jenny primary trainer in flight over hangars of the Aviation Armorers' School and armament testing station, Wilbur Wright Field, 1918

Signal Corps aviation school flying cadets stand roll call behind Wilbur Wright Field hangars, spring 1918.

sectors, one for dual-control flights and one for solos. When this proved impractical, a more efficient and safer plan was adopted in which the entire field was devoted to dual instruction in the morning and solo instruction in the afternoon, with a small portion of the field permanently set aside for cross-country flying and radio airplanes.

By May 31, the faculty had increased to 25 instructors, and 180 cadets were involved in various stages of instruction, including courses in military studies, gunnery, radio, photography, airplanes, engines, poison-gas defense, and aerial navigation. The flying instruction proceeded from dual-control to solo instruction, then to cross-country. Cadets flew a cumulative average of 66.6 hours per day, with an average of 22.5 airplanes in commission. One cadet had graduated as a reserve military aviator, 16 were ready to graduate, two had been discharged because of flying deficiencies, and the school had suffered its first flying fatality when one cadet died in a crash.

AIRPLANE TESTING AT WILBUR WRIGHT FIELD

The gunnery program was not the first military testing function at Wilbur Wright Field. Airplane testing was inaugurated by a March 1, 1918, request from McCook Field to provide hangar space for experimental flying. McCook Field operated under the Airplane Engineering Department of the Signal Corps Equipment Division. Established October 13, 1917, it was located near downtown Dayton, about 10 miles by road from Wilbur Wright Field.

McCook's mission was to research, develop, test, and evaluate U.S. military aircraft and, occasionally, to test airplanes designed or manufactured by allied nations. Although McCook had its own flying field for airplane testing, space was limited. The special macadam-and-cinder accelerated runway—a definite improvement over the bumpy grass strips to which most pilots of the day were accustomed—was 1,000 feet long and 100 feet wide. In order to take advantage of the prevailing winds, however, the runway had to be laid across the short expanse of the field. This resulted in extremely short takeoff distances due to surrounding obstacles (trees, the river, etc.). It also led to the coining of McCook Field's motto, "THIS FIELD IS SMALL— USE IT ALL," which was emblazoned on the front of one of the hangars. As aircraft grew in size and power, this constraint became one of the major factors that forced McCook's activities to relocate to Wright Field in 1927.

From McCook's inception, it was understood that a certain amount of both hangar space and maintenance support might be available from Wilbur Wright Field. McCook's first request for assistance was in conjunction with the brief testing of three Italian airplanes: one SVA (Societa Verduzio Ansaldo) single-place "Scout," one Pomilio two-place fighter with a Fiat engine, and one SIA (Societa Italiana Aviazione) two-place fighter with a Fiat engine. The airplanes' arrival on March 19, 1918, marked the debut of Wilbur Wright Field as a test site for modern military aviation. After two months of testing, the airplanes were disassembled and returned to Italy.

By that time, flight-testing aspects of three other McCook programs had been moved to Wilbur Wright. The first of these began and ended on the same day. An American Morse pursuit [fighter] airplane, equipped with a small, eight-cylinder, Liberty engine, was trucked from McCook and assembled for testing. On March 28,

Aerial view of McCook Field, the "cradle of aviation," along the eastern bank of the Great Miami River in Dayton

BARLING BOMBER

The U.S. Army Air Service gave little attention to large airplanes prior to the end of World War I, instead spending time and effort on the production of the Liberty engine. Soon after the war, however, Brigadier General William "Billy" Mitchell, assistant chief of the Army Air Service, recognized the need for development of a large bomber. Consequently, historians often accredit General Mitchell for the construction of the Barling bomber. In 1919, General Mitchell discovered a young Englishman by the name of Walter J. Barling, Jr., who had previously worked for the Royal Air Force. He asked Barling to design a new style of airplane that could carry an adequate number of bombs to sink a battleship. Mitchell's goal was to prove his theories to non-believers on the effectiveness of bombers in aerial warfare. For two prototypes, Mitchell projected the cost of this large bomber at $375,000.

As designed, the Barling bomber was considered a triplane. It was 28 feet in height and 65 feet in length; the wingspan was 120 feet. The gross weight of the bomber totaled 42,569 pounds. Military specifications required a 5,000-pound bomb load, but with a larger bomb, the Barling's flight time was limited to two hours. The specifications also required a speed not less than 100 mph and a service ceiling of 10,000 feet. The Barling bomber was powered by six, 12-cylinder, 400-horsepower, Liberty motors (two tractor engines and one pusher on each side). It had a gasoline capacity of 2,000 gallons and an oil capacity of 181 gallons.

The Witteman-Lewis Company of New Jersey manufactured the bomber components. Only six airfields in the country, however, were adequate for the flight tests of the huge airplane: Mitchel Field, Long Island, New York; Teterboro Field, Hasbrouck Heights, New Jersey; Ellington Field, Texas; Kelly Field, Texas; Langley Field, Virginia; and Wilbur Wright Field, Ohio. While the airfields in Texas were considered good sites, most of the others were rejected because of their size and congested surroundings. In the end, Wilbur Wright Field was chosen based on its proximity to the aeronautical engineering function at McCook Field.

After manufacturing, the bomber's components were shipped by rail for assembly at McCook Field. The Army's Engineering Division insisted on having a hand in the project, an arrangement that resulted in mismatched parts and drove the cost to $525,000 for one prototype. A hangar large enough to protect the plane from rainwater added $700,000 to the cost of the big bomber program.

On August 22, 1923, the Barling bomber, the largest airplane in the world at that time, made its first flight at Wilbur Wright Field. It was piloted by Lieutenant Harold R. Harris and Lieutenant Muir S. Fairchild, future U.S. Air Force vice chief of staff. The Barling was the first airplane to require a flight engineer to assist with flight operations. Engineer Douglas Culver performed this duty, and Walter Barling flew as a passenger. The large bomber lifted from the grass after a 13-second, 960-foot run, thus silencing critics who said the plane would roll into Dayton before taking off. During their 28-minute flight, Lieutenants Harris and Fairchild flew the bomber over Wilbur Wright Field at an altitude of 2,000 feet.

In October 1923, the bomber set records in both duration and altitude at Wilbur Wright Field. On October 25, it carried a load of 4,409 pounds to an altitude of 6,772 feet. Two days later, it carried 6,612 pounds to an altitude of 5,344 feet. Lieutenant Harris piloted the bomber on each of these record setting flights.

Despite the Barling bomber's remarkable performance, the six Liberty engines could not lift the bomber high enough to safely cross the Appalachian Mountains, thus canceling its appearance at a Washington, D.C., air show. In 1927, after authorizing well over $1 million on one prototype, Congress cut further funds for the Barling bomber. It was subsequently dismantled and put in storage at Fairfield Air Depot. In 1929, Major Henry H. "Hap" Arnold ordered the bomber burned. The hangar remained in use until 1942 when it was dismantled.

Sources: Engineering Division, Air Service, Specification for Barling Bombardment Airplane, May 15, 1920; "Notes on Barling Bomber;" Captain Earl H. Tilford, Jr., "The Short, Unhappy Life of the Barling Bomber," *AIR FORCE Magazine*, February 1978, pp 68-70; "Description of the Barling Bomber;" Marshall Lincoln, "The Barling Bomber," *Air Classics* 2(5) February 1965; "From the Editors Desk," *Aerospace Historian* 27(1), Spring/March 1980, pp 55-57. All sources in the Barling Bomber File, Aeronautical Systems Center History Office.

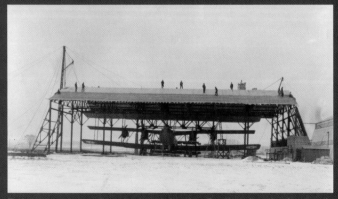

Rainwater was often trapped in the Barling bomber's poorly sealed wings, compromising weight measurements during flight tests. To protect it from rain, a specially designed hanger was constructed to shelter the huge bomber. The hangar, shown here in February 1924, was demolished in 1942 to make room for the new Patterson Field Air Transport Terminal (Building 146).

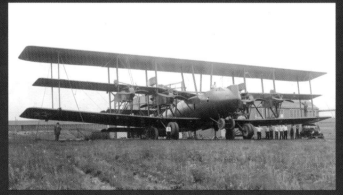

The Barling bomber at Wilbur Wright Field, 1923. Two trucks of four wheels each formed the main gear, with an additional pair called "nose-over" wheels farther forward. These front wheels prevented the nose from hitting the ground during takeoff and landing. The tires on the main wheels were 60 inches in diameter and 12 inches wide.

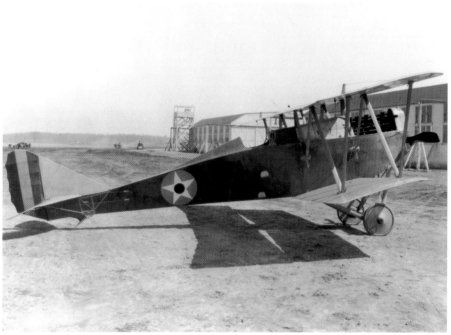

McCook Field engineers tested this Italian SVA with U.S. Army Air Service markings at Wilbur Wright Field in 1918. This airplane was the Italian Army's outstanding fighter of World War I. (*United States Air Force Museum*)

DeHavilland DH-4, an all-wood machine designed for the dual functions of observation work and light bombing. The British-designed airplane, modified for American production methods and mounted with the Liberty 12-cylinder engine, was America's greatest engineering contribution to World War I. DH-4s were common sights around McCook Field in the postwar years.

Aerial view of Wilbur Wright Field. Visible in the center left is the 42,000-pound Barling bomber, which was tested at the field between August 1923 and May 1925. This photograph was taken before a special hangar was constructed to house the big bomber. Wilbur Wright Field was one of only six fields in the nation large enough to accommodate the huge airplane.

the first test flight ended abruptly when the airplane crashed from a height of about 50 feet and was destroyed.

Another short-term effort produced better results. On May 15, 1918, a French LePere pursuit equipped with a Liberty engine arrived for a series of tests that were to be conducted by three French pilots. At the time of Major Wilbourn's May 31 annual report, the airplane had flown almost daily.

A longer-range program began April 20 when Wilbur Wright Field agreed to furnish McCook with accommodations and limited logistical support for eight airplanes, including British DeHavilland DH-4 reconnaissance and Bristol pursuit aircraft. The support included not only hangar and shop space, but also a force of enlisted mechanics to both assemble and maintain the airplanes, particularly the engines. For its part, McCook Field agreed to furnish two Liberty engines and two instructors to assist in training Wilbur Wright Field mechanics. McCook also promised to provide flight time on the DH-4 airplanes being tested in order to upgrade Wilbur Wright Field pilots for this new type of airplane.

By April 24, 1918, one DH-4 and one Bristol fighter had arrived. The career of the Bristol fighter was brief; it crashed and was destroyed May 7. The DH-4 enjoyed more success, and was soon joined by seven more DH-4s manufactured by the Dayton Wright Airplane Company. To support these aircraft, 40 to 50 enlisted mechanics received 20 days of on-the-job training on the Liberty engine. Despite the fact that McCook did not follow through in setting aside a DH-4 specifically for upgrading Wilbur Wright pilots, by the end of May, at least 45 pilots had gained considerable experience in handling the new airplane. Moreover, every engineering officer on station had taken the initiative to study the structure of both the DH-4 and its engine.

OVER THERE AND BACK

At 11:00 a.m., November 11, 1918, the fighting ceased, and World War I ended. Records maintained between June and November 1918 indicated that both the flying instruction program and the armorers' school at Wilbur Wright Field contributed significantly to the Air Service record in Europe. Four of the original aero squadrons at the field earned combat credits in France. The 12th Aero Squadron

MCCOOK FIELD DURING WORLD WAR I

While Wilbur Wright Field focused its efforts on training pilots and mechanics for the war, another military establishment in the Dayton area focused on airplane and equipment research and development. Edward Deeds, who had been instrumental in the establishment of Wilbur Wright Field earlier in the year, also looked to the Dayton area as a location to centralize aviation research. The chosen tract of land was a 254-acre plot between the Miami River and present-day Keowee Street and was owned by Deeds and Dayton inventor Charles F. Kettering. A committee appointed by the Aircraft Production Board inspected the site in late September 1917, and immediately accepted it for the Signal Corps' purposes.

On October 13, 1917, the establishment of McCook Field was formalized to serve as the headquarters of the Airplane Engineering Department; operations officially commenced on December 4, 1917. McCook's mission was to research, develop, test, and evaluate U.S. military aircraft and, occasionally, to test airplanes designed or manufactured by allied nations. The installation was called an "experimental field" and the work conducted there consisted of the test and evaluation of materials and structures for the Army and industry. Over the next several years, McCook Field was assigned functional responsibility for the design of all airplanes and accessories, including engines, cameras, bomb sights, bomb racks, parachutes, and clothing.

Throughout 1918, the Signal Corps underwent a number of organizational changes and, in May, the Army Air Service was created. Most significant for McCook Field was the September merger of the Engineering Production Department and the Airplane Engineering Department to form the Airplane Engineering Division of the new Air Service, removing them from the control of the Bureau of Aircraft Production. The title "Engineering Division, Air Service," became the permanent designation for the organization at McCook Field until 1926.

Maintenance on a Huff-Daland LB-1 Pegasus bomber, March 24, 1926, at McCook Field. This photograph illustrates the method of hoisting the wing to remove the wheel.

Trucking Office building being readied for the move from McCook Field to the new Wright Field, May 1926

During the war, engineers at McCook Field developed one of America's most significant contributions to Allied air power—the Liberty engine mated with the British DeHavilland DH-4 biplane. A few weeks after the United States entered World War I, the Aircraft Production Board decided that one of America's most expedient contributions to victory would be the large-scale production of a high-powered airplane engine. By June 4, 1917, an American engine design was approved and the first Liberty engine was completed late that month and tested, appropriately, on July 4. It was an eight-cylinder, 200-horsepower model and was shortly joined in production by more powerful 12-cylinder, 300-horsepower versions and, later, by 12-cylinder models with 440 horsepower. By the end of the war, six American automobile manufacturers had produced 15,600 Liberty engines, and another 60,000 were on order by the United States and its allies for use on a wide range of aircraft.

The American-designed Liberty engine was mounted on the British DeHavilland DH-4, all-wood, two-place biplane, the only foreign machine produced in quantity in the United States. A sample airframe arrived in the United States on August 15, 1917, and was rushed by rail to McCook Field. Over the next two months, the machine underwent "extensive detail redesign" to accommodate American

"The Baby Blimp" landing at McCook Field. This observation balloon burned at Wilbur Wright Field on September 30, 1921, following a lightning strike. The watchman was injured.

Propeller Unit at McCook Field where all wooden propellers were made. The wood was laminated and glued to the desired thickness, then shaped, carved, and finished, ready for use on the airplane.

production methods and was fitted with the new 12-cylinder, 400-horsepower, Liberty engine specifically designed for it. The modified English biplane was renamed the "Liberty Airplane." On October 18, 1917, the Dayton Wright Airplane Company received an initial order for 250 redesigned DH-4s equipped with Liberty engines. The first production model was test flown at McCook Field on October 29 and the first American model DH-4 reached France in May 1918. By the end of November 1918, American manufacturers had produced more than 4,500 Liberty airplanes (DH-4s); Dayton Wright had constructed 3,106 of them.

Although the DH-4 represented America's major contribution to the war effort in Europe, other significant projects were also under way at McCook Field, including modification and production of the most combat-effective British, French, and Italian warplanes. As the war progressed, McCook Field engineers began to apply their efforts toward developing original American designs as well. The first of these developments was the USAC-1 (U.S. Army, Combat-1) two-seat biplane modeled after the British Bristol fighter. The development of the USAC-1 firmly established an engineering discipline at McCook that permitted the rapid development of other aircraft, such as the redesign and standardization of the DeHavilland DH-9 (a later model of the DH-4) for production. The McCook Field version was named the USD-9 and a reconnaissance version named the USD-9A.

For the decade after the war, the Airplane Engineering Division continued to serve as the center of all Army aviation research and development. Throughout the early 1920s, McCook engineers concentrated on developing standards unique to military aircraft, reviewing designs, modifying and testing procured machines, and developing ancillary equipment to enhance military aircraft. Flight testing continued from both McCook Field and Wilbur Wright Field. In 1919 alone, McCook's Flight Test Section recorded 1,276 test flights. It was soon recognized, however, that McCook Field was inadequate for its mission because of its size limitations. Consequently, in 1927, McCook's research and development program became a function of the newly established Wright Field.

Source: ASC History Office, *Splendid Vision, Unswerving Purpose: Developing Air Power for the United States Air Force during the First Century of Powered Flight* (Wright-Patterson Air Force Base, Ohio, 2003).

Wilbur Wright Field, winter 1918

was cited for its participation in aerial operations in the Lorraine, Ile-de-France, Champagne-Marne, Champagne, St. Mihiel, and Meuse-Argonne battles. The 13th and 20th Aero Squadrons also flew in the Lorraine, St. Mihiel, and Meuse-Argonne engagements. The 19th Aero Squadron flew liaison missions in France between January and March 1918.[53] Wilbur Wright Field's first commanding officer, Major Arthur R. Christie, subsequently promoted to the rank of lieutenant colonel, served as chief of Air Service, V Corps, in the St. Mihiel offensive, August 10 to September 16, 1918.[54]

After the war, priorities shifted. Just as America had clamored for instant armament in April 1917, it now demanded immediate relief from the burden of supporting nearly five million men in arms.[55] Demobilization soon brought drastic changes at installations across the nation. Although available sources do not specifically state Wilbur Wright Field's military population as of November 1918, the installation probably operated near its planned peak of 1,700 personnel. All training at the facility ceased by the end of November with flying limited to experimental and test aircraft participating in McCook Field programs. Wilbur Wright Field's mission shifted abruptly from training pilots and armorers to serving as a temporary repository for surplus war materiel.

Effective January 10, 1919, Wilbur Wright Field merged administratively with the Air Service Armorers' School and nearby Fairfield Aviation General Supply Depot (FAGSD).[56] The new unit was named the Wilbur Wright Air Service Depot (WWASD) and FAGSD's commander assumed control over all three organizations. The designation "Wilbur Wright Field" remained in use until 1925, although functions of the field were administered by WWASD and its successors. By the time the designation "Wilbur Wright Field" was dropped in 1925, the depot function had changed names several times. In brief, these changes were: Wilbur Wright Air Service Depot, January 10, 1919; Aviation General Supply Depot, November 3, 1919; Air Service Supply and Repair Depot, Fairfield, Ohio, September 20, 1920; and Fairfield Air Intermediate Depot, January 14, 1921 (see Chapter 3: The Story of Air Force Logistics).

Military strength at WWASD initially consisted of 70 officers and 830 enlisted men, but as demobilization continued, uniformed strength declined rapidly. On February 20, 1919, the flying school, armorers' school, and eight squadrons demobilized. A growing civilian workforce assumed the continuing function of the depot. By the end of 1918, this force peaked at 1,000 employees.[57]

As Army Air Service training fields and stations throughout the nation and overseas closed, supplies and equipment were shipped to major air depots such as WWASD for storage, inventory, and disposal. At the peak of this activity, 77 buildings under depot management, including some hangars, were used to house 2,500 aeronautical engines, 700 airplanes of various series, and thousands of instruments, compasses, watches, altimeters, and gauges of all descriptions. Four of the 12 Unit Two hangars were relocated from the flightline to the interior of the installation and converted to other uses (one garage, one gymnasium, and two warehouses). Final disposition of all wartime surplus items dominated functions at the Fairfield, Ohio, depot for nearly eight years.

The end of World War I also slowed the tempo of activities at McCook Field, but did not diminish the mission itself. During the immediate postwar years, McCook Field's organization underwent some changes. The creation of the Engineering Division through the merger of the Airplane Engineering Division, the Technical Section of the Division of Military Aeronautics, and the Testing Squadron at Wilbur Wright Field represented one of the most important of these changes.

Despite the unsettled ambiance and restrictive funding of the immediate postwar period, McCook's engineers and pilots continued to make significant progress in many areas. When, in 1919, the Engineering Division accepted responsibility for all experimental aircraft activities previously conducted at Langley Field, Virginia, McCook became the nerve center of aircraft and engineering activities for the Air Service. In addition to the development and test of airplanes, the Engineering Division devoted its efforts to engines and associated equipment such as parachutes, leak-proof tanks, flotation gear, various types of machine guns, flexible mounts, synchronizing devices, bombs, and bombing equipment.[58] By 1924, the

The WRIGHT IDEA

WOW! OFF WENT THE LID ON VICTORY DAY!

(From the front page of *THE WRIGHT IDEA,* November 16, 1918)

VOL. 1 NO. 19.　　　　　　　　　　　　　　　　　　　　　　　PRICE FIVE CENTS.

Wow! Off Went Lid On Victory Day!

It was a day of days! It was a night of nights! Has there ever been a night like it since the dawn of history! On November 11th, 1918, the war of wars was over. The tremendous upheaval which had cost 10,000,000 lives was at an end. The mad monarch had been forced to abdicate. Ungodly autocracy was overthrown. While millions had been saddened, there were millions of mothers, fathers, wives, sweethearts and sisters whom were made infinitely happy by the wonderful news. Who could measure their happiness? The terrible uncertainty—the terrible tension were over. And what an excuse for a hectic, hilarious celebration! America went wild with joy. From Maine to Alabama—from Virginia to California, the nervous American temperament energetically rejoiced—cut loose—went the limit! And it was a great day that night for the Irish! It was a day of days! A night of nights!

On Broadway it was a wild night at sea. The Battle of Champaigne was reproduced.

In Charge of Gunnery at Wright Field

Wright Field vs. Camp Sherman

One of the biggest football games of the season to be played by military organizations in this section of the country will be staged at Redland Field (National League baseball park), Cincinnati, Ohio, on Saturday, November 23, when Camp Sherman meets Wilbur Wright aviation field for the championship. Both of these teams will be in the pink of condition, as neither of them will play a game the week preceding their meeting.

Last year when Camp Sherman boys played the Camp Sheridan team at Cincinnati the largest crowd that was ever in Redland Field viewed the game and many of those who happened to come late were unable to even get near the gates. Special trains brought people from the surrounding country and there were more automobiles than there are usually people at the park.

It was a day of days! It was a night of nights! Has there ever been a night like it since the dawn of history? On November 11, 1918, the war of wars was over. The tremendous upheaval which had cost 10,000,000 lives was at an end....

Out here in Dayton it was a dry but nevertheless wonderful celebration. Mayor [J.M.] Switzer apparently had decided that the occasion required no other stimulant than the mighty stimulating news of victory. Early in the day John Barleycorn had been throttled and sat upon until the great demonstration was over, although many thought that the Gem City would celebrate in a liquidly sparkling way.

Workmen and work ladies pushed aside their airplanes, shells and cash registers to march through the town in impromptu parades.

The streets were one mass of swaying humanity and machines. There was a continuous snowstorm of confetti and torn paper, and a Bedlam of roaring, cheering, whistling, ringing and honk-konking....

A half-holiday had been declared at [Wilbur] Wright Field and few were the men that remained at the post. Some journeyed to Springfield, while the majority came flocking to Dayton....

But so far as painting the town red was concerned, Springfield had it on Dayton. For there John Barleycorn was unfettered, which seems to bear out their contention that the residents of that friendly town celebrated with more spirit....

Christmas will be here shortly—the time for rejoicing and the most appropriate time to render thanks and praise to the Prince of Peace—the God of Justice. And with our hearts hoping—

> God grant that a newer season,
> Be the dawn of a happier age,
> When men shall return to reason
> And peace shall extinguish rage.
> —J. J. McIntyre

Melodious ladies from the villages of Fairfield and Osborn join amateur Air Service musicians in entertaining at a World War I victory dance at Wilbur Wright Field.

Dayton soldiers enjoy a victory dinner at Hills and Dales Park. (*Montgomery County Historical Society, NCR Archives*)

Wilbur Wright Field, July 1923. The balloon hangar at the northern end of the Unit One hangars (at extreme left) was a major post-World War I addition to the hangar line. In the far upper portion of the photograph are the villages of Fairfield and Osborn, which merged in 1950 to become Fairborn, Ohio. (*United States Air Force Museum*)

Wilbur Wright Field civilian work crew

Engineering Division at McCook had ceased all aircraft construction and concerned itself solely with monitoring airplane design and aircraft produced by private manufacturers.

In 1926, the Air Service became the Air Corps. McCook Field became home to the Materiel Division, one of three major branches of the new organization. The Materiel Division was an expansion of the Engineering Division and included not only engineering but also supply, procurement, and maintenance of aircraft. Included in the new Materiel Division were six air depots, including Fairfield Air Depot (see Chapter 3: The Story of Air Force Logistics).

In the postwar period, McCook Field's airplane flight testing also continued at Wilbur Wright Field. On November 1, 1920, Wilbur Wright Field's role as a test site was formally recognized by Special Order 178 activating a "Department of Testing and Flying."[59] In 1921, the testing facilities expanded to include a high-altitude bombing range; a two-mile, electrically timed, speed course; and equipment for testing machine-gun butts. Perhaps the most impressive test flight at Wilbur Wright Field occurred on August 22, 1923, when Lieutenant Harold R. Harris (later brigadier general), chief of the McCook Field Flight Test Section, and Lieutenant Muir Fairchild (later general) flew the XNBL-1 Barling bomber on its maiden flight of 28 minutes. The giant bomber, operating solely from Wilbur Wright Field, set four very

significant world records, thereby firmly establishing the concept of the heavy bomber.

Until 1924, the United States government leased the site of Wilbur Wright Field. By that time, it was clear that the site would continue to be used as a center of aviation activity, maintaining the flying heritage begun there some 20 years before. McCook Field did not fare as well. The reorganization of 1926 and its consequent shift in personnel made evident, more than ever, the inadequacy of the facilities at the small field. Fortunately, definite plans for relocation of the Materiel Division had been approved by 1926 and implementation was under way. In August 1924, a group of prescient Dayton citizens was responsible for the donation of more than 4,500 acres of land, including the site of Wilbur Wright Field, to the federal government. On August 21, 1925, the War Department discontinued the designation "Wilbur Wright Field" in anticipation of the establishment of the new and larger reservation, to be known as "Wright Field" in honor of both illustrious brothers. Wright Field would become the new home of the Air Corps Materiel Division. Operations moved from McCook Field over the course of 1927, with some functions remaining at McCook into 1929 as facilities at the new location were completed. On October 12, 1927, the expanded reservation was formally dedicated in an elaborate ceremony.

Over the next two decades, Wright Field, encompassing the old Wilbur Wright Field and the Fairfield Air Depot, became the scene of engineering development and procurement, as well as the heart of Army Air Corps/Air Forces logistical support. Wilbur Wright and McCook fields had established the foundation for Wright Field's success in just 10 short years. The visionary engineers, scientists, artisans, and pilots of the early Signal Corps and Air Service facilities in the Dayton area had established traditions of excellence that stand to this day.

THE MATERIEL DIVISION AT WRIGHT FIELD—1926-1939

The Materiel Division was established at McCook Field on October 15, 1926, nearly one year prior to the dedication of Wright Field. The division was one of three major activities of the newly designated Air Corps, as approved July 2, 1926. Brigadier General William E. Gillmore became the first chief of the Materiel Division, assuming responsibility for all functions previously performed by the Engineering Division, as well as several other divisions of the now defunct Air Service. The Supply Division of the Air Service moved to Dayton in the fall of 1926, thereby centralizing all materiel activities of the Air Corps in Dayton, with temporary headquarters at McCook Field pending completion of the permanent headquarters at Wright Field. Beginning in the spring of 1927 and concluding in the spring of 1929, Materiel Division operations moved to the new field.

Under the direction of General Gillmore, operations of the Materiel Division were conducted by six major sections: Procurement, Engineering, Administration, Field Service, Industrial War Plans, and Repair and Maintenance. In addition to materiel and engineering functions, the chief of the division also directed the operations of the Air Corps Engineering School and the Army Aeronautical Museum at Wright Field. The Engineering School, established as the Air Service School of Application at McCook in 1919, was the forerunner to the Air Force Institute of Technology. The creation of the museum at McCook in 1923 heralded the beginning of a lasting tradition that eventually became the United States Air Force Museum.

Through its various sections, the Materiel Division developed and purchased practically all equipment and supplies used by the Air Corps, from new airplanes to necessary operating and maintenance supplies. While the Procurement Section secured bids and awarded contracts for materiel, the Engineering Section designed, tested, and developed airplanes, engines, propellers, accessories, and associated ground equipment. The Engineering Section consisted of seven main engineering branches: Aircraft, Power Plant, Equipment, Materials, and Armament, each of which was responsible for a specific category of equipment; and the Engineering Procurement and Shops branches. The Shops Branch acted as the Engineering Section's service laboratory, inspecting, modifying, and repairing experimental and prototype airplanes submitted by manufacturers. The Shops Branch also manufactured equipment not readily available in the private sector.

The Field Service Section of the Materiel Division managed all supply and maintenance operations at six Air Corps depots, located at Fairfield, Ohio; Little Rock, Arkansas; Middletown, Pennsylvania; Rockwell, California; San Antonio, Texas; and Scott Field, Illinois. The Flying Branch conducted all authorized flight testing of experimental airplanes, engines, aircraft accessories, and miscellaneous equipment. Responsibility for publicity, public relations, and photographic services for the Materiel Division fell to the Technical Data Branch. The Industrial War Plans Section held primary responsibility for planning for national defense, including preparing guidance for movement of supplies to front lines, standardizing equipment, and ensuring a healthy industrial complex capable of rapid mobilization.

Throughout the late 1920s and 1930s, the engineering personnel at Wright Field put their collective efforts into developing improved power plants and equipment, and aircraft with improved aerodynamic and structural characteristics. They researched and developed various airfoil designs; retractable landing gear; all-metal, monocoque, monoplane construction; pressurized cabins; air-cooled, radial engines; superchargers and turbosuperchargers; controllable-pitch and full-feathering propellers; high-octane fuels; blind-flying instrumentation; and free-fall parachutes.

Engineering work carried on by the Materiel Division at Wright Field resulted in great benefit to civil and commercial aeronautics as well. The standards maintained by the Materiel Division ensured the high quality of equipment for military application, and also engendered improvements in civil aeronautics.

Technical Data Branch film library. Film production facilities were established at Wright Field in November 1938, with the Training Film Field Unit No. 2 charged to produce historical and technical films for the Air Corps. The first sound film, *Wings of Peace*, told the story of the 1938 flight of six Air Corps B-17s to South America.

Boeing's Model 299, forerunner of the B-17 bomber, arrived at Wright Field to undergo competitive tests, August 20, 1935. This airplane crashed and burned on October 30, 1935. (*United States Air Force Museum*)

Wright Field in 1926, looking southeast over the flying field. Huffman Dam is visible in the center foreground. (*Darrell R. Larkins Collection*)

Source: ASC History Office, *Splendid Vision, Unswerving Purpose: Developing Air Power for the United States Air Force during the First Century of Powered Flight* (Wright-Patterson Air Force Base, Ohio, 2003).

Chapter Three

THE STORY OF AIR FORCE LOGISTICS

WORLD WAR I ORIGINS

While the Signal Corps negotiated with the Miami Conservancy District in Dayton, Ohio, to lease acreage for Wilbur Wright Field, discussions also were under way to purchase land for a centrally located, aviation general supply depot. Such an institution was essential for logistics support of the Signal Corps Aviation Schools planned for Wilbur Wright Field, Scott and Chanute fields in Illinois, and Selfridge Field in Michigan. Located near Fairfield, Ohio, the depot would provide everything from airplane parts and engines to laces for the mechanics' shoes.

Fairfield Aviation General Supply Depot, 1918. Troops of the 246th Aero Squadron occupied the tents on the left. To the right were the post exchange, bachelor officers' quarters, and Building 1, headquarters of the depot.

Due to time restrictions, rapid negotiations between the Signal Corps Construction Department and the Miami Conservancy District culminated on June 10, 1917. The Signal Corps Equipment Division paid $8,000 to purchase 40 acres of wheat fields. The triangular tract lay about nine and one-half miles north of Dayton, bordering Bath Township on the east and south and Wilbur Wright Field on the north and west. While the depot and Wilbur Wright Field were neighbors, they operated independently and reported to separate divisions within the Signal Corps.

Following the grain harvest in September 1917, work crews that had labored during the summer months to build Wilbur Wright Field began constructing the depot. The major building of the Fairfield Aviation General Supply Depot, Fairfield, Ohio (FAGSD), "opened its doors for business"

on January 4, 1918, less than four months later. This brick-and-concrete main structure, now known as Building 1, Area C, was constructed at a cost of $981,000. A unique feature of the 234,000-square-foot, U-shaped building was a 600-foot-long double rail spur that ran between the two wings. A large roof spanned both wings. The spur, or "government switch" as civil engineers called it, connected the depot with the Big Four Railroad Company whose main lines recently had been relocated from the village of Osborn to the eastern side of Fairfield. The main building housed the depot headquarters, a Signal Corps weather office, and thousands of square feet of storage space for freight and supplies. Six other buildings, including three steel storage hangars and the depot garage, also were constructed as part of FAGSD.

Interior view of the Building 1 trainway, Fairfield Air Depot, 1940. The signs extending over the platform were color-coded to aid individuals in finding specific articles. Blue signs indicated the location of hardware storage; orange, engine spares; red, airplane spares; and yellow, miscellaneous.

Building 1, with its covered trainway, is the oldest, permanent military building at Wright-Patterson. Built in 1917, it continues to be a vital part of the business of the base. Although the interior has undergone considerable modification, the rail spur still exists and remained in use until railroad operations at Wright-Patterson ceased in 1993.

Tents of the 246th Aero Squadron, 1918. Each tent could hold eight soldiers in cramped accommodations. (*United States Air Force Museum, Howard S. Mitchell Collection*)

"Tent City" of the 246th Aero Squadron, Fairfield Aviation General Supply Depot, during World War I. The mess hall was located at the far end of the "street." Flower gardens in front of the tents were carefully planted and tended by privates. (*United States Air Force Museum, Howard S. Mitchell Collection*)

Lieutenant Colonel James A. Mars, a 1903 graduate of the U.S. Military Academy, became the Fairfield Aviation General Supply Depot's first commanding officer on January 4, 1918. The initial station complement included 150 troops each from the 612th, 669th, and 678th Aero Squadrons that arrived from Kelly Field, Texas. Later in 1918, the depot hired its first civilian employees: six female clerk-stenographers and a male janitor.

The depot's primary mission was to provide supply support for wartime training operations. In particular, it received, stored, and issued equipment and supplies to the Signal Corps' aviation, mechanics', and armorers' schools, as well as other programs at Wilbur Wright, Chanute, Scott, and Selfridge fields, and other Army installations (such as McCook Field) as directed by higher headquarters. The depot fell under the responsibility of the Signal Corps Equipment Division in Washington,

D.C., and operated independently of the various Army airfields it supported. (The airfields reported to Headquarters, Central Department, Chicago, Illinois.)[1] In May 1918, all aviation responsibilities were reassigned from the Signal Corps to the Air Service.

Early in 1918, when it became obvious that the Allies would be victorious, the Air Service surveyed its existing installations and began making plans for their use after hostilities ended. Two factors had immediate implications for Dayton-area facilities: the need for aviation training programs declined, and the disposal of war surplus materiel assumed great importance. Air Service headquarters consolidated the two installations at Fairfield, Ohio; terminated the training mission of Wilbur Wright Field; and shifted control and the vacated space to the Fairfield depot.

Accordingly, Wilbur Wright Air Service Depot (WWASD) was formed January 10,

1919, by consolidating Wilbur Wright Field, the Air Service Armorers' School, and the Fairfield Aviation General Supply Depot.[2] Major Charles T. Waring, who had assumed command of FAGSD on December 24, 1918, remained as commander of the newly designated installation.

Demobilization began in earnest shortly after WWASD was formed. On February 20, 1919, the following Wilbur Wright Field organizations were demobilized: Signal Corps Aviation School; Armorers' School Squadrons D (4th Provisional Squadron), E (5th Provisional Squadron), and F (2nd Provisional Squadron); Squadrons A (231st Aero Squadron), B (851st Aero Squadron), I (42nd Aero Squadron), K (44th Aero Squadron), L (246th Aero Squadron), and M (342nd Aero Squadron); and Squadrons N and O, which were casual organizations formed at Wilbur Wright Field on October 1, 1918, to aid in the projected demobilization.[3]

CHRONOLOGY OF FAIRFIELD INSTALLATION

Designation	Date Established	Commander(s)	Assumed Command
Fairfield Aviation General Supply Depot, Fairfield, Ohio	Jan. 4, 1918	Lt. Colonel James A. Mars	Jan. 4, 1918
		Major Charles T. Waring	Dec. 24, 1918
Wilbur Wright Air Service Depot	Jan. 10, 1919	Major Charles T. Waring	Continued
		Major Prince A. Oliver	May 1, 1919
		Lt. Colonel George E.A. Reinburg	Aug. 2, 1919
Aviation General Supply Depot, Fairfield, Ohio	Nov. 3, 1919	Lt. Colonel George E.A. Reinburg	Continued
Air Service Supply and Repair Depot	Sept. 20, 1920	Captain George E. A. Reinburg	Continued
Fairfield Air Intermediate Depot	Jan. 14, 1921	Captain George E. A. Reinburg	Continued
		Major Augustine W. Robins	Aug. 27, 1921
Fairfield Air Depot Reservation	June 21, 1927	Major Augustine W. Robins	Continued
		Major J. Y. Chisum	July 4, 1928
		Major Henry H. Arnold	June 25, 1929
		Lt. Colonel L. E. Goodier	July 1, 1930
		Major Albert L. Sneed	Aug. 15, 1930
Fairfield Air Depot, Patterson Field	July 1, 1931	Major Albert L. Sneed	Continued
		Major Fred H. Coleman	Mar. 13, 1933
		Lt. Colonel Junius H. Houghton	July 28, 1936
		Lt. Colonel James F. Doherty	Aug. 1, 1939
		Lt. Colonel Merrill G. Estabrook	Sept. 11, 1939
Fairfield Air Depot Control Area Command	Feb. 1, 1943	Colonel Merrill G. Estabrook	Continued
Fairfield Air Service Command	May 17, 1943	Colonel Merrill G. Estabrook	Continued
		Colonel Clifford C. Nutt	Jan. 1, 1944
		Brig. General Clarence P. Kane	Feb. 12, 1944
		Colonel Clarence H. Welch	Apr. 29, 1944
		Colonel Elmer H. Jose	June 8, 1944
Fairfield Air Technical Service Command	Dec. 6, 1944	Brig. General Harold A. Bartron	Dec. 6, 1944
		Brig. General Joseph T. Morris	Nov. 6, 1945

(right) Lieutenant Colonel James A. Mars, a 1903 graduate from the U.S. Military Academy, was commissioned a lieutenant in the cavalry before transferring to the Air Service.

(far right) Lieutenant Colonel George E. A. Reinburg, August 3, 1918. Reinburg commanded the supply depot at Fairfield between 1919 and 1921.

As military personnel at the depot left active duty, new civilian employees filled their positions. The 50-man guard section was one of the first to become "civilianized." A civilian personnel office, with a staff of four, opened on October 24, 1919, in the office of the post adjutant. Civilian employment increased to nearly 1,000 in the immediate postwar period before leveling off in March 1920 to about half that number.

In November 1919, two significant changes occurred at the Fairfield facility. First, on November 3, the installation formally transferred to the Air Service's list of permanent depots and was renamed the Aviation General Supply Depot, Fairfield, Ohio.[4] Second, an Air Service Stockkeepers' School, with a staff of one officer and 65 enlisted men, moved to the depot from Washington, D.C., to train the rapidly expanding civilian workforce. When the flow of surplus materiel slowed in 1921, the need for stock keepers and the stockkeepers' school diminished. On August 17, 1921, the school relocated to Chanute Field, Illinois.

POSTWAR DEMOBILIZATION AND REORGANIZATION

As demobilization gained momentum, buildings originally used to train flying cadets, mechanics, and armorers at Wilbur Wright Field became storage facilities. Immediately following the armistice, an Air Service Liquidation Board was created in Paris to dispose of war materiel. The board completed the bulk of its work in a record six months. The steady stream of war surplus property flowing into the Fairfield supply depot from Europe, as well as from closed Air Service installations in the continental United States, became a flood. Storage and disposal of this property became a major project for depot personnel during the next eight years.[5]

Keenly mindful that austere appropriations had constrained the Air Service before World War I and cognizant that funds might be extremely limited in the postwar years, the Air Service supply authorities at first attempted to save nearly everything for future use. Mountains of materiel soon swamped storage facilities. Seventy-seven buildings at the depot were crammed with more than 2,500 aircraft engines of all types, 700 airplanes of various makes, and thousands of instruments of all descriptions, as well as general items

such as lumber, clothing, and personal equipment. The sheer mass of materiel overwhelmed the classification and storage systems.

Between 1919 and 1922 at the Fairfield depot alone, millions of dollars of property were classified and disposed of, either by direct sale locally or through Air Service Supply Division headquarters in Washington, D.C. The Supply Division coordinated the sale of airplanes, engines, and equipment through advertisements in national magazines. Tons of clothing were sold to merchants, and 1.5 million feet of hard lumber—cherry, mahogany and walnut—were sold through sealed bids.

By 1922, most of the nonstandard and obsolete materiel at the aviation depot had been distributed to other government agencies, sold, salvaged, scrapped, or otherwise discarded. Serviceable airplanes and engines were inventoried and set aside for future use. Smaller serviceable supplies were cataloged into a comprehensive storage system and consolidated in Building 1 and three adjacent structures. This monumental disposal effort was carried on in addition to the depot's normal supply function (i.e., furnishing parts to repair shops on the post and to other Air Service organizations and installations).[6]

Major General Charles T. Menoher, commander of the famous 42nd "Rainbow" Division in France during World War I, became chief of the Air Service on December 23, 1918. The next month he drew up plans for a postwar Army aviation force in conjunction with the War Department General Staff. The plans projected a force of 24,000 officers and enlisted men, with a fleet of 1,000 modern airplanes. Supporting such a force required 26 flying fields, including some that had operated during World War I. In April 1919, the Air Service further specified an active-duty force of 2,000 officers and 21,850 enlisted men, an inventory of 1,700 active airplanes, and 3,400 reserve aircraft. Active-duty forces would be organized into 87 "service" squadrons, with each assigned 18 airplanes. Moreover, plans called for 42 balloon companies. If allowed, this total strength in personnel and equipment would have put the Air Service "on a par with the infantry, cavalry and artillery divisions of the Army."[7]

Unfortunately, these grand plans ran afoul of congressional frugality and less than one-third the desired budget was approved. The Army Reorganization Act of 1920 provided for only 280,000 officers and enlisted men overall. The Air Service was designated a combatant arm with an

authorized strength of 1,516 officers and 16,000 enlisted men, including 2,500 flying cadets. Consequently, this reduced the active-duty officer corps to those men with Regular Army commissions, and released reservists from active duty. Temporary commissions also were terminated, so most officers who remained on active duty reverted to permanent grades. To preserve needed skills and expertise in managerial positions, the Air Service encouraged discharged officers to accept jobs as civil-service employees.[8]

Other significant parts of the Army Reorganization Act of 1920 authorized flight pay amounting to 50 percent of base pay and required that tactical units be commanded by "flyers," a term generally construed to mean officers rated as airplane pilots. The chief of the Air Service was assigned the rank of major general and the assistant chief given the rank of brigadier general. The "blueprint" of tactical organizations called for 27 squadrons in seven groups under two wings. The squadrons were further designated by function: observation squadrons (15), surveillance or reconnaissance squadrons (4), pursuit or fighter squadrons (4), and bombardment squadrons (4). The bombardment squadrons flew Martin MB-2 "heavy" bombers. The 27 squadrons also possessed 32 balloon companies.

AIR SERVICE SUPPLY AND REPAIR DEPOT

In 1919, the War Department requested $55 million for Air Service operations, but Congress authorized only $25 million. Not only did this lesser amount preclude, according to General Menoher, the purchase of even one new airplane, it also forced the Air Service to consolidate organizations. One of these mergers affected the Aviation Repair Depot at Indianapolis, Indiana. This facility had functioned before and during World War I as a regional center for major repairs to airplanes and engines. On July 16, 1920, the chief of the Air Service ordered the Speedway Aviation Repair Depot to move from Indianapolis to Fairfield, Ohio, and merge with the Aviation General Supply Depot. The combined activity would be named the Air Service Supply and Repair Depot.[9]

Personnel at the repair depot apparently did not view the move favorably. A few days before publication of the official order, the Speedway depot's engineering officer, Captain Shiras A. Blair, asserted that the move could not occur "due to the fact that the flying land at Fairfield consisted of swamp land and would not make a fit flying field." Upon receipt of the order, however, an advance crew of eight civilian workers went to Fairfield to begin converting buildings for use as engineering shops. "Locks were broken from deserted hangars, and tractors and trucks, stored after the World War of 1918 [*sic*] were revamped in order that facilities might be installed for the shops."[10]

Relocating the repair facility generated the usual sparring between the two commanders. On August 9, Captain Joseph H. Rudolph, commanding officer of the Speedway Aviation Repair Depot, said he received permission from Air Service headquarters to delay the move until the Fairfield depot completed the promised modifications to buildings allocated to house repair materials and equipment. The Fairfield commander, Captain George E. A. Reinburg, disagreed. (Captain Reinburg had assumed command on August 2, 1919, as a lieutenant colonel but, in the spring of 1920, he reverted back to his permanent rank of captain.) He replied that the interior floor plans were the responsibility of the Speedway's engineering officer because of his familiarity with his own requirements.

Captain Reinburg asserted his intention to operate the aero-repair function as merely a department of the Fairfield facility and not necessarily as the facility's *raison d'être*. In fact, Captain Reinburg stated that he could provide only three carpenters for the renovation process, implying that any additional help would have to come from Indiana. Moreover, Captain Rudolph would have to provide his own pine lumber for the renovations, his own office furniture, "steam radiators, piping, etc., connected with the internal heating of the departments," because the Fairfield depot was short of these items.

Living quarters, however, were not a problem. The Aviation General Supply Depot at Fairfield had 52 bachelor officer quarters, barracks for 2,100 enlisted men, family housing for 72 married officers and 48 married noncommissioned officers, and four sets of family quarters for "commanding officers." These facilities became virtually empty after World War I demobilizations. There were also "about 15 sets of [married] quarters for high-class mechanics and civilian employees…whose services [were] required at all times." These vital employees, including the "post plumber, electrician, engineer, shop foreman and [11] department heads," were allowed to reside in on-post quarters. It was expected that 300 other civilian

employees transferring from the repair depot in Indianapolis would find adequate housing in the nearby villages of Fairfield and Osborn where most of the depot workers lived.[11]

Captain Joseph Rudolph (shown here as a major) commanded the Speedway Aviation Repair Depot in Indianapolis, Indiana, before its move to the Aviation General Supply Depot in Fairfield, Ohio. In 1924, Major Rudolph served as the engineering officer in charge of flying operations at the International Air Races held at Wilbur Wright Field.

Staff officers of the Speedway Aviation Repair Depot, Indianapolis, 1919. Salvation Army and YMCA buildings (such as the one pictured here) were common on military installations of the day and sometimes used for official military activities.

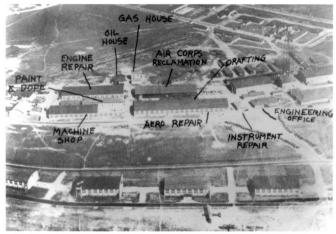

Center of aero-repair activities near the flightline, Air Service Supply and Repair Depot, Fairfield, Ohio, 1921

Members of the aero-repair function, Air Service Supply and Repair Depot, 1921. Many of these employees transferred with the Speedway Aviation Repair Depot to Fairfield in the summer of 1920.

Mindful of the approaching winter and the consequent need to relocate the repair facility before bad weather, Air Service headquarters ended the sparring between the merging installations. Headquarters ordered Captain Reinburg at Fairfield to identify the exact buildings to be used by the aero-repair facility and to renovate the structures quickly. Altogether, eight buildings were modified. The first building was constructed in the vicinity of present-day Building 207 (Area C) by adding a wooden hangar to an existing structure. Near this building were facilities for the "Drafting, Reclamation and the Gas House;" "the Machine Shop;" and the "Paint and Dope, Engine Repair and the Oil House." A small "Instrument Building," and another structure for drafting and blueprint work completed the aero-repair complex.[12]

On September 20, 1920, the former Speedway Aviation Repair Depot reopened for business as the Engineering Repair Section of the Air Service Supply and Repair Depot at Fairfield, Ohio. Captain Blair headed the section, whose mission remained the "repair and maintenance of aircraft and the overhaul of engines."[13]

Engineering Repair Section

During 1917 and 1918, Allied airplane manufacturers produced 19,600 British-designed, DeHavilland DH-4, single-engine, observation biplanes and about 15,600 Liberty airplane engines. After the war, thousands of engines and airplanes were stockpiled at supply depots where they were drawn upon for spares. By 1921, the various engineering repair installations supported an active aircraft inventory of

1,108 DH-4 biplanes powered by Liberty engines; 721 Curtiss JN-4 and 800 Curtiss JN-6 trainers equipped with Curtiss OX-5 engines; 170 Standard SE-5 pursuit (fighter) airplanes; and 12 Martin MB-2 bombers. Additionally 38 free balloons, 250 observation or captive balloons, and 250 non-rigid airships were in the inventory.[14]

From the end of World War I until the late 1930s, the Engineering Repair Section at Fairfield (soon simplified to "Engineering Department," but not to be confused with the Engineering Division at McCook Field) remained a relatively compact organization with four functions: aero repair, metal manufacture and repair, engine repair, and final assembly and inspection. Engine overhaul was a rather simple process in the 1920s. After an airplane engine had been removed from the fuselage, a mechanic and his assistant disassembled the power plant and placed the components in a single, specially designed parts truck. The components were dipped into a cleansing

solution, and inspectors then closely checked them, sending some for reconditioning and replacing others. (Cylinders and related parts were reconditioned by "sub-assembly" and "accessories" departments.) Finally, the elements made their way back to the parts truck and for reassembly by the mechanic. Final inspections were conducted intermittently as the engine resumed its identity during the rebuilding process. The average monthly production using this procedure totaled 50 overhauled engines.

When first established, the Fairfield repair facility provided this service for JN-4 and JN-6 trainers from the 15th Squadron, Chanute Field, Illinois; DH-4 observation airplanes from the 1st Squadron at Chanute Field, the 5th, 11th, and 49th Squadrons at Langley Field, Virginia, and the 17th Squadron, Selfridge Field, Michigan; and MB-2 bombers from the 11th, 20th, and 49th Squadrons at Langley Field. Later in the 1920s, when more powerful, faster, and

Rebuilding an airplane fuselage at Fairfield Air Intermediate Depot, 1925. The employee in the center is the foreman, Mr. Erwin F. Boger.

WACO—"ASK ANY PILOT"

Troy, Ohio, 20 miles north of Dayton, is Waco territory—the home, for many years, of the Weaver Aircraft Company, which for a time was the largest producer of aircraft in the United States. While Waco production ended long ago, a group of volunteers, determined not to let the Waco name be forgotten, manage the Waco Historical Society, the Waco Museum, and the Aviation Learning Center—a three-prong effort to recall Wacos of the 1920s, 1930s, and 1940s, and encourage young people to become involved and enthused about aviation.

The Weaver Aircraft Company (Waco, for short) was founded in 1920 in Loraine, Ohio. George E. "Buck" Weaver collaborated with friends, Clayton Brukner and James Elwood Junkin, to manufacture high-quality civil aircraft. Within three years, Weaver departed the company and Clayton Brukner built a new manufacturing center, flight test site, and sales office in Troy, Ohio, approximately 20 miles north of the Army's McCook Field. The Waco aircraft product name was retained because the young company had quickly established a reputation for high-quality pleasure aircraft. Waco engineers hand-built a variety of models featuring open cockpits and cabins, bi-wings, and air-cooled engines favored by pleasure flyers—especially the Waco 10. The company's slogan answered anyone who questioned their performance: "Ask any pilot."

With the advent of World War II, Waco converted to war production. Soon, the company's 2,600 employees also produced UPF-7 trainers and cargo-carrying gliders for the United States government. Ironically, the aircraft manufacturer with a reputation for meticulously handcrafted pleasure aircraft made its mark on military aviation with its design for a cheap, quickly built, piloted glider. The glider's ties to today's Wright-Patterson were real. They met specifications set by officials at Wright Field, and they were dubbed, somewhat affectionately, "jeeps with wings" by General Henry "Hap" Arnold, commander of the Army Air Corps, who had been assigned to the installation.

General Arnold's "jeeps with wings" were acquired through the Materiel Center's Aircraft Laboratory, Experimental Engineering Section, at Wright Field, which, in 1942, issued a requirement for a glider capable of carrying a fully loaded, one-quarter-ton truck with three crewmen or a maximum of 15 troops. The Aircraft Laboratory's Glider Branch tested all submissions, and the Waco-designed model (XCG-4) best satisfied the requirement for an air-transportable jeep with a nose section that could be hoisted upward, allowing a fully loaded truck to drive into the fuselage. By war's end, the 13,909 Waco-designed gliders—designated CG-4A's—were purchased. The Troy Waco plant could produce only a fraction of the CG-4A's and other manufacturers, such as Babcock, Cessna, Commonwealth, Ford, and Gibson, among others, produced the bulk of the gliders. Newsreel footage of the D-Day invasion of France showed hundreds of Waco-designed CG-4A's delivering thousands of Allied troops to the Normandy coastline.

By 1946, Waco's tally of engine-driven aircraft produced was 3,723 and, before the war ended, they had produced 1,075 CG-3A, CG-4A, CG-13A, and CG-15A gliders used as inexpensive troop and cargo carriers.

Production at Waco declined following the war, but the old Waco biplanes retained a loyal following and many of the pre-World War II-era models still fly, based at Waco Field in Troy. Waco replicas are still manufactured today, although not in Troy.

Today's three-pronged Waco activity is one of volunteers including Jack Waters, a retired U.S. Air Force major general who had been assigned to Wright-Patterson. Waco's Learning Center personnel have conducted programs with local schools for 10 to 12 years, have worked with the Civil Air Patrol and, for 21 years, have hosted an annual Fly-In which draws about 40 privately owned Waco planes and 3,000-5,000 spectators. For six years, the group also has staged a five-day Aviation Summer Camp for youth.

Clayton Brukner's contributions to the Miami Valley extend beyond the production of aircraft. Long a lover of nature, in 1934, he bought 165 acres of land near Troy for a wildlife refuge. The Brukner Nature Center was incorporated in 1967 by the state of Ohio for educational purposes and, today, is one of the state's leading natural history sites with a nationally recognized animal-rehabilitation center and an extensive environmental-education program.

Wright State University's Special Collections and Archives house the papers of Clayton J. Brukner and documentation on Waco's involvement in World War II production. In 1997, Brukner was enshrined in the National Aviation Hall of Fame, an organization that recognizes America's outstanding air and space pioneers.

Sources: Joe Balmer and Ken Davis, *There Goes a Waco: An American Classic Aircraft* (Troy, Ohio, 1992); Clayton J. Brukner Biographical Sketch, viewed online June 19, 2003, at http://www.libraries.wright.edu/special/manuscripts/ms109biosketch.html; Jim Beisner, Waco Learning Center, personal communication with Helen Kavanaugh-Jones, April 2003.

Waco biplane

Waco CG-4A glider, demonstrating its loading method. CG-4A's could also carry up to 15 troops.

Liberty V-12 airplane engines in storage at the Fairfield Air Depot. Designed for the DeHavilland DH-4 observation airplane, these engines remained in the active Air Corps inventory until the mid-1930s.

heavier airplanes came into the inventory, maintenance standards prescribed that all pursuit (fighter) aircraft be overhauled every 10 months "regardless of flying time." Primary trainers were overhauled every 15 months and bombers every 18 months. All other types of airplanes underwent overhaul every 12 months. This schedule remained flexible, allowing for more frequent overhauls when particular types of airplanes were subjected to increased flying hours or above-normal stress and strain.

The Fairfield Engineering Department gained additional functions as time passed. On November 1, 1920, it began testing airplanes and engines that had been repaired at the depot.[15] The following spring, the Airplane Engineering Division at McCook Field expanded the separate testing facilities it operated at Fairfield with the opening of a testing site for machine-gun butts, a high-altitude bombing range, and a two-mile, electrically timed speed course to accurately record flying speeds of experimental aircraft. In 1925, Fairfield assumed responsibility for drop-testing new parachutes and repairing and maintaining those in service. This allowed McCook Field's experimental engineers to concentrate their efforts on developing new and better parachutes.

Wilbur Wright Field, as part of the Fairfield depot, was ideal for both experimental and test programs because of its size and the absence of flying hazards surrounding it. Consequently, until the new Wright Field opened in late 1927,

practically every history-making airplane owned by the Army was tested at the same location where the Wright brothers flew their pioneer aircraft.

FAIRFIELD AIR INTERMEDIATE DEPOT

The year 1921 brought major changes to the Fairfield depot, beginning with a new name. At the end of 1920, the depot commander asked the chief of the Air Service for a permanent unit designation to cease the confusing variety of names

and titles that had identified the Fairfield installation since its establishment in June 1917. The facility had been known, in turn, as the Fairfield Aviation General Supply Depot; Wilbur Wright Air Service Depot; Aviation General Supply Depot, Fairfield, Ohio; and the Air Service Supply and Repair Depot. Because the facility also had consolidated Wilbur Wright Field, the Signal Corps Aviation School, and the Aviation Armorers' School, and served as a test function for McCook Field programs, confusion lingered over those names as well.

The War Department clarified the situation by establishing "air intermediate depots" (AIDs), to serve as centers for both supply and repair. On January 14, 1921, the Ohio facility became the Fairfield Air Intermediate Depot (FAID), one of four such installations.[16] The others were located at San Antonio, Texas; Rockwell, California; and Middletown, Pennsylvania.

The Middletown AID supplied units in New England, New York, Pennsylvania, and Virginia. The San Antonio AID supported organizations in Texas, Oklahoma, Colorado, New Mexico, and Arizona. Units in California, Washington, Oregon, Nevada, Utah, Montana, Idaho, and Wyoming were serviced by the Rockwell AID (later Sacramento Air Depot at McClellan Field, California). FAID supported 24 bases scattered across 23 states: Ohio, Kentucky, West Virginia, North Carolina, South Carolina, Tennessee, Georgia, Alabama, Florida, Louisiana, Mississippi, Arkansas, Missouri, Indiana, Illinois, Kansas, Nebraska, North Dakota, South Dakota, Minnesota, Iowa, Wisconsin, and Michigan.

Vought pursuit biplane displaying the insignia of the Fairfield Air Intermediate Depot. Major Augustine Warner Robins approved the insignia in 1923. The pack represents the supply depot, the tool kit stands for the repair depot, and the pen behind the ear symbolizes the Property, Maintenance, and Cost Compilation Section.

BRIGADIER GENERAL AUGUSTINE WARNER ROBINS

Major Augustine Warner Robins served as commander of Fairfield Air Depot from August 1921 to July 1928.

Augustine Warner Robins is generally recognized as the father of Air Force logistics. He was born September 29, 1882, in Gloucester County, Virginia, the eldest son of Civil War veteran Colonel William Todd Robins and his wife, Sally Nelson Robins, both descendents of early Virginia colony settlers. His family called him Warner, a family name on his mother's side. To his close friends, however, he was known simply as "Robby."

General Robins began his long and distinguished military career at the U.S. Military Academy at West Point. Following graduation in 1907, he spent the first 10 years of his Army life with the cavalry, where his duties included mapping assignments in China and the Philippines, teaching mathematics at West Point, maintaining civil order during the Colorado coal-mine strikes, patrolling the United States-Mexico border in New Mexico, and serving with the U.S. Punitive Expedition in Mexico in 1916.

In September 1917, Robins transferred to the Aviation Section of the Signal Corps as a major. He earned his wings in 1918 following flight training at Scott Field, Illinois, and Park Field, Tennessee. A seasoned Army officer at the age of 35, Robins served briefly as district supervisor of the Northern District of the Air Service at Indianapolis and then was summoned to Washington at the conclusion of World War I as deputy chief of the Supply Division of the Air Service from 1919 to 1921. In January 1921, he was involved in a serious aircraft accident. His broken jaw and facial injuries were so severe that he required six months to recuperate at Walter Reed Hospital in Washington, D.C.

After his release from the hospital, Robins completed the first of three tours at Wright and Patterson fields. During his first tour, Robins served as commander of Fairfield Air Intermediate Depot (August 27, 1921 to June 21, 1927) and the Fairfield Air Depot Reservation (June 21, 1927 to July 3, 1928) with headquarters in what is today Building 1 at Wright-Patterson. He also served concurrently as commander of Wilbur Wright Field until that designation was dropped in 1925.

In 1927, Robins became chief of the Army Air Corps Field Service Section. While at the depot, Robins made two very significant contributions to the growth and development of Air Corps logistics. The Robins Board on Supply Accountability, convened in 1922, developed the basic system of inventory and supply accountability that was used by the Air Corps, and subsequently by the Air Force, for the following 30 years. He was, likewise, instrumental in creating a specialized training system for supply officers assigned to Air Corps depots. During this period, Robins, his wife Dorothy, and their three daughters resided on Patterson Field in Building 8, now the Arnold House Heritage Center.

After a number of tours to other fields, Robins returned to Dayton in November 1931 as the executive to the chief of the Materiel Division, headquartered in Building 11 at Wright Field. Four years later, he became chief of the Materiel Division and assistant chief of the Air Corps, with the temporary rank of brigadier general. During his next four years at Wright Field, he directed the supply, repair, research, development, and purchase of Army Air Corps equipment.

In February 1939, Robins resumed his Regular Army rank of colonel and entered his last assignment as commanding officer of the Air Corps Training Center at Randolph Field, Texas. At Randolph, he supervised flight training for the expanding pre-World War II Air Corps. Colonel Augustine Warner Robins died suddenly of a heart attack at his home at Randolph Field on June 16, 1940, and was buried at Arlington National Cemetery. General Henry H. Arnold, a West Point classmate and good friend, remembered Robins as "every inch a soldier." The Air Corps considered him "one of its ablest officers and most gallant gentlemen."

On January 23, 1942, Robins Field near Macon, Georgia, was named in General Robins' memory. The air depot located there, originally known as Wellston Air Depot, was renamed Warner Robins Air Depot in October 1942 after the town of Wellston changed its name to Warner Robins. The memory of General Robins has been perpetuated in the title Warner Robins Air Logistics Center.

Brigadier General Augustine Warner Robins and his family in Quarters 1 (Building 700), 1935. Robins and his family were the first occupants of Wright-Patterson's senior-officer quarters. On October 2, 1989, the house was officially designated the Robins House in the general's honor.

Major Augustine Warner Robins, commanding officer of the Fairfield Air Intermediate Depot, poses with his staff in front of FAID headquarters during the 1924 International Air Races. Front row, left to right: Captain W. E. Donnelly, Captain J. B. Powers, Major J. H. Rudolph, Commander Robins, Major George H. Brett, and Major Hugh J. Knerr. Back row, left to right: Captain Edward Laughlin, Captain C. O. Thrasher, Lieutenant L. H. Dunlap, Lieutenant C. E. Thomas, Lieutenant Harold A. Bartron, Lieutenant L. E. Sharon, and Captain F. F. Christine

All four intermediate depots initially provided supplies, repairs, and regularly scheduled overhauls to aircraft stationed at bases within their control areas. Over the course of 1921, however, FAID assumed the repair function of the Middletown control area, thereby expanding its services to the entire region east of the Mississippi River and a considerable portion of the region west and north, including regular Army Air Service bases, National Guard installations, and organized Reserve Corps sites.

At the time Fairfield Air Intermediate Depot was designated in January 1921, Captain George Reinburg had served as the commander through three name changes. Major Augustine Warner Robins succeeded Captain Reinburg on August 27, 1921.[17] Fresh from the Office of the Chief of the Air Service (OCAS), Major Robins remained in command of FAID until July 3, 1928, holding the position longer

than any preceding officer in the history of the facility.[18] (In 1927, Major Robins also was detailed as chief of the Army Air Corps Field Service Section.[19] While working in this capacity, he was instrumental in devising important policies and procedures for overall handling of Air Corps materiel.)

Property, Maintenance, and Cost Compilation Section

In July 1921, a month before Major Robins assumed command, the Army Air Service relocated physical control of its entire depot system from Washington, D.C., to the Fairfield Air Intermediate Depot. Directed by Captain Elmer E. Adler, the Property, Maintenance, and Cost Compilation Section (PMCCS) of the Air Service Supply Division was a "sub-office" of the Property Requirement Division of

the Office of the Chief of the Air Service (OCAS). Though collocated at the depot, the PMCCS was not a subordinate organization and continued to report functionally to the OCAS. This new agency was the earliest antecedent of today's Air Force Materiel Command (AFMC), currently headquartered at Wright-Patterson Air Force Base.

As host of the PMCCS, FAID assumed a more prominent national role and was designated the central control depot for all paperwork pertaining to depot operations. All correspondence concerning requisition, issue, distribution, and storage of Air Service materiel had to be channeled through the PMCCS.[20] The PMCCS received daily reports on stock levels of materiel from the four air intermediate depots; from the additional repair functions at Fairfield, Ohio; Dallas, Texas; and Montgomery, Alabama; and from the smaller supply depots at Little Rock,

Commanding officer and original staff of the Property, Maintenance, and Cost Compilation Section (PMCCS) of the Supply Division, Army Air Service, at the Fairfield Air Intermediate Depot, 1921. Left to right: Lieutenant R. V. Ignico (later brigadier general); Lieutenant Omar Niergarth (later colonel); Captain Elmer E. Adler, commanding officer (later brigadier general); Lieutenant J. L. Stromme (later colonel); Lieutenant Edwin R. Page (later brigadier general); and Lieutenant F. P. Kenney. In 1924, the PMCCS changed its name to Field Service Section.

Fairfield Air Depot and Wilbur Wright Field, April 1922. At left, a line of World War I wooden hangars borders the flightline. In the center, maintenance and engineering shops mingle with warehouses and outdoor storage areas managed by the depot. Building 1 is visible at center right. (*Montgomery County Historical Society, NCR Archives*)

function over all supply activities within the First through Seventh Corps areas. Air Service customers within these areas forwarded requisitions to FAID where the Supply Department balanced requests against consolidated ledger reports. This procedure determined whether the requisitioned materiel was in stock at FAID, at one of the smaller depots, or on hand at some other Air Service installation. Shipping instructions were issued, with close attention to manifesting materiel from the source nearest the requester. For example, a requisition from Mitchel Field, Long Island, New York, would likely be filled from the Middletown, Pennsylvania, depot, and not from Fairfield or any further point.

Lieutenant McPike estimated that, as of May 2, 1922, FAID had in storage $250 million worth of government property. He pointed out that materiel was listed on 120,000 stock record cards that were continuously updated to assure a current inventory. This painstaking process involved identifying, describing, cataloging, and indexing entries for "hundreds of thousands of supplies of a technical nature."[22] This mammoth job occupied about 350 civilian employees for four years.

As described above, two separate functions existed simultaneously at Fairfield Air Intermediate Depot during the mid-1920s: a combined repair and supply function, and a central control function. The repair and supply function served the entire region east of the Mississippi River, a few areas to the west and north, and also made extensive shipments to the Panama Canal Zone, Hawaiian Islands, and the Philippine Islands. The central control function (PMCCS) was a national function that monitored and directed all supply and repair functions.

On January 26, 1924, at the request of the PMCCS director, the name of the Property, Maintenance, and Cost Compilation Section was changed to the Field Service Section.[23] The section reported to the Supply Division, OCAS, until October 15, 1926, when it became one of six major sections of the Materiel Division, a new organization operating at McCook Field. On June 21, 1927, the Field Service Section moved from FAID to newly constructed buildings at Wright Field (now Area B, Wright-Patterson Air Force Base) and joined the Materiel Division, which had also relocated to the new facility from McCook Field. From 1924 to 1939, the Field Service Section continued to supervise the

Arkansas, and Scott Field, Illinois. Based on this information, the PMCCS directed the flow of supplies between the various depots nationwide and, thus, indirectly controlled supplies to all Air Service bases and organizations.

The depot function at FAID operated in the same fashion as the other depots and was serviced by the PMCCS in the same way. According to Lieutenant George V. McPike, the Fairfield supply officer, FAID distributed and coordinated supplies so that "a well balanced stock level [would]

be maintained for every type of organization in the area."[21] When supplies dropped below a fixed minimum, the depot forwarded requisitions to the PMCCS for restocking. The PMCCS then either directed shipment of required materiel from one of the other control depots to Fairfield Air Intermediate Depot or recommended to the OCAS that the needed materiel be purchased.

Fairfield Air Intermediate Depot, in turn, provided the same kind of service to its own "customers," exercising a control

In 1927, the Field Service Section offices were at the intersection of today's Pearson Road and Allbrook Drive (Area C) opposite the Base Exchange Service Station and Fire Station No. 1 (Building 163). The area occupied by these offices now forms a wide lawn adjacent to 88th Air Base Wing headquarters (Building 10).

operations of the four intermediate depots and, in 1927, acquired formal control of the departmental depots in Panama, Hawaii, and the Philippines.

1924 Round-the-World Flight

Both the Fairfield Air Intermediate Depot and McCook Field played major, albeit largely unsung, roles in one of the Air Service's most spectacular achievements during the two decades between the world wars—the 26,345-mile circumnavigation of the globe by four single-engine Douglas World Cruisers. Eight young fliers, dubbed "Magellans of the Air," departed Seattle, Washington, at 8:47 a.m. on April 6, 1924.[24] When two of the four World Cruisers touched down in Seattle on September 28, they had logged about 363 hours in actual flight, suffered no fatalities, and had sustained only minor injuries on their epochal flight. This was a most remarkable record, considering their itinerary: Canada; Alaska; the Aleutian chain; the North Pacific; Siberia; Japan; China; Southeast Asia; India; the Middle East; Europe; Iceland; the North Atlantic; Nova Scotia; Massachusetts; New York; Washington, D.C.; Ohio; and Washington.

Secretary of War John W. Weeks proclaimed in November 1923 that the flight's official purpose was "to demonstrate the feasibility with which aerial communication may be established between the various continents, and to obtain much valuable information concerning the operation of present type

aircraft in various climates of the world."[25] Other official motives for the flight, according to one source, included economics and patriotism. "The purpose . . . is to point the way for all nations to develop aviation commercially and to secure for our country the honor of being the first to

encircle the globe entirely by air," giving the United States the position of "leading power in the peace-time application of flying."[26]

Since 1919, nations competed to become the first to circle the earth by air. The United States had set the pace on May 17, 1919, when three U.S. Navy Curtiss Flying Boats took off from Newfoundland to the Azores. One of the NC-4s, piloted by Lieutenant Commander Albert C. Read, continued on to Lisbon, Portugal, and Plymouth, England. The NC-4 thus became the first airplane to bridge the Atlantic, with an en route stop.[27]

Great Britain eclipsed the United States' achievement the next month and scored an impressive first in aviation history. On June 14 and 15, 1919, John Alcock and Arthur Whitten-Brown, both Royal Flying Corps veterans, flew nonstop from Newfoundland across the stormy North Atlantic to Ireland in a Vickers-Vimy biplane.[28]

U.S. Army Air Service scientists, designers, engineers, pilots, "mechanicians," craftsmen, and support personnel maintained an unfaltering desire to expand aeronautical science and strengthen American air prowess. Between

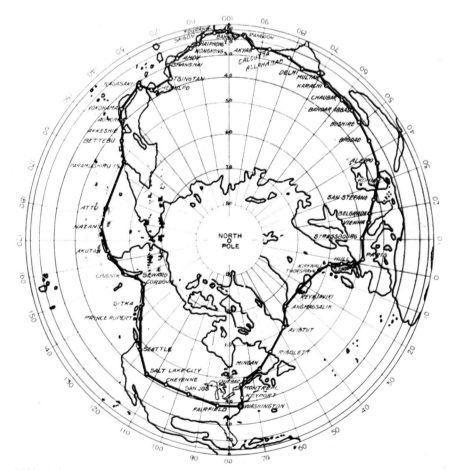

Route of the 1924 round-the-world flight. Two of the original four Douglas World Cruisers—*New Orleans* and *Chicago*—completed the 26,345-mile global odyssey in a cumulative flying time of 15 days 3 hours 7 minutes, with an average speed of 79.5 mph. (*United States Air Force Museum*)

1918 and 1923, some of the Air Service's most significant domestic and world records in aviation occurred at or above McCook and Wilbur Wright fields (including the airspace over Fairfield Air Intermediate Depot).

By 1923, British, French, Italian, Portuguese, and Argentinean airmen, generously supported by their own governments, challenged the United States' position in the world of aviation.[29] Not only was the external competition formidable, but a keen rivalry also developed between the air services of the U.S. Army and Navy for the biggest share of congressional appropriations. Something exceptionally daring—near the point of impossibility—was needed to earn honor, prestige, and money.

According to several writers of Air Service history, the World Flight concept originated with Brigadier General William "Billy" Mitchell, assistant chief of the Air Service. Along with Major General Mason M. Patrick, chief of the Air Service, General Mitchell "had promoted a series of bold aerial ventures that, they hoped, would sell military aviation to the people, the Army and the Congress."[30] Based on careful planning and analysis, they designed each venture to test new concepts, new procedures, and new equipment; to blaze air routes; and to expand the Air Service's operational capabilities. The World Flight advanced all of these objectives in one effort.[31] The ultimate success of the World Flight resulted from an extraordinary team effort. Officers and men who held no hope of glory or direct participation in the flight "worked to exhaustion preparing charts, gathering worldwide weather data, and planting fuel, oil and spare parts in remote areas around the globe."[32]

Meticulous planning for this aerial odyssey began more than a year before the four heavily laden Douglas World Cruisers struggled off the waters of Lake Washington, near Seattle. The Air Service headquarters' staff carefully selected a World Flight committee chaired by Lieutenant Robert J. Brown, Jr., and composed of four other members: Captain W. Volandt, and Lieutenants Clarence E. Crumrine, Erik Nelson, and St. Clair Streett. All were veteran pilots with hundreds of hours in the air traversing thousands of miles.

Lowell Thomas, the flight's unofficial historian, became acquainted with the principals involved in the historic voyage

Major General Mason M. Patrick, chief of the Air Service, 1924. Patrick was a 1896 graduate of the U.S. Military Academy and a veteran of World War I, during which he served as chief of the Air Service, American Expeditionary Forces, France. Patrick commanded the Army Air Service/Air Corps from 1921 to 1927. He retired in December 1927. Despite spending his military career in the air arm, General Patrick did not earn his pilot wings until the age of 60, a feat not duplicated before or since by a general officer in the U.S. Army or Air Force.

Brigadier General William Mitchell (1879-1936) began his military career in the infantry before earning his pilot wings in 1917 at a civilian flying school. He served with the American Expeditionary Forces, France, during World War I and was appointed assistant chief of the Air Service in 1920. In 1926, after being court-martialed for his advocacy of theories that ran contrary to Army doctrine, he resigned. In 1947, Mitchell was posthumously promoted to major general and awarded a Special Congressional Medal of Honor for his pioneer service and foresight in the field of American military aviation. (*United States Air Force Museum*)

Lieutenant Erik H. Nelson, a Swedish immigrant, joined the Royal Canadian Air Force in July 1917, but three months later was reassigned to the U.S. Signal Corps as a bomber pilot. Nelson had experience with distance flying prior to the World Flight. In 1919, he was one of four pilots who completed a 7,000-mile circuit of the United States. In 1920, Nelson was the chief engineering officer on an endurance flight headed by Captain St. Clair Streett from Mitchel Field, New York, to Nome, Alaska, and return. Nelson later rose to the rank of brigadier general. (*United States Air Force Museum*)

and reported that the Air Service staff studied three essential factors: aircraft, logistics, and aircrew. The Air Service had to obtain several rugged aircraft in the hopes that at least one would survive the perilous voyage and, thus, justify the expenditure of man-hours and money. The logistical task, setting up supply bases around the globe, required the maximum support of all other branches of the national defense establishment, including the U.S. Coast Guard; the closest cooperation of the State Department and the Bureau of Fisheries; and the assistance of generous American corporations with offices abroad. The final task was the selection of pilots and mechanics possessing the highest degree of expertise and stamina, both physical and mental, to complete the hazardous mission.

After surveying newspapers, magazines, and books of that period, two contemporary historians concluded that, "To the man in the street the project seemed doomed to disaster. He surmised that the fliers who engaged in it would stand little chance of returning alive." When foreign governments revealed they were sponsoring pilots in similar ventures, however, the American public responded with national pride. Public attitude changed to a forthright, "However slim the chances of success American fliers must lead the field."[33]

Logistical Support Plans

Logistical support was the most complex of the three requisite factors. A route was selected to avoid dangerous climatic conditions in Alaska, the North Pacific, Japan, China, and the Middle East during the six-month journey.[34] For flight and logistical planning, the world was divided into six regions and a project officer selected for each division. About the first of July 1923, Lieutenant Clifford Nutt of McCook Field and Lieutenant Clarence E. Crumrine stationed in the Philippine Islands pioneered separate path-finding trips. Lieutenant Nutt surveyed England, Greenland, Iceland, the Faroe Islands, France, Italy, and Canada. Lieutenant Crumrine visited Japan, the Kurile Islands, the Aleutian Islands, and Alaska.

Through American embassies in the respective host nations, the Air Service made arrangements for the reception and accommodation of airplanes and their crews; storage of spare parts and equipment; and availability of maintenance

Douglas World Cruiser outfitted with twin pontoons for water takeoffs and landings and oceanic flights

facilities and tools, as well as maps, charts, photographs, and sketches of landing fields and water areas for pontoon operations. American corporations with offices and facilities in the various countries became instrumental in obtaining services from foreign governments for the World Flight participants. For example, the Standard Oil Company made arrangements for caching aviation gasoline, oil, and lubricants in remote areas.[35]

Air Service logisticians headed by Captain Elmer E. Adler of the Field Service Section, Fairfield Air Intermediate Depot, initiated the support plans for the flight. McCook Field, assisted by FAID, was the main logistics base with subordinate supply depots in each of the six global regions. The Field Service Section procured, packaged, and distributed the necessary spare parts and equipment for the World Cruisers.

Part of Lowell Thomas' account, based on Air Service records, reads:

Spare parts for planes and engines, a fairly complete outfit of tools, small quantities of standard utility parts, and material, such as tubing, shock-absorber cord, plywood and items other than spare parts, were sent to each station on the route of the Flight. The spare parts and tools were packed in specially constructed boxes designed at the Fairfield Depot and built in the repair shops. The boxes themselves were constructed of ash, spruce, and plywood so that they might be used for the furnishing of

wood for emergency repairs. Carpenter tools for working up the wood were sent in the tool chests.

Tubing and other items which could not readily be bent were packed with the propellers in lengths of six feet or more. The weight, cubic contents, and dimensions of every article were carefully considered, about four hundred and eighty separate items being sent to each station, so arranged that the Fliers could find spare parts or repair material even in the dark. On the outside of each crate a diagram showed exactly where each item was located inside.[36]

All shipments to points east of Calcutta, India, were sent from FAID to Seattle for steamship transportation, and all shipments to Karachi, India, and points westward went to New York for oceanic shipment.[37] Spare parts sent overseas included 15 Liberty, 12-cylinder, 400-horsepower engines; 14 extra sets of pontoons; and approximately 200 percent of airframe replacement parts.[38]

A maintenance schedule developed for the trip called first for engine changes in Japan. At Calcutta, new wings were to be fitted, new engines installed, and pontoons replaced with landing gear. Final engine changes were scheduled at Hull, England, where landing gear would also be replaced with pontoons.[39] Each of the four airplanes carried a set of tools "deemed sufficient to maintain the aircraft along the way," plus a screwdriver, hammer, wrench, a pair of pliers, and a flashlight.[40]

The four World Cruisers on display. *Chicago* (No. 2) and *New Orleans* (No. 4) successfully completed the flight. *Seattle* (No. 1) crashed in Alaska and *Boston* (No. 3) ditched in the Atlantic Ocean near Iceland.

Douglas World Cruiser

When Air Service headquarters surveyed its current aircraft inventory to find a machine capable of circumnavigating the globe, it found no suitable vehicle. In October 1922, the Engineering Division at McCook Field had recommended the purchase of an airplane similar to the Douglas Airplane Company's new DT-2 being produced in large quantities for the U.S. Navy. On June 24, 1923, the War Department instructed the Air Service to obtain all available data on the Fokker F-5 transport and on the Davis-Douglas Cloudster, both similar to the DT-2. The Douglas Company, located at Santa Monica, California, submitted specifications for a modified DT-2. The new design was named the D-WC (Douglas World Cruiser). The company promised delivery of the $23,271 prototype within 45 days of receiving a signed contract.

Lieutenant Erik Nelson, age 35, an aeronautical engineer assigned to McCook Field and a veteran pilot, was placed on temporary duty at the Douglas factory in California. He worked closely with Donald Douglas, assisted by John Northrop, in designing the reconfiguration of the DT-2 into a World Cruiser. On August 1, 1923, General Patrick approved the reconfiguration plans and a contract was let for construction of the prototype D-WC. The delivery date was met and Lieutenant Nelson flew the new airplane to the Engineering Division at McCook Field for a series of tests. The Air Service chief flew to the installation to inspect the prototype and recommended a few changes to increase performance. The airplane then flew to Langley Field, Virginia, where it was equipped with pontoons and successfully completed additional tests. Satisfied with the aircraft's performance, the War Department gave final approval for the World Flight on November 9, 1923. Eight days later, the Douglas Airplane Company received a $192,684 contract for the manufacture of four World Cruisers and spare parts.[41]

World Flight Crews

As expected, the response to the call for volunteer pilots and mechanics for the exciting, perilous world flight was overwhelming. The Air Service chief, General Patrick, after "test-hopping" the prototype World Cruiser at McCook Field, purportedly remarked that he wished he was "young enough to go on this great undertaking."[42] The only stipulations, aside from meeting General Patrick's stringent requirements of expertise, character, courage, and initiative, were that the individual be unmarried and write his rationale for volunteering on the application. Although Air Service headquarters never disclosed the exact criteria used in selecting the eight crewmembers, Lowell Thomas noted the pilots had all "clocked many hundreds of hours in either cross-country or endurance flying." Some also were exceptionally proficient in both theoretical and practical aeronautical engineering.[43]

The Air Service chief personally selected Major Frederick L. Martin to command the flight, although he was married. Other airplane commanders included Lieutenants Lowell H. Smith, Leigh Wade, and Erik H. Nelson of McCook Field. Lieutenants Leslie P. Arnold and LeClaire Schultze served as alternates. These pilots arrived at Langley Field, Virginia, a few days before Christmas, 1923. They underwent concentrated courses in weather phenomena and forecasting, world geography and climatology, aerial navigation, and "enough medical and surgical knowledge to pull them out of mishaps in case of a forced landing in some region remote from civilization."

"Modern Magellans," the crews of three of the World Cruisers gathered at the Douglas Aircraft Company in Santa Monica, California, before the flight. Left to right: Technical Sergeant Arthur Turner, Staff Sergeant Henry H. Ogden, Lieutenant Leslie P. Arnold, Lieutenant Leigh Wade, Lieutenant Lowell H. Smith, Major Frederick L. Martin (Flight Leader), and Sergeant Alva Harvey. Turner did not make the flight. (*United States Air Force Museum*)

At Langley they worked closely with 10 of the Air Service's best, enlisted mechanics who had been assigned to work on the prototype World Cruiser. At the end of the six-week indoctrination period, the aircraft commanders selected the mechanics they wished to accompany them on the global flight: Sergeant Alva Harvey, Technical Sergeant Arthur Turner, Staff Sergeant Henry Ogden, and John Harding, an Engineering Division civilian engineer from McCook Field who was also a Reserve lieutenant.[44]

At the end of February 1924, the officers and enlisted mechanics went to the Douglas factory in Santa Monica, where they watched the daily progress in manufacturing the World Cruisers to familiarize themselves with every detail of the airplanes' components. The Cruisers were completed by March 11, 1924, and the pilots named the four aircraft. Major Martin chose *Seattle*, Lieutenant Smith picked *Chicago*, Lieutenant Wade selected *Boston*, and Lieutenant Nelson opted for *New Orleans*. From the Douglas factory the aircraft flew to the Sand Point flying field on Lake Washington near Seattle, where wheels were exchanged for pontoons. The final selection of crews occurred at this juncture:

Seattle—Major Frederick L. Martin and Sergeant Alva Harvey

Chicago—Lieutenant Lowell H. Smith and Lieutenant Leslie P. Arnold

Boston—Lieutenant Leigh Wade and Staff Sergeant Henry Ogden

New Orleans—Lieutenant Erik H. Nelson and Lieutenant John Harding

Success

The quartet of Douglas World Cruisers began their historic odyssey at 8:47 a.m., April 6, 1924, lifting off from Lake Washington. Of the original four World Cruisers, only the *New Orleans* and the *Chicago* completed the circumnavigation of the world, landing at Seattle, Washington, on September 28, 1924. They had flown 26,345 miles in 175 days.

No lives were lost or major injuries sustained when the *Seattle* crashed against a mountain side near Dutch Harbor, Alaska, on April 30, or when the *Boston*, with no oil pressure in the engine, ditched in the North Atlantic between the Orkney Islands and Iceland on August 2. However, General Patrick "felt that Wade and Ogden [crew of the *Boston*] deserved to enjoy and

The World Cruisers landed at McCook Field on September 14, 1924, during their return leg to Seattle, Washington. *Chicago* (No. 2) led the flight, flanked by *Boston II* (the prototype airplane) and *New Orleans* (No. 4). They were met in Columbus by the Barling bomber, flown by McCook Field pilot Lieutenant Harold Harris. (*Montgomery County Historical Society, NCR Archives*)

General Billy Mitchell met with some of the World Flight pilots at the 1924 Air Races at Wilbur Wright Field.

participate in the homecoming celebrations."[45] He directed the *Boston II*, the 1923 prototype World Cruiser, to join the *Chicago* and *New Orleans* in Nova Scotia and complete the triumphal journey with them.

President Calvin Coolidge and his cabinet members, along with thousands of spectators, greeted the six crewmen on their arrival at Bolling Field, D.C., from Nova Scotia. The final transcontinental portion of the World Flight from the nation's capitol to Seattle, Washington, included a stopover at McCook Field on September 14-16, 1924.

As the aircrews neared Columbus, Ohio, they were joined by a special escort—

McCook Field test pilot Lieutenant Harold Harris in the huge, six-engine, Barling bomber. According to Lieutenant Nelson, "As we passed over Wilbur Wright airdrome we saw 'Welcome World Fliers' painted in huge letters on the ground, and between fifty thousand and a hundred thousand people cheered us a moment later as we came gliding down over McCook."

Airplane mechanics and specialists, working in shifts during the next two days, minutely examined "every bolt and wire to find out exactly how the Cruisers had mechanically withstood the strain of the flight." The aircraft were judged capable of flying the remaining 3,000 miles to Seattle via Chicago. The aircrew members underwent thorough medical examinations by McCook Field flight surgeons. All men "were passed as very 'paragons' of physical fitness."[46]

Chicago's wheels touched down on Sand Point Field, Seattle, Washington, at 1:28 p.m. local, September 28, 1924. A crowd of 50,000 people greeted the World Fliers. The airplane and its sister, *New Orleans*, had flown the 26,000 miles at an average speed of 79.5 mph. Flying time was the equivalent of 15 days 3 hours 7 minutes.[47]

Several days later, other crews returned the World Cruisers to McCook Field, ending their magnificent odyssey. A few months following the flight, Congress voted the Distinguished Service Medal "never before awarded except for services in war," to the six World Flight crewmen. Each received several foreign decorations (with congressional permission), including appointments as Chevaliers of the French Legion of Honor. All of the World Cruiser officers were advanced "five hundred files each" on promotion rosters.[48]

In April 1957, the United States Air Force Museum at Wright-Patterson Air Force Base obtained the *New Orleans* on loan from the Natural History Museum, Los Angeles, California, rejuvenated it, and placed it on permanent display. The *Chicago* is on display at the National Air and Space Museum, Smithsonian Institution, Washington, D.C.

1924 International Air Races

Less than a week after the Douglas World Cruisers made their final landing near Seattle, Washington, the Dayton area was center stage in the spotlight of aviation. On October 2-4, 1924, two decades and 12 days after aviation pioneer Wilbur Wright flew the first complete circle in the history of powered airplanes at Huffman Prairie, Fairfield Air Intermediate Depot and Wilbur Wright Field hosted the International Air Races. This aviation extravaganza offered prestigious prizes, such as the Pulitzer Trophy and $80,000 in prize money (payable in U.S. Government Liberty Bonds) to the winners of 12 events. The National Aeronautic Association (NAA), whose president that year was aviation enthusiast Frederick Beck Patterson, president of Dayton's National Cash Register Company, sponsored the event. As it turned out, the "international" aspect of the races did not materialize. The expected foreign contestants cancelled a few weeks prior to the meet, ostensibly because they knew

More than 100,000 visitors attended the 1924 International Air Races. A fence and wide road separated the crowds from the bustling flightline with its noise, dust, fumes, and whirring propellers. (*Montgomery County Historical Society, NCR Archives*)

The Wright brothers' Flyer, the airplane that made the world's first, controlled, powered flight on December 17, 1903, was reassembled for the air races and exhibited in the Wright 1910 hangar that still stood on Wilbur Wright Field. Signs announced that proceeds from the display would go to the nonprofit National Aeronautic Association. This was the last time the airplane was shown before shipment to England in 1928. (*Montgomery County Historical Society, NCR Archives*)

their European airplanes could not match the speed, maneuverability, and general performance of American civilian and military airplanes.[49]

The three Air Service installations in the Dayton area had major responsibilities in the thrilling aerial drama that attracted more than 100,000 spectators during three days of exhibits, demonstrations, and races.

Major Augustine Warner Robins, Wilbur Wright Field/FAID commander, worked closely with Frederick Patterson and

Charles H. Paul, general manager of the air races organization. Major Robins' personal assistants were Lieutenants Elmer E. Adler and C. E. Thomas, the latter serving as executive officer. Other officers in charge of air race activities at Wilbur Wright were Major George H. Brett, housing and entertainment; Major J. H. Rudolph, engineering officer in charge of flying operations; Major Hugh J. Knerr, in charge of ferrying airplanes and furnishing aircraft for aerial photography; Captain Edward

AIR RACING IN AMERICA

In August 1909, just six years after the Wright brothers' first flight, the world's first air meet was held in Rheims, France. Prizes were offered for selected events, such as the highest altitude reached, longest flight, most passengers carried, and the fastest one, two and three laps over a 10-kilometer course. Glenn Curtiss won two of the biggest prizes: the James Gordon Bennett Trophy for the quickest two laps and the Prix de la Vitesse for the quickest three laps. Curtiss received $5,000 in prize money, which he used to buy out his partner in the Herrington-Curtiss Company of Hammondsport, New York, and created his own aircraft company. This set the pattern for air racing in the early years of aviation: many of the racers were not only pilots, but designers and builders as well.

The first meet in the United States occurred in Los Angeles, in January 1910. A second meet held in October 1910, in Belmont Park, Elmont, New York, attracted dozens of pilots from Europe and the United States to compete for the coveted Bennett Trophy. The outbreak of World War I in 1914 caused the cancellation of the Bennett Trophy race. After the race resumed in 1917, French pilots won the trophy three consecutive years. The French retired the trophy permanently in 1920.

In addition to the Bennett Trophy, pilots competed for several other awards during the early years of aviation. Some of these included:

· The Michelin Cup, given to the pilot who made the longest flight between sunrise and sunset in a single day. Orville Wright won the first competition in 1908.
· The London Daily Mail prize, given to the pilot who completed the first flight over the English Channel. In 1909, Louis Bleriot of France won this coveted trophy.
· The William Randolph Hearst prize, with a payment of $50,000 to the first pilot to fly across the North American continent in 30 days or less. The Hearst prize, first offered in 1911, inspired the first transcontinental flight by Cal Rodgers. Unfortunately, Rodgers did not receive the prize because the flight took him 49 days. His flight was also marred by 19 crashes.
· The Jacques Schneider Trophy, initiated in 1913, was given to the winner of seaplane races over open water. The first 150-nautical-mile race, held in Monaco, was won by the French pilot, Maurice Prevost.

In 1919, new trophy races emerged, including prizes for the first nonstop flight between England and the United States (1919), and the first solo, nonstop flight between New York and France (1919). In 1927, this latter prize led to one of the most famous flights in history, when Charles Lindbergh flew across the Atlantic Ocean in 33 hours 32 minutes. Many more awards were proposed, and the

Advertising poster for the 1924 Air Races

interwar years soon became known as the "Golden Age of Air Racing." One of the trophies contributing to the popularity of air racing was the Pulitzer Trophy, established in 1920 by American publisher Ralph Pulitzer. The first Pulitzer race, a speed trial, was held in 1920 on Long Island, New York. An American military pilot, Major Corliss Moseley, took home the trophy after setting a new speed record of 156 mph.

The National Air Meet was established in 1921 and, in October 1924, it became known as the International Air Races, the first of which was held at Wilbur Wright Field in Dayton, Ohio. Prior to 1924, almost all race events were individual time trials—the individual with the fastest time won the race. Beginning at the 1924 races in Dayton, the John L. Mitchell Trophy Races, during which the planes flew simultaneously in a closed-circuit course around pylons, changed the method of determining the winner. As the Mitchell trophy races increased in popularity, more events were added. For the next five years, the races were held in various venues, including Long Island, New York; Philadelphia, Pennsylvania; Spokane, Washington; and Los Angeles. Cleveland hosted the air races every year between 1929 and 1939, except in 1930 (Chicago), 1933, and 1936 (both in Los Angeles).

With the United States entry into World War II, the races were cancelled, but the enormous changes in air power resulting from the war created new public interest in aviation. With the backing of the War Department, the National Air Races resumed in Cleveland in 1946, with competitors primarily flying retired military airplanes. Three years later, during the Thompson Trophy race in Cleveland, a World War II P-51 fighter crashed into a private residence, killing a woman, her small child, and the pilot, Bill Odom. The following year, with the outbreak of the Korean War, the Department of Defense refused to allow military aircraft to participate. The National Air Races folded.

The Professional Race Pilots Association, later renamed the National Air Racing Group, rekindled the sport in 1964 with the establishment of the National Championship Air Races in Reno, Nevada. The Reno National Air Races have been held annually ever since. While other air shows often include racing on their programs, Reno continues to hold the worlds only true air races.

Sources: Ralph Hickok, "History of the International Air Races," viewed online July 16, 2003, at http://www.hickoksports.com/history/airrace.shtml; European Sport Pilot Association, "The History of Air Racing," viewed online July 17, 2003, at http://www.esparacing.com/Air%20Racing%20History/MAIN%20MENU.htm; "The Major Trophy Races of the Golden Age of Air Racing," U.S. Centennial of Flight Commission, viewed online July 17, 2003, at http://www.centennialofflight.gov/essay/Explorers_Record_Setters_and_Daredevils/trophies/EX10.htm; Bill Meixner, Society of Air Racing Historians, personal communication with Lori S. Tagg, July 28, 2003.

Timer's stand at Fairfield Air Races, October 12, 1927

Pulitzer Trophy awarded annually between 1920 and 1925 for the fastest time over a closed-course

NBS-1 in flight at the 1924 Air Races, Wilbur Wright Field. Note the mock-up display of New York City in the background. It was constructed to demonstrate the tactics, accuracy, and power of aerial bombardment. Held on the third day of the races, the bombing attack on New York was repelled by pursuit aircraft defending the "city." In a final demonstration, the display was destroyed to show "the power of the bomb."

Captain Burt E. Skeel, assigned to Selfridge Field, Michigan, died in the crash of his Curtiss R-6 racer during the Pulitzer Trophy race on October 4. Skeel Avenue, named in his honor, parallels today's Area C flightline.

Lieutenant Alexander Pearson of McCook Field died on September 2, 1924, during a practice flight for the International Air Races. Pearson Avenue in present-day Area C of Wright-Patterson Air Force Base is named in his honor

Major Augustine Warner Robins and Lieutenant Charles Thomas welcome Chief of the Air Service Major General Mason Patrick and his sister, Miss Patrick, to the 1924 International Air Races.

Laughlin (McCook Field), assistant operations officer; Captain F. F. Christine, photographic officer; Captain J. B. Powers, surgeon; Lieutenant Samuel G. Eaton, Jr., communications officer; Lieutenant L. E. Sharon, publicity and press relations; and Lieutenant Harold A. Bartron, in charge of the baggage and check room.[50]

For several weeks before the races, FAID and Wilbur Wright Field personnel put final touches on 130 racing planes and prepared to receive scores of distinguished guests. They erected a huge grandstand over one mile in length at Wilbur Wright Field; converted the Officers' Club into a dining hall; and drained the swimming pool to turn it into a sunken garden. The post gymnasium was the site of a gala reception and aviation ball held the evening of October 2, attended by about 500 military personnel and guests. In local newspaper accounts, the Air Service Information Division described the evening's sparkling glamour:

> Long lines of flaunting colors which bordered the room represented every nation in the world, conveying the international appeal and importance of aviation. At the farther end of the ballroom was the American crest, and in the center was erected in huge form the emblem of the major general of the United States Army.
>
> Flags of various nations were placed at intervals about the huge room to form stalls, one being allotted to each government flying field in the United States and its possessions which sent delegates to the international classic.
>
> In direct contrast to the color scheme were placed paintings at either end of the ballroom. Both of the works of art were executed at [Wilbur] Wright Field and depicted a group of Curtiss pursuit planes in battle formation and the other, the epoch-making flight of Lieut. Maughan across the continent. To enhance these elaborate decorations profusion of fern and other greenery were used about the boxes and autumn flowers were banked on all tables.
>
> Despite the fact that America is a republic, the scene rivaled closely the brilliance of European court functions.
>
> Distinguished army officers, marines and representatives of foreign countries mingled with the handsomely gowned women upon the dance floor, producing a scene that will be unequalled for many years to come.[51]

In Dayton, enthusiasm for the races was epidemic. On Friday, October 3, all public schools closed. Over the weekend, city and county offices shut down, and all factories, including the National Cash Register Company (NCR), Dayton Engineering Laboratories Company (Delco), and the General Motors Corporation, stopped operations to allow their workers to attend the event. The banquet of the National Aeronautic Association, held Friday evening at the NCR dining hall with 1,000 persons attending, included Frederick Patterson, who as president of the NAA presided over the dinner; Charles F. Kettering, head of the General Motors Research Corporation, who served as toastmaster; and Frederick H. Rike, president of the Rike-Kumler Company, who was honored as vice-chairman of the race officials.[52]

The first event of the weekend was an On-To-Dayton race, which attracted 51 civilian entries from New York, Illinois, Texas, Kansas, Michigan, and a dozen other states, plus Canada. The brightly painted airplanes of all makes created a wondrous spectacle. The events held on October 2-4 encompassed 12 categories of major races, skywriting exhibitions, free balloon flights, parachute demonstrations, freak flying, aerial combat, and formation flying.

A nostalgic highlight occurred with the display and flight of early Wright brothers' aircraft. The 1903 Flyer *Kitty Hawk* was displayed but not flown.[53] The 1911 Model B, however, was removed from storage in the FAID Supply Department and flown by McCook Field's famous test pilot, Lieutenant John A. Macready. Macready thrilled throngs of spectators by circling Wilbur Wright Field at an altitude of 1,000 feet for 15 minutes at a speed of about 45 mph.

Another highlight was the release of the tiny Sperry Messenger. Suspended by hooks in its upper wing, the 862-pound biplane hung from a trapeze below the TC-5 non-rigid dirigible. At an altitude of 2,500 feet, the TC-5 released the Messenger into powered flight.

The highlights of the event were the races, each sponsored by a different newspaper, manufacturer, labor union, civic organization, or prominent citizen. The chief of the Air Service personally selected 23 officers from among the Army's most courageous and skillful pilots to fly in the four most prestigious trophy races. For the Liberty Engine Builders Trophy race on October 2, nine principals and three alternates were chosen to fly DeHavilland DH-4 and Corps Observation airplanes. Lieutenant C. A. Cover from FAID was one of the participants, but the winner was Lieutenant D. G. Duke, assigned to the Office of the Chief of the Air Service, who flew a DH-4 observation biplane powered by a 12-cylinder, 400-horsepower, Liberty engine. He averaged 128 mph over the 12-lap, 15-mile course.

Six Martin MB-2 twin-engine bombers and one Curtiss-built, Martin NBS-1 bomber entered the Dayton Chamber of Commerce Trophy race on October 3. The race featured "large capacity airplanes" (capable of carrying useful loads of 2,000 pounds or more) and offered $4,000 in Liberty bonds as prizes. Captain George C. Kenney of McCook Field participated in the race, but Lieutenant D. M. Myers of Phillips Field, flying an MB-2 bomber, won the 150-mile race with an average speed of 109.85 mph.

Airways DH-4 airplane at McCook Field. Fairfield Intermediate Air Depot in 1923. The Model Airway moved from McCook Field to more spacious accommodations at

Night-landing light developed at McCook Field. Mounted on a DH-4B observation airplane, the light illuminated the ground directly in front of the aircraft. The lamp was a 12-volt, 500-watt, concentrated filament, incandescent bulb. The entire light weighed 11 pounds. (*United States Air Force Museum*)

The John L. Mitchell Trophy race on October 4 featured pursuit-type airplanes and offered $5,000 in Liberty bonds as prizes. Major Carl Spaatz, commanding officer of Selfridge Field, chose 11 Air Service candidates from his command to compete in the race. Lieutenant Cyrus Bettis won the three-lap, 200-kilometer (124.27-mile) competition with a speed of 185.45 mph in a Curtiss PW-8 biplane with a 460-horsepower, Curtiss D-12 engine.[54]

For the renowned Pulitzer Trophy speed race on October 4, which offered $10,000 in prizes, General Patrick chose Captain Burt E. Skeel, commanding officer of the 27th Pursuit Squadron at Selfridge Field, Michigan, and Lieutenant Wendell H. Brookley, McCook Field, to fly Army-type, Curtiss R-8 racers. Lieutenant Alexander Pearson, Jr., McCook Field, was picked to fly the Navy version of the Curtiss R-8. Lieutenant H. H. Mills, McCook Field, served as an alternate pilot. Lieutenant Mills won the Pulitzer Trophy race, flying a Verville-Sperry racer over the 200-kilometer course at an average speed of 216.55 mph. This race, however, was marred by two deaths. Lieutenant Pearson, age 29, died September 2 during a practice flight for the race at Wilbur Wright Field when the left wing of his Curtiss racer failed at an altitude of about 300 feet, causing the airplane to crash. Captain Skeel, age 30, died during the race when the wings broke away from the fuselage of his Curtiss racer at 2,000 feet. He went into a dive at about 275 mph. (Skeel and Pearson avenues in present-day Area C, Wright-Patterson Air

Force Base, are named in memory of these two intrepid, young pilots.)

Model Airway

In 1920, General Billy Mitchell, chief of training and operations for the Air Service, traveled from Washington, D.C., to McCook Field for an inspection. Upon his return to Washington, Mitchell wrote to Colonel Thurman Bane, commander of McCook Field:

> After leaving the other day, I drove continuously until I reached Washington. I had to change tires eight times en route; covered 502 miles and reached here in 23 1/2 hours. I think the time has come when we should establish a model air route between here and Dayton....[55]

The Model Airway System, an experimental Air Service airline, was thus established in June 1921, becoming the first in the nation to operate regularly scheduled flights between fixed points. Initially, the airway offered regular service for both passengers and cargo between McCook Field, Ohio; Bolling Field, D.C.; Langley Field, Virginia; and Mitchel Field, Long Island, New York. Increased operations and the need to expand facilities prompted the Air Service to relocate the airline headquarters from crowded McCook Field to more spacious accommodations at Fairfield Intermediate

Air Depot. The three coordinating functions of the airway—administration, communications, and meteorology—operated at FAID under the direct supervision of the commanding officer, Major Robins.

In July 1923, the Air Corps' weather station also moved from McCook Field to the loft of Building 1 at FAID. By the end of the year, the Model Airway expanded to cover Selfridge Field, Michigan; Chanute Field, Illinois; and FAID, Fairfield, Ohio. Several other facilities were integrated into radio nets for communications and weather reporting.

Fairfield Intermediate Air Depot had become interested in the airway and promoted its growth by conducting a cost analysis of both cargo and passenger flights. Major Robins reported his findings in August 1925: between June 1922 and June 1925, DH-4 aircraft in the program made 546 flights covering 1,046,610 miles. They carried 868 passengers and 46,179 pounds of cargo. Major Robins estimated that 600 of the passengers had flown on official orders. He concluded that, based on an average trip of 2,000 miles, the Army had saved $600,000 in passenger travel expenses during the three-year period. Cargo savings amounted to $4,263, calculated at 1,000 miles per load.[56]

Major Robins was pleased with the progress of the Model Airway because it strengthened a vital link in the national defense chain. He also applauded the many benefits of the operation to the future of both military and commercial aviation, especially in areas of instrumentation, navigation, standardization of facilities and procedures, and communications. For example:

• A special night air route was established between McCook Field and Norton Field, Ohio, for a three-month test of night-flying equipment, both on airplanes and on the ground. (Norton Field, Columbus, Ohio, dedicated on June 30, 1923, had no connection with Norton Air Force Base in California.)
• The Airways Branch of the Engineering Department, FAID, experimented with "instrument-only" flying. Cockpits were entirely enclosed to simulate fog and poor visibility and to force pilots to fly by navigation instruments alone.
• The Radio Branch at FAID developed direction-finder radio beacons. In tests, radio beacons guided aircraft between McCook and Langin Field, Moundsville, West Virginia; and between McCook and Chanute Field, Illinois.

DISTINGUISHED VISITORS TO FAIRFIELD AIR DEPOT

According to the official history of the Fairfield Air Depot written in 1943, personnel were busy feting several distinguished visitors in 1927 and 1928:

Colonel Charles A. Lindbergh, flying back home from his trans-Atlantic hop in the monoplane "Spirit of St. Louis", passed over this station June 17, [1927]. He was joined here by the First Pursuit Group of Selfridge Field and several observation planes which formed an escort on the last leg of the journey to St. Louis. Five days later, Colonel Lindbergh returned to the field in the company of Major Lamphier and Lieutenant Street and spent the night as the guest of Orville Wright....

Arthur Goebel, winner of the Dole prize for his Hawaiian flight [a race between Oakland, California, and Hawaii sponsored by the Dole Pineapple Company], stopped here in March. A group of engineering students from Yale University inspected the shops in April. A new commercial tri-motor Fokker monoplane, "the last word" in aircraft, stopped at the field for servicing in May and was inspected with interest by personnel of the post. On June 6, members of Commander [Richard E.] Byrd's crew flew the "Floyd Bennett" to Fairfield where tests on radio apparatus were conducted before the plane left on the Antarctic expedition. An enthusiastic group of 600 4-H boys and girls from Logan County toured the field in June. Amelia Earhart, first woman to fly the Atlantic Ocean, landed for a brief stop on September 2, at which time also the Goodyear airship "Puritan" visited the post and took many officers and civilians for rides.

A number of these visitors became record-holding aviators, their names well known in the annals of aviation history. Lindbergh's 33-hour, solo flight in 1927, from Long Island, New York, to Paris, France, earned him a place in the Aviation Hall of Fame in 1967. Earhart and Byrd also were inducted into the Hall of Fame in 1968. Earhart was not only the first woman to fly solo across the Atlantic Ocean in 1932; she also completed a solo flight between Hawaii and California before she and her navigator, Fred Noonan, disappeared during an attempted round-the-world flight in 1937. Byrd and his mechanic and friend, Floyd Bennett, completed the first airplane flight over the North Pole in 1926, and in 1929, Byrd flew over the South Pole in the airplane he named in honor of Bennett, who passed away the previous year.

Source: *History of the Air Depot at Fairfield, Ohio, 1917-1943* (Fairfield Air Service Command, Patterson Field, 1944), pp 56, 59-60.

Amelia Earhart, pictured in front of the Lockheed Electra she flew across the Atlantic Ocean in 1932. Earhart landed for a brief stop at Fairfield Air Depot on September 2, 1928. (*Smithsonian Institution*)

Commander Richard E. Byrd, pilot of the "Floyd Bennett," who completed a flight over the South Pole in 1929 (*Hill Aerospace Museum*)

Brigadier General William E. Gillmore, chief of the Air Corps Materiel Division, and Orville Wright visit with Charles Lindbergh at Wright Field on June 22, 1927, one month after Lindbergh's famous transatlantic flight.

- A new system of aviation maps showed private, as well as military, airfields.
- Names of small towns and villages along flight routes were marked on hangars and flat roofs of major buildings. Together with a program of installing rotating beacons, this identification system saved the lives of scores of military and civilian pilots during times of poor visibility and bad weather. During 1934, when the Air Corps assumed responsibility for flying the airmail, these markers proved invaluable.
- FAID outlined a standardized plan for equipping emergency landing fields with hangars, radio towers, fuel tanks, etc. A "logical plan of expansion" specified dimensions of runways and standard field markers for future field development.
- FAID also standardized procedures for filing flight plans, invaluable for both preflight planning and in the event of in-flight emergency. Information "filed" prior to departure included airplane type and identification number, names and addresses of crewmembers and passengers, description and weight of cargo, volume and weight of fuel aboard, route to be flown, scheduled intermediate stops, and final destination with estimated time of arrival. Moreover, a reservation system for passengers and cargo was formulated.

Major Henry H. Arnold, the most famous commander of the Fairfield Air Depot, served as commander from June 1929 to July 1930. He also headed the Field Service Section of the Materiel Division at this time. Arnold became chief of the Army Air Corps (later Army Air Forces) in 1938 and led the service during World War II. On December 5, 1944, he was promoted to the five-star supergrade of General of the Army. General Arnold retired in 1946 and, three years later, was named General of the Air Force, the only person to hold that title.

- A radio net embraced all major stations of the airway, furnishing complete meteorological reports three times daily. This ended previous dependence on sketchy local weather reports and newspaper forecasts.

In Major Robins' judgment:

The Airway has increased the general interest in cross country flying because of excellent maps, marked towns, emergency fields, aeronautical bulletins, radio communications, and other conveniences which have been instituted by the Model Airway over the entire country. The aeronautical bulletins, which include the latest information on all landing fields throughout the United States are published with a complete sketch and information regarding facilities at each field. These aeronautical bulletins are furnished to Air Service personnel and to all persons engaged in aeronautics.[57]

Due to federal legislation restricting competition with the newly contracted commercial airmail service, the airway project was discontinued in 1926. The Model Airway, however, became the prototype for a growing commercial industry.

CREATION OF THE FAIRFIELD AIR DEPOT RESERVATION

In 1924, a chain of events began that soon culminated in the establishment of the Fairfield Air Depot Reservation (FADR). That year, the citizens of Dayton donated land to the government for the establishment of a new and expanded Air Service installation, Wright Field, which was formally dedicated in October 1927. In response, War Department General Orders No. 20, issued in August 1925, directed that the designation "Wilbur Wright Field" be discontinued. The original flying field, together with the newly donated acreage, was designated "Wright Field" in honor of both Wright brothers. The area occupied by the Fairfield depot (including the 40 acres purchased by the government in 1917) also became part of the larger Wright Field, although the depot's mission remained unchanged.[58]

More widespread changes affected the Fairfield depot with implementation of the Air Corps Act signed on July 2, 1926. It authorized the formation of the Air Corps Materiel Division, originally located at McCook Field but which would have permanent facilities at the new Wright Field.[59] As mentioned previously, the Field Service Section at FAID performed a materiel function in its management of the Air Service depots. The section became a major component of the new Materiel Division, physically removing it from the Fairfield depot. This move clarified the often confusing distinction between the Air Corps' depot command and the Fairfield depot that had been its temporary home. During certain periods after the move, however, the depot commander directed the Field Service Section; from May to September 1927, Major Robins commanded both functions, and from September 1929 to June 1930, Major Henry H. "Hap" Arnold was in command.[60]

Building 13, constructed in 1930, was FAD's primary engine overhaul and aero-repair facility.

Concurrent with the removal of the Field Service Section on June 21, 1927, the depot changed names once more. War Department General Orders No. 9, dated the next day, designated the depot as the Fairfield Air Depot Reservation (FADR).[61] Again, the mission of the depot remained unchanged, although as part of the new Wright Field, an increase in construction and activities occurred.

At FADR, Building 13 was built to keep pace with modern repair and overhaul functions. The structure was completed in June 1930, with new machinery and equipment installed in less than a month, causing no lost time in production. The hangar system of overhaul was reinstated and replaced the assembly-line method used since 1926. In November 1930, Building 54 was completed by bricking up

an old hangar. This building housed parachute repair, the fabric department, the wood mill, and the local issue unit.[62] FADR also provided temporary space for several functions of the Materiel Division moving from McCook Field, but whose facilities on the new section of the field (present-day Area B) had not been completed.

1931 Air Corps Maneuvers

In May 1931, FADR and Wright Field hosted the 1931 Air Corps maneuvers, which were much larger than previous field exercises held at the depot in 1926 and 1929. The maneuvers were carefully designed to test the capabilities of all facets of the Air Corps. Brigadier General Benjamin D. Foulois, assistant chief of the Air Corps, flew his staff to Fairfield from Washington and personally organized the First Provisional Air Division. This temporary organization consisted of one bombardment wing, one pursuit wing, two observation wings, one attack group, and one transport group. It mustered 24 of the Air Corps' 25 tactical squadrons in the continental United States. In addition, National Guard units from 19 states attended. In all, 672 airplanes participated. Orville Wright served as honorary chairman of the Dayton committee for the maneuvers.

Initial practice formations flew over Ohio from May 15-19 to rehearse an "aerial war" to be staged on the East Coast the following week. Foulois' division flew a full-scale review over Chicago on May 20, and then headed east. On May 22, the division landed at five different fields in New York and General Foulois set up division headquarters at Mitchel Field on Long Island. Operations were carried out over New York; Hartford; Springfield, Massachusetts; Boston; Atlantic City; Trenton; Jersey City; Philadelphia; and Washington, D.C. This required perfect coordination of supply lines, maintenance facilities, and staff support agencies. Orders for the maneuvers called for air transport of sufficient food, medical items, and clothing to supply 1,400 men for 15 days as they moved over a dozen states. This experience under actual field conditions gave the Air Corps a precise picture of its ability to handle a large, mobile air force.

The maneuvers closed with an aerial review over Washington, D.C., on Memorial Day 1931, with President Herbert C. Hoover observing from the White House

Service-type parachute inflated by a Douglas observation airplane at the Fairfield Air Depot in 1928. The 175-pound dummy used for ballast in drop-testing parachutes is visible on the running board of the truck.

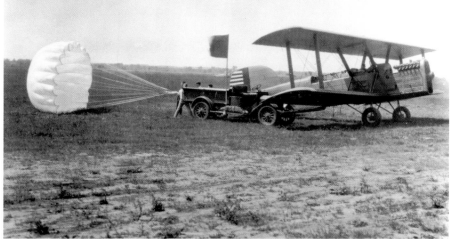

Pilots discuss flying operations during the refueling of a bomber aircraft, Fairfield Air Depot.

Soldiers on the flightline at Fairfield Air Depot Reservation, 1926 (*Darrell R. Larkins Collection*)

AIR CORPS CARNIVAL—1930

On June 7, 1930, Wright Field and Fairfield Air Depot Reservation (FADR) sponsored a gala Air Corps Carnival. The carnival was a benefit air exhibition to raise funds for the Army Relief Association, the Enlisted Men's Fund, and the Civilian Welfare Association at Wright Field and FADR, as well as the Wilbur Wright Officers' Club. Major "Hap" Arnold, commander of the depot, served as general chairman of the event, and widespread advertising attracted a large crowd. Civilian employees were admitted free; the general public paid an entrance fee of 25 to 50 cents. Food and souvenir stands, a merry-go-round, and airplane displays were part of the attraction.

The 40-piece FADR band provided morning entertainment. The afternoon program offered a pursuit race, simulated aerial combat, a balloon bursting exhibition, parachute jumping (during which eight men simultaneously jumped from a single airplane), glider target towing, pursuit formation flying, balloon races, model airplane contests, demonstration of a radio-controlled airplane, and airplane aerobatics. Visitors were welcome to take dollar airplane rides and to dance in the gymnasium. The evening program opened with a display of fireworks during which airplanes carried lights as they flew overhead. A display of night photography and an illuminated smoke screen closed the spectacle.

Source: *History of the Air Depot at Fairfield, Ohio, 1917-1943* (Fairfield Air Service Command, Patterson Field, 1944), pp 62-63.

Thousands of spectators attended a 1930 air show and carnival on the Fairfield Air Depot Reservation flightline. The "modern" steel hangar in the foreground is now Building 145, Area C. FADR headquarters (Building 1) is at center top; completed Building 13 is at center left.

Fairfield Air Depot Reservation during the 1931 maneuvers. Part of the pursuit wing, First Provisional Air Division, is parked beside the flightline of World War I wooden hangars. A total of 672 airplanes participated in the maneuvers. The large, dark structure in the center is the hangar specially constructed for the Barling bomber.

Major General Benjamin D. Foulois. While a brigadier general and assistant chief of the Air Corps, Foulois organized the First Provisional Air Division for the 1931 Air Corps maneuvers. Previously, he served for one year (1929-1930) as chief of the Materiel Division at Wright Field and, in December 1931, Foulois was promoted to major general and named chief of the Air Corps. He retired in 1935.

lawn. During the division's mobilization, operation, and demobilization, its airplanes flew over all 48 states and accumulated four million miles without any fatalities or serious injuries.[63]

PATTERSON AND WRIGHT FIELDS

From August 1925 to July 1931, the area occupied by the Fairfield Air Intermediate Depot and, subsequently, the Fairfield Air Depot Reservation technically remained part of Wright Field. This situation changed on July 1, 1931, when Wright Field was divided into two separate installations. All of the land west of Huffman Dam, including much of the area known today as Area B, Wright-Patterson Air Force Base, retained the Wright Field designation. All of the land east of Huffman Dam was designated Patterson Field, in honor of Lieutenant Frank Stuart Patterson, per War Department General Orders No. 5 (see below). Patterson Field consisted of the area now known as Area A, Wright-Patterson Air Force Base, and the whole of

the Fairfield Air Depot Reservation, now Area C. The reservation incorporated Huffman Prairie and the World War I sites of the Fairfield Aviation General Supply Depot and Wilbur Wright Field. Although located on Patterson Field, the Fairfield Air Depot retained its title (dropping "Reservation") and continued as a major function of the new installation until the depot function inactivated in 1946.

Patterson Field, with its logistics mission, and Wright Field, dedicated to engineering advancement, remained separate installations until the end of World War II. By 1945, the functions and identities of the two fields had become so intertwined that most people considered them a single installation. Their official merger, however, did not occur until January 13, 1948.

The Patterson Name

Frank Stuart Patterson was born in Dayton, Ohio, on November 6, 1897, the son of Frank Jefferson Patterson and Julia Shaw Patterson. The elder Patterson and his brother, John H. Patterson, founded the National Cash Register Company and

Patterson Field, May 5, 1931

figured prominently in local history. Young Frank Stuart, as his family called him, descended from a long line of military officers. His great grandfather, Colonel Robert Patterson, was a veteran of the Revolutionary War, the War of 1812, and skirmishes with Native Americans in Ohio, Indiana, and Illinois. Three uncles were Union Army captains during the Civil War.

Frank Stuart carried on the family's military tradition. After schooling at the Florida-Adirondack preparatory school, he enrolled at Yale University, from which he received his bachelor of arts degree *in absentia* in the spring of 1918, as did many of his fellow classmates who had joined the Signal Corps. On May 21, 1917, he enlisted as a private in the Aviation Section of the Enlisted Reserve Corps, U.S. Army.[64] He received ground training at the Massachusetts Institute of Technology, Cambridge, Massachusetts. After completing primary training at Buffalo, New York, and advanced flying training at Mineola, Long Island, New York, Frank Stuart trained in aerial observation at Post Field, Fort Sill, Oklahoma. Upon completing this training, Private First Class Patterson was honorably discharged from the Enlisted Reserve Corps on September 14,

1917, at Fort Sill, and commissioned the following day as a first lieutenant in the Officers Reserve Corps, U.S. Army, with the aeronautical rating of pilot.

Lieutenant Patterson was assigned to the 137th Aero Squadron at Wilbur Wright Field, near his hometown of Dayton, on May 10, 1918.[65] Orders issued May 9 at the new station assigned him to a board of officers "for the purpose of conducting tests of the DeHavilland Four and Bristol Fighter airplanes."[66] On June 19, 1918, little more than a month after arriving at Wilbur Wright, Lieutenant Patterson and his aerial observer, Lieutenant LeRoy Amos Swan, went aloft in their DH-4 (serial number 32098) to test newly installed machine guns synchronized by Nelson, interrupter-gear equipment. Their instructions were to fire about 100 rounds into the field from 6,000 feet, 10,000 feet, and 15,000 feet. They completed the first two trials successfully, firing the guns through the propeller as they dived. Lieutenant Patterson then climbed to 15,000 feet and pointed the airplane downward in a steep dive. Just as reports of the guns reached the earth, the wings of the airplane collapsed and separated completely from the fuselage, leaving it to travel across the field at full power during

its fall. The machine was completely wrecked and the crew crushed.[67]

The initial report of the accident wired to Washington indicated that it was not clear whether the wings folded up or were swept back. The aircraft accident investigation board subsequently determined that

Patterson's accident believed to be due to shearing of tie rod that passes through fuselage near radiator and connects the two fittings to which are attached nose drift wires. Accident occurred while diving hence considerable strain put on those two fittings. Tie rod sheared but fittings show very little elongation showing that weakness lay in tie rod. There being nothing else to take backward strain on wings, wings probably folded back and separated from fuselage.[68]

This report contradicted the erroneous assumption that bullets from the machine gun shattered the propeller blades, which flew back and tore the wings from the airplane.

Lieutenant Patterson was buried next to his father in Dayton's Woodland Cemetery

Lieutenant Frank Stuart Patterson (standing sixth from left) with fellow pilots, May 1918

Following a military funeral on June 21, 1918, Lieutenant Frank Stuart Patterson was buried next to his father in Woodland Cemetery, Dayton.

Test pilot Lieutenant Frank Stuart Patterson died in the crash of his DH-4 biplane while testing a machine-gun synchronizer over Wilbur Wright Field, June 19, 1918.

A Logistics Heritage

Patterson Field's principal function from July 1, 1931, through 1945 was that of logistics. The field was assigned to the Air Corps Materiel Division until October 17, 1941, when the Air Service Command assumed the logistics functions of the Materiel Division. The Fairfield Air Depot (FAD) occupied the major portion of Patterson Field. By the early 1930s, FAD's area of control extended over 23 states, serving 28 of the approximately 50 Air Corps stations in the United States.[72] Fairfield was linked to other depots in the system by an air transport supply service, a descendant of the earlier Model Airway. The transport service was staffed at FAD by the 1st Provisional Air Transport Squadron, constituted March 1, 1935, and redesignated the 1st Transport Squadron on June 25, 1935. The 2nd, 3rd, and 4th Transport Squadrons were stationed at the San Antonio, Middletown, and Rockwell depots, respectively. These squadrons comprised the transport system within the Air Corps, an idea proposed by Lieutenant Colonel Augustine Warner Robins while serving as executive officer to the chief of the Materiel Division.

Statistics from 1931 show that personnel strength at Patterson Field ranged between 12 and 15 officers and 500 to 550 civilian employees. The monthly civilian payroll approximated $67,000. During June 1931,

on June 21, 1918, following a military funeral. He was survived by his mother, Julia Shaw Patterson Carnell; a brother, Jefferson, who was a lieutenant in the field artillery serving with the 83rd Division in France; a sister, Mary, who was active in wartime Red Cross work; and other members of the illustrious Patterson family, including his first cousin, Frederick Beck Patterson, who later headed the Dayton Air Service Committee in its drive to donate land for the creation of Wright Field.

LeRoy Amos Swan was born June 5, 1894, in Norwich, Connecticut. After attending public schools and the Norwich Free Academy, he graduated in 1917 with a bachelor of science degree in mechanical engineering from the Massachusetts Institute of Technology. Swan enlisted in the Air Service on July 14, 1917, receiving a commission in the Officers Reserve Corps the following September. He attended the aerial observers' school at Post Field, Oklahoma, as had Lieutenant Patterson. During his assignment to Wilbur Wright Field beginning in April 1918, Swan participated in a number of flights involving machine-gun testing. Lieutenant Swan was buried with military honors in his hometown. According to the local newspapers, he was the first military man from Norwich to die during the war.[69]

The Patterson-Swan accident was not the first involving a DH-4 aircraft in the local area. The earliest recorded crash in a DH-4 occurred May 2, 1917, and killed Lieutenant Colonel Henry J. Damm and Major Oscar Brindley.[70] Perhaps the earliest fatality at the site occurred before it became Wilbur Wright Field. On May 21, 1912, newspaper accounts claimed, "Fred Southard stole an airplane...and went aloft for his solo flight," falling 100 feet to his death when the engine failed.[71]

PATTERSON FIELD COMMANDERS

Major Albert L. Sneed (later lieutenant colonel)	Aug. 30, 1931 – Mar. 13, 1933
Major Fred H. Coleman (later colonel)	Mar. 13, 1933 – July 28, 1936
Lt. Col. Junius H. Houghton	July 28, 1936 – Aug. 1, 1939
Lt. Col. James F. Doherty	Aug. 1, 1939 – Sept. 11, 1939
Lt. Col. Merrick G. Estabrook, Jr. (later brig. gen.)	Sept.11, 1939 – Feb. 12, 1944
Colonel James A. Woodruff	Feb. 6, 1943 – Dec. 13, 1943
Colonel Raymond E. Culbertson	Dec. 13, 1943 – Feb. 12, 1944
Brigadier General Clarence P. Kane	Feb. 12, 1944 – Apr. 29, 1944
Colonel Clarence H. Welch	Apr. 29, 1944 – June 8, 1944
Colonel Elmer H. Jose	June 8, 1944 – Dec. 6, 1944
Brigadier General Harold A. Bartron	Dec. 6, 1944 – Nov. 5, 1945

Colonel Estabrook continued, as of February 6, 1943, to command Fairfield Air Service Command (FASC), while Colonel Woodruff and then Colonel Culbertson commanded Patterson Field. On November 27, 1944, FASC and Patterson Field base command again merged, with Colonel Jose as commanding general of Fairfield Air Technical Service Command (FATSC), Patterson Field Army Air Base, and the 4100th Army Air Forces Base Unit.

General Bartron, as commander of FATSC, was the last commander of Patterson Field as a facility separate from Wright Field. On November 6, 1945, a combined staff was appointed for the concurrent operations of Wright and Patterson fields under Brigadier General Joseph T. Morris, then commander of Wright Field and the 4000th Army Air Forces Base Unit. General Morris continued in command on December 15, 1945, when Wright and Patterson fields combined for administrative purposes under an umbrella organization designated the Army Air Forces Technical Base (AAFTB), and on December 9, 1947, when the AAFTB was redesignated the Air Force Technical Base (AFTB). On January 13, 1948, when Wright and Patterson fields were redesignated Wright-Patterson Air Force Base, General Morris became the first commander of the permanently merged fields.

Lieutenant Colonel Junius H. Houghton (shown here when a brigadier general) served as Patterson Field commander from 1936 to 1939.

Colonel Elmer H. Jose, Jr. (shown here as a major), served as commander of Patterson Field until November 1945, when the Fairfield Air Technical Service Command was created.

General Harold A. Bartron, commander of Fairfield Air Technical Service Command and Patterson Field until November 1945

depot supply received 784 shipments totaling 875,815 pounds and dispatched 1,379 orders totaling 1,099,277 pounds. The repair shops received 44 airplanes and 67 engines for overhaul. Airplane overhauls completed that month totaled 36, while engine overhauls numbered 85.[73]

In 1931, 437 airplanes of all types were assigned in the FAD control area. These included attack, bomber, observation, photographic, pursuit, basic training, primary training, and cargo aircraft. FAD was responsible for maintaining these airplanes in flying condition. The overhaul schedule in force at that time called for bombers to be overhauled every 18 months, primary trainers every 15 months, pursuits every 10 months, National Guard airplanes every two years, and all other types once each year. The FAD Engineering Department accomplished this work in Building 13 and two small adjacent buildings, 52 and 54.

Administrative personnel were originally housed in Building 1 at the depot. In March 1933, however, a new headquarters, Building 11, was completed and accepted.[74] A government radio station, WYD, was first located at the depot and served both the depot and Wright Field. In September

Fairfield Air Depot (FAD), Patterson Field. A major portion of Patterson Field was occupied by FAD until the depot closed in 1946. The large building at lower right is Building 1, the original depot headquarters.

C-33 cargo transport, able to carry 5,000 pounds of cargo at a top speed of more than 200 mph. Along with C-27 and C-39 transports, C-33s were flown by the 1st Provisional Air Transport Squadron assigned to FAD in July 1935. (*United States Air Force Museum*)

Building 11, Area C, served as Patterson Field headquarters from 1933 to 1948.

The Transient Camp, located at the south end of Patterson Field, stood approximately where Buildings 262 and 266, Area A, stand today.

1933, the radio station and telegraph office were consolidated with the message center at Wright Field and subsequently moved their operations there.

Depression Years

As Fairfield Air Depot entered the 1930s, the Air Corps was in the midst of the five-year expansion program outlined by the 1926 Air Corps Act. This program, which got off to a late start, lasted until 1932. Goals of the program included 1,800 serviceable airplanes, 1,650 officers, and 15,000 enlisted men. In terms of quality, the goal was continual improvement of aircraft systems. World records for altitude, speed, and distance flying were routinely broken as the Air Corps struggled to come of age.

Notwithstanding progress made under this five-year expansion program, the economic conditions of the Great Depression were reflected in operations at the Fairfield Air Depot. Noncritical departments were reorganized to enable employment of personnel in mission-essential departments. In some cases, positions held by male employees were reassigned to women, who worked for lower

Transient work crews graded the landscape for construction of the Brick Quarters.

salaries. In February 1930, the station supply department consolidated with depot supply to eliminate duplicate receiving, shipping, and inspecting departments. In April 1930, the finance and personnel departments also combined in the interest of efficiency.[75] By July of 1933, funds to meet the civilian payroll declined to the point that it became necessary to declare an occasional administrative furlough of personnel without pay.

Beginning in 1934, the workforce at the depot was augmented by temporary laborers employed under Depression-era programs. In May, the depot set aside several acres of land for men forced to live as transients by the economic conditions of the Depression. By 1935, according to one account, 527 men lived in this camp. Because many of these men possessed skills useful to the depot, they received lodging and meals at the camp in exchange for 20 hours of work per week. Transient workers also received a subsistence pay of approximately $1.50 per week in cash.[76]

One valuable service performed by the transient workers involved the renovation of buildings, particularly those frame structures hastily built during World War I. For example, Buildings 2, 3, and 4, south of Building 1, were essentially "shells" of corrugated metal with cinder floors. Although still in use as storage buildings, they had deteriorated greatly through the years. Building 2 was renovated by a commercial contractor and Building 3 by the Civil Work Administration (CWA), a federal agency created to help relieve the nation's unemployment situation. The men living at the transient camp, however, renovated Building 4. (The three buildings were later linked together to form what is today Building 2, Area C.) The same group of men performed most of the landscaping around the brick officers' quarters in 1934.

In January 1936, the federal Works Progress Administration absorbed large numbers of unemployed laborers, and the transient camp was discontinued. The site of the camp, however, retained the unofficial name "Transient Camp" for many more years.[77]

The Civilian Conservation Corps also maintained a camp on Patterson Field. The young men in this organization had no official relationship with the depot, but they did assist in sodding lawns, landscaping, and other tasks during 1935 and 1936.

Airplane engine storage, Fairfield Air Depot, 1933

The Brick Quarters complex on Patterson Field was constructed between 1933 and 1935. It consisted of 92 units of officer quarters and the Wilbur Wright Officers' Club (today the Officers' Open Mess, Building 800). The complex was arranged in the form of a horseshoe with the Officers' Club centered at the top of the curve.

Quarters 1 (Building 700), the largest home in the Brick Quarters, was the residence of the installation's senior commander. In 1989, the residence was named the Robins House in honor of Brigadier General Augustine Warner Robins, who resided in the home while commander of the Air Corps Materiel Division at Wright Field from 1935 to 1939.

Keystone B-6A, 1931. Keystone B-4A and B-6A bombers were modified at Fairfield Air Depot for use by the Air Corps in flying the United States airmail in 1934. (*United States Air Force Museum*)

Kellett YG-1 autogiro. The autogiro school, established at Patterson Field in April 1938, was the first military school in the United States to train pilots and mechanics with rotary wing aircraft.

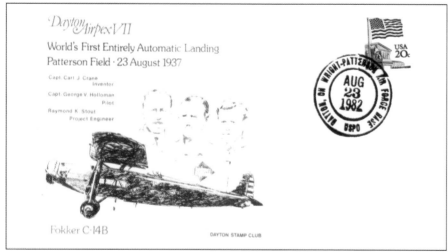

Dayton Airpex VII

World's First Entirely Automatic Landing
Patterson Field · 23 August 1937

Capt. Carl J. Crane
Inventor

Capt. George V. Holloman
Pilot

Raymond K. Stout
Project Engineer

USA 20c

AUG 23 1982

Fokker C-14B

DAYTON STAMP CLUB

A cachet issued by the Dayton Stamp Club in 1982 commemorated the 45th anniversary of the world's first, entirely automatic landing.

FAD Activities During the 1930s

In April and May of 1933, FAD supported the Air Corps Anti-Aircraft Exercises based at Patterson Field. These exercises were designed primarily to perfect new techniques and tactics being developed in bombardment aviation. FAD personnel installed and maintained special equipment on airplanes arriving from bases in Virginia, Texas, California, and New York. The squadrons not only came from widely separated stations, they operated many different types of airplanes. For example, the bombardment section consisted of squadrons of twin-engine, Curtiss B-2 biplane bombers; Douglas B-7 monoplane bombers; and all-metal, Boeing B-9 monoplane bombers. Upon their arrival at Patterson Field, the squadrons and their aircraft underwent one month of intensive training and equipment conditioning. Final exercises were held May 15-24. The entire proceedings received the commendation of Air Corps Chief Brigadier General Benjamin D. Foulois, who was in attendance.[78]

In 1934, FAD tackled a new project as the Air Corps assumed responsibility for flying the United States mail. During the winter of 1933, the federal government cancelled its airmail contracts with commercial airlines and delegated the job to the Air Corps. In February 1934, FAD began to furnish supplies for the aircraft supporting the mission. Although the Air Corps was poorly equipped and unprepared to assume responsibility for the airmail on such short notice, the Engineering Department at Fairfield did its best to meet the challenge. In 11 days, the Engineering Department modified enough airplanes to support operations. Rear controls in the cockpits of "stick and wire"

Colonel Don McNeal and his staff of the Air Corps Weather School, Patterson Field, 1937

Keystone biplanes were removed to make room for baggage compartments and the airplanes' structures were reworked. The airplanes were equipped with instruments to enable blind flying so that airmail schedules could be met day and night.

Supply personnel at Fairfield took turns sleeping in the depot to fill emergency requests for stock received during the night. Depot supervisors worked diligently with commercial businesses to ensure that supplies, including oil and gasoline, remained available. In addition, supply personnel went to various Air Corps and commercial installations throughout the country as FAD liaisons to facilitate operations.

Despite these Herculean efforts, the mission was doomed to failure, the victim of insurmountable difficulties. By June 1934, in the face of numerous fatalities and tragic accidents, the federal government abandoned the project and returned contracts for flying the mail to commercial airlines. Although Air Corps facilities proved inadequate and inappropriate for such service at the time, valuable experience was gained and lessons were learned.[79] Perhaps most important, the Air Corps' inadequate funding was dramatically brought to public attention and

strengthened the argument for increased Air Corps appropriations.

The popular, long-distance, 1934 Alaskan flight initiated by Lieutenant Colonel Henry H. "Hap" Arnold and Major Ralph Royce also increased public awareness. Ten new Martin YB-10 twin-engine bombers were specially equipped for the journey to Fairbanks, Alaska, and return. In June and July of 1934, FAD personnel prepared the airplanes for the flight. Trial flights were conducted from Patterson Field to March Field, California, and to Dallas, Texas, although the Alaskan trip itself was launched from Bolling Field in Washington, D.C. The complete flight occupied one month and included a brief stop at Patterson Field on August 19 during the return trip.[80]

A number of developments in military meteorology can be traced to activities at Patterson Field in the 1930s. On March 26, 1935, the Army's adjutant general explained a new War Department policy designed to improve the meteorological service furnished to the Air Corps. It involved closer technical supervision of the Army service, more contact and closer cooperation with the larger services of the U.S. Weather Bureau and the Department of Commerce, and the introduction of air-

mass-analysis weather forecasting. These changes were expected to yield great benefits to military pilots. A Signal Corps officer implemented the new policy in areas of considerable military air activity.

A radio transmitter, installed at Patterson Field on January 8, 1936, broadcast local weather conditions at hourly intervals. Pilots and operations officers received this weather data on a frequency of 379 kilocycles under the call letters WXA.[81] On July 1, 1937, the Meteorological School at Fort Monmouth, New Jersey, transferred to Patterson Field and its name changed to the Air Corps Weather School. The Fairfield Air Depot commander served as the school commandant. The first class of 20 students graduated from the six-month course at Patterson Field on January 28, 1938. The school operated until its transfer to the Air Corps Technical School at Chanute Field on June 1, 1940.[82]

Another significant event in Patterson Field history was the opening of the first military autogiro school in the United States on April 20, 1938. The school trained officers as pilots and enlisted men as maintenance crews for the service testing of autogiros. It started with three new, Kellett YG-1B, direct-control autogiros, and augmented its small training fleet until

The Officers' Club, Building 800, was completed October 13, 1934, and opened to all officers stationed at both Wright and Patterson fields. The swimming pool complex was a gift from Lieutenant Frank S. Patterson's mother, Julia Shaw Patterson Carnell. The pool was dedicated in memory of the test pilot in 1936. It was completed in 1937 and remained in service until 1997.

BRIGADIER GENERAL MERRICK G. ESTABROOK, JR.

Brigadier General Merrick G. Estabrook, Jr., oversaw many of the dramatic changes that occurred at Patterson Field immediately prior to World War II and during the early years of the war. He served as commander of both Patterson Field and Fairfield Air Depot from 1939 to 1943.

According to many who served under him, General Estabrook spent long days and nights, at the expense of his own health, overseeing the massive construction projects at Patterson Field and supervising a steadily increasing number of staff members. His right-hand men were Chief Warrant Officer Charles M. "Smitty" Smith and Post Adjutant Major Eugene M. Becher, veterans of the operations at Fairfield Air Depot. The relatively quiet and closely-knit Patterson Field community of the 1930s disappeared almost overnight as the post became a center of wartime activity.

General Estabrook first served at Wright Field from September 1927 to January 1928, when he was assigned as an assistant chief of the Supply Branch, Materiel Division. He graduated from the Air Corps Engineering School at Wright Field in June 1932 and from the Army Industrial College in Washington in 1936. He then returned to Wright Field as chief of the Engineering and Shops Branch, Materiel Division. He subsequently became assistant chief of the Contract Section at Wright Field.

On September 11, 1939, Lieutenant Colonel Estabrook assumed command of Patterson Field and Fairfield Air Depot. In 1943, he served as commander of the nine-state Fairfield Air Depot Control Area Command, supervising all Air Service Command (ASC) units contained therein. On May 21, 1943, he became the first commander of the newly created Fairfield Air Service Command, one of the 11 ASC "Keep 'Em Flying" organizations in the United States. In September 1943, Estabrook was promoted to the temporary rank of brigadier general in this position.

On February 12, 1944, General Estabrook transferred to ASC headquarters at Patterson Field. He retired from active service due to disability on August 31, 1944. Three years later, on December 19, 1947, General Estabrook passed away.

Brigadier General Merrick G. Estabrook

Chief Warrant Officer Charles M. "Smitty" Smith, General Estabrook's right-hand man

Patterson Field, 1943, showing the buildup that occurred in the early years of World War II. In the foreground are numerous depot maintenance and storage buildings essential to wartime operations.

seven were on hand, the largest assembly of such machines ever before gathered in one location. Hangar 5, erected during World War I, served as classroom and workshop for the four-week course. During the first and second sessions, the autogiro school trained 12 officer pilots and 15 enlisted mechanics. These graduates then went to Fort Monroe, Virginia; Fort Sill, Oklahoma; and Fort Bragg, North Carolina, to conduct further tests with the ground services. Lieutenant H. F. Gregory, one of three faculty members for the autogiro pilot course, subsequently went to Wright Field to become project officer for the Army's rotary-wing aircraft.[83]

The world's first, entirely automatic landing occurred August 23, 1937, at Patterson Field. Captain George V. Holloman piloted the Fokker C-14B transport airplane used in the test. Captain Carl J. Crane, inventor of the system, and Raymond K. Stout, project engineer, were also present. All three were assigned to the Materiel Division at Wright Field. The landing was successfully completed without any assistance from the human pilot or from the ground. Captains Crane and Holloman received the Mackay Trophy

and the Distinguished Flying Cross for their achievements.[84]

The foregoing special activities provided occasional breaks in the normal routine of the Fairfield Air Depot. As the 1930s came to a close, however, the days of such routine activity were numbered. Fairfield soon became one of scores of depots in the United States charged with maintaining and distributing the largest stockpile of wartime materiel in history.

WORLD WAR II EXPANSION

Prior to World War II, the depot supply operation at Patterson Field was modest in size, even though it provided service to Air Corps installations nationwide. In 1939, FAD's Supply Department occupied approximately 7,500 square feet of office space, 306,000 square feet of warehouse floor space, and employed less than 200 workers. Operations were housed almost entirely in Building 1, which had been constructed during World War I. Additional workers employed by the Engineering Department, the Signal Corps

radio section, the fuels and lubricants unit, the lumber yard, the airfreight terminal, and other related units brought total employment at the depot to nearly 500 people. A standard, 40-hour workweek was in effect, with six to eight civilians remaining on duty over weekends to handle emergency shipments. Railroad service consisted of two tracks, which entered Building 1 from the east, and a single track that connected Patterson Field to Osborn.[85]

In light of engineering advancements made during the 1920s and 1930s in the United States and abroad, Air Corps officials recognized that a rapid expansion of its tactical capabilities would be central to the nation's rearmament program. Prodded by President Franklin D. Roosevelt, Congress authorized $300 million for Air Corps development in April 1939. This allowed the Air Corps to schedule production of unprecedented quantities of warplanes. At the Air Corps' depots, the need for supply and repair materials and associated equipment to keep new airplanes in flying condition increased dramatically. These changes brought a new way of life to the depot at Fairfield, Ohio.

Buildings 2, 3, and 4 at Patterson Field were originally constructed as separate buildings. During 1940, they were joined together with annexes to form one large structure, known today as Building 2, Area C.

In May 1935, Colonel Fred H. Coleman, commanding officer of Fairfield Air Depot, had prepared for the Materiel Division a detailed, 45-page outline of construction needed at Patterson Field, as well as a revised station plan showing proposed locations. For five years the program was studied, proposed, bypassed, and reconsidered by the Materiel Division, the Field Service Section, the Quartermaster Corps, and the War Department. In 1940, funds amounting to $1,970,000 were finally earmarked for these projects as part of the Air Corps Expansion Program.[86] It was the beginning of a tremendous wartime buildup at the depot. From 1938 to 1945, the face of Patterson Field changed drastically as shops, warehouses, and military and civilian housing complexes began to dominate the landscape.

During a special visit to the depot in late 1940, Brigadier General Henry H. Arnold outlined in graphic terms the scope of physical growth and increased depot activities FAD could expect.[87] His

Base Operations (Building 206), completed in December 1941 on the Patterson Field flightline. The FADO (Fairfield Air Depot Operations) Hotel for transient pilots was located in the center portion of the building.

Operations, Building 206, Patterson Field

Pilots register at the FADO Hotel. (*Dick Cull*)

The FADO Hotel provided comfortable but crowded quarters for pilots visiting Patterson Field.

predictions soon became reality. Supplies arrived at the Fairfield depot—by rail, mail, truck, and airplane—in ever-increasing quantities. The Receiving Department in Building 1 soon proved inadequate and was greatly expanded.[88] Plans were made to construct additional buildings and relocate some of the various Supply Department sections. Other buildings were joined by annexes to form larger structures. Barracks, mess halls, and other buildings were constructed rapidly to house mushrooming numbers of newly recruited and transferred military personnel.

In 1940 and 1941, Building 206 was constructed. It housed an airplane repair facility, offices for Patterson Field Operations, and a hotel for transient pilots. The hotel became known as the FADO (Fairfield Air Depot Operations) Hotel, a break in the Air Corps' tradition of referring to all airfield hotels as "DeGink."[89] In 1943, a new air terminal, Building 146, was constructed adjacent to Building 206 (on the site of the old Barling bomber hangar).

A repair warehouse set up in Building 6 in the fall of 1942 accommodated the huge mass of reparable materiel arriving from overseas. In June 1943, the newly completed Building 80 further augmented storage space and relieved the already crowded conditions in Building 6. It was constructed by combining eight demountable steel hangars, thus placing 32,000 square feet under a single roof.[90]

A new complex of warehouses constructed between 1941 and 1943 in the vicinity of Buildings 1 and 2 accommodated the steady deluge of

The Patterson Field Air Transport Terminal served the field's needs during the 1930s and early 1940s. In 1943, a new terminal, Building 146, was constructed on the site of the old Barling bomber hangar.

Warehouse operations in Building 258

FAD leased storage space in 14 buildings in Springfield, Ohio, to augment warehouse space available at Patterson Field.

Maintenance on a B-17 bomber, Building 13, during World War II

Constructed in 1943 as an engine overhaul and repair facility, Building 89 became home to the United States Air Force Museum from 1954 to 1971.

Building 219, completed in 1942, was one of two hospitals serving Wright and Patterson fields during the war years. The second hospital was a cantonment-type, temporary facility in Wood City.

Patterson Field control tower, 1940

supply materiel at the depot. These new structures included Buildings 70, 71, 72, 114, 174, 252, 253, 254, 255, 257, and 258. FAD leased additional warehouse space, totaling more than 561,000 square feet, in 14 buildings in Springfield, Ohio. Eventually, a packing and shipping department established in Springfield facilitated operations and saved time and expense.

Building 13, used for engine overhaul and repair, was expanded by consolidating and connecting several existing buildings. Building 89, also used for overhaul and repair, was completed in 1943, and Building 95, erected during 1943, housed the installation's salvage and disposal branch.

Building 10, which serves today as headquarters for the 88th Air Base Wing, was constructed in 1943 as headquarters for the Fairfield Air Service Command, a designation begun in May 1943 to refer to the command of the depot (see below). Building 11 continued to serve as Patterson Field headquarters. Extension wings added to both ends of Building 11 in 1942 provided additional offices for personnel.

As Patterson Field operations expanded, so did the field's administrative and support needs. The power plant expanded in 1942 to service increased energy needs. The post hospital, Building 219, was completed and activated on June 17, 1942, with a staff of 15 medical officers, 550 enlisted personnel, and six civilians. By December 1942, even this new facility proved inadequate to minister to the ills of all personnel at both Wright and Patterson fields. A cantonment-type hospital, activated December 21 in Wood City, soon became the main base hospital.[91]

Permanent runway construction took place on Patterson Field during 1942 and 1943 with concrete Runway "A" (now Taxiway No. 8) completed in December 1942. New concrete taxiways and an extension to Runway "B" (no longer in use) were built the following summer and accepted on September 3, 1943. An extension to Runway "C" (now Taxiway No. 12) was accepted on October 11, 1943.

Land acquisition during World War II became an important factor in Patterson Field's development. The 1943 and 1944 purchases of 851 acres in Greene County from the Miami Conservancy District and private landowners expanded Patterson Field to the northeast, allowing the 1947 construction of the Very Heavy Bomber (VHB) runway (see below).

The headquarters complex for the Air Service Command was a major construction project at the south end of Patterson Field in 1942 and 1943. The central structures,

Building 262 and its annex, 262A, were completed and occupied in 1942. Numerous warehouses, as well as barracks for Air Service Command personnel, surrounded the headquarters building. Area A, as the command headquarters area came to be known, was connected to the existing complexes on Patterson Field by an extension of Skeel Avenue, but remained, in effect, a separate area. In fact, from 1944 until 1948, Area A was administratively considered part of Wright Field.

As a result of the military and industrial buildup during World War II, a housing shortage arose in Dayton, Fairfield, and Osborn for military and civilian personnel working at Wright and Patterson fields. To remedy this situation, temporary wood-frame quarters were constructed throughout the base in 1941 and 1942. The largest of these were Wood City (renamed Kittyhawk Center in 1972, see Chapter 4: Fulcrum of Base Support) and the hilltop portion of Wright Field (Area B). Wood

Air traffic controllers direct traffic around the Patterson Field runway and flightline.

Completed Patterson Field runway complex, 1945

Civilian employees received guidance from "Patty Patterson" in Patterson Field's Civilian Handbook.

A large Air Service Command complex was constructed at the south end of Patterson Field between 1941 and 1943. The headquarters buildings (262 and 262A), at center, are flanked by rows of warehouses and barracks. Large buildings at center right are Buildings 287 and 288.

The Wood City complex at Patterson Field, located east of Route 4, contained a large cantonment hospital, military housing, and recreation facilities. Wood City was renamed Kittyhawk Center in 1972.

Construction of temporary enlisted barracks on Patterson Field, June 1941

City was erected across the highway from Patterson Field in an area that had served as a radio antenna farm and recreation area between 1924 and 1940. Its name derived from the wooden buildings and its street names, all of which referred to trees (Ash, Locust, Buckeye, etc.). This typical housing complex included living quarters, chapel, post hospital, education center, library, and recreation facilities. It also contained a 48-space trailer camp and the First Sergeants' Family Quarters.

Large barracks complexes also existed on Wright Avenue across from Building 1, on Wright Avenue west of Pearson Road, and on Skeel Avenue at the far north end of Patterson Field. Hebble Homes, a wartime housing development constructed near the base in Fairfield, housed civilian employees. Other housing areas were located between Building 219 (the hospital), Building 88 (the Foulois House), and along Communications Boulevard behind Building 262. Splinter City stood near Patterson Field's Xenia Avenue exit (Gate 33C). Sherwood Forest was located between today's Gate 15A and the Twin Base Golf Course. It also derived its name from the streets serving the housing complex (Friar Circle, Lear Street, Sherwood Street, etc.). On Patterson Field and Wright Field, housing and mess halls were racially segregated until President Harry S. Truman ended the practice in 1948.

One additional housing area, known as Skyway Park, was constructed at the intersection of present-day Kauffman Avenue and Colonel Glenn Highway during the summer of 1944. A model wartime village, it provided housing for civilian workers and the families of military personnel. Skyway Park included 546 family-housing units and Skyway Lodge, a dormitory complex complete with its own cafeteria, medical clinic, and housing for 640 individuals. The lodge became home to a large number of single women who worked on the installation.

On October 25, 1947, Skyway Park transferred from the National Housing Agency to the U.S. Air Force and was designated "Area D," Wright-Patterson Air Force Base. By 1957, the complex was no longer needed and some of its buildings were razed, while others were purchased and moved. In 1963, the General Services Administration officially transferred 190 acres of Skyway Park property to the state of Ohio. It now comprises a portion of the Wright State University campus (see Chapter 5: Wright-Patterson and the Miami Valley).

Splinter City was a barracks complex at the north end of Patterson Field across the street from today's Building 22, Area C.

Skyway Park, left, was constructed as housing for Air Service Command's civilian employees. Located at the intersection of present-day Kauffman Avenue and Colonel Glenn Highway, this property is now part of the Wright State University campus.

The extensive physical growth of Patterson Field was accompanied by a significant increase in the military and civilian workforce. By July 1940, civilian employment at FAD rose to 1,013, double the average of the preceding decade. By January 1942, employment had soared to 9,041; it peaked in March of 1943 at 19,433.[92]

The demographics of the workforce also changed significantly. Prior to 1939, women made up less than 10 percent of employees at the depot. Female workers primarily held office positions because they were prohibited from working in storerooms and industrial areas. The critical need for workers in the early months of the war, however, rapidly altered this situation. By

During World War II, women filled jobs in storage and maintenance previously reserved for men.

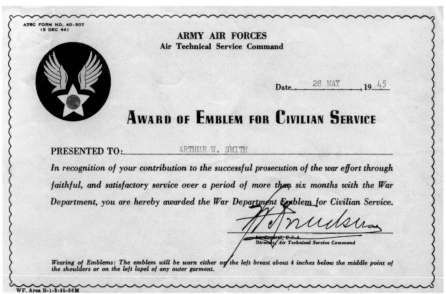

Civilian award certificates were issued upon completion of six months of satisfactory government service.

1942, women worked alongside men in warehouses as storekeepers and later as tug and truck drivers. By 1944, women comprised more than 50 percent of the workforce at the depot.[93] Many of them had been actively recruited to fill essential swing and graveyard shifts. Female employees were recruited in the Dayton area, as well as the hills and hamlets of Appalachia.

By March 1941, to serve the increasing number of employees at the depot, a system of payroll checks replaced cash disbursements. By 1945, the total military and civilian payroll at both Patterson and Wright fields reached $131.5 million, equivalent to 35 percent of Dayton's total industrial payroll.[94]

When wartime operations at Patterson Field shifted to a 24-hour, 7-day-a-week schedule, the nature of civil-service appointments altered as well. Under the former, standard probation system, new employees achieved permanent, civil-service status after a six-month to one-year period. Under war-service appointments, workers were not hired permanently, but were placed on the civil-service payroll "for the duration and six months."[95]

To acknowledge the real contribution that civilian employees made to the war effort, the War Department directed that a special recognition ceremony be held at each military establishment in the nation. At Patterson Field, the ceremony took place on December 8, 1943, and 11,400 civilian employees with at least six consecutive months of satisfactory government service received a special ribbon. The attractive service ribbon featured dark blue fabric

with horizontal silver stripes on which was centered the Army Air Forces insignia.[96]

The work of depot personnel was complicated by prewar systems and procedures not designed to handle the massive increases in materiel activities resulting from the war effort. Existing forms and procedures needed to be analyzed, simplified, and streamlined. A more efficient stock-control system needed to be devised and implemented. Methods had to be perfected for receiving and screening tremendous quantities of materiel from overseas, and the perpetual problem of insufficient warehouse and storage space had to be tackled. Fairfield Air Depot personnel devised and tested many new concepts at the facility. In fact, because of FAD's past history and experience, the Army Air Forces (AAF) often used the installation as a proving ground for new ideas before adopting them for use throughout the supply system. Thus, in a very real sense, FAD pioneered advances that presaged the modern logistics system.

The emergency in the supply field was echoed in maintenance and repair. The war provided a skyrocketing volume of maintenance and repair work linked to the warplane production program. The Engineering Department at Fairfield met this challenge with a like amount of innovation. More complete and detailed work-order records were kept on airplane and engine operations. Work-order and airplane parts-handling procedures were standardized and published. Cost accounting was discontinued and replaced with job control, which recorded man-hour expenditures for the various operations. Even these streamlined procedures required a dramatic

RABBIT PATCH

When the activities of the Army Air Corps expanded beyond the available land at McCook Field, new facilities for research, development, assembly, and testing were built at Wright Field. The Power Plant Section was one of the earliest organizations to receive a new facility, and construction of Building 18 was completed in 1928.

This facility housed a large number of aircraft engine assembly "units" where engineers and technicians designed and fabricated many of the special, experimental components of new aircraft power plants. While McCook Field hosted the nation's first aircraft engine dynamometer (an instrument that measures power or force), by 1931, a new world-class dynamometer was designed and under construction to allow Power Plant engineers the ability to "spin-up" and test any size of aircraft engine within their imagination's grasp.

These new, truly world-unique facilities, however, were limited in terms of the number of engines that could be operated simultaneously, and testing focused on measurements of power plant performance. Over the next 10 years, as the potential of air power in war became increasingly apparent, the need arose to perform long-term testing on numerous engines to evaluate engine durability and the effects of "foreign object damage"—the military term for any form of debris (even birds) that might fly into the vicinity of the engine and hinder its operation. By the late 1930s, a remote area at Wright Field called the "Rabbit Patch" was fully developed to run these tests.

Because simultaneously running a large number of aircraft power plants produced significant noise, the Rabbit Patch was placed in the south-central portion of Wright Field where small, rolling hills provided natural sound barriers. Concrete pads served as engine-stand locations, each with a small, state-of-the-art control room. While dark and dismal by today's standards, these control rooms contained all the instrumentation required to monitor power plant operation and utilized that era's state-of-the-art data collection capabilities.

The Rabbit Patch was in full operation, testing piston-based power plants at the onset of World War II. Power Plant Laboratory engineers utilized the Rabbit Patch for development and testing the "highly modified," Allison-manufactured V-1710 engines used in the P-38 Lightning, the P-40 Warhawk (made famous by the Flying Tigers in the days following Pearl Harbor), and an early version of the P-51 Mustang. Power Plant Lab engineers solved overheating problems that plagued the P-38H and modified new engines with advanced turbosuperchargers to give American pilots improved high-altitude performance. The P-38J also had redesigned Prestone coolant scoops on the tail booms. All P-38Js retained the V-1719-89/91 engines of the P-38Hs, but their more efficient cooling installations increased engine performance from 1,240 to 1,425 horsepower at 27,000 feet. The technical knowledge, skill, and perseverance of the Power Plant Laboratory personnel and their use of the Rabbit Patch test facility provided the United States with the necessary edge in battle in both the Pacific and the Atlantic during World War II.

The Rabbit Patch also was used to test the earliest versions of the revolutionary new aircraft power plant, the turbojet, invented independently by Dr. Hans Von Ohain, a German scientist, and Sir Frank Whittle, an English engineer. The Power Plant Laboratory merged the exciting advancements of both configurations. The J33, the United States' earliest representation of gas turbine technology, became an important wartime engine. In June 1944, the Air Corps' first operational turbojet fighter, the Lockheed P-80 Shooting Star powered by a J33/I-40 engine rated at 4,000 pounds thrust, made its first flight. In 1947, it set a world speed record at 620 mph.

Located in an area near the current Interstate 675 gate in Area B, the Rabbit Patch continued to be used through the early 1950s to test and evaluate aircraft engine technology developed by the Power Plant Laboratory. Shortly after the Page Manor housing complex was constructed in 1953, the facility closed because engine operation, on a 24-hour, 7-day-a-week basis, disturbed the military families who lived there. The control rooms were dismantled, and the engine test stands removed. Today, all that remains of the Rabbit Patch are a few concrete pads, but the contributions it made to the U.S. Air Force will never be forgotten.

The Power Plant Section at Wright Field constructed the Rabbit Patch in a remote location to perform test and evaluation on multiple engines simultaneously. Each engine test stand had a dedicated control room to monitor engine operations during the long-term tests.

The V-1710 piston-driven aircraft engine was tested extensively at the Rabbit Patch. The test shown was conducted in June 1944 on a V-1710-89 version for the P-38J Lightning.

FAD's packing and crating shop. Employees constructed custom wooden boxes for equipment shipments.

Employees of the FAD Supply Department in front of Building 1, 1941

Women employees were recruited by Air Service Command to fill many critical jobs at the Fairfield depot. Here, women work side by side with men overhauling engines on Patterson Field.

increase in personnel to handle the load. By December 1942, the civilian workforce in the engineering shops had increased to 600 workers.[97]

Transportation of supplies and equipment presented another major problem. Railroads and truck lines became so overburdened that supply missions were jeopardized. Rail and motor transportation, formerly under the Quartermaster's control, were assigned to a specialized Transportation Corps under the jurisdiction of the post commander. A new tug pool moved materiel and equipment between the various storerooms and the shipping department at the depot. A truck pool moved supplies between the various depots. Finally, a Traffic Section was established to facilitate more efficient rail service and to supervise the receipt of all commercial rail shipments to Patterson Field.[98]

A summary of the monthly activity report submitted September 30, 1944, by the Supply Division of the Fairfield Air Technical Service Command to Air Technical Service Command headquarters illustrated the scope of the work the depot had capably organized and accomplished. During the month of September, the Supply Division received 28,791 individual requests involving 96,511 different items of stock. These requests resulted in the preparation of 41,060 individual shipping tickets. The materiel involved filled 460 freight cars with 6,165 tons of domestic and 2,350 tons of overseas shipments. During the same period, warehouse personnel received 630 freight cars loaded with 17,622 tons of supplies and equipment, of which 3,187 tons were gasoline and oil shipments.[99]

These engines are carefully wrapped in Pliofilm and crated to survive shipment to other AAF depots and overseas.

The Landis-Shank Cemetery, originally the Hebble Creek Cemetery, was included in the 4,520 acres donated to the government in 1924 for the construction of Wright Field. Around 1941, the eight graves of the Landis and Shank families were moved to accommodate construction of Building 262. The cemetery is currently located off Hebble Creek Road near the tenth hole of the Prairie Trace Golf Course.

Thus, Patterson Field and the Fairfield Air Depot exemplified the drive and enthusiasm that existed throughout the Army Air Forces in the national effort to "Keep 'Em Flying" during the war.

Command Assignments during the War

The story of the Fairfield Air Depot's command assignments from 1941 until its inactivation in 1946 is complex but interesting. Changes in command occurred both prior to, and during, the war with some frequency to reflect organizational changes pursued at the national level. The changes were necessitated by fluctuating wartime needs and continuous efforts to improve materiel management in support of AAF programs. As part of the Air Corps' depot system, Fairfield came under the jurisdiction of the Field Service Section of the Materiel Division from 1926 until 1941. As supply and maintenance requirements intensified, the logistics functions of the Materiel Division separated from those of engineering and procurement. Effective March 15, 1941, logistics became the responsibility of the Provisional Air Corps Maintenance Command, established as a "service test" of the maintenance command system under the direction of the Materiel Division. The Provisional Maintenance Command consisted of its headquarters, the Field Service Section, the 50th Transport Wing, and the six major depots,

including Fairfield. Subordinate to this command, two Provisional Maintenance Group Areas (PMGAs) were established: the Fairfield PMGA and the San Antonio PMGA. The Fairfield PMGA encompassed the Fairfield Air Depot and four sub-depots at Selma, Montgomery, and Maxwell Field in Alabama, and Eglin Field in Florida.

On April 29, 1941, the War Department dropped the designation "Provisional" and officially established the Air Corps Maintenance Command, although it remained under the guidance of the Materiel Division for some time. The new command located its headquarters at Patterson Field. The Field Service Section, however, retained its quarters at Wright Field. The awkwardness of this arrangement soon became apparent and, on June 19, the Maintenance Command offices moved to Wright Field also. On August 28, 1941, the Maintenance Command was instructed to establish sub-depots at all stations under the direct control of the Air Corps. FAD played a large role in this program (see below).

The Air Corps Maintenance Command, however, had been established to perform a specific, limited job—supplying the Air Corps and maintaining its equipment. By the summer of 1941, the nature of the Air Corps had changed with the establishment of the combined Army Air Forces and the rapid expansion of Air Force Combat Command (formerly General Headquarters [GHQ] Air Force). An expanded maintenance organization was needed— one that would assume supply and

maintenance functions for both branches of the Army Air Forces.

In October 1941, Maintenance Command was reorganized as Air Service Command (ASC) to meet this need. New Air Service Command Areas originally followed the boundaries of the numbered Air Forces (i.e. 8th Air Force), but later followed depot control area boundaries. On December 11, 1941, ASC separated from the Materiel Division entirely and became directly responsible to Army Air Forces headquarters in Washington. The Field Service Section remained at Wright Field until the new ASC headquarters building (Building 262) was completed at Patterson Field (Area A). In September 1942, Field Service Section headquarters moved to the new Patterson Field location. Meanwhile, in March 1942, the Materiel Division was elevated to command status. Materiel Command, headquartered temporarily in Washington, retained responsibility for the engineering and procurement work of the Army Air Forces. On April 1, 1943, Materiel Command headquarters returned to Wright Field.

As the war progressed, further changes occurred in Air Service Command. In December 1942, ASC headquarters moved from Washington to the newly constructed Building 262 on Patterson Field. Coinciding with the move, the Field Service Section was officially discontinued. On December 19, Air Service Command disbanded the Air Service Command Areas and activated Air Depot Control Areas. On February 1, 1943, ASC outlined a new organization of 11 Air Depot Control Areas. At that time, the depot at Fairfield was redesignated the Fairfield Air Depot Control Area Command (FADCAC), with responsibility for a nine-state area (Ohio, Iowa, Michigan, Wisconsin, Minnesota, Illinois, Indiana, Missouri, and Kentucky), including 21 storage depots, 16 sub-depots, three servicing detachments, four air-depot detachments, two air-cargo depot detachments, and two overhaul detachments.

Officer strength at FADCAC on February 22, 1943, stood at 1,013, with 41 assigned to headquarters, 197 to the Fairfield Air Depot, 81 to sub-depots, 90 to storage depots, 348 to tactical units, and the remainder to smaller organizations. Offices for FADCAC were established in the Patterson Field headquarters, Building 11, side by side with base offices. To an extent, functions of the FADCAC offices and the base administrative offices intermingled, with depot and base officers

PRISONER OF WAR MURALS

During World War II, more than 400,000 German, Italian, and Japanese prisoners of war (POWs) were brought to the United States and interned in camps throughout the country. Wright Field, while not a major POW camp, housed several hundred prisoners during the war. Available evidence suggests that the prisoners, mostly Germans, worked in warehouses and mess halls and helped maintain the grounds of both Wright and Patterson fields. Building 280, Area A, served as a dining hall for these men. Several of the German POWs who had worked as artists or had artistic abilities requested paints so they could decorate their dining hall with murals. The prisoners created three wall murals. Two of the paintings were lost in the early 1950s when the building was refurbished. One wall mural, however, was preserved in place as a cultural treasure by order of the base commander. These fanciful paintings of creatures from German folklore remain, mute witnesses to a nearly forgotten part of the myriad wartime roles and missions of Wright and Patterson fields.

Portion of the German prisoner of war murals in Building 280

Intersection of Wright Avenue and Pearson Road, 1944. Patterson Field headquarters (Building 11) is at right. Fairfield Air Service Command headquarters (Building 10) is at left, along with the Patterson Field Officers' Dining Room.

performing duties of either headquarters. This situation resulted from a shortage of available officers to staff both offices and the familiarity of most personnel with base operations. A clear and distinct separation of the two offices did not occur until March 8, 1944, with the completion of Building 10 as a separate headquarters for the FADCAC.[100]

Meanwhile, a conference of all control area commanding officers convened at Air Service Command headquarters on March 1-2, 1943, to explain the new ASC organizational structure, how the various ASC divisions were to operate within it, and how the system extended down through the depots and into the sub-depots. The internal reorganization of Air Service Command was completed in slightly more than four months. It had been ably implemented under the direction of Major General Walter H. Frank and his staff at Patterson Field, with special guidance provided by retired Brigadier General Robert E. Wood. General Wood, then president of Sears Roebuck Corporation, assisted General Frank in applying modern, industrial, organizational concepts to the military supply and maintenance system.

As the system of depot control areas was further refined, General Frank recommended that each control area be identified as "Air Service Command," preceded by the name of the town in which the respective headquarters were located. Thus, effective May 17, 1943, the Fairfield depot command was redesignated the Fairfield Air Service Command (FASC).[101] The command's responsibilities remained unchanged.

On August 31, 1944, Materiel Command at Wright Field and Air Service Command at Patterson Field were combined and redesignated Army Air Forces Technical Service Command, once again centralizing control of all logistics and engineering operations. Headquarters for the new command was set up in the former ASC headquarters facility, Building 262.[102]

The name of FASC was changed to the Fairfield Air Technical Service Command (FATSC) on December 6, 1944, reflecting the organizational change. (Other depot control areas also changed from "Air Service Commands" to "Air Technical Service Commands.") Just prior to this name change, on November 27, 1944, FASC headquarters and the Patterson Field base command again merged, so that the commanding general of FATSC also commanded the Patterson Field Army Air Base and the 4100th Army Air Forces Base Unit, activated April 1, 1944. The designation Fairfield Air Technical Service

Air Service Command (ASC) headquarters moved from Washington, D.C., to Building 262, Patterson Field, in December 1942. The ASC emblem, as shown on the command flag, was displayed above the main entrance of the headquarters.

General Harold A. Bartron and staff in front of FATSC headquarters (Building 10), 1945

WRIGHT FIELD AND WORLD WAR II

On April 3, 1939, Congress authorized $300 million for the creation of an Air Corps force of 5,500 military airplanes. Chief of the Air Corps Major General Henry H. Arnold was mandated to build the best air force money could buy. By March 1941, the Air Corps, later a subordinate element of the Army Air Forces, planned for a force of 84 combat groups, 7,800 combat airplanes, and 400,000 personnel. This air power ultimately determined the course of World War II, in both the European and Pacific theaters. The logistical support for this wartime force originated at Wright and Patterson fields. By June 1943, nearly 50,000 men and women, military and civilian, worked at the two fields, laboring day and night to provide the materiel support for an Allied victory. Wright Field's Materiel Division was the "nerve center of air power," the place where the airplanes responsible for Allied victories were conceived and put into production.

Gearing up for the war was a monumental task for the Materiel Division, which had managed its experimental engineering and procurement functions with limited appropriations since 1926. Now central to the rearmament process, the Materiel Division expanded all phases of its activities many months before the rest of the nation, and by 1939, managed more than $110 million in contracts. The division, elevated to command status in March 1942, was enlarged and realigned to sustain the procurement of airplanes in production quantities and, at the same time, provide facilities to sustain a large-scale program of testing and development.

In 1942, the Materiel Division became the Materiel Command. Two major divisions of the Materiel Command, the Production Division and the Engineering Division, conducted the primary work at Wright Field during the war. In the Production Division, skilled engineers and attorneys prepared purchasing contracts at Wright Field for thousands of warplanes and vast quantities of supplementary equipment. At aircraft production facilities in the private sector, technical experts from the Production Division supervised the details of contract administration and helped speed production schedules.

It was the Engineering Division, however, that provided the design work contracted out to manufacturers. Beginning in 1935, the division worked on designs that, less than a decade later, drastically changed world history and introduced a new age of air power. Some of the more significant airplanes conceived and developed at Wright Field for the war included the Douglas A-20 Havoc, a low-level attack bomber; the Vought-Sikorsky R-4, the first full-production helicopter for the American military forces; the four-engine Boeing B-17 Flying Fortress bomber, known as "the plane that carried the war to the enemy's homeland"; the Consolidated-Vultee B-24 Liberator bomber, of which more than 18,000 were manufactured; the Boeing B-29 very heavy bomber, one of the largest wartime

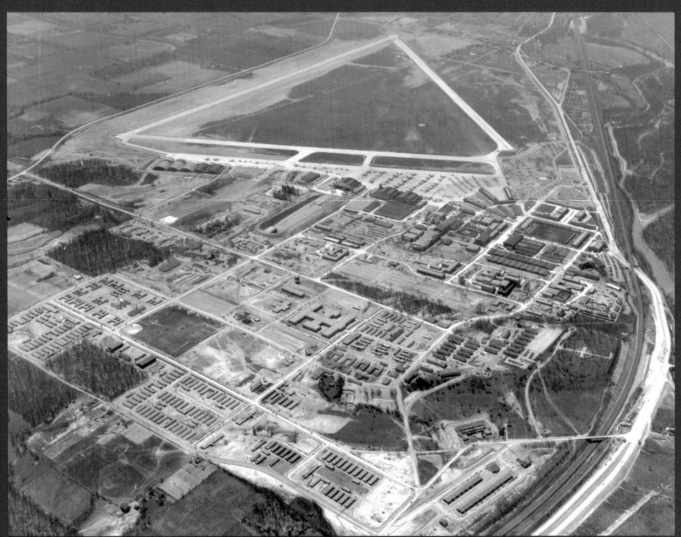

Aerial view of Wright Field in 1944 showing expansion during the war. The size of the field doubled with the purchase of additional acreage surrounding the flightline. The runways were extended and paved in 1941. Additional facilities were constructed for the laboratories, and barracks were built in the hilltop area to house the influx of military personnel.

projects at the field and the airplane that dropped on Japan two atomic bombs, the first and only such weapons used in warfare; the C-45, C-46, C-47, and C-54 transports; the P-35, P-36, and P-39 pursuit airplanes and the Curtiss-Wright P-40 Warhawk, of which 13,738 were produced; Lockheed's P-38 Lightning, which was credited with shooting down more Japanese airplanes than any other fighter in the Pacific; the Republic P-47 Thunderbolt, "the roughest, toughest fighter of the war, with the ability to take a tremendous amount of punishment"; the North American Aviation P-51 Mustang, the best all-around, American-built fighter of World War II; and the Northrop P-61 Black Widow night fighter.

In addition to airplanes, Wright Field engineers developed a number of other products for the war, including nylon fabrics and synthetic silk materials for use in parachutes and corded tires; camouflage paints for warplanes; mobile, field repair outfits; portable steel mats to convert soggy fields into useable runways; and jacks for hoisting bombers off the ground so that tires or landing gear could be serviced.

One historian summarized the Wright Field engineers' efforts:

They don't build airplanes at Wright Field. They don't teach men to fly them. But they dictate the size and shape and number of every plane we use for military purposes. They buy it, follow it into production. They test it to see if it meets with required standards. And they do the same thing with every piece of equipment that pertains to military aviation.

Following the war, in October 1945, Wright Field hosted an Army Air Forces Fair, which showcased the airplanes and technology that helped the Allies win the war. Originally intended as a local weekend event, the fair attracted more than 500,000 visitors from 27 countries in the first two days. With this huge interest, the fair was extended for an additional week. It was a spectacular end to an exhausting but successful war effort by the personnel of Wright and Patterson fields.

Sources: ASC History Office, *Splendid Vision, Unswerving Purpose: Developing Air Power for the United States Air Force during the First Century of Powered Flight* (Wright-Patterson Air Force Base, 2002); Douglas J. Ingells, *They Tamed The Sky* (New York, 1947), p 20.

In addition to developing and testing aircraft and equipment, Wright Field personnel maintained a prisoner of war camp for about 600 German and Italian POWs.

A week-long AAF Fair in October 1945 attracted over 500,000 visitors to Wright Field. In addition to Allied aircraft, captured German and Japanese planes were also on display.

The B-29 bomber program was the largest wartime effort at Wright Field. Pictured is the XB-29 on Wright Field, July 1943.

P-38 at Wright Field

Command remained in effect until the depot's inactivation in January 1946.

Major Organizations

Because Patterson Field was a logistics hub, its subordinate tactical organizations were mostly transport units. The 10th Transport Group, activated May 20, 1937, at Patterson Field, consolidated the 1st Transport Group and the 10th Observation Group. The 10th Transport Group subsequently was reassigned to Wright Field on June 20, 1938, but returned to Patterson on January 16, 1941. The group transferred to General Billy Mitchell Field, Wisconsin, on May 25, 1942. The 10th Transport Group trained at both Patterson and Wright fields with C-27 and C-33 aircraft, and consisted of five subordinate squadrons: the 1st (1937-1943), 2nd (1937-1943), 3rd (1937-1940), 4th (1937-1940), and 5th (1937-1944).[103]

The 1st Provisional Transport Squadron, as mentioned previously, was activated July 15, 1935, at Fairfield Air Depot and assigned to the 10th Transport Group on May 20, 1937. While at Patterson, the squadron flew C-27, C-33, and C-39 aircraft, as well as various civilian and military modifications of the DC-3.[104] The 5th Transport Squadron activated at Patterson Field on October 14, 1933, and operated C-33 and C-39 aircraft.

A second tactical organization located at Patterson was the 63rd Transport Group.

It was constituted November 20, 1940, activated December 1 of the same year at Wright Field, and transferred to Patterson on February 17, 1941. While at Patterson, the 63rd operated C-33, C-34, and C-50 aircraft to transport supplies, materiel, and personnel in the continental United States and in the Caribbean area.[105] On September 9, 1941, the group moved to Brookley Field, Alabama. The 63rd Transport Group had three subordinate squadrons: the 3rd (1940-1944), 6th (1940-1942), and 9th (1940-1943), all of which moved to Alabama in 1941.

For a short period, a third transport group was located at Patterson Field. The 316th Transport Group was constituted February

CBS radio program, "Cheers From the Camps," broadcast live from Patterson Field (*Dick Cull*)

Bob Hope performed before standing-room-only crowds at Patterson Field to bolster morale and entertain the troops. (*Dick Cull*)

Frances Langford performs for the troops at Patterson Field. (*Dick Cull*)

2, 1942, and activated February 14 at Patterson Field, where it was equipped with C-47 aircraft. On June 17, the organization and its subordinate transport squadrons, the 36th, 37th, 38th, 44th, and 45th, were reassigned to Bowman Field, Kentucky.[106]

Four other squadrons activated at and were assigned briefly to Patterson Field as the Air Corps began expanding for World War II. The 11th Transport Squadron, assigned to the 60th Transport Group, activated December 1, 1940; the 13th Transport Squadron, assigned to the 61st Transport Group, activated December 1, 1941; the 19th Bombardment Squadron (Medium), assigned to the 2nd Bombardment Group (Medium) at Mitchel Field, New York, activated February 1, 1940; and the 33rd Bombardment Squadron (Medium), assigned to the 22nd Bombardment Group (Medium), also activated February 1, 1940.[107] The 19th and the 33rd Bombardment Squadrons both flew Douglas B-18 medium bombers.

Military Training Programs

During World War II, Fairfield Air Depot conducted an amazing array of training programs. One of the most impressive military training efforts involved teams of military specialists known as air depot groups, formed and trained at Patterson Field to perform all the functions of a miniature, mobile depot at remote field locations. Other military programs included training for service groups and squadrons, and depot repair squadrons. During 1942 and early 1943, military personnel were recruited from the Fairfield Air Depot Control Area Command (FADCAC) in maximum numbers, frequently approximating 10,000 men, and trained for duty with these combat support functions.[108]

Fairfield Air Depot was responsible for Phase I (activation and unit training) and Phase II (group training). Final or Phase III training normally was conducted at specific, field training stations, although a certain amount of Phase III training did take place at Fairfield. In February 1943, FADCAC assumed training responsibility for the service groups at Lockbourne Army Air Base, Ohio; Kellogg Field, Michigan; and Baer Field, Indiana. The special staff at Fairfield also supervised the instruction of other arms and services units.

Two special efforts of the Fairfield

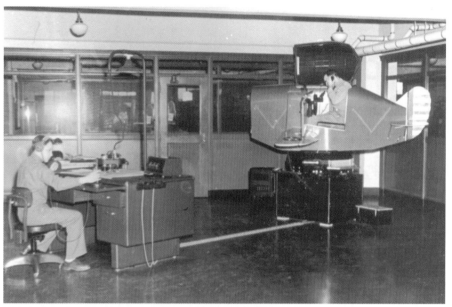
Link trainer in use at base operations, Building 206

Aero Repair Training School held in Building 1084 of Fairfield Air Service Command, Patterson Field

Women's Army Corps officers assigned to the WAC Supply School at Patterson Field, January 1944. The women were all classified as lieutenants. H. T. Sears and David A. Mote instructed the class.

ORGANIZATIONS AND UNITS TRAINED BY FAD PERSONNEL

The available list of organizations and units trained by Fairfield Air Depot personnel during World War II included:

Headquarters 2054th Ordnance Company (Aviation)	Patterson Field
1157th Signal Depot Company (Aviation)*	Springfield, Illinois
912th Engineer Headquarters Company	Patterson Field
Detachment 905th Quartermaster (Aviation)	Patterson Field
Detachment 859th Signal Service Company (Aviation)	Patterson Field
838th AAF Specialized Depot Detachment	Marion, Ohio
555th Service Squadron*	Springfield, Illinois
315th Depot Repair Squadron	Patterson Field
97th Depot Supply Squadron	Patterson Field
96th Service Group**	Oscoda, Michigan
88th Depot Repair Squadron	Patterson Field
85th Depot Repair Squadron	Patterson Field
55th Air Depot Group	Patterson Field
18th Medical Supply Platoon	Patterson Field
1st Mobile Rubber Repair Detachment	Patterson Field
18th Air Depot Group	Patterson Field
407th Service Squadron*	Patterson Field
345th Aviation Squadron**	Patterson Field
1916th Quartermaster Truck Company (Aviation)**	Patterson Field
2007th Quartermaster Truck Company (Aviation)**	Patterson Field

* Chinese-American squadrons
** African-American squadrons

Source: *History of Fairfield Air Depot Control Area Command and Fairfield Air Service Command, 1 February 1943 – 1 October 1944* (Fairfield Air Service Command, Patterson Field, Ohio), p 125.

Members of the 407th Service Squadron, Chinese-Americans from around the United States, received their technical training from Fairfield Air Depot personnel at Patterson Field in 1943 and 1944.

Air depot repair squadrons were trained at Patterson Field to perform all functions of a mobile, miniature depot at remote field locations.

training program resulted in the preparation of the 96th Service Group, consisting of black officers and enlisted personnel, and the training of three Chinese-American service squadrons. The 96th was the only such service group trained at Fairfield.[109] In another special effort at Patterson Field, Army Air Forces Nurses Training Detachment No. 6 activated on November 10, 1943, to train nurse recruits.[110]

Special training facilities for officers at FAD included the U.S. Army Materiel Division Supply School, established in October 1940 at the direction of the Materiel Division chief, and the Engineering Maintenance Officers' Training School, established July 1942 at the request of Air Service Command.[111] The supply school also conducted training of enlisted personnel from various supply squadrons and service groups.

The Civilian Club on Patterson Field (Building 274, Area A) opened on December 1, 1944, to the strains of Lawrence Welk and his orchestra. The club continued to be used as a recreation facility until 1979, when it was converted to a conference center.

Civilian Training Programs

Equally important at Fairfield was the training of civilian depot workers to support operations. Prior to 1939, little need for training programs for new workers existed because skilled employees remained readily available in the Dayton area. When job training was necessary, it was handled on an individual basis. Wartime expansion demanded that an ever-increasing number of workers be trained for jobs requiring a multitude of technical and mechanical skills. By February 1943, the civilian employment office at Patterson Field hired 300 new employees per day.[112] FAD was responsible for creating training programs not only for its own employees, but for thousands of employees scheduled to work in many sub-depot supply and engineering departments activated across the country.

A shops training school was activated in June 1941 (replacing the older aircraft training unit) to provide instruction in all phases of aero repair, engine overhaul, machine-shop operations, sheet-metal manufacturing, and welding.[113] Instructors at the school were often challenged to create innovative programs to compensate for the lack of current textbooks. Engines, starters, and generators built from rejected parts were used for demonstration purposes. A visual aids unit provided a variety of technical films for instructional purposes.[114]

In the Supply Division, civilian workers received job-related training beginning in

Civilian workers in the Supply Division at Fairfield Air Service Command

March 1941.[115] Classroom instruction was given in warehousing, stock records, inspection, and shipping and receiving.

Although originally separate, all training activities conducted by the Engineering and Supply departments eventually combined under a separate post training department, and later the Personnel and Training Division. A special curriculum unit assumed responsibility for issuing study guides, manuals, worksheets, and other instructional materials for use in the Fairfield training programs. Within a short time, artists, photographic illustrators, draftsmen, veritypists, and other professionals augmented the curriculum writers, and together produced a number of quality training publications. Many of these manuals received wide distribution in other air depots in the United States and served as the basis for national standardization of Air Service Command manuals.[116]

As classes grew larger, the makeshift classrooms at FAD became inadequate. For a short time, education of civilian supply personnel moved to Osborn High School during after-school hours and during the summer of 1943. In September 1943, the supply school classes moved to new buildings in Wood City (Buildings 50, 1044, 1045, and 1046).[117] The move ensured that more space and permanent, fixed equipment were available to accommodate larger supply and maintenance classes. Engineering Department courses took place in a special building erected as a theater and auditorium in Area C. This building remained in use until 1943, when the engineering training school also moved to new quarters in Wood City.[118]

Recruiting for the depot training programs at Fairfield remained a continual challenge. In order to ensure a high number of qualified trainees at Fairfield and other depots, Air Service Command initiated pre-induction or pre-service training programs at various high schools and junior colleges. Patterson Field cooperated with the Ohio State Board of Vocational Education and local boards of education in Ohio to establish this pre-service instruction. The first course started at Springfield High School on March 1, 1943, with a class of 22 senior boys in a program of aircraft engine repair. Other Ohio centers offering pre-service training in engine repair were set up at Norwood, Portsmouth, Zanesville, Washington Court House, Cambridge, Franklin, Ironton, and Hamilton. In addition, Portsmouth and Zanesville offered courses in aircraft repair. During

Enlisted mechanics in training at Patterson Field were subsequently transferred to other Army Air Forces depots in the United States and abroad.

the spring of 1944, 12 classes had an enrollment of 190 students, and by June 20, the graduates entered the armed services or accepted employment at Patterson Field.[119]

Additional off-reservation schools were instituted in October of 1943 as the local employment situation became acute. In cooperation with the State Board of Vocational Education, training centers were established at nine selected schools in Ohio and four in Kentucky. Machine-shop and sheet-metal courses were offered in four-week sessions. A two-week engine course also was offered. The off-reservation schools soon supplied trainees to the civilian training branch at the rate of approximately 75 new workers per week. The Maintenance Division at Fairfield depended almost wholly upon this source for personnel replacements.[120]

A 1944 report to the depot's commanding officer revealed that, for the period May 1, 1943, to May 1, 1944, $676,065 had been invested by the government in the civilian training program at Patterson Field, including the off-reservation training mentioned above and a radio-mechanics training program conducted in part under contract at the University of Wisconsin.[121]

Assisting with the Creation of New Depots

As the American air fleet grew, decentralization of repair and maintenance facilities became imperative. FAD, as one of the oldest, permanent repair depots in the country, was called upon to apply lessons learned from its many years of experience to support this program. Providing its expertise in the establishment, layout, and manning of new depots and sub-depots became one of FAD's most important missions during the war.

The Provisional Air Corps Maintenance Command, with headquarters at Patterson Field, coordinated the organization of new depots in the spring of 1941. Service and training at depots in the eastern part of the country were the responsibility of the Fairfield Provisional Maintenance Group. Key personnel for new sub-depots were recruited from among the engineering and supply employees at FAD. FAD established 41 sub-depots by September 1942, with 21 remaining under its control.[122]

Once new depots were established, the crucial need for trained, experienced personnel had to be met immediately. FAD

SUB-DEPOTS UNDER FAIRFIELD AIR DEPOT CONTROL—1942

OFFICIAL NAME	OFFICIAL ADDRESS	DATE ACTIVATED
Alliance Sub-Depot	Alliance, Nebraska	September 2, 1942
Baer Sub-Depot	Fort Wayne, Indiana	March 21, 1942
Berry Sub-Depot	Nashville, Tennessee	January 10, 1942
Bowman Sub-Depot	Louisville, Kentucky	March 21, 1942
Chanute Sub-Depot	Rantoul, Illinois	September 16, 1942
Coffeyville Sub-Depot	Coffeyville, Kansas	August 2, 1942
George Sub-Depot	Lawrenceville, Illinois	June 16, 1942
Kellogg Sub-Depot	Battle Creek, Michigan	September 1, 1942
Lockbourne Sub-Depot	Lockbourne, Ohio	July 2, 1942
Madison Sub-Depot	Madison, Wisconsin	June 19, 1942
Rapid City Sub-Depot	Rapid City, South Dakota	July 20, 1942
Salina Sub-Depot	Salina, Kansas	September 7, 1942
Scott Sub-Depot	Belleville, Illinois	August 15, 1941
Sedalia Sub-Depot	Sedalia, Missouri	September 2, 1942
Selfridge Sub-Depot	Mt. Clemens, Michigan	March 12, 1942
Shaw Sub-Depot	Sumter, South Carolina	October 1, 1941
Sioux City Sub-Depot	Sioux City, Iowa	August 1, 1942
Sioux Falls Sub-Depot	Sioux Falls, South Dakota	June 19, 1942
Smyrna Sub-Depot	Smyrna, Tennessee	May 15, 1942
Topeka Sub-Depot	Topeka, Kansas	August 1, 1942
Wayne County Sub-Depot	Romulus, Michigan	January 10, 1942

Source: *History of the Air Depot at Fairfield, Ohio, 1917-1943* (Fairfield Air Service Command, Patterson Field, 1944), Exhibit 64.

instructors trained these depot employees both at Fairfield and on site. FAD-trained personnel often remained at these depots as key supervisors.

Everything was rush, both at Fairfield and at the sub-depots as each was established. Thousands of new employees were being hired and all departments of the Fairfield Air Depot were being rapidly expanded. The relative handful of trained and qualified employees were urgently needed in each department to train and supervise hundreds of new workers…Gradually however the training program at Fairfield produced sufficient trained personnel to ease the situation.[123]

During 1942 and 1943, emphasis was given to the concept of specialized depots, with each depot assigned one or more classes of spare parts, equipment, and supplies for specified aircraft. By the end of 1943, 68 specialized depots had become the backbone of the Air Corps' supply system.[124]

PATTERSON FIELD AND THE END OF THE WAR

The small air depot at Fairfield, Ohio, opened in 1918 as the Fairfield Aviation General Supply Depot to serve World War I Signal Corps aviation schools. By 1945, it had greatly expanded, ably serving the needs of the U.S. Army Air Forces stationed around the world. Its host, Patterson Field, also expanded from a small, closely knit installation to a major logistics center and Midwest hub of World War II activities.

In the months following the war, Patterson Field became an active separation center for military personnel. In September 1945, the 4265th AAF Base Unit Separation Center activated under the command of Colonel Richard Gimbel. The Separation Center was located in buildings formerly used for civilian training at Wood City. In the early weeks, up to 150 men were separated daily, with the processing period averaging nearly 36 hours per man. By the end of December, the 4265th accelerated the separation process to nearly 1,000 men per day. As of November 13, 1945, records indicate that 14,675 enlisted men and 3,508 officers had been processed. In total, the 4265th separated more than 35,000 men.

The Separation Center was established primarily to discharge officers and enlisted men arriving from other Army Air Forces posts, but did everything possible to expedite the discharge of men stationed at Patterson Field. Initially, members of the Women's Army Corps (WAC) were not processed at Patterson, but were sent to the nearest service and ground force point of separation. Base newspapers indicated, however, that WACs eventually separated from the Army Air Forces at Patterson as well.[125]

Depot functions at Patterson Field underwent substantial reductions, and were eventually discontinued entirely in the months following V-J Day (August 15, 1945). Postwar reorganization called for supply and maintenance depot functions to be concentrated in selected air materiel areas (the successors to Air Technical Service Command control areas). Fairfield was one of the depots selected for inactivation, and the long history of the air depot at Fairfield thus came to an end. On the first of January 1946, Fairfield Air Technical Service Command officially inactivated and its functions were reassigned to other air materiel areas. Most ATSC personnel transferred to other agencies on Wright and Patterson fields or to other air materiel areas.

The physical appearance of Patterson Field also changed as the base settled into its postwar mission. Experience gained during the war emphasized the importance of coordinated planning. From 1943 on, construction at Wright and Patterson fields, as well as land acquisition, had been handled through coordinated planning boards. Master plans were drawn up by city planners of national reputation, approved by the Army Air Forces, and executed under the direction of local installation planning boards. In 1945, the master plans for both Wright and Patterson fields were integrated into one master plan for the Wright-Patterson complex.[126]

One of the major projects of the new master plan was the construction of the Very Heavy Bomber (VHB) runway at Patterson Field in 1946-1947. Initially popularized as "the B-36 runway," it was designed to service the very heavy bombers and jet-powered aircraft anticipated in the postwar period. It was 8,000 feet long and 300 feet wide, with an additional overrun of 1,000 feet at each end, and a load capacity of 300,000 pounds per square foot. Patterson Field, rather than Wright Field, was chosen as the site of the VHB runway because of the more flexible limits on its military reservation boundaries, and because the existing topography was more adaptable to the physical proportions of a VHB runway.[127]

By late 1945, Wright Field and Patterson Field had begun to merge functions and services. The two fields, along with two satellite organizations, were consolidated into the Army Air Forces Technical Base, Dayton, Ohio, under Air Technical Service Command headquartered in Building 262. On December 15, 1945, Brigadier General Joseph T. Morris, previously the Wright Field commander, took command of the new administrative organization.

On March 9, 1946, Air Technical Service Command was renamed Air Materiel Command (AMC). Headquarters for the command remained in Building 262. On December 9, 1947, the AAF Technical Base, Dayton, Ohio, was redesignated Air Force Technical Base, reflecting the independent status of the U.S. Air Force. One month later, on January 13, 1948, the installation's designation was finalized as Wright-Patterson Air Force Base, culminating more than 30 years of development. Air Materiel Command (predecessor of today's Air Force Materiel Command) remained the parent command for both Wright and Patterson fields until their merger, and subsequently assumed command authority over Wright-Patterson Air Force Base.

Colonel Joseph T. Morris (seated), commander of the Air Force Technical Base and former Wright Field commander

Construction of the Very Heavy Bomber (VHB) runway at Patterson Field, 1947

Patterson Field, 1947, prior to its merger with Wright Field to form Wright-Patterson Air Force Base. Another merger was also eminent—the adjacent towns of Fairfield and Osborn merged to become the city of Fairborn in 1950.

BRIDGE OF WINGS

While commanding Air Service Command, Major General Walter H. Frank asked a 34-year-old private assigned to the art department at Wright Field to paint a mural for his new headquarters, Building 262. Stuyvesant Van Veen, already a noted artist, had painted a major mural for the Hall of Pharmacy at the 1939 New York World's Fair. He had studied at several art schools, including the National Academy of Design and the Art Student's League, and his instructors included famed American artist Thomas Hart Benton.

Bridge of Wings became Van Veen's 23rd mural. The 8-by-40-foot canvas was painted on the west wall of the second floor lobby in Building 262, now Gillmore Hall. Working between his military duties, Van Veen took more than a year to complete the mural, although he spent only six months actually painting. During that time, he rose in rank from private to sergeant.

The mural captures Van Veen's postwar vision of the Western Hemisphere united through air power. It contains a series of images representing the flow of culture and commerce between North and South America fashioned in a blend of social realism and allegory. "I had as great a freedom of expression and creativeness in executing the mural," Van Veen later wrote, "as in any of my more than thirty other mural works." Van Veen sketched his images from live models and incorporated several elements that were controversial for the period. Among these were images of the emancipated American woman and a female scientist (the latter being a portrait of Van Veen's wife). They portrayed women as independent, professional, and equal to their male counterparts. Equally noteworthy were two musicians—a white bass fiddler and a black saxophonist. These metaphors represented the artist's belief that the two races could live and work together in harmony. Van Veen added a personal note by painting himself as the bass player.

Several artists assisted Van Veen with the design and execution of the mural. Corporal Eva Mirabal, a member of the Pueblo Nation from Taos, New Mexico, was an artist in her own right and creator of the wartime cartoon strip "G.I. Gertie." Corporal Edward "Red" Sachs from San Francisco also worked on the project.

Since the artwork had to set for six months before being coated with a permanent preservative, the artist placed only a temporary buttermilk preservative on it. Unfortunately, Van Veen was reassigned during the waiting period. He returned in 1956 as a contractor to complete the task. He also repaired the area around the coffee plant that had been damaged by fire. "When I painted that, I had never seen a coffee plant," he later explained. "It wasn't too realistic."

Wright Field, a second Van Veen mural, was installed in Building 126 on Wright Field until removed during renovation and placed in safe storage in 1998. This 6-by-15-foot canvas portrayed the field's World War II flightline complemented by an allegorical figure of "Flight" rising up to the sky.

While at Wright Field, Van Veen won the 1945 Ohio Valley Exhibition and the 1945 Wright Field Army Art competitions. After the war he worked as supervisor of the Art Academy at the Cincinnati Art Museum and, from 1949 to 1974, taught art at the City College of New York. His works reside in many museums and permanent collections including the Cincinnati Historical and Philosophical Society, Ohio State University, and the Lincoln Center of Performing Arts. In 1972, Van Veen received one of the highest honors conferred on living American artists, composers, and writers, when he was elected to the American Academy and Institute of Arts and Letters. He died in 1988 at the age of 77.

Van Veen's two cultural legacies from World War II were formally accessioned to the Air Force Art Collection in 1998.

Bridge of Wings

Bridge of Wings mural detail

Wright Field

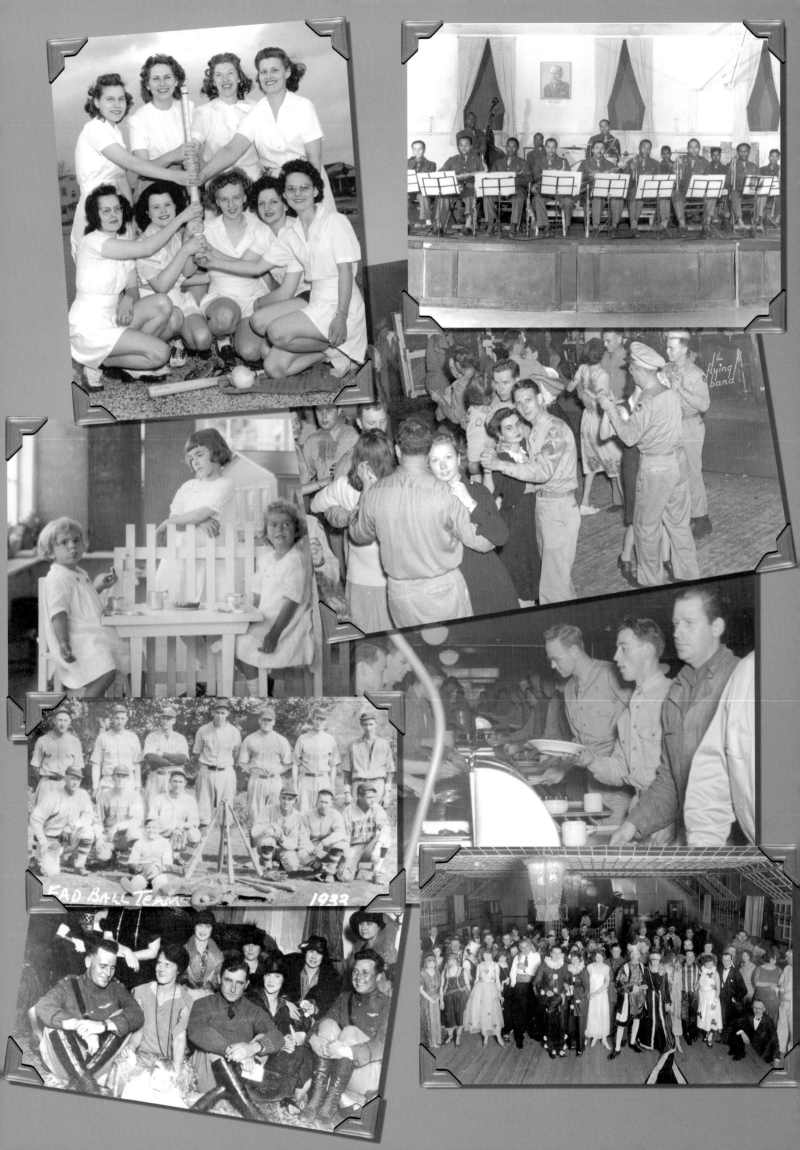

E & D BALL TEAM 1932

Chapter Four

FULCRUM OF BASE SUPPORT

WRIGHT-PATTERSON ENTERS THE COLD WAR

Wright-Patterson Air Force Base came into existence on January 13, 1948, when the newly created U.S. Air Force officially merged Wright Field and Patterson Field into a single installation.[1] The new "base" designation conformed with an Air Force policy of renaming its numerous fields as bases. This action also ended a long succession of designations applied to the Dayton installations. Wright-Patterson was assigned to Headquarters Air Materiel Command (AMC).

To facilitate daily management, refinements were made to the designations of the combined installation's various

Wright-Patterson Air Force Base, 1950

internal components. The south end of Patterson Field, including Headquarters AMC and the officers' Brick Quarters became Area A of Wright-Patterson Air Force Base (since 1944 it had been Area A of Wright Field). The former Wright Field became Area B, Wright-Patterson (it had been Area B of Wright Field). The north end of Patterson Field, including the former Fairfield Air Depot and Wilbur Wright Field, became Area C. Finally, Skyway Park, located across Kauffman Avenue and encompassing the area that is now part of Wright State University, became Area D. In April 1952, the installation's address officially changed from Wright-Patterson Air Force Base, Dayton, Ohio, to Wright-Patterson Air Force Base, Ohio.

Shortly after receiving its new designation, the installation displayed both its pride and its success as "crossroads of the Air Force" at a daylong Air Force Day celebration held on

September 18, 1948. To commemorate the first birthday of the autonomous U.S. Air Force, the base highlighted recent technological advances in weapons and equipment on the Area B flightline. More than 150,000 spectators attended the celebration, featuring Strategic Air Command's new Consolidated-Vultee B-36 six-engine bomber (unofficially called the Peacemaker) on central display. Attendees also enjoyed flight demonstrations by AMC's Flight Test Division and the Ohio National Guard's 162nd Fighter Squadron. A six-foot birthday cake was served in celebration of the Air Force's first anniversary.[2] The Air Force Day open house conveyed the message that AMC, through its Wright-Patterson Air Force Base offices, laboratories, and personnel stationed around the world, played a key role in making the Air Force a valuable instrument for international peace. The 1948 Air Force Day evolved into an annual

Crowds lined the apron on the Area B flightline during the gala Air Force Day celebration held September 18, 1948. The base celebrated the Air Force's first anniversary with a six-foot birthday cake.

General Joseph T. Morris, popularly recognized as the "father of Wright-Patterson Air Force Base"

CHRONOLOGY OF WRIGHT-PATTERSON AIR FORCE BASE AND ITS ANTECEDENTS

Installation	Date Established
Wilbur Wright Field	June 6, 1917
McCook Field	October 13, 1917
Fairfield Aviation General Supply Depot	January 4, 1918
Wilbur Wright Air Service Depot	January 10, 1919
Aviation General Supply Depot, Fairfield, Ohio	November 3, 1919
Air Service Supply and Repair Depot, Fairfield	September 20, 1920
Fairfield Air Intermediate Depot	January 14, 1921
Fairfield Air Depot Reservation	June 21, 1927
Wright Field	October 12, 1927
Patterson Field	July 1, 1931
Army Air Forces Technical Base, Dayton, Ohio	December 15, 1945
Air Force Technical Base, Dayton, Ohio	December 9, 1947
Wright-Patterson Air Force Base	January 13, 1948

Armed Forces Day held in May. Over the years, participants included the Army, Navy, Civil Air Patrol, Ohio National Guard, Civil Defense Organization, and Air National Guard.

Overseeing Wright-Patterson Air Force Base during its earliest years was Brigadier General Joseph T. Morris, popularly recognized as "father of Wright-Patterson

Air Force Base." General Morris, former commander of Wright Field and subsequently of Air Force Technical Base, continued to serve Wright-Patterson in a leadership capacity until March 1952. He served not only as base commander, but also as wing commander of the 4000th Air Force Base Unit, which had supported Patterson and Wright fields prior to the

merger. (Patterson Field's 4100th Army Air Forces Base Unit merged with Wright Field's 4000th on February 21, 1946.) On August 28, 1948, the 4000th and its subordinate units were redesignated as Headquarters and Headquarters Squadron 2750th Air Force Base.[3] The following year, on October 4, 1949, the 2750th Air Force Base was redesignated as Headquarters and

BRIGADIER GENERAL JOSEPH T. MORRIS—
"FATHER OF WRIGHT-PATTERSON AIR FORCE BASE"

Joseph Theodore Morris was born in Punxsutawney, Pennsylvania, on April 17, 1894. He graduated from Pennsylvania State College in 1917 with a bachelor of science degree in electrical engineering. He also later received a master of science degree from Yale University and an honorary doctorate of science in military aeronautics from Bowling Green State University.

Enlisting in February 1918, Morris was commissioned June 13, 1918, as a second lieutenant in the Air Service and became a radio officer with the First Provisional Wing at Mineola, Long Island, New York. He served as a communications and maintenance officer at a variety of posts before arriving at Wright Field in 1931 to attend the Air Corps Engineering School. He graduated the following June and reported to Rockwell Air Depot, California, where he served until 1940.

General Morris returned to Dayton in 1941, as assistant chief of the Maintenance Division, Air Service Command, at Patterson Field. He served in various positions at both Wright and Patterson fields until July 1943, when he became commander of the 8th Air Force Service Command in England. Morris also served as chief of Maintenance for the U.S. Strategic Air Forces in the European theater and commander of the 12th Air Force Service Command in Italy before returning stateside in late 1945.

After the war ended in Europe, Morris returned to Ohio once more, this time as commander of Wright Field. He served subsequently as commanding officer of the Army Air Forces Technical Base, Dayton, Ohio (November 1945 to December 1947), and

"Uncle Joe" Morris at work in his 2750th Air Base Wing headquarters

of the Air Force Technical Base (December 1947 to January 1948). Once Wright and Patterson fields merged, Morris served as the first commander of Wright-Patterson Air Force Base from January 13, 1948, to March 28, 1952.

A master planner with an impressive breadth of vision, General Morris skillfully guided the installation from wartime to peacetime, managed the transfer of its property and facilities from the Army to the Air Force, directed the racial integration of the base, and oversaw base operations support for the Berlin Airlift and the Korean War. His brilliant and sensitive stewardship earned him the nickname "Uncle Joe" and recognition as the "father of Wright-Patterson Air Force Base."

Brigadier General Morris left Wright-Patterson in March 1952 to become commander of the Spokane Air Depot, Washington. He finished his career with a temporary assignment to the United Nations Command truce team in Korea.

Upon his retirement on July 31, 1953, General Morris returned to Dayton as vice president of United Aircraft Products, Inc., until 1959. The general resided in Fairborn until his death, on May 21, 1980, in the U.S. Air Force Medical Center, Wright-Patterson. He was buried at Arlington National Cemetery on May 27, 1980. In his honor, the air base wing headquarters, Building 10, Area C, was officially dedicated as Morris Hall on August 28, 1981.

Colonel Morris inspects the Wright-Patterson Air Force Base Police Department, April 1948.

Headquarters Squadron 2750th Air Base Wing. The "Headquarters" and "Headquarters Squadron" were subsequently deleted, leading to the designation of Wright-Patterson's base operating unit as the 2750th Air Base Wing. (The wing was redesignated the 645th Air Base Wing in 1992 and then became the 88th Air Base Wing in 1994.)

The 2750th Air Base Wing (ABW) provided services (civil engineering, airfield operations, transportation, logistics, billeting, personnel management, finance, recreational services, etc.) to many associate/tenant organizations located at Wright-Patterson and limited services to other government agencies and components of the Department of Defense located off base. This support encompassed the operation and maintenance of the base's airfields, aircraft, buildings and grounds, communications systems, automotive equipment, and supply and medical facilities, as well as housing, messing, and training for military personnel. The Directorate of Base Air Installations, an essential part of this support system, handled a multi-million-dollar housekeeping operation comprising the care and maintenance of the base's many buildings, surfaced areas, railroads, and utility plants. It monitored the work of numerous private contractors, as well as personnel attached to a resident office of the Army Corps of Engineers.

Wright-Patterson Air Force Base's unification coincided with the onset of the Cold War. The Berlin Airlift, Korean War, and other Cold War activities directly impacted operations at the base. Operation Vittles, more popularly known as the Berlin Airlift, was the first event of global import to pose a significant challenge to the effectiveness of Air Materiel Command. On June 22, 1948, a Soviet-imposed rail and highway blockade of West Berlin isolated American, British, and French occupied zones of the city, including more than two million German citizens. The United States' response to the crisis demonstrated to the world the flexibility of the Air Force and called attention to the intense logistics planning necessary to maintain a long-term, aerial supply line. Over a 15-month period, from June 26, 1948, to September 30, 1949, Operation Vittles airlifted more than two million tons of food, fuel, and supplies into West Berlin.

Headquarters AMC at Wright-Patterson provided parts and supplies necessary to maintain the Air Force fleet involved in the airlift. The only transport airplanes available to the Air Force overseas at the

Four-engine C-54 transports (shown) and twin-engine C-47s maintained the round-the-clock supply line into Berlin.

Headquarters Air Materiel Command provided aircraft parts and supplies for airplanes involved in Operation Vittles (popularly known as the Berlin Airlift) during 1948-1949. Here supplies are loaded aboard a C-54 Skymaster.

beginning of Operation Vittles were twin-engine, Douglas C-47 Skytrains. On June 27, AMC began shipping parts and supplies to Germany for four squadrons of larger and faster four-engine, Douglas C-54 Skymaster aircraft. AMC first supervised the transfer of C-54 transports to Germany and then ensured sufficient spare parts were shipped to Europe to keep the airplanes in operation.

As the blockade continued, Headquarters AMC established a priority requisition system with the airlift nerve center at Rhein-Main Air Base in West Germany. Daily cables received at Wright-Patterson, many as long as 20 teletyped

pages with 20 items to the page, listed parts needed to keep the cargo transports airworthy. Headquarters AMC disseminated high-priority orders to its various depots, which then immediately filled requisitions for air shipment to Germany. With the support of Headquarters AMC personnel, the airlift's round-the-clock schedule remained uninterrupted by maintenance or parts delays.[4]

Although the Soviet surface blockade ended in May 1949, the airlift continued for an additional four months to allow reserve stocks in Berlin to reach satisfactory levels and ensure that the international political situation had stabilized. By the time of the final flight in September 1949, Wright-Patterson Air Force Base had laid a firm foundation for its enduring role in the history of the modern United States Air Force. During the year and a half of this first event of the Cold War, installation manpower, which had reached a postwar low of 21,000 in 1947, rose to 25,000.

Not long after the end of the Berlin Airlift, on June 27, 1950, the United States entered combat in Korea. Although better prepared for this conflict than it had been for World War II, the Air Force still found most of its 20,000 aircraft were of World War II vintage, and a significant percentage of these, especially combat aircraft, were in storage. Congressional emphasis over the previous three years had been on reducing military expenses and, accordingly, acquisition of new airplanes came very low on the list of priorities. Although the Far East Air Forces (FEAF) in Korea and Japan had a variety of fighters and bombers in its combat fleet, only one model, the Lockheed F-80 Shooting Star, was a jet.

XB-45 (top) and XF-86, two jet aircraft developed by engineers at Wright Field in the late 1940s. The North American F-86 Sabre became the first successful American swept-wing jet fighter, earning renown in combat against Russian MiG-15s in Korea. Three North American RB-45C's also arrived in Korea in late 1950 for combat suitability tests but were never officially committed to the war.

Area C flightline, June 1953, located on the site of the earliest military flying field in the Dayton area, Wilbur Wright Field

Support aspects of the conflict in Korea rested heavily on AMC at Wright-Patterson. To meet immediate demands, AMC's depots overhauled and modified aircraft in storage and sent them into action. By November 1950, five months after the start of the war, the depots had modified and reconditioned more than 400 aircraft for use in the Far East. What the Air Force really needed, however, were thousands of aircraft incorporating the latest developments in technology delivered to the front as fast as possible. Congress appropriated $10 billion to buy new airplanes, and AMC set goals for the industrial effort, scheduled output, and evaluated the impact of aircraft programs on basic national resources. In December 1950, AMC became the sole procuring agency within the Department of the Air Force. Under the impetus of wartime support and rearmament, its role became mammoth. AMC's workforce grew from 93,600 in 1948 to 137,000 in 1951, reaching 224,000 by the late 1950s. Fiscal year expenditures during the Korean conflict further reflected the extent of the overall logistics mission: $1.7 billion in 1950; $3.6 billion in 1951; $8.1 billion in 1952; and $10.5 billion in 1953.[5] As the result of significant efforts at Wright-Patterson, new airplanes at the front when the Korean conflict ended on July 27, 1953, included the Republic F-84 Thunderjet and the North American F-86 Sabre, both capable of engaging and defeating Communist MiG aircraft in aerial combat.

In addition to the procurement of new aircraft, AMC ably managed the flow of supplies to Korea, a difficult assignment under the best of circumstances. The long pipeline to the Far East called for extraordinary efforts in maintaining an effective and reliable logistical support system. As the Korean War continued, Air Force activities, including those at Wright-Patterson, expanded on a global basis to meet wartime demands. As the pace of base functions escalated, the base commander faced numerous quality-of-life issues. New organizations added to the base significantly increased the military and civilian workforce. Manpower surged to 34,000 in 1951 before beginning a decade-long decline to 25,000 in 1959. Additional personnel compounded the chronic family-housing shortage the base had faced throughout the 1940s, and the growing volume of automobile traffic forced the base to institute staggered work shifts to relieve rush-hour congestion.

Wartime activities affected all aspects of base operations. Due to increased flying missions, the Ground Controlled Approach (GCA) on the Area C flightline accelerated to 24-hour operations. The Air Reserve Training Branch, established to continue military training of reserve officers in civilian life, was inactivated (its activities having ceased almost entirely as reserve officers were recalled to active service). The Wright-Patterson hospital received its first combat casualties in October 1950 and continued to treat wounded personnel throughout the conflict. Furthermore, the American Red Cross designated the hospital as a special, blood-collection center to meet the increased need for blood products. The base chaplain's office even sponsored a clothing drive for the aid of Korean war victims.[6]

In early 1952, in the midst of the Korean War, General Edwin W. Rawlings, the AMC commander, initiated the process of decentralization at AMC. Under his direction, Headquarters AMC's primary mission shifted from operations to program management. Field commanders at various air materiel areas (the depots) became responsible for selected aspects of supply, maintenance, and procurement. The respective depots also specialized in handling specific commodities and received exclusive responsibility for computing requirements, purchasing, receiving, storing, shipping, and maintaining particular items assigned to them. Once relieved of the voluminous load of paperwork these processes involved, headquarters concentrated on the most important phase of procurement—purchasing complex and expensive aerial weapon systems and their supporting subsystems.

As Wright-Patterson's base support organization, the 2750th Air Base Wing assisted in moving many AMC divisions to outlying areas. Although the workload of each wing component was affected in some manner by this major effort, the

The introduction of staggered work schedules eased traffic congestion surrounding the base in the early 1950s.

Plans for the final organization of Air Research and Development Command (ARDC) were made at Wright-Patterson in early 1951. Reviewing the new organizational structure are Dr. M. J. Kelly (left); Brigadier General Ralph P. Swofford, Jr., deputy commanding general for operations, ARDC (standing, left); Major General David M. Schlatter, commanding general, ARDC (seated, center); and Mr. Donald A. Quarles. Kelly and Quarles, civilian scientists and vice presidents of Bell Telephone Laboratories, served as advisors to General Schlatter.

Civilian Personnel Branch carried the heaviest load. It processed all paperwork involved in the transfer of employees and their positions to 15 AMC installations throughout the United States. The Employee Utilization Section dealt with special problems in connection with the decentralization program. An Out-Service Placement unit assisted people unable to transfer with their positions and who could not be reassigned at Wright-Patterson. A Transportation unit arranged travel for employees and their dependents to new locations and transportation of their household goods.[7] By 1957, the command completed the decentralization process.

In July 1954, General Douglas MacArthur helped dedicate AMC's first computer, a Remington Rand UNIVAC (Universal Automatic Computer), signaling a new age in the field of logistics. The complex work of computing logistics requirements soon became automated, allowing supply, maintenance, and procurement information to be integrated quickly and accurately, in ways previously unimagined.

Although decentralization and automation of AMC remained dominant themes during the early 1950s, perhaps the most significant change in the command structure was separation of the research and development (R&D) function of AMC into a distinct command. R&D had been one of three pillars of AMC's World War II antecedent organization, the Materiel Division, and dated from early experimental work at McCook Field. (The other Materiel Division pillars were procurement-acquisition and maintenance-logistical support.) In the closing stages of World War II, it had become increasingly apparent that science and technology would determine America's future air supremacy and the nation's security.

The R&D program of Air Materiel Command largely focused on development of new and improved equipment and, hence, toward service and production engineering. Basic research usually took second place to applied research, leading to the inherent danger that, over a period of time, the technological base crucial to future military superiority suffered. In 1949, Air Force Chief of Staff General Hoyt S. Vandenberg appointed a special committee of the U.S. Air Force Scientific Advisory Board to study the capabilities and future requirements of the Air Force R&D program. This committee, chaired by Dr. Louis N. Ridenour, recommended that research and development be divorced from production engineering and placed in a separate command. The Air Force, subsequently, established such a command drawn from elements of AMC. Air Research and Development Command (ARDC), created in 1950 and activated in April 1951 with headquarters at Wright-Patterson, became responsible for all research and development engineering on aircraft and aeronautical equipment.

Wright-Patterson remained an acquisition, logistics, and research and development center, but the scope and management of these activities changed substantially. The laboratories in Area B were reorganized on April 2, 1951, to form Wright Air Development Center (WADC), the largest of 10 R&D centers under ARDC. WADC consisted of four elements: the Engineering, Flight Test, and All-Weather Flying divisions, and the Office of Air Research. The 12 laboratories under WADC's jurisdiction supervised development of most weapon systems, airborne components, ground equipment, and materials.[8] In June 1951, ARDC relocated its headquarters from Wright-Patterson to Baltimore, Maryland. The

Wright Air Development Center functions in Area B, however, remained essentially the same. In January 1958, ARDC moved from Baltimore to new headquarters at Andrews Air Force Base, outside Washington, D.C. Three years later, it was redesignated Air Force Systems Command (AFSC).

Research remained a critical part of life at Wright-Patterson in the early Cold War period. With the success of jet-engine technology, Area B laboratories announced the end of Air Force piston-engine research in 1949. A new hypersonic wind tunnel began operations, and a microscope that could display pictures of electronic conditions on the surface of materials was added to the Materials Laboratory's tool kit. In 1950, Area B technicians conducted the first tests with bluff-bomb shapes, and the Photographic Laboratory determined, from test flights, the feasibility of taking strip photographs from high-altitude balloons. E. R. Ballinger conducted the first, manned-aircraft, weightlessness experiments with instrumented humans at Wright-Patterson. Another milestone occurred on December 13, 1956, when Major Arnold I. Beck attained a simulated altitude of 198,770 feet, the highest on record, in an ARDC altitude chamber at Wright-Patterson Air Force Base.

Flight Operations

The 2750th Air Base Wing monitored daily flight operations of the installation. All flying activities were consolidated in Area B in July 1952. At the same time, WADC's Flight Test Division transferred to Area C, where longer and heavier runways were better suited for the types

WADC headquarters was established in Building 14, Area B, in 1951.

A WADC engineer calibrates the wing aileron on a flutter model of the F-89 Scorpion.

AIR BASE WING EMBLEM

Until 1969, the 2750th Air Base Wing (ABW) used the emblem of its predecessor organization, the 4000th Army Air Forces Base Unit (the World War II host organization for Wright Field). In 1968, the wing decided it needed a new, distinctive emblem and motto that more accurately represented its service to a combined Wright-Patterson Air Force Base. It hosted a month-long contest to solicit proposals and offered $25 United States savings bonds as prizes for the winners. Because the wing served the entire Wright-Patterson community, the contest opened to all military and civilian personnel at the installation. A panel of experts examined more than 60 entries before selecting the winning design of Craig W. Gridley, an engineer with the Aeronautical Systems Division. Lieutenant Colonel Clifford T. Manlove, the 2750th ABW's executive officer, submitted the winning motto.

Headquarters U.S. Air Force officially approved the motto and a slightly modified version of the emblem on April 4, 1969. The emblem bore the Air Force colors—golden yellow to denote the sun and the excellence of personnel in assigned duties, and ultramarine blue to symbolize the sky, the primary theater of Air Force operations. The Wright Flyer in the upper right corner indicated the aviation heritage of Wright-Patterson Air Force Base, the birthplace of military aviation and the only base the 2750th Air Base Wing had ever supported. Stylized aircraft with red contrails at the lower left represented modern weapon systems. A mach cone or shock wave in the center of the shield denoted the aerospace mission of the unit. Finally, a Lamp of Knowledge on the cone signified the research imparted by wing personnel.

Lieutenant Colonel Manlove based his winning motto—"Strength Through Support"—on the fundamental concept that any organizational goal could be more effectively and economically achieved through the combined and harmonious efforts of the group working as a single unit.

The wing used its new emblem until 1993, when General Merrill A. McPeak, Air Force chief of staff, personally reviewed all Air Force emblems and ordered changes that conformed to his perceptions of correct emblem design and appearance. Concluding that the wing emblem was cluttered with too many symbols and colors, the general ordered the design simplified. As a result, the wing eliminated the Lamp of Knowledge and the clouds around the Wright Flyer. The aircraft contrails changed from red to white, and the red detailing on the Wright Flyer was eliminated. Finally, the shield's two-toned blue background changed to a single shade of blue. General McPeak approved the revised design in November 1993.

4000th Army Air Forces Base Unit emblem

Original Air Base Wing emblem approved in 1969

Air Base Wing emblem revised at the direction of General Merrill McPeak, Air Force chief of staff, 1993

Activity at the Air Terminal, Building 146, Area C, increased in 1954 with the introduction of LOGAIR, enabling the transportation of high-priority supplies and equipment.

of aircraft used by the division. Using Area C's runways also eliminated the danger of mishaps occurring in housing areas adjacent to the Area B flightline. In connection with the move, all base aircraft operations moved from Building 206, Area C, to Building 8, Area B. The only aircraft left on the Patterson side of the base not associated with the Flight Test Division were two B-17s, 12 C-47s, and one C-54 employed for administrative flights and instrument training.

The mission of maintaining flight proficiency among pilots assigned to Wright-Patterson fell to the 2750th Air Base Wing. The wing's Flight Training Branch fulfilled this task through extensive training programs in instrument flight, jet transition flight, and conventional transition flight. During 1954, Wright-Patterson joined other Air Force bases in establishing a special, jet transition program for its pilots using F-80 fighter and T-33 trainer aircraft. Ground school instructors from the 2750th completed the jet indoctrination course at Craig Air Force Base, Alabama, before

opening the ground school phase of the Wright-Patterson jet program on July 27, 1954. The Pilot Transition Branch eventually transferred from the main training branch in Area C to Building 8, Area B.

In July 1953, the Area C control tower shut down temporarily for rehabilitation. A modern FRC-19 console was installed in the tower and approach control moved from the tower to a new, radar, traffic-control room in Building 206. Normal control of air traffic resumed on August 16. Transmitters for the tower were located in Building 199, with receivers off Sand Hill Road east of the flightline.[9] The installation of ultra-high-frequency (UHF) equipment also broadened the facilities available in the tower.

On September 9, 1957, the Air Traffic Control Division of the Civil Aeronautics Administration (CAA) accepted operational responsibility for controlling air traffic within a six-airport area, including Wright-Patterson Air Force Base. This organization, designated Dayton RAPCON

(radar approach control), handled civilian and military air traffic for Wright-Patterson, Clinton County Air Force Base (located near Wilmington, Ohio), and the Dayton, Springfield, and Richmond (Indiana) municipal airports.

Wright-Patterson in the 1950s became an active component of the Air Force's transportation network. To ensure mail delivery during the 1951 national rail strike, the 2750th Air Base Wing initiated administrative flights linking Wright-Patterson (Headquarters AMC) and Washington, D.C. (Headquarters U.S. Air Force). This shuttle service, known as the Kittyhawk flight, proved so satisfactory that, in November 1952, the shuttle's schedule was extended to six days a week, with 10 dedicated crews.[10]

On October 6, 1952, two AMC courier flights, known as Dixieland and Alamo, were inaugurated at Wright-Patterson to provide air service for AMC personnel on official business. The flights also expedited mail service between headquarters and its outlying air materiel areas. On Mondays,

From the 1950s to the 1990s, Wright-Patterson stored and shipped tires worldwide.

Wednesdays, and Fridays, Dixieland departed for points east and south and Alamo for points west and south. Beginning in 1954, weekly passenger and cargo flights began between Wright-Patterson and Brookley Air Force Base, Alabama.

Creation of the Logistics Airlift (LOGAIR) system in 1954 opened a new era at Wright-Patterson. AMC had long been in need of an efficient air transportation network to support its logistics distribution operations. It recognized airlift capability as a key factor in constructing a modern logistics management system capable of global mobilization. Mercury Service, as the AMC airlift system originally became known, was approved in February 1954 as a scheduled airlift within the continental United States (CONUS). Its purpose was two-fold: to move materiel quickly to CONUS-based operational units, and to shuttle materiel between the AMC air depots. Aircraft and services to support this system were contracted from civilian airlines, much as they had been in emergencies during World War II, the Berlin Airlift, and the Korean War.

Mercury Service had an eastern and western trunk operated, respectively, by Capital Airlines of Nashville, Tennessee,

and American Export and Import Company of Miami, Florida. Wright-Patterson Air Force Base was included in the Eastern Zone, which consisted of five-day-a-week, roundtrip service from the Air Force bases of Kelly, Texas, to Tinker, Oklahoma; Wright-Patterson; Olmsted, Pennsylvania; Westover, Massachusetts; Robins, Georgia; and Brookley, Alabama. The Western Zone provided flights from Kelly to Tinker; Hill, Utah; Travis, California; Norton, California (flag stop only); and McChord, Washington.[11]

Since American Airlines also used the Mercury designation for part of its fleet, AMC changed the name of its system to LOGAIR in August 1954. When initial LOGAIR contracts expired on October 31, 1954, a second phase began in November and included a new transcontinental operation. Daily service was established from principal air materiel areas to aerial ports of embarkation to overseas destinations.[12] Resort Airlines conducted transcontinental operations, while the original contractors retained the eastern and western trunks. Twin-engine C-46 and four-engine C-54 aircraft were used throughout the system.

The LOGAIR system placed priority on transporting items urgently needed or that would achieve significant savings, such as

aircraft engines and spare parts. LOGAIR also provided a means for moving sensitive items, such as hazardous materials, that civilian airlines were not permitted to carry. During the last few months of 1954 and the first half of 1955, the young airlift system grew impressively. One year after its beginning, the number of route miles flown and tonnage transported had doubled and, a year later, doubled again. Operations at key locations were running 24 hours a day, seven days a week. As service expanded, other Air Force commands, including Strategic Air Command, Air Defense Command, and Air Research and Development Command, negotiated for use of LOGAIR services. LOGAIR's capacity for rapid response and flexibility soon established the system as an essential element in America's combat readiness.

In addition to courier services, Wright-Patterson, with WADC assistance, also furnished materiel support and services to Strategic Air Command for 11 Boeing B-52 Stratofortress aircraft during 1957. These *Red Scramble* heavy bombers, committed to first-response missions in the event of a Soviet attack, had to be maintained in combat-ready status. In November, Wright-Patterson provided support and services to five Strategic Air Command Boeing KC-135 Stratotankers during Operation Sun Run, a transcontinental speed and effectiveness test.[13]

By 1959, between courier services and test flights, annual takeoff and landing operations at Wright-Patterson's two fields rivaled commercial movements at New York's International Airport, Idlewild. Area C's flightline logged 139,276 takeoffs and landings that year. In Area B, where the flightline closed to jet aircraft operations on February 27, 1958, flight controllers recorded 44,699 takeoffs and landings.

Base Activities

While wartime activities dominated the early part of the decade, a wide variety of other events ensured that Wright-Patterson remained a vibrant community throughout the 1950s. As home of Headquarters AMC, Air Force Institute of Technology (AFIT), Foreign Technology Division (FTD), and a number of other high-profile organizations, Wright-Patterson frequently hosted senior military officers, government officials, and business leaders from across the globe. Perhaps the most touching of these visits came on October 1, 1951, when General E. Marras, chief of staff for Defense, Italian Air Command, visited the base and placed flowers on the memorial to Lieutenant Giovanni Pirelli, an Italian test pilot who died in a crash at Wilbur Wright Field on February 4, 1919.

In addition to distinguished visitors, Wright-Patterson hosted numerous summer encampments for the Air Force Reserve Officer Training Corps (AFROTC), Civil Air Patrol, U.S. Military Academy, and U.S. Air Force Academy cadets, as well as Explorer and other scout groups. Thousands of AFROTC cadets from Midwestern colleges and universities received special preflight briefings and other instruction in preparation for flight experience in C-45, C-47, and T-33 aircraft at Wright-Patterson Air Force Base. The Wright-Patterson Non-Commissioned Officers Academy, established in 1955, provided a concentrated, four-week course in leadership and management for selected noncommissioned officers from the 2750th ABW and base tenant organizations. The academy received official accreditation in December 1957.[14]

On several occasions throughout the 1950s, the base served as a safe haven for aircraft evacuated from other installations experiencing severe weather, such as Hurricane Florence that hit the northwestern Florida coast in 1953. Weather significantly impacted operations on base as well. An unusually heavy snowstorm struck the state of Ohio over Thanksgiving weekend in 1950. Ten inches of snow accumulated on Saturday night and another two fell the following morning. The Dayton weather bureau declared it the deepest, 24-hour snowfall ever recorded in the area. Drifts up to five feet halted all city, county, and state traffic. Wright-Patterson employees were unable to report to work until the following Wednesday.

The biggest weather challenge at the end of the decade, however, was flooding. Between January 20 and 25, 1959, Mad River floodwaters, impounded by Huffman Dam, inundated all but 6,000 feet of the Patterson Field runway, closing the airfield to jet aircraft operations. Night aircraft operations were suspended and transient air traffic diverted from Area C to Area B. Despite problems on base, personnel from Wright-Patterson helped evacuate marooned civilians along the shores of the Miami River and in the Springfield area. On January 21, 1959, a Bell H-13 and two Army helicopters, one local and one transient, flew nine rescue missions.

In another example of base support to the community, Annual American Radio

Armed Forces Day celebration, Area B. This May event became an annual tradition during the 1950s.

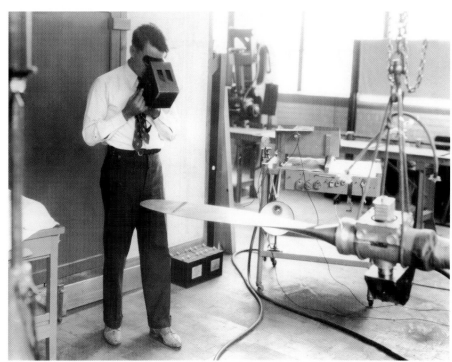

A student at the Air Corps Engineering School studies the effects of propeller vibration, 1930s. The line of dust across the blade indicates an area of weakness.

Relay League "Field Day" exercises were held on the hilltop in Area B on June 19 and 20, 1954. Military Amateur Radio System (MARS, later renamed Military Affiliate Radio System) members and radio amateurs from Wright-Patterson participated in the event. The exercises tested the capabilities of MARS radio stations operating in the field under conditions approximating those encountered during an actual emergency. During the two-day event, the group made radio contact with other amateur radio

stations in all 48 states and in the Canadian Amateur Radio District, some of which were beyond the Arctic Circle. They also worked stations in Alaska, Puerto Rico, and North Africa. The success of the exercises led to the formation of an official Wright-Patterson Air Force Base MARS and Amateur Radio Club. The club participated in subsequent contests and field exercises and provided backup communications during actual emergencies in the Dayton area.

GIOVANNI PIRELLI

Lieutenant Giovanni Pirelli, a member of the Italian Military Mission, was an Italian pilot with decorations from his government for shooting down four enemy airplanes on the Italian-Austrian front during World War I. After the war, in January 1919, Pirelli came to Wilbur Wright Field to test Italian-made airplanes. On February 4, 1919, Pirelli took off from McCook Field in his Ansaldo Balilla airplane headed for Wilbur Wright Field. His mechanic, Lodovico Fusari, stated that his aircraft was "apparently working in perfect condition and his controls were performing satisfactorily." At an altitude of 200 feet over Wilbur Wright Field, Pirelli's airplane suddenly nose-dived straight into the ground, crashing in the vicinity of Building 25 (on present-day Area C). Lieutenant Pirelli was killed instantly. At an honorary ceremony in 1930, General Benjamin D. Foulois officiated the placement of a brass plaque on the grass ramp in front of Hanger 152. (The field's runways had not yet been paved.) The marker was enclosed by a white picket fence. When the runways were paved in the 1940s, the plaque was flush-mounted in place on the apron.

Italian-Americans associated with the Dayton chapter of the Sons of Italy paid for Pirelli's funeral expenses. In appreciation, Pirelli's father, an Italian senator and a wealthy rubber manufacturer in Milan (Pirelli Tires), sent a check for $1,000 to the chapter, which used the money to purchase a lodge hall at Fifth and Wayne avenues in Dayton. The group dedicated the lodge in the anglicized name of the pilot, John Pirelli. The lodge was later torn down, but the chapter is still known as the John Pirelli Lodge #1633.

General Benjamin D. Foulois, in his opening remarks at the 1930 dedication, stated:

Eleven years have passed but [Pirelli's] sacrifice has not been made in vain. His example while amongst us here and his passing at a time when he was so actively engaged in service to two great nations cannot soon be forgotten. It is one more important link in the friendship bond between Italy and America.

In 1986, the plaque honoring Pirelli was removed from its location on the Area C flightline because it was endangered by construction in the area. The plaque is currently held in safekeeping in the office of the base historian pending rededication.

Sources: "A Dedication to Aviator and Test Pilot Lieutenant Giovanni (John) Pirelli," John Pirelli Lodge #1633 of Dayton, Ohio, viewed online July 21, 2003, at http://www.johnpirelliosia.org/johnpirelli_dedication.htm; various articles in file: Giovanni Pirelli, on file at Aeronautical Systems Center History Office.

Lieutenant Pirelli's Ansaldo Balilla airplane crashed at Wilbur Wright Field on February 4, 1919.

Lieutenant Giovanni Pirelli (*John Pirelli Lodge #1633, of Dayton, Ohio*)

WRIGHT-PATTERSON IN THE VIETNAM ERA

In 1961, the Air Force underwent a major restructuring that stemmed from an extensive search for the most efficient method to acquire and maintain weapon systems. The restructure also responded to the effects of new technology that were making weapon systems more capable, more complex, and more expensive, and thus resulted in acquisition of fewer units. On April 1, 1961, the Air Force transferred Air Materiel Command's procurement and production functions for new systems to Air Research and Development Command. AMC was then redesignated Air Force Logistics Command (AFLC) and ARDC became Air Force Systems Command (AFSC). This restructuring made Air Force Systems Command, with headquarters at Andrews Air Force Base in Maryland, responsible for new weapon systems from the research and development phase through initial deployment. As part of this reorganization, in Area B of Wright-Patterson, procurement and production duties of AMC's Aeronautical Systems Center, established in 1959, combined with R&D responsibilities of Wright Air Development Division (WADD)—1959 successor to Wright Air Development Center. The new organization became Aeronautical Systems Division (ASD) under AFSC. In a concurrent move, the former WADD laboratories were separated from engineering development and assigned to Headquarters AFSC, allowing them to fully concentrate on advanced technology research. Once new systems deployed, AFLC, with headquarters in Area A of Wright-Patterson, assumed responsibility for supporting them throughout their operational lifetimes.

The 2750th ABW, now assigned to Air Force Logistics Command, continued its mission as host organization for Wright-Patterson and its many associate/tenant organizations. Approximately 100 tenant-support agreements, representing more than 150 diverse organizational units, made the wing's mission quite complex. Headquarters AFLC, in keeping with command management and organization policies, exempted the 2750th from normal air materiel area command jurisdiction. Instead, the air base wing reported directly to Headquarters AFLC. The installation's real property, however, was aligned within the 10-state Mobile (Alabama) Air Materiel Area of responsibility. Wright-Patterson's fixed capital assets at the end of fiscal year 1961 totaled more than $208 million on base and nearly $6 million off base.

The 2750th's flight operations function continued to expand in the 1960s. Late in 1960, Headquarters Air Force directed the wing to provide mission support airlift for organizations assigned or attached to Wright-Patterson Air Force Base. On November 3, 1960, 10 tenant units transferred 16 aircraft and responsibility for scheduling, operations, and maintenance to the wing. The base Transport Flight office also received responsibility for the operation of staff aircraft on October 11, 1960.[15] Sixteen new T-39 aircraft joined the 2750th's fleet of Convair T-29s in 1962. These aircraft supported advanced facilities for jet pilot checkout on base and replaced older C-47s for transporting passengers. By 1962, the base air terminal handled an average of 472 flights and 1,310 tons of cargo per month, while the flightline averaged 92,000 takeoffs and landings per

Air Materiel Command was redesignated Air Force Logistics Command on April 1, 1961. Headquarters remained in Building 262, Area A.

Building 14 served as headquarters for Materiel Command(1943-1944), Wright Air Development Center (1951-1959), Wright Air Development Division (1959-1961), Aeronautical Systems Division (1961-1992), and Aeronautical Systems Center (1992-2003).

Handling cargo at the base airfreight terminal was part of the 2750th Air Base Wing mission.

Building 10, headquarters for the 2750th Air Base Wing, about 1960

Line of C-5 Galaxy cargo aircraft await departure in support of combat operations. (*U.S. Air Force photo by Staff Sergeant P. J. Farlin*)

year. The opening of a new control tower on the Patterson Field runway in 1963 facilitated the gradual shift of flight operations away from Wright Field and initiation of plans to close the Area B runway.

In addition to successful flight operations, consolidation and automation continued as trends in base-support operations. The air base wing automated its base-supply activities by installing an electronic data-processing system and implementing a micro-mechanized Engineering Data Automated Logistics Program. In February 1966, a newly acquired UNIVAC 1050-II computer system, one of only a dozen such systems in the Air Force, was programmed. Air Force Systems Command had developed the specialized system to store more than 100,000 supply and 30,000 equipment items. The UNIVAC had functions for standardization, requisitioning, purchasing, receipting, storage, stock control, issue, shipment, reporting, disposition, identification, and accounting.

The mid-1960s found Wright-Patterson attending to escalating support requirements for the nation's military operations in Vietnam. Beginning in the autumn of 1964, the mission of the 2750th ABW and AFLC included providing materiel support for major Air Force commands engaged in combat zones. The bulk of this support came through the Sacramento Air Materiel Area. In addition to massive airlifts of spare parts, supplies, munitions, and other materiel, highly skilled teams of depot maintenance technicians, known as Rapid Area Maintenance (RAM) teams, deployed to Southeast Asia to repair weapon systems that had sustained crash or battle damage. The teams often salvaged valuable airplanes and equipment at crash sites under very dangerous conditions.

The 2750th ABW provided certain essential services and a limited number of personnel in support of the AFLC mission in Southeast Asia. In September 1964, the wing shipped flight and hangar equipment to the 2851st Air Base Group at Kelly Air Force Base, Texas. The 2750th soon

became the prime procurer of loaders, revetments, and shelters used to protect resources in the combat theater. The first prototype revetments were erected and tested at Wright-Patterson in May 1965. Subsequently, the 2750th ABW Procurement Division let contracts for more than $18 million to manufacture revetments. Standard revetments consisted of rectangular, 16-gauge steel bins, 10 feet by 7.7 feet, 16 feet high. Once they reached their destination, the bins, filled with dirt, sand, or gravel, provided aircraft with maximum protection from enemy mortar fire and accidental ground explosions.

The air base wing performed other activities in support of combat operations. Some 2750th ABW agencies processed civilian and military personnel bound for Vietnam. Others, such as Wright-Patterson's first, 15-man, maintenance-specialist team, left for temporary duty in Southeast Asia. In November 1965, the 15 airmen, all from the 2750th ABW, became AFLC's first contribution to the Air Force Prime BEEF (Base Engineering Emergency

Prisoners of war returned to Wright-Patterson during Operation Homecoming included (left to right) Major Donald Heiliger, Lieutenant Colonel Robert Purcell, Captain Burton Campbell, Captain Edward Mechenbier, Colonel Ronald Byrne, Colonel Nick Apple.

FB-111, developed in the 1960s, was the Air Force's strategic bomber variant of the F-111 fighter. The aircraft served until 1991.

SR-71 Blackbird strategic reconnaissance aircraft. Throughout its nearly 24-year career, the SR-71 remained the world's fastest and highest-flying operational aircraft. (*U.S. Air Force photo by Technical Sergeant Michael Haggerty*)

Force) concept of a mobile, military, civil engineering force.

The wing also conducted flight training, small arms weapon training, and vehicle operator training, as well as laundry management courses. It handled shipments from Lexington Blue Grass Army Depot, Kentucky; Ravenna Ordnance Plant, Ohio; the USS *Enterprise*; and the National Cash Register Corporation in Dayton. The wing supplied vehicles and weapons in accordance with levies and provided for offloading, temporary storage, and reloading of Southeast Asia-bound materiel. The 2750th even supported a limited maintenance mission. Stock number user-directory reconciliations were handled

on base, and the wing furnished supplies, in-flight lunches, quarters, and rations as tasked.[16]

On the Area B side of the installation, the research laboratories and Aeronautical Systems Division (ASD) invented and improved systems used by warfighters. The laboratories developed jet fuels and lubricants for all services, worked on phased-array radar and airborne lasers, explored the use of composite materials in structures, and pursued stealth and fly-by-wire technology. ASD set up a special Limited War Office to respond to combat requirements and quickly evaluate new hardware ideas. The office conducted hundreds of rapid-response programs ranging from development of tactical, electronic-warfare systems and guided bombs to modification of cargo aircraft into side-firing gunships. Its efforts produced the AC-47 and AC-130 gunships, intrusion alarms, and a mobile, tactical air-control system. ASD deployed F-4C, F-111, and F-15 fighters; C-141 and C-5A transports; the SR-71 Blackbird; and A-10 attack aircraft during these years and conducted extensive research on XB-70 and B-1A bombers.

Wright-Patterson's tenant organizations also had roles in the Southeast Asia conflict. The Air Force Institute of Technology established extension courses in the combat zone and participated in Project Corona Harvest to glean lessons learned. Wright-Patterson's most public contribution came in 1973 when the U.S. Air Force Medical Center Wright-Patterson became one of 10 Air Force medical facilities selected to receive and process former prisoners of war. Operation Homecoming eventually brought 30 Air Force officers to Wright-Patterson for processing and reorientation.

Base Activities

Throughout the 1960s and early 1970s, Wright-Patterson rarely missed a chance to celebrate or support its community. On June 15, 1960, Wright-Patterson Air Force Base held official ceremonies commemorating a landmark in aviation history by recognizing Brigadier General Frank P. Lahm as "Father of Air Force Flight Training." Lieutenant Lahm, a member of the Aeronautical Board, was Orville Wright's passenger during the first, successful test flight of a military airplane on July 27, 1909, at Fort Myer, Virginia. Lahm later served as commander of the Air Corps Training Center at Kelly Field, Texas, and as assistant chief of the Army Air Corps. The general traveled from his home in Mansfield, Ohio, for the Wright-Patterson ceremony. Six hundred Air Force Academy cadets attended the celebration, which included an aerial demonstration by pilots of the Wright Air Development Division. The inscription on the plaque presented to General Lahm read in part:

> Presented to Brigadier General Frank P. Lahm...in recognition of his lifelong devotion to aviation and aeronautical science. Taught to fly by Wilbur Wright in the first military aeroplane, Signal Corps No. 1, at College Park, Md., in 1909. Awarded by "The Early Birds," an organization of those who flew solo before December 17, 1916.[17]

Wright-Patterson celebrated another landmark with the first, official reunion of World War I flyers held at the United States Air Force Museum in late June 1961. More than 400 World War I aviators attended the event, their first reunion in 43 years. This

OPERATION HOMECOMING

One of the most touching scenes in Wright-Patterson history unfolded between February 15 and April 1, 1973, during the return of the nation's Southeast Asia prisoners of war to American soil. Operation Homecoming concluded a series of plans begun in June 1968 and alternately known as Sentinel Echo (June 1968) and Egress Recap (September 1972). On the eve of the prisoners' release, Secretary of Defense Melvin R. Laird changed the project's title to "Homecoming."

Many local residents waited on the flightline for the arrival of the former prisoners of war.

Based upon World War II and Korean War experiences, Department of Defense officials expected that the returnees might require significant medical and psychological assistance. Many had been held captive for more than five years, some for as long as eight years. Therefore, the POWs were placed under military medical auspices as soon as possible after their release and remained in medical channels for transportation to the continental United States. After initial examination, treatment, and processing overseas, the men were evacuated by air to military medical facilities in the United States. The U.S. Air Force Medical Center Wright-Patterson was one of 10 Air Force medical facilities to receive and process the returnees.

A tearful reunion marks the Operation Homecoming welcome of a repatriated prisoner of war on the Wright-Patterson flightline.

The 2750th Air Base Wing (ABW) furnished logistical support for the repatriation program. Wing agencies provided family quarters for next of kin, as well as operational facilities for the Air Force debriefing team, the Wright-Patterson processing team, and the press center. Dependents stayed in visiting officers' quarters (Buildings 832 and 833), temporarily reconfigured as family accommodations. A reception room for visitors was located in Building 833, and the debriefing team used rooms in the north wing of the medical center for administration and consultation. The Dodge Gymnasium (Building 849) housed the Operation Homecoming News Center. All returnees were quartered in the north wing of the medical center.

Actual repatriation began February 12, 1973, when 143 American servicemen landed at Clark Air Base, Republic of the Philippines, in the first of 12 release increments. Immediately following their arrival at Clark, invitational orders authorized dependents to travel at government expense to the stateside hospitals receiving the repatriates. The initial phase of Operation Homecoming ended April 4, 1973, at Clark Air Base, by which time 597 former American captives had returned to freedom.

Thirty of the repatriated prisoners were flown to Wright-Patterson between February 15 and April 1. Upon disembarking, Wright-Patterson's most honored guests were warmly greeted by either General Jack J. Catton, the Air Force Logistics Command (AFLC) commander, or Lieutenant General Richard M. Hoban, AFLC vice commander, and by Brigadier General Irby B. Jarvis, the 2750th ABW commander. The returnees walked down a red carpet extending from the aircraft and smartly saluted the American flag held by a four-man color guard. Their families then greeted them on the flightline, in view of media representatives and spectators, before being transported to the medical center.

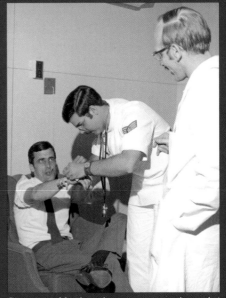

Sergeant Meske places a name bracelet on Captain Michael Burns, while Dr. Harry Delcher looks on. Burns was one of the POWs returned to Wright-Patterson during Operation Homecoming.

Processing of the former POWs involved intelligence debriefs, medical examinations and evaluations, counseling sessions, updates to fiscal affairs, a chaplain's visit, family assistance, and public affairs activities. Families held five news conferences in the news center. Once processing concluded, the returnees received 90 days of convalescent leave. President and Mrs. Richard M. Nixon invited all repatriates, along with their wives or mothers, to a White House reception and formal dinner on May 24, 1973.

Operation Homecoming concluded a period of particularly stellar performance by the 2750th Air Base Wing. The Air Force recognized its accomplishments during the period from February 1, 1972, to January 31, 1974, by awarding the wing its first Air Force Outstanding Unit Award. The Air Force cited the unit's support of national and international operations, as well as the milestones it established in community relations and employee morale and welfare projects. Most noteworthy, however, was the wing's successful handling of Operation Homecoming.

Lieutenant Colonel William Breckner receives a plaque from a local boy scout troop following Operation Homecoming.

extraordinary group included Captain Eddie Rickenbacker, the country's "Ace of Aces;" Douglas Campbell, the first American ace; and George Vaughn, the second-ranking living ace. Other dignitaries in attendance included General Carl Spaatz, and Brigadier Generals Frank P. Lahm and Benjamin D. Foulois, two of the first three military pilots taught to fly by the Wright brothers.[18]

In September 1961, the air base wing received its first Air Force Flying Safety Award, covering the period from January to June 1961, during which base pilots flew more than 25,000 hours in a wide variety of aircraft without a single accident or incident. The air base wing subsequently received the AFLC Flying Safety Award in recognition of more than 54,000 hours of accident-free flying over the year.[19]

In July 1964, the sixth annual Dayton Soap Box Derby took place on the accelerated runway in Area B. Seventy-five boys from the Dayton area competed. The derbies, initiated in 1933, continued as a base and community tradition for several decades.

Personnel at Wright-Patterson also turned their attention toward environmental issues. Between 1928 and 1954, base personnel planted more than 7,500 trees on the installation, primarily for ornamental purposes. Yet, this activity paled when compared to the greening of Wright-Patterson that took place in the 1960s, when an aggressive, tree-planting campaign permanently enhanced Wright-Patterson's aesthetics. During 1960, the base planted 10,000 multiflora rose trees in conjunction with the state of Ohio's fish and wildlife conservation program. Four years later, it planted another 51,000 trees as part of Operation Green Rush, a base beautification program.

Then, in 1966, Wright-Patterson entered into a 10-year, cooperative, timber management program with the Ohio Division of Forestry and Reclamation to reforest 420 acres on base. Volunteers from various base organizations, most notably members of the Twin Base Rod and Gun Club, planted tens of thousands of trees. The overall scheme provided for adequate road and firebreak development, erosion control, insect and disease control, and wildlife habitat conservation. A survey of existing base timberlands identified sellable saw timber and timber products. The inspection led to harvesting 75,000 board feet of saw logs and 200 tons of pulpwood in the fall of 1966. Conservation actions like these helped Wright-Patterson win the Air Force's General Thomas D. White Fish and Wildlife Conservation Award four times during the 1960s. (The tree planting also did not go unnoticed. In the 1990s, the National Arbor Day Foundation recognized Wright-Patterson seven times as a Tree City USA.)

In 1967, Wright-Patterson Air Force Base celebrated its Golden Anniversary. A half-century of dedication and progress separated the establishment of Wilbur Wright Field, as a Signal Corps training school for World War I pilots, and the sophisticated research and flight operations that characterized Wright-Patterson Air Force Base on the horizon of the aerospace age. Celebrations on base were accompanied by publication of a pictorial history of Wright-Patterson's 50 years of accomplishments.

Several 1971 events again sparked interest in Wright-Patterson's heritage. In March, a new tombstone marker was placed at the grave of Private Hiram Honaker, a Civil War veteran of the 5th Regiment, U.S. Colored Cavalry. Private Honaker was buried in the Cox family cemetery that had become part of Wright-Patterson Air Force Base in 1950. Also in 1971, the half-acre pylon site on Huffman Prairie Flying Field was listed on the National Register of Historic Places.

Soap box derbies have been a part of Dayton life since the 1930s. On occasion, they are held on the abandoned accelerated runway in Area B.

Area C flightline, looking south, in the early 1960s. Facilities for the 17th Bombardment Wing are visible on the right side of the photograph. Alert hangars for the 58th Air Division, which inactivated in 1958, were at the north end of the runway.

Headquarters
2750th Air Base Wing (AFLC)
United States Air Force
Wright-Patterson Air Force Base, Ohio 45433

15 May 1967

Office of the Commander
All Personnel
Wright-Patterson Air Force Base, Ohio

Dear Friend

This marks the 50th anniversary of military aviation at Wright-Patterson Air Force Base. During this span of fifty years, our contribution to the growth of aviation and to our nation's air power has been enormous.

Only at Wright-Patterson can one trace all aspects of an aircraft system literally from the cradle to the grave. That is, from the original concept of research and development, through the entire operational life phase supported by the Air Force Logistics Command, to its final resting place, the Air Force Museum. As a consequence, Wright-Patterson is today one of the greatest and best known air bases in the world.

The contribution this base has made to air power and the significant role we have played in Air Force history was made possible only through the individual efforts of the many thousands of military and civilian personnel who were and are stationed and employed here.

You, as a member of this great team, should be justly proud of the role you have played; for the true history of any organization is really written in the combined efforts of its people.

JOWELL C. WISE
Brigadier General, USAF
Commander

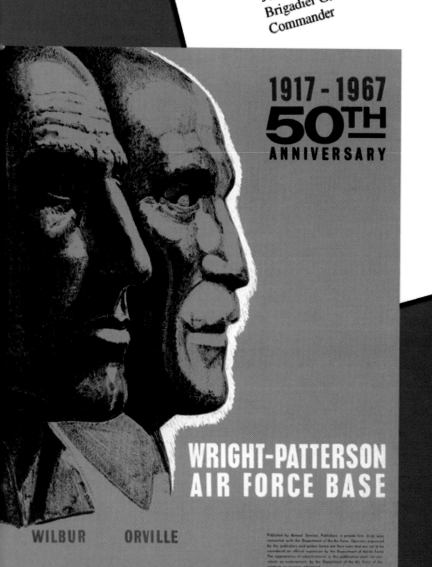

1917 - 1967
50TH
ANNIVERSARY

WRIGHT-PATTERSON
AIR FORCE BASE

WILBUR   ORVILLE

Published by Armed Services Publishers a private firm in no way connected with the Department of the Air Force. Opinions expressed by the publishers and writers herein are their own and are not to be considered an official expression by the Department of the Air Force. The appearance of advertisement in this publication does not constitute an endorsement by the Department of the Air Force of the products or services advertised.

WRIGHT-PATTERSON AND THE END OF THE COLD WAR

The post-Vietnam War period was one of rapid transition and uncertainty within the defense establishment. First came the postwar drawdown and a series of A-76 cost comparison studies, the latter of which converted many activities at Wright-Patterson Air Force Base from government to contract operations. The base population dropped to around 23,000 in 1976. Inflation and rapidly rising fuel prices further aggravated competition for declining funds. Postwar austerity gave way in the early 1980s to the nation's largest increase in peacetime defense spending as the Cold War heated up. Base manpower rose again, peaking at about 30,500 in 1989. Wright-Patterson underwent organizational changes as the Air Force moved away from its Vietnam-era structure toward one better suited to the Cold War and the dynamic progress of technology. Another period of cutbacks in the late 1980s coincided with climactic events leading to the tearing down of the Berlin Wall in November 1989. This symbolic act marked the collapse of communism and brought to a conclusion the Cold War that had lasted for almost half a century.

In December 1974, Wright-Patterson had a workforce of 25,731 and a payroll of more than $402 million. On the base flightline, three units (17th Bomb Wing, 2750th Air Base Wing, and 4950th Test Wing) operated 72 bomber, tanker, test, cargo, and training aircraft. When the Cold War concluded in 1989, Wright-Patterson's workforce stood at 30,543; its annual payroll was almost $900 million; and its economic impact on 18 counties within a 50-mile radius of the base exceeded $3.1 billion. Three units still

The Electronic Warfare Research Center, Building 620, Area B, with its distinctive twin towers, was dedicated in June 1967.

operated 65 aircraft from the base flightline, but only the 4950th Test Wing had been there the previous 15 years.

Despite postwar cutbacks, work on cutting-edge technologies continued in Wright-Patterson's laboratories. Research covered a variety of projects, such as "all glass" cockpits, very-high-speed integrated circuits, variable-camber wings, short takeoff and landing technology, advanced structural alloys and composite materials, artificial intelligence, and doubling the performance of gas turbine engines. The rapid pace of technological development, however, also contributed to organizational instability for the laboratories. In 1975, the Aerospace Research Laboratories (previously the Aeronautical Research Laboratory) inactivated, and its personnel and resources transferred to the newly activated Air Force Wright Aeronautical Laboratories (AFWAL), assigned to Air Force Systems Command. AFWAL included the Materials, Avionics, Aero Propulsion, and Flight Dynamics laboratories.

In October 1982, Air Force Systems Command again reorganized its laboratories. This time it ordered them either to merge or affiliate with their product divisions to improve the transition of new technologies into operational systems. The Air Force Wright Aeronautical Laboratories, assigned to Aeronautical Systems Division (ASD), evolved into Wright Research and Development Center (WRDC) in October 1988, adding an Electronic Technology Laboratory and directorates for Manufacturing Technology, Cockpit Integration, and Signature Technology. Two years later, WRDC acquired the Armament Laboratory at Eglin Air Force Base. In December 1990, the organization was redesignated Wright Laboratory.

Two B-1B Lancers soar over Wyoming in formation. (*U.S. Air Force photo by Staff Sergeant Steve Thurow*)

F-117A Nighthawk stealth fighter

WRIGHT-PATTERSON AIR FORCE BASE COMMANDERS

Lieutenant General Richard M. Scofield (shown here as a major general) commanded Aeronautical Systems Center and Wright-Patterson Air Force Base during the Balkan Proximity Peace Talks in November 1995. ASC and the 88th Air Base Wing received an Air Force Organizational Excellence Award and an Air Force Outstanding Unit Award, respectively, for their handling of the peace talks.

Commanders	From	To
Brigadier General Joseph T. Morris	November 6, 1945	March 29, 1952
Colonel C. Pratt Brown (later brig. gen.)	March 29, 1952	October 2, 1953
Colonel Paul L. Barton (later brig. gen.)	October 2, 1953	August 10, 1957
Brigadier General Donald L. Hardy	August 10, 1957	July 1, 1958
Brigadier General John D. Howe	July 1, 1958	May 20, 1960
Colonel James C. Cochran	May 20, 1960	August 10, 1960
Brigadier General Elbert Helton	August 10, 1960	August 3, 1962
Colonel Glen J. McClernon (later brig. gen.)	August 3, 1962	August 1, 1964
Colonel Arthur E. Exon (later brig. gen.)	August 1, 1964	December 21, 1965
Colonel Jowell C. Wise (later brig. gen.)	December 21, 1965	July 9, 1968
Colonel Colman O. Williams (later brig. gen.)	July 9, 1968	September 3, 1970
Brigadier General Edmund A. Rafalko	September 3, 1970	June 15, 1972
Colonel Irby B. Jarvis, Jr. (later brig. gen.)	June 15, 1972	January 31, 1975
Colonel Robert W. Clement (later brig. gen.)	January 31, 1975	January 13, 1976
Colonel Titus C. Hall	January 13, 1976	January 14, 1977
Colonel Rano E. Lueker (later brig. gen.)	January 14, 1977	April 23, 1979
Colonel James H. Rigney, Jr.	April 23, 1979	June 26, 1981
Colonel Leonard R. Peterson	June 26, 1981	June 28, 1984
Colonel Charles E. Fox, Jr.	June 28, 1984	March 28, 1987
Colonel Stephen F. Kollar	March 28, 1987	July 25, 1989
Colonel Dennis P. Tewell	July 25, 1989	July 28, 1990
Colonel William B. Orellana	July 28, 1990	July 1, 1992
Lieutenant General Thomas R. Ferguson, Jr.*	July 1, 1992	May 26, 1993
Lieutenant General James A. Fain, Jr.	May 26, 1993	October 14, 1994
Lieutenant General Richard M. Scofield	October 14, 1994	May 24, 1996
Lieutenant General Kenneth E. Eickmann	May 24, 1996	May 9, 1997
Brigadier General Robert P. Bongiovi	May 9, 1997	May 29, 1997
Lieutenant General Kenneth E. Eickmann	May 29, 1997	May 9, 1998
Brigadier General Robert P. Bongiovi	May 9, 1998	June 1, 1998
Lieutenant Robert F. Raggio	June 1, 1998	June 29, 2001
Lieutenant General Richard V. Reynolds	June 29, 2001	December 8, 2003
Lieutenant General William R. Looney III	December 8, 2003	Present

Colonel William B. Orellana guided the 88th Air Base Wing through realignment under Aeronautical Systems Center and two redesignations of the wing.

* When Air Force Systems Command (AFSC) and Air Force Logistics Command (AFLC) were disestablished and reorganized as Air Force Materiel Command in 1992, Wright-Patterson Air Force Base was realigned under the Aeronautical Systems Center (ASC), with the center commander becoming the commander of the base. Previously, the base commander had reported directly to Headquarters AFLC, and ASC's predecessor (Aeronautical Systems Division) had been a tenant of the base. The new arrangement was a culmination of Air Force Chief of Staff General Merrill A. McPeak's desire to streamline all elements of the Air Force. He directed the policy of "one base, one wing, one boss." Colonel Orellana continued to command the 2750th Air Base Wing. As of July 1, 1992, in addition to managing its acquisition programs and the 2750th Air Base Wing (now the 88th Air Base Wing), ASC gained responsibility for the U.S. Air Force Medical Center at Wright-Patterson.

Colonel Dennis Tewell, 2750th Air Base Wing commander, operates the base locomotive. Railroad operations ended in 1993.

Aeronautical Systems Division continued to modernize the Air Force's tactical and strategic forces. It upgraded the F-15 and F-16 fighters, improved avionics, fielded the B-1B bomber and the stealthy F-117 fighter, and began work on the Advanced Tactical Fighter (today's F/A-22), C-17 wide-body transport, T-1A Tanker-Transport Training System aircraft, and two new presidential aircraft. It also returned to hypersonic research with the X-30A National Aerospace Plane (NASP).

Air Force Logistics Command sought to support and equip the force despite inflation and stiff competition for government dollars. The command focused on improving its cost effectiveness through computer technology and better management techniques. It remodeled its archaic, World War II depot system and physical plants and initiated support for the Rapid Deployment Force. In 1976, AFLC established an Acquisition Logistics Division to plan the integration of operational support in the acquisition process. Over the next 10 years, automation and microchip technology revolutionized its supply and distribution business, while modular electronics challenged aircraft inventory management and support. In 1987, AFLC began emphasizing reliability and maintainability as solutions to its increasingly complex weapon systems. Two years later, the command initiated major procurement reforms to counter ethics problems that had arisen with its defense contractors. These reforms focused on price, quality, a

Aviation and motor fuels became primary targets for conservation measures at Wright-Patterson during the Arab oil embargo in the 1970s.

streamlined source-selection process, and improved acquisition strategy and oversight. AFLC also worked on promoting artificial intelligence, bolstering the United States' industrial base for mobilization, and developing quality management techniques.

The last years of the Cold War saw the 2750th Air Base Wing involved in a host of activities, including development of a self-guided tour of historic base sites, support of annual encampments of the Air Force Reserve Officer Training Corps (beginning in 1980), responses to local and national emergencies, and decreased flight operations.

National and Local Emergency Responses

The November 1973 energy crisis had lasting effects on Wright-Patterson Air Force Base. Energy conservation, long a matter of concern at Wright-Patterson, suddenly intensified when Arab nations stopped oil shipments to the United States in retaliation for U.S. support of Israel during the Yom Kippur war. This embargo precipitated AFLC's Pacer Energy fuel-conservation program that continued through the end of fiscal year 1974. The 2750th Air Base Wing established a Pacer Energy Task Force to implement and administer a comprehensive energy-conservation program. The Task Force outlined three general target areas for fuel savings: a 14-percent reduction in aviation fuels, a 15-percent reduction in motor fuels, and a 15-percent reduction in utilities, especially heating fuels and electricity. Guidance from higher headquarters directed aviation-fuel conservation practices of Wright-Patterson's three flying wings (the 2750th ABW, 17th Bomb Wing, and 4950th Test Wing). Programmed flying hour reductions for the wings went into effect in January 1974. Through reduced flying hours and greater economies in ground and air operations, the 2750th ABW used 12.3 percent less jet fuel and 22.7 percent less aviation gasoline in the first eight months of fiscal year 1974 than it had the previous year, representing a monetary savings of $743,000 and $114,600, respectively.

The base achieved similar reductions in consumption of motor fuels and utilities. Between July 1, 1973, and March 31, 1974, through the vigilance and cooperation of Wright-Patterson's workers and on-base residents, the base saved $1,347,018 in

energy consumption compared to the previous year.

Energy-conservation efforts continued even after the embargo lifted. Plans covering future contingencies aimed to achieve up to 75-percent curtailment of specific energy sources. By September 1976, the wing had attained a 6.8-percent reduction in energy consumption (i.e., electricity, natural gas, coal, motor-vehicle fuels, and fuel oil for heating purposes) over fiscal year 1974 figures. This decrease was greater than had been anticipated when the campaign began. Throughout the remainder of the decade, especially during the severe winters of 1976-1977 and 1977-1978, the base continued to enlarge its energy-conservation program.

The 2750th Air Base Wing responded to a local emergency on April 3, 1974, when a devastating tornado hit Xenia, Ohio, just 12 miles from base. That day entered United States' history as the "Day of 100 Tornadoes." Slightly more than 10 percent of the deaths resulting from these natural disasters occurred in and around the city of Xenia, a quiet, progressive community southeast of Wright-Patterson. Residents of the city included 1,297 Wright-Patterson employees (1,064 civilian and 233 military). The killer tornado struck Xenia at 4:40 p.m., carving a swath of destruction four miles long and one-half mile wide. In its wake, 34 people died and 500 were injured. The tornado destroyed more than a thousand homes (including those of 293 base employees), heavily damaged 660, and slightly damaged another 904. Insurance adjusters placed losses at $500 million.

At 5:00 p.m., Brigadier General Irby B. Jarvis, Jr., the 2750th ABW commander, activated the base Disaster Preparedness Control Center. The medical center and the air base wing responded quickly to two of Xenia's most pressing needs: medical aid for the injured and assistance in sorting through tons of debris to recover casualties. At 6:30 p.m., a medical team arrived, in addition to three on-scene commanders from the air base wing, to direct Wright-Patterson's assistance. From then until 1:00 p.m. the next day, roads between Wright-Patterson and Xenia filled with a steady stream of traffic from the base to the disaster scene. The most seriously injured were admitted to the U.S. Air Force Medical Center. A 4950th Test Wing CH-3 helicopter flew eight sorties to airlift medicine and equipment from St. Elizabeth's Hospital in Dayton to Greene Memorial Hospital in Xenia. Five hundred base volunteers contributed blood at the medical center. Base civil engineers dispatched

Aftermath of the Xenia tornado. The violent tornado struck the city of Xenia, Ohio, on April 3, 1974, killing 34 people and causing more than $500 million in damages. The medical center and the 2750th Air Base Wing immediately dispatched medical teams and convoys of heavy equipment to the scene.

The Wright-Patterson Disaster Preparedness Control Center coordinated the assistance offered by virtually every organization on base to help sort through the debris and return order to Xenia.

heavy equipment to assist in search and recovery operations and to open traffic arteries. As the first long night ended, other supplies and assistance arrived, including generators for emergency lighting, floodlights, gasoline, 7,000 gallons of water, box lunches, and 30 gallons of coffee for volunteer rescue workers.

Wright-Patterson's support continued throughout the following week, as personnel from nearly every base organization volunteered their services. The 2046th Communications Group moved its Military Affiliate Radio System (MARS) van to downtown Xenia, where it established VHF, UHF, and radio-telephone communications linking on-scene civil defense, the Wright-Patterson Fire Department, and base Security Police with the Disaster Preparedness Control Center. The 2046th also opened the Springfield Municipal Airport tower to assist Ohio National Guard helicopters ferrying emergency supplies. Meanwhile, aerial photography was provided by the 4950th Test Wing along with the 155th Tactical Reconnaissance Group, Air National Guard, at Lincoln, Nebraska, and assistance from the 2750th ABW Operations and Training Division.

On April 6, a Federal Disaster Assistance Team established temporary offices in Building 89, Area C, and assembled a staff of 30 people. A Fifth U.S. Army liaison officer from Fort Knox, Kentucky, joined the team to coordinate all military assistance efforts. By April 8, once the team assumed full control of the operation, Wright-Patterson concluded its major support role in disaster operations.

On April 9, President Richard M. Nixon arrived on base via Air Force One to survey the disaster area. James T. Lynn, secretary of Housing and Urban Development; Thomas J. Donne, chief of the Federal Disaster Assistance Administration; and Presidential Press Secretary Ronald Ziegler accompanied the president. After viewing the disaster area from the air, the president's helicopter landed at an elementary school on the outskirts of Xenia and the party drove into town, where the president conferred with Greene County and other officials. That afternoon, about 500 spectators greeted the president upon his return to Wright-Patterson. Two days later, the air base wing inactivated its Disaster Preparedness Control Center.

In 1977, the base again implemented its disaster-preparedness plans. The winter of 1976-1977 entered Ohio history as one of the worst on record. General F. Michael Rogers, AFLC commander, described the winter as "the most unforgiving weather this region has ever seen." The average temperature of 11.6 degrees F for January 1977 set a record, and an additional 13 days registered temperatures at or below zero. Only the record snowfall of 34.4 inches recorded in January 1918 bested the total of 20.2 inches that fell in January 1977. February's temperatures registered 2.9 degrees below normal. As temperatures descended, energy usage increased and a statewide crisis in natural gas supplies ensued.

Paced by the 2750th Civil Engineering Squadron and the 2750th Logistics Squadron, the air base wing exerted extraordinary effort keeping Wright-Patterson fully operational and helping distressed local communities as they struggled with blocked roads and frozen water lines. The 2750th Air Base Wing delivered 1,500 gallons of fresh water to Trotwood, Ohio, where many homes had frozen pipelines. It dispatched snow blowers to Clark, Greene, Preble, Clinton, and Fayette counties, and loaned ice-thawing equipment to the cities of Fairborn,

The winter of 1976-1977 exposed Ohio to extraordinarily severe weather. The month of January recorded 13 days with temperatures at or below zero.

The 2750th Logistics Squadron was cited for its efforts to sustain vital base functions during the 1978 blizzard. The squadron received an Air Force Outstanding Unit Award for the period April 1977 to March 1978.

New Carlisle, and Xenia. It supplied water containers to Miamisburg, West Milton, and the American Red Cross. The Civil Engineering Squadron's performance earned the unit the Air Force Outstanding Unit Award for the period April 1, 1976, to March 31, 1977.

As the winter of 1977-1978 approached, early indicators pointed toward a repeat of the previous year. The forecasts proved accurate, and the 1977-1978 season was almost as harsh, and decidedly more dramatic. A severe blizzard with 75-mph wind gusts and seven to 12 inches of snow whipped the Miami Valley from January 26 to January 29, 1978. On-base activities were

reduced to minimum essential operations and the flightline closed to all aircraft traffic from 4:33 a.m. January 26 through 4:00 p.m. January 27. According to a local newspaper, Dayton's 1977-1978 snowfall totaled 62.7 inches, bettering the 1976-1977 winter total of 38.8 inches. Both years set new records for the Dayton area.

Although the 2750th Air Base Wing had made preparations for another severe winter, the magnitude of the blizzard and subsequent snowfalls stretched its resources. Emergency measures notwithstanding, snow removal on base was hampered by insufficient equipment in the active inventory, especially front-end

PRESIDENTIAL VISITS TO WRIGHT-PATTERSON AIR FORCE BASE

Over the years, numerous presidents visited Wright-Patterson Air Force Base to fulfill a variety of obligations at the base and within the Dayton community.

President Richard M. Nixon and Congressman Clarence J. Brown, right, arrived at Wright-Patterson on April 9, 1974, to survey tornado damage in the city of Xenia. Base Commander Brigadier General Irby Jarvis and AFLC Commander General Jack J. Catton, left, briefed the presidential party.

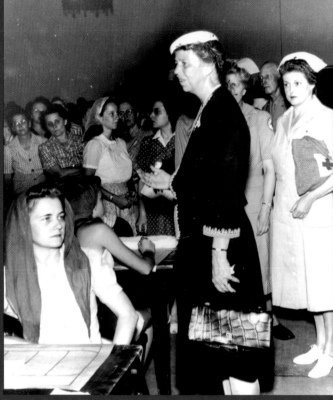

During World War II, First Lady Eleanor Roosevelt addressed officers' wives involved in Red Cross work at Patterson Field.

President George W. Bush checks Technical Sergeant Rob Bowling's haircut as he greeted military personnel at Wright-Patterson Air Force Base, April 24, 2003. President Bush was in Ohio to speak to the workers of the Lima Army Tank Plant in Lima. He returned to the United States Air Force Museum on July 4, 2003, to kick off the centennial of flight celebration and to dedicate the new display hangar at the museum. (*White House photo by Paul Morse*)

loaders and dump trucks. The services of commercial contractors were needed into mid-February to supplement base efforts to clear and haul snow from Wright-Patterson's streets and parking lots. Once again the 2750th Civil Engineering Squadron assisted beleaguered communities within a seven-county area with the loan of snow-removal equipment and military operators. The 2750th Logistics Squadron earned the Air Force Outstanding Unit Award for the period from April 1, 1977, to March 31, 1978, most specifically for sustaining vital base functions during the January 1978 blizzard.

Airfield Operations

In 1975, the 2750th Air Base Wing received the Air Force Flight Safety Certificate recognizing three years of accident-free flying. This was the last such award the wing received. By June of that year, the 2750th complied with Air Force directives to transfer its 24 support aircraft to other units. For financial reasons, Headquarters Air Force decided, in late 1974, to drop nearly 400 aging, administrative support aircraft from its active inventory. It released a tentative disposition schedule affecting 343 aircraft, including both reciprocating engine aircraft and jet engine T-33s. Among the airplanes involved were six T-33s from Wright-Patterson, scheduled for transfer to the Military Aircraft Storage and Disposition Center (MASDC) at Davis-Monthan Air Force Base, Arizona, in February 1975.

In early December 1974, a more comprehensive, phase-out schedule covered the remainder of the 2750th ABW's support fleet. The wing transferred its remaining aircraft, except for five T-39s, to MASDC by the end of the fiscal year. Meanwhile, Military Airlift Command (MAC) was designated single manager for pooled T-39 aircraft located in the continental United States. Wright-Patterson became one of 15 host bases, and MAC subsequently announced that the base would bed down nine T-39s, including the five transferred to MAC from the 2750th ABW. On April 21, 1975, MAC initiated central scheduling for a portion of its new T-39 fleet, with scheduling for Wright-Patterson starting on June 20. The newly established Detachment 2, 1401st Military Airlift Squadron (MAS), operated the five T-39s reassigned to Military Airlift Command. Initial personnel for this detachment were reassigned from the

Although the 2750th Air Base Wing dispersed all of its administrative support aircraft during 1974 and 1975, the wing's Operations and Training Division retained management of the Wright-Patterson Air Force Base aerodrome.

In 1975, Wright-Patterson Air Force Base was one of 15 bases hosting Military Airlift Command (MAC) T-39 aircraft. The MAC fleet was assigned to Detachment 2, 1401st Military Airlift Squadron.

2750th ABW's Flight Operations Branch.

Although it did not possess aircraft, the air base wing continued operating the airfield and supporting the 4950th Test Wing; Detachment 2, 1401st MAS; and transient aircraft with supplies, petroleum, oil, and lubricants. Effective July 1, 1975, the 4950th Test Wing assumed base-level maintenance, including the transient alert function and such responsibilities as chief of maintenance, quality control, maintenance control, organizational maintenance, field maintenance, survival equipment maintenance, avionics maintenance, and the Precision Measurement Equipment Laboratory (PMEL).

The transfer of support aircraft also

affected the air base wing's Simulator Training Branch, which ceased operations, except that of the T-40 trainer, as of July 1, 1975. The branch's equipment transferred to the 4950th Test Wing in September, and instructor and maintenance personnel followed in October. The 2750th also transferred facility responsibility for a number of Area C buildings, including Buildings 13, 105, 148, 152, 169, 206-North, 256, and 884, as well as Building 188 in Area B, to the 4950th Test Wing.

For the first time since 1948, the 2750th Air Base Wing did not possess its own aircraft or have an active flying mission. It had also lost most support functions associated with flying operations. Although the wing continued to operate

Firefighters practice putting out an aircraft fire.

the airfield and support tenant flying organizations, the scope of these operations shrank significantly on June 1, 1976, when the Air Force officially closed the Area B aerodrome, ending 50 years of flight operations at the former Wright Field.

Meanwhile, Wright-Patterson secured the base's future flying operations through development of an Air Installation Compatible Use Zone (AICUZ) around the installation. Due to the growth of the base and its surrounding communities, air base wing commanders had been concerned for many years of potential adverse impacts of business and residential encroachment on Wright-Patterson's flying operations. As early as 1962, Base Commander Brigadier General Glen J. McClernon held meetings with community leaders concerning land use and development in areas adjacent to the installation. In October 1966, a Wright-Patterson Air Force Base Airport Zoning Regulation became law. Although subsequently challenged and rescinded, this basic document provided a firm foundation for the AICUZ concept. Its key element was an atmosphere of mutual trust and helpfulness between the base and surrounding communities.

AICUZ was a community interface

program designed to coordinate the needs of the Air Force with the development of surrounding communities to assure continuance of Wright Patterson Air Force Base as a center of flying operations. Through the program, the Air Force hoped to achieve compatible land uses around military installations. While Air Force officials were concerned about the development of land near Air Force bases, they also recognized that the land presented an intrinsically attractive area for development and they had a responsibility to protect public areas surrounding their airfields from noise, pollution, and flight hazards. AICUZ prevented on-base and off-base development from adversely impacting flight operations at the airfield.

In May 1975, after years of careful planning, the final AICUZ study was released to the public. The base commander invited state and local officials and Ohio's U.S. congressional representatives to attend a special briefing on the study. In July, the installation and its four adjoining counties enacted a Wright-Patterson Air Force Base Airport Zoning Regulation. This regulation defined noise and accident zones around the air base and suggested compatible land use for the zones. Some areas remained open

space, while more densely settled areas were available for limited construction, with noise-reduction features designed into buildings and other planning measures.

Another milestone in Wright-Patterson's airfield management occurred on September 20, 1978, with the decommissioning of the precision approach AN/FPN-16 radar in Area C after 26 years of continuous operations. A new, solid-state, dual instrument landing system assisted pilots using Runway 5L23R. With the new equipment, pilots received control tower permission to land and then automatically received required data to make instrument landings without further assistance.

In 1978, the Federal Aviation Administration (FAA) also relocated the Dayton approach control facility, activated in 1957, from Wright-Patterson Air Force Base to Dayton International Airport at Vandalia, Ohio. A memorandum of understanding between the 2750th Air Base Wing; the 2046th Communications Group; Detachment 15, 15th Weather Squadron; and the FAA terminated in October. The FAA also released space it occupied in Buildings 206 and 841, Area C. The air base wing's Operations and Training Division continued to manage the

THE ORIGIN OF FFO

Within the continental United States, the airfield at Wright-Patterson Air Force Base, Ohio, is referred to simply as FFO. Within the international aviation community, and on most aviation maps and charts, the letter K, which represents an airport located in the United States, is prefixed to form KFFO. Aviators around the world are familiar with KFFO and FFO, but the origin of the designation is cloudy.

The earliest military flying field associated with Wright-Patterson was Wilbur Wright Field (1917-1925), located on what later became Areas A and C. Wilbur Wright Field served as a training field for pilots. Fairfield Air Depot (1917-1946) was adjacent to the flying field. The two important facilities were given the airfield designator FFO.

At the time of the assignment of the FFO designator (about 1917), the towns of Fairfield and Osborn were in close proximity to the field. Fairfield was located directly east of the airfield and Osborn was to the west. The little town of Fairfield was growing due to the airfield's flying operations and the influx of jobs. Osborn, which had boomed in the mid-1800s from the railroad industry, faced a major problem. Most of the town was located on the Mad River floodplain west of the flying field and was prone to flooding. To solve the problem, in the early 1920s, Osborn literally moved directly east of the town of Fairfield, placing Fairfield between Osborn and Wilbur Wright Field. Fairfield continued to prosper from its desirable location next to the flying field and the depot, but it could not expand. Osborn, on the other hand, had growth potential but needed renewed prosperity. Consequently, in 1950, Fairfield and Osborn merged to become the village of Fairborn.

The FFO designator was issued around 1917, long before the merger of the two towns. According to the Wright-Patterson airfield supervisor, the designator FFO was based on the proximity of the two towns of **F**airfield and **O**sborn to the airfield before Osborn moved. According to the written history of Fairborn, FFO derived from **F**airfield, **O**hio, because it was directly affected by the flying industry. Both sources are reliable, but further documentation has not been discovered to substantiate either theory.

Side note: The designator for the airfield located on Wright-Patterson's Area B, adjacent to the United States Air Force Museum, was (K)DWF, which stood for Dayton-Wright Field. The DWF airfield closed in 1976, and the International Civil Aviation Organization no longer recognized the designator. Wright-Patterson Air Force Base, however, continued to use the field unofficially when an aircraft was delivered by air to the museum.

Sources: Chamber of Commerce, Fairborn, Ohio, "Community Profile," viewed online July 18, 2003, at http://www.nationjob.com/showcomp.cgi/faoh.html; Gary R. Chandler, Wright-Patterson Air Force Base Airfield Supervisor, personal communication with Master Sergeant David L. Wolf, Aeronautical Systems Center History Office, May 13, 2003.

Air traffic controllers from the 2046th Communications Group supervise runway operations from the base control tower.

Wright-Patterson aerodrome in coordination with other Air Force and federal government agencies.

Contracting Operations

The post-Vietnam drawdown spawned a series of A-76 cost comparison studies that converted many activities at Wright-Patterson Air Force Base from government to contractor operations. This wave of studies continued into the twenty-first century. Cost comparison studies were designed to determine the most cost-efficient way to deliver government services. Some resulted in the government continuing to perform services, albeit with greatly reduced resources; others led to operations being contracted out to businesses in the private sector. In either case, base operations support was directly affected. The 2750th Air Base Wing conducted several studies of its administrative and logistics functions. As a result, the base gradually ceased performing its own laundry, fuels management, audiovisual services, publications distribution, unclassified mail distribution, commissary shelf stocking, and custodial services, and turned them over to contractors.

Environmental Concerns

An active public concern with environmental matters in the 1980s gave impetus to an aggressive environmental management program at Wright-Patterson Air Force Base. In 1981, following congressional passage of the Comprehensive Environmental Response, Compensation and Liability Act, the installation's environmental managers adopted a restoration program focused on identifying, controlling, and eliminating environmental contamination from past base operations and disposal practices. The program began with an extensive survey that identified 27 potentially contaminated sites on base. Studies determined the presence, nature, magnitude, and potential migration of any contaminants in the sites. After analyzing the results, the base developed remediation action plans to clean up and monitor the sites. In one case, for instance, Hadden Park, a 65-acre facility complete with hiking trails and sport facilities, had to be closed in 1985 due to environmental

Recycling became a major activity at Wright-Patterson in the 1980s.

concerns associated with adjacent landfills. As part of its ongoing work, environmental managers from the air base wing worked closely with the U.S. and Ohio Environmental Protection Agencies to develop standards and agreements conforming to regulatory requirements.

Environmental concerns at Wright-Patterson also extended to issues of radiation, asbestos, lead paint, water quality, and recycling. Air quality became another area of great interest. Wright-Patterson invested $37 million to modify and modernize the base's coal-fired heating plants. Six new, large-capacity boilers replaced 17 antiquated units that dated from the 1930s, and three of the base's five heating plants closed. Together with new electrostatic precipitators that removed nearly all particulate matter exhausted from the heating plants, Wright-Patterson met federal and state environmental regulations affecting particulate and sulfur-dioxide emissions. Overall, the new system provided Wright-Patterson with the most modern, solid-fuel boiler plants and fuel-handling facilities in the Air Force.

Computer Technology

While computer technology had already made an imprint at Wright-Patterson, the post-Vietnam era witnessed rapid growth of automation as a tool for conducting routine business. In late 1974, the AFLC deputy chief of staff for Procurement and Production announced that the 2750th ABW had been selected as the command's "lead base" for implementing the Customer Integrated Automated Procurement System (CIAPS). The Air Force Data Systems Design Center developed this automated system to handle computer-produced delivery orders for the Federal Supply Schedule for all Air Force base procurement activities. It used a Burroughs B-3500 computer as an automated link with the Base Supply UNIVAC 1050-II computer system and the Medical Supply B-3500 computer system. Wright-Patterson finished implementing the new system in 1976.[20]

Digital technology seemed to be everywhere. In 1982, the base transitioned from its 1940s electro-mechanical telephone switching mechanism to the Digital Multiplex System (DMS-100), a computer-controlled, electronic, digital telephone switch system. This event coincided with the establishment of the first, Air Force-managed, Defense Metropolitan Area Telephone System (DMATS). Within the 2750th Air Base Wing, a word-processing center replaced the old typing pool in 1980. By mid-decade, the wing activated the Base Headquarters Automated Notification Network that linked wing agencies via a package of office automation and communication tools. Wright-Patterson had e-mail. As the decade progressed, microcomputer technology was increasingly integrated into the wing and base work centers.

Base Celebrations

Several major celebrations highlighted these years at Wright-Patterson. Among the most exciting activities in the 1970s included those associated with celebrating the nation's bicentennial. Throughout America, the bicentennial observance was divided into three themes: Heritage '76,

An Employees' Monument, located at the corner of Skeel Avenue and Novick Road, was dedicated during the bicentennial to honor all military members and civilian employees who have ever worked at Wright-Patterson.

Flag-lined streets of Brick Quarters in Area A

Wright-Patterson was recognized as an official Bicentennial United States Air Force Installation in 1975. The base hosted a wide range of activities in connection with the nation's 200th birthday, including a People for People Festival.

Festival USA, and Horizons '76. The American Revolution Bicentennial Commission officially recognized Wright-Patterson as a "Bicentennial United States Air Force Installation" on August 19, 1975. Wright Patterson established separate committees to carry out programs based on all three themes. The Wright-Patterson Air Force Base Planning and Coordinating Committee, chaired by the base commander, coordinated overall direction of the base-wide program. During autumn and winter of 1975-1976, the committees coordinated with each other and with their counterparts in local communities to assure the best possible local and area observance of the nation's 200th birthday.

Wright-Patterson opened its schedule of events with a Community Day on May 24, 1975, one week after Armed Forces Day. An acrobatic demonstration by the U.S. Air Force Thunderbirds highlighted the six-hour program. The Ninth Virginia Regiment, attired in Revolutionary War uniforms and regalia, gave two demonstration drills with muskets. (The original regiment was organized November 19, 1776, and fought in many major campaigns, including incursions into the Ohio River Valley, during the Revolutionary War.)

The United States Air Force Museum hosted activities in July 1976. From July 11 to 13, an Ohio Region Bicentennial Boy Scout Jamboree on the museum's grounds attracted nearly 10,000 scouts. On July 23, the museum dedicated its new $1 million Visitors Reception Center, a gift from the Air Force Museum Foundation. Senator Barry M. Goldwater was principal speaker at the ceremonies. Secretary of the Air Force Thomas C. Reed cut the symbolic ribbon to open the center. The scissors he used had belonged to Orville Wright and were loaned for the occasion by Mr. Wright's niece, Mrs. Ivonette Wright Miller.

The highlight of the base's bicentennial celebration, however, was the grand People for People Festival held September 20, which attracted nearly 16,000 visitors. The festival's purposes were to "promote human relations through awareness, communication, and understanding," and to provide a suitable program to represent the heritage of America "through arts,

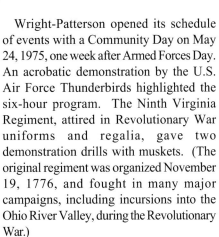

crafts, drama, display, dance, cuisine, fashion, music, and song."

Another initiative honored Air Force members of past decades. Wright-Patterson established a Base Memorialization Committee to name appropriate streets, buildings, recreational areas, and medical facilities in honor of distinguished members of the Air Force. Over the next few years, numerous facilities across the installation were memorialized (see Appendix 7).

The final observance in the yearlong bicentennial celebration was the dedication of the Employees' Monument to honor all military and civilian employees who had worked at Wright-Patterson Air Force Base throughout its long history. The monument was a stainless steel sculpture, six feet high and approximately 20 feet tip to tip, mounted on a reinforced concrete pedestal. Wright State University art student Ray Williams designed the sculpture, described as an abstract of upswept wings symbolizing man's reach toward outer space. The monument was erected in the spring of 1977 at the corner of Skeel Avenue and Novick Road in Area A, overlooking Huffman Prairie. Sadly, it was removed a decade later due to serious weather corrosion.

In 1978, the base and the Dayton community celebrated the 75th anniversary of powered flight. A special steering committee of the Greater Dayton Area Chamber of Commerce coordinated many local activities. Retired Lieutenant General James T. Stewart, who had commanded Aeronautical Systems Division from June

Hot air balloon rally on the grounds of the United States Air Force Museum

1970 to August 1976, chaired the committee. Wright-Patterson played a crucial role in the observances and adopted a special anniversary logo, which it displayed on all letters posted during the year.

On September 9, 1978, the Miami Conservancy District transferred three parcels of land that had been intimately associated with the Wright brothers to the United States Air Force and the 2750th Air Base Wing. The first was a 0.52-acre plot on Pylon Road in Area A, on which the Wilbur and Orville Wright Commission, in cooperation with the Miami Conservancy District, had erected a concrete pylon in 1941. The marker commemorated the starting point of the oval flight path flown by the Wright brothers. The other two parcels of land comprised a 27-acre park and memorial on Wright Brothers Hill dedicated to the memory of the Wright brothers. The park was home to the Wright Memorial and several Adena culture burial mounds dated between 500 B.C. and A.D. 200. (It is located in Area B near the intersection of State Route 444 and Kauffman Avenue.)

The Wright Memorial was originally designed, planned, and constructed between 1922 and 1944 by the Wright Memorial Committee, the Miami Conservancy District, and the National Park Service, in consultation with the Olmsted Brothers landscape architectural

On May 18, 1983, area residents were treated to the sight of the Space Shuttle *Enterprise* mounted atop a modified Boeing 747. The 747 made a low-level flight over the area so people could see the piggybacked aircraft before it landed for an overnight refueling stop at Wright-Patterson. Thousands of spectators came to the base for a public viewing of the shuttle, en route to the Paris Air Show.

firm and Orville Wright. The Dayton community funded the project. At the center of the memorial was built a 17-foot-high obelisk of pink, North Carolina granite from the same quarry used for the construction of Wright Brothers National Memorial in Kitty Hawk, North Carolina. A bronze plaque commemorating the Wrights' accomplishments adorns the shaft, and four smaller, bronze plaques on the memorial's

inner wall list the names of the 119 flyers trained at Huffman Prairie; the contributions Wright-Patterson has made to aviation; the Wrights' work at Huffman Prairie; and directions to the burial mounds on the hill. An African-American camp of the Civilian Conservation Corps provided the grading, roadwork, and landscaping for the area. After completion, the Miami Conservancy District cared for the memorial and its

EVOLUTION OF BASE NEWSPAPERS

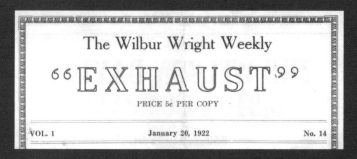

Since 1917, weekly newspapers have been communicating information to the installation's workers and residents. Wilbur Wright Field started publishing *Aviation Weekly* on September 21, 1917, for personnel and students at the aviation school. The paper covered news of the school, in addition to general aviation news, technical discussions submitted by readers, and social and athletic events taking place on the field. The following year, the field introduced a new weekly, *The Wright Idea*, to meet the needs of its postwar workforce. The Wilbur Wright Field *Exhaust*, a non-government, subscription publication produced in Fairfield, made its appearance in 1921. This 5-cent-per-copy newspaper targeted the workforce at both the airfield and Fairfield Air Depot. Yet another paper emerged later in the 1920s with the publication of the *Wright Field Supercharger*, aimed at the personnel at Wright Field.

World War II fostered a massive personnel expansion at Wright and Patterson fields, and a series of newspapers served the large, diverse community. Air Service Command published *The Mascot* for civilians working in Area A of Wright Field, *The Wright Flyer* for civilians in Area B, and *Take-Off* for Wright Field's enlisted personnel. *The Mascot* and *The Wright Flyer* combined in 1944 to create a single civilian newspaper, *The Wright Flyer*. This endeavor briefly discontinued in 1947, but based on popular demand, Headquarters Air Materiel Command began publishing it again under the sponsorship of the Wright-Patterson Welfare Fund Council. The *Patterson Field Postings* served the logistics community. Air Technical Service Command also published *Wright Field Air-Tec Skylines* in 1945 for all members of the command, and the Technical Data Section of the Maintenance Division published its own *Skyliner* beginning in 1942.

The Wright Flyer had to be renamed in 1947 because of a War Department circular stating that the name of an installation could not be used in civilian newspapers. Consequently, the *Skywriter* began distribution on December 19, 1947. The *Skywriter* continued to be a base-sponsored venture before being merged, on June 18, 1948, with the *Post Script*, a newspaper designed initially to serve the military community of Wright-Patterson Air Force Base. The *Post Script* remained the only base newspaper until late 1955, when it was replaced by *Wingspread* magazine. *Wingspread*, also a commercial publication, met the needs of all employees and organizations on base. The last phase in this journalistic evolution arrived in February 1960, when the Wright-Patterson Air Force Base *Skywrighter*, a base-sponsored, commercial paper, made its appearance. It became the longest running base newspaper, continuing to publish into the twenty-first century.

The half-acre plot of land on Pylon Road in Area A commemorating the site of the Wright brothers' first hangar was one of three parcels transferred from the Miami Conservancy District to Wright-Patterson Air Force Base in September 1978.

While a bugler sounded taps, two T-39 aircraft flew over the Wright Memorial during the December 17, 1978, ceremony to commemorate 75 years of powered flight. The 27-acre Wright Brothers Hill park and memorial were transferred to the Air Force on September 9, 1978.

Ivonette Miller and Horace Wright, niece and nephew of Wilbur and Orville Wright, placed ceremonial wreaths on the Wright Memorial at the 1978 dedication. Wreaths are laid at the memorial each December 17 in honor of the Wrights' achievements.

surrounding park until it was transferred to the Air Force in 1978.

About 150 guests and visitors attended the formal ceremony marking the transfer of the properties. Robert S. Oelman, president of the Miami Conservancy District Board of Directors, conveyed the original copy of the special warranty deed to General Bryce Poe II, AFLC commander. In his remarks, General Poe stated that it was fitting that the memorial, "which for 38 years has rested on Miami Conservancy District land is now a part of Wright-Patterson—just as Huffman Prairie—also once Conservancy land, is now part of the Base." Among the distinguished attendees were two U.S. congressmen: Clarence J. Brown of Urbana and Charles W. Whalen

of Dayton, as well as Ivonette Wright Miller of Dayton and Horace A. Wright of Xenia, niece and nephew of Wilbur and Orville Wright.

Wright-Patterson received an unusual honor later in the month. On September 23, the Dayton Stamp Club and the 75th Anniversary of Powered Flight Committee, supported by the United States Air Force Museum, hosted the first-day sale of two commemorative stamps honoring the Wright brothers. U.S. Postmaster General William F. Bolger spoke at the ceremony preceding initial sale of two differently designed 31-cent airmail stamps.

On December 16, the eve of the official anniversary, 700 persons attended a First Flight Banquet at the Dayton Convention

and Exhibition Center. Participating dignitaries included Lowell Thomas, famous newscaster and author; retired Lieutenant General James Doolittle; former astronaut Neil Armstrong, first man to walk on the moon; and Milton Caniff, a nationally known cartoonist (creator of the Steve Canyon series).

The morning of December 17 marked the actual anniversary of the first flight. Base employees, local citizens, and other visitors began arriving early on Wright Brothers Hill. General Bryce Poe and Lieutenant General James Stewart began the formal ceremonies with appropriate remarks. Ivonette Miller and Horace Wright then laid large wreaths at the base of the granite Wright Memorial. At 10:35 a.m., the precise hour and minute of Orville Wright's historic first lift-off at Kitty Hawk, North Carolina, on December 17, 1903, a bugler sounded taps and two T-39 aircraft flew overhead. The ceremony served as a fitting tribute to the spirit and accomplishments of the Wright brothers and to the thousands who followed them in the intervening 75 years. It was also the first of what became a tradition at Wright-Patterson Air Force Base—the annual First Flight Ceremony.

From September 18-20, 1987, Wright-Patterson Air Force Base also hosted official observance of the United States Air Force's 40th anniversary celebration. Wright-Patterson was chosen for this honor because the event coincided with the 70th anniversary of McCook Field and the 60th anniversary of the dedication of

A flag and aircraft display erected in front of AFLC headquarters to celebrate the command's 45th anniversary in 1989

The nation's prisoners of war and missing in action are remembered at base ceremonies.

FROM DESERT STORM TO IRAQI FREEDOM

United States President George H. W. Bush and Soviet President Mikhail Gorbachev signed the Strategic Arms Reduction Treaty on July 31, 1991. The treaty, coupled with the collapse of communism, marked the end of the Cold War. Wright-Patterson and the nation quickly returned to a traditional peacetime status for the first time since World War II. This transition, like previous ones, caused dynamic restructuring, consolidations, personnel reductions, budget pressures, and new modes of business. Before this transition began, however, the nation found itself at war with a new enemy in the Middle East. For the next decade, Wright-Patterson supported contingency operations around the globe, hosted international peace talks, and continued to perform its traditional missions of research, development, acquisition, logistics, and education.

At the conclusion of previous wars, America's armed forces had demobilized. The 1990s experienced similar cutbacks with the end of the Cold War. Plans for major defense restructuring started well before the Persian Gulf War (Operation

Wright Field. The Air Force's Festival of Flight opened with a Friday afternoon ceremony on Huffman Prairie Flying Field punctuated by Wright B Flyer and F-16 flyovers. On Saturday, the festival moved to the site of Wright Field in Area B where the weekend's activities were conducted under the theme "Carrying Out the Grand Tradition." Events included an aerial parade of 30 active and retired aircraft from the Wright B Flyer to the F-15 and aerial demonstrations by F-15 and F-16 fighter aircraft. A special ceremony honored those Americans who died in war or were missing in action. A fair with amusements, athletic contests, and entertainers ranging from the Air Force's Tops in Blue talent troupe to singer Glenn Campbell captivated the crowds. The celebration concluded with a formal banquet at the Dayton Convention Center.

The *Spirit of Texas* was the first B-2 bomber to visit Wright-Patterson Air Force Base, June 16, 1995. (*U.S. Air Force photo by Spencer P. Lane*)

A C-17 Globemaster III from Charleston Air Force Base, South Carolina, leads an airdrop formation. The C-17 was capable of rapid strategic delivery of troops and cargo to main operating bases or forward bases in deployment areas. (*U.S. Air Force photo by Staff Sergeant Jeffrey Allen*)

command, was redesignated Aeronautical Systems Center (ASC) on July 1, 1992. ASC remained responsible for research, development, test, evaluation, and initial acquisition of aeronautical systems, munitions systems, and related Air Force equipment. As part of a wider Air Force reform to simplify base management by establishing standard organizational frameworks across Air Force bases, however, ASC became more than a product center. It became host organization for Wright-Patterson Air Force Base, and its commander assumed the duties of installation commander. The 2750th Air Base Wing, which had been host organization, was assigned to ASC, as was the 74th Medical Group.[21]

Other restructuring actions included inactivation of the Wright-Patterson Contracting Center and assignment of its functions to ASC's Contracting Directorate; activation of the Joint Logistics Systems Center; and reassignment of the Foreign Technology Division from AFSC to Air Force Intelligence Command. Weather support resources were decentralized and placed under the direct command of operational and support units. The Logistics Communications Division inactivated, as did the Air Force Contract Maintenance Center, whose operations moved to the Defense Contract Management Center International also located at Wright-Patterson. The base's flight test mission also officially ended in 1994, when the 4950th Test Wing inactivated and the function moved to Edwards Air Force Base, California. This brought to a close 90 years of flight testing activity in the Miami Valley.

Several restructurings affected the laboratories. First, the Air Force consolidated its 14 laboratories into four "super" laboratories: Armstrong Laboratory, Phillips Laboratory, Rome Laboratory, and Wright Laboratory. Wright Research and Development Center became Wright Laboratory and its seven laboratories were assigned as directorates. The unit remained at Wright-Patterson and was assigned to Aeronautical Systems Center. The Harry G. Armstrong Aerospace Medical Research Laboratory on base became a detachment of the Armstrong Laboratory headquartered at Brooks Air Force Base, Texas. Another laboratory reshuffling came in April 1997, when AFMC activated the Air Force Research Laboratory (AFRL), headquartered at Wright-Patterson Air Force Base. AFMC then reassigned the four Air Force laboratories, together with the Air Force

Desert Storm), which was too brief to have an effect on downsizing. Dollar savings primarily drove the drawdown, with the armed services increasingly looking to technology and business-model efficiencies to save money and maximize the effectiveness of their shrinking forces. In early 1990, the Department of Defense proposed saving the Air Force $10.9 billion over a five-year period through job reductions and command mergers. A new Air Force soon took shape as the Cold War-era commands yielded to new organizations. Strategic Air Command, Tactical Air Command, and Military Airlift Command were consigned to oblivion. Air Combat Command rose to manage the bomber and fighter fleets; Air Mobility Command assumed control of tankers and cargo aircraft; and Space Command inherited intercontinental ballistic missiles.

Wright-Patterson also experienced major changes when the Air Force restructured

its acquisition and logistics missions on June 30, 1992. The Air Force disestablished Air Force Systems Command and Air Force Logistics Command and activated in their stead Air Force Materiel Command (AFMC), headquartered in Building 262 at Wright-Patterson Air Force Base, on July 1, 1992. The new command represented the Air Force response to national policies mandating greater control of defense acquisition by civilians and the Department of Defense, and fewer layers of bureaucracy. AFMC created a single, streamlined organization for managing all Air Force research, development, test, acquisition, and support functions. It restored "cradle to grave" responsibility for weapon systems to a single command, as had existed prior to 1950.

The establishment of AFMC profoundly affected units at Wright-Patterson. Aeronautical Systems Division, one of four product centers assigned to the new

Office of Scientific Research, to AFRL. Wright Laboratory, likewise, inactivated as AFRL absorbed its assets.

Loss of Wright Laboratory was not the only change Aeronautical Systems Center experienced in 1998. That same year, ASC transferred its munitions systems management to a newly activated Air Armament Center at Eglin Air Force Base, Florida. In turn, ASC acquired the human systems development and procurement missions from Human Systems Center at Brooks, when it was reorganized as the 311th Human Systems Wing and assigned to ASC.

The work traditionally performed at Wright-Patterson Air Force Base continued despite the restructurings. Research, development, acquisition, logistics, and training remained the installation's major activities. Wright-Patterson success stories in the 1990s included deployment of the B-2 stealth bomber, C-17 transport, and T-1A trainer. Several C-135 airframes were modified to support the Open Skies Treaty mission and the F/A-22 Raptor jumped from drawing board to test flight. With the RQ-1 Predator and RQ-4 Global Hawk programs, the base began managing a new generation of aircraft—unmanned aerial vehicles. Wright-Patterson also became home to one of four major, Department of Defense, high-performance computing centers and the world's eighth largest supercomputer.

The manner in which Wright-Patterson performed its traditional work changed significantly. Acquisition reform reached a fevered pitch in the 1990s as federal law, coupled with Department of Defense and Air Force initiatives, transformed the acquisition process to produce systems faster, cheaper, and with greater reliability. Air Force Materiel Command developed long-term strategic plans, focused on core missions, and freed up resources for other defense priorities. AFMC adopted business practices and began operating more like a commercial enterprise when assessing cost and value in the development of its products. It contracted out to commercial businesses those activities that could be performed cheaper or better by the private sector. It turned to communications-computer technology and electronic commerce to economize and accelerate the acquisition process. The command reengineered its inventory and maintenance management functions to reduce the number of spares in the pipeline and accelerated delivery times. Adoption of Integrated Weapon System Management (IWSM)—a philosophical

The RQ-1 Predator was a medium-altitude, long-endurance, unmanned aerial vehicle. The fully operational system consisted of four air vehicles (with sensors), a ground control station, a Predator primary satellite link communication suite, and a 55-member ground crew.

cornerstone of the new command—integrated all functional, life-cycle requirements into the development of weapon systems. Greater agility, speedier decisions, and more effective handling of multiple issues enabled AFMC to better support operational forces and achieve the Air Force's new tenet of "better, faster, and cheaper."

Air Force Materiel Command even invigorated and systematized its technology transfer process to promote transition of technology from the laboratory to advanced weapons. These efforts challenged a perceived adversarial relationship between the nation's military technology community and its industrial base. AFMC actively encouraged transferring military technology to the industrial sector and increasingly aimed military investment strategies toward technologies that had commercial applications.

Force restructuring played a major role in acquisition management as well. The Air Force scaled back resources and eliminated excess capacity. AFMC reduced its test and evaluation infrastructure, including aircraft and personnel. The command closed several of its depots and other facilities. The San Antonio Air Logistics Center at Kelly Air Force Base, Texas, and the Sacramento Air Logistics Center at McClellan Air Force Base, California, were closed. Newark Air Force Base, Ohio, closed in 1996 following privatization of the Aerospace Guidance and Metrology Center.

The last decade of the twentieth century

A maintenance technician works on a C-21 assigned to the 47th Airlift Flight. (*U.S. Air Force photo by Spencer P. Lane*)

also brought formidable changes to the air base wing and base operations support. These changes began with several redesignations. On October 1, 1992, the 2750th Air Base Wing became the 645th Air Base Wing; its subordinate units also received the "645" designator. The renumbering occurred at the direction of General Merrill McPeak, U.S. Air Force chief of staff, who wanted to eliminate four-digit unit designations and preserve the Air Force's most historic and prestigious unit designators. The change did not alter the wing's distinguished lineage and honors. On October 1, 1994, another redesignation made the unit the 88th Air Base Wing. Again, General McPeak directed the renumbering to eliminate unnecessary three-digit designators. The "88" designator also applied to the wing's subordinate units but did not alter the wing's lineage and honors.

The organization and mission of the wing also changed. During 1990, the AFLC Command Center was realigned from the AFLC Logistics Operation Center to the air base wing, as was the 2046th Communications Group. As mentioned previously, the 2750th Air Base Wing was reassigned to Aeronautical Systems Center effective July 1, 1992. At the same time, the duties of installation commander, performed by the wing commander since World War II, transferred to the ASC commander. The ASC commander, in turn, delegated many of the responsibilities back to the wing commander. As a result of the transfer, however, many inherently base operations and support functions moved from the air base wing to ASC, including finance, contracting, inspector general, protocol, public affairs, human resources, and for brief periods, communications and the chaplaincy. Despite these changes, the air base wing managed to garner its fifth and sixth Air Force Outstanding Unit Awards. The fifth award, for the period January 1, 1990, to December 31, 1991, recognized the wing's role in supporting the Persian Gulf War (see below). The sixth award (January 1, 1995, to December 30, 1995) honored the wing's outstanding work hosting the Balkan Proximity Peace Talks (see below).

All base organizations contended with serious manpower reductions and their consequences. Force restructuring, mission changes, and a greatly expanded series of A-76 cost comparison studies powered the reduction. Beginning in 1993, a steady stream of civilians, spurred by cash incentive programs, left government service at Wright-Patterson. These voluntary retirements eliminated, or greatly reduced, involuntary reductions-in-force. For those employees who remained, constant personnel turnover, reassignments, and heavier workloads became routine.

The A-76 cost comparison studies had a great impact on the wing and base support operations as they reduced wing resources, transferred work to contractors, and cut back the scope of wing operations. The studies officially began in 1996 and continued through 2000. They affected Aeronautical Systems Center and the 74th Medical Group, as well as the 88th Air Base

Wright-Patterson Air Force Base became a forward operating location for the National Airborne Operations Center's E-4B in 1994.

U.S. Air Force team chief Lieutenant Colonel Den Hovatter and Russian team chief Colonel Fedor Semkin discuss their Open Skies Treaty joint flight by the Russian Tu-154 Open Skies aircraft on the Wright-Patterson flightline, July 27, 2000. (*U.S. Air Force photo by Spencer P. Lane*)

Wing. Following the studies, most of the air base wing's logistics work (transportation, supply, and equipment maintenance) converted to contract operations. The 88th ABW's communications group, civil engineering directorate, and environmental management office experienced substantial manpower reductions, with the latter two losing all of their military personnel.

Defense reductions and consolidations directly influenced base-level operations. The 88th Air Base Wing gained customers and expanded the geographic area it serviced as installations and units closed.

The wing also lost some responsibilities. Wright-Patterson closed its Correctional Custody Facility in 1990. In 1992, the Defense Finance and Accounting Service absorbed the base's payroll and other financial work, and the base hosted its last Air Force ROTC summer training program. With the growth of the Internet, the Air Force Personnel Center in Texas gradually assumed direct control of many personnel management services. An August 17, 1993, ceremony at Building 143 marked the closure of the base railroad after 75 years of operation. In 1996, the Aircraft Tire Storage and Distribution Point ceased

LOGAIR operations at Wright-Patterson ended in 1992 after 38 years of service.

Technicians install the world's eighth largest supercomputer in Wright-Patterson's Major Shared Resource Center. The Compaq ES-45, with 836 processors operating at 1,000 MHz, had a peak performance capability of 1.67 trillion calculations per second. (*U.S. Air Force photo by Spencer P. Lane*)

operations after 30 years. Three years later, the Command Publications Distribution Center closed.

Other Wright-Patterson organizations had similar experiences. In 1999, the Wright-Patterson Medical Center closed its aerospace physiological training unit, ending its physiological training for aircrews. The center also terminated operation of its hyperbaric altitude chamber after 47 years. Simultaneously, Wright-Patterson Medical Center and Kettering [Ohio] Medical Center implemented the first-ever, medical service, cooperative research and development agreement in the Air Force. Under its terms, Kettering Medical Center assumed operation of the hyperbaric chamber.

The Air Force Research Laboratory also turned over some functions to contractors. In June 2000, the laboratory transferred its maintenance and operations responsibility for six base wind tunnels and a water-flow tunnel to Ohio State University under terms of a five-year contract.

Yet, the arrival of new customers, services, and capabilities offset the loss of old ones. The Persian Gulf War, global contingency operations, the Peace Talks, and Operations Enduring Freedom, Noble Eagle, and Iraqi Freedom kept the air base wing and Wright-Patterson hopping. From 1993 to 1997, the air base wing even regained an active flying mission with assignment of the 47th Airlift Flight. The flight was an administrative support unit

that flew C-12 (until 1994) and C-21 aircraft. It was reassigned to Air Mobility Command in 1997 but remained at Wright-Patterson as a tenant unit. Wright-Patterson also became a forward operating location for the National Airborne Operations Center's E-4 aircraft in 1994.

A reorganization of meteorological units in 1991 made the weather unit an organic component of the air base wing. Activation of an integrated, automated, weather communications system further enhanced Wright-Patterson's weather capabilities. In addition to operating the base weather station, the 88th Weather Squadron performed specialized weather support. It provided scientific and technical support to AFRL, ASC, and a host of other military and civil agencies. It evaluated and exploited environmental factors for current and future aerospace weapon systems and components. Its services varied from training Baltic nation teams on weather support for the Open Skies mission to assisting acquisition managers with the Global Hawk unmanned aerial vehicle (UAV) program. The squadron also served as lead agency for collecting weather sensitivity data for Air Force weapons, sub-systems, and components.

Signing of the 1992 Open Skies Treaty brought an international mission to Wright-Patterson Air Force Base. The treaty permitted unarmed reconnaissance flights by specially configured aircraft over more than 25 North Atlantic Treaty Organization (NATO) and former Warsaw Pact nations, including the United States and Russia. Almost immediately, the On Site Inspection Agency began using Wright-Patterson as a staging area for mock certification and trial flights. ASC's Developmental Manufacturing and Modification Facility converted two WC-135B's into OC-135B treaty aircraft. The National Air Intelligence Center (NAIC, successor of the Foreign Technology Division) processed media gathered by the flights. In 1995, Wright-Patterson Air Force Base was designated primary sensor target for Open Skies missions, an official Open Skies airfield, and the media processing facility (operated by NAIC). It hosted the first foreign aircraft to fly an Open Skies surveillance mission in June 1995.

In 1992, LOGAIR operations that had begun on the base airfield in 1954 ended. The next year, the air base wing assumed operations of the Precision Measurement Equipment Laboratory (PMEL) and transient alert functions from the departing 4950th Test Wing. During Operation Uphold Democracy in Haiti in 1994, the 47th

Airlift Flight stood alert and launched two aircrews to support the airlift into Port-au-Prince.

Airfield personnel responded to an unusual emergency on November 7, 1996, when a Delta Air Lines MD-11 commercial jet made an emergency landing at the base. The flight en route to Cincinnati from London had been diverted due to thunderstorms. Flap irregularities forced it to make an emergency landing. The aircraft, carrying 210 passengers, landed just as a tornado warning was issued for the base. Its crew and passengers were unharmed.

Wright-Patterson's emergency forces went into action the evening of September 20, 2000, following reports of a tornado touchdown at 7:30 p.m. in Xenia, Ohio. Although less severe than the 1974 tornado, one person was killed and 100 injured. Thirteen base firefighters and paramedics responded with two pumper trucks and one rescue vehicle. One unit staged at the Xenia Wal-Mart store where 30 employees were trapped. Another unit established a triage center at a nearby elementary school. A base team, assisted by a crisis response team from the 74th Medical Center's mental health flight, searched a 44-home area to aid injured and help where needed.

Environmental management and historic preservation remained major issues in the 1990s. At Wright-Patterson, an aggressive campaign attacked pollution and preserved the base's historic structures and cultural property. Wright-Patterson Air Force Base and the U.S. Environmental Protection Agency (USEPA) signed a Federal Facility Agreement on March 21, 1991, to facilitate cleanup of hazardous-waste sites on base. On the state level, the base adhered to provisions of an Administrative Orders on Consent signed by the base commander and the Ohio Environmental Protection Agency in 1988. The base also signed agreements with the city of Dayton to protect the city's wellheads and to treat contaminated groundwater that had migrated from the base. Responsibility for tracking all hazardous waste generated on base transferred from the Defense Reutilization and Marketing Office (DRMO) to the air base wing. Wright-Patterson constructed a low-level, radioactive-waste storage facility, a hazardous-materials warehouse, and a recycling center. It cleaned up and recapped old landfills, demolishing several Woodland Hills housing units in the process. By 2000, the base had completed final remediation action on 67 of its 68 identified restoration sites. The skills of the 88th Air Base Wing's

environmental managers became so highly regarded, that its members were invited onto the international stage. Under the Nunn-Lugar Exchange Act, base-level environmental experts began advising former Warsaw Pact nations on environmental matters.

On November 29, 1995, a construction crew excavating a sewer line along 13th Street between "M" and "P" Streets in Area B unearthed a cache of M-114 cluster bombs. Initially, investigators erroneously identified the ordnance as biological weapons with live explosives containing potentially live anthrax or *Brucella suis.* After extensive investigation, the cluster bombs proved to be inert simulation devices containing properly sterilized, dead *Brucella suis* bacteria. The site was then cleaned and closed.[22]

Wright-Patterson's commitment to historic preservation was symbolized in 1990 when the 2750th Air Base Wing received the Air Force's General Thomas D. White Historic Building Preservation Award. Over the next 10 years, wing environmental managers worked closely with the Ohio Historic Preservation Office to preserve and manage Wright-Patterson's historic properties. They wrote a Historical Resources Management Plan and supervised a major Historic American Engineering Record project documenting Wright-Patterson's significant engineering facilities. They surveyed historic structures

and established four historical districts on base: Fairfield Air Depot, Wright Field, Brick Quarters, and World War II. The base railroad system also was documented prior to its closure. World War II prisoner of war murals in Building 280, Area A, were cleaned and preserved. Rather than demolishing obsolete historic structures, base engineers restored their original façades and adapted their interiors for new uses. Buildings 32 and 11A, original Wright Field facilities, were saved in this manner. Environmental managers also preserved eight archeological sites and performed remote sensing on the installation's Native American burial mounds and Huffman Prairie Flying Field.

Computer technology became a fact of life at Wright-Patterson Air Force Base in the 1990s. Computer workstations sprouted everywhere. Mainframes, personal computers, and laptops became essential workforce tools, streamlining and improving productivity. Networking enhanced communication and made electronic commerce a reality. Military and civilian personnel management, financial operations, and most other aspects of work on base gradually shifted from local, people-intensive operations, to computer-based, self-help operations managed from centralized locations at other installations.

The computer explosion, coupled with rapid equipment obsolescence, quickly generated stockpiles of outmoded

During Operation Desert Storm, the 2750th Air Base Wing operated an aerial port of embarkation that deployed Army units to the combat theater. Here, soldiers wait in the base mobility center for deployment to the Middle East.

equipment. The situation became so serious that, in 1994, the base conducted a "Computer Roundup" and collected more than 30,000 excess computer items. Under an agreement between Wright-Patterson and Dayton's Alliance for Education (a coalition of local corporate, foundation, and education leaders), the base distributed its excess equipment to schools in 11 neighboring counties.

While computers became part of the base's daily fabric, Wright-Patterson also developed its resources in high-performance computing. Supercomputers, with their ability to rapidly perform complex calculations, were essential for the cutting-edge work of engineers and scientists at Wright-Patterson and throughout the defense community. Wright-Patterson's critical role and capabilities in research and development led the Department of Defense to select it as the site for one of four supercomputers. On May 12, 1997, Major General Richard R. Paul, commander of the Air Force Research Laboratory, and

U.S. Congressman David Hobson officially opened the Major Shared Resource Center (MSRC) at Wright-Patterson Air Force Base. The high-performance supercomputer aided researchers from the Department of Defense, AFRL, academic world, and industry in solving complex, technical problems. Scientists and engineers also used it for modeling and simulations. MSRC's capabilities increased substantially in late 2001, when four semi-trailers delivered the world's eighth largest computer. The $34 million Compaq ES-45 with a Tru 64 Unix operating system gave MSRC a peak performance capacity of 1.67 trillion calculations per second.

Combat Operations

The Persian Gulf War began in August 1990 when Saddam Hussein, president of Iraq, ordered his army to invade Kuwait. Within days, a coalition of 32 nations led

by the United States moved to protect the Middle East's oil reserves and liberate Kuwait. On August 7, President George H. W. Bush ordered U.S. forces to the Persian Gulf and announced the beginning of Operation Desert Shield, defense of the Middle East. When Saddam Hussein refused to withdraw from Kuwait, the United States-led coalition launched Operation Desert Storm, liberation of Kuwait. A massive air campaign against Iraq began on January 16, 1991. Main coalition ground forces invaded Kuwait and southern Iraq on February 24. Within four days, they encircled and defeated Iraqi forces, liberating Kuwait. President Bush declared a cease-fire on February 28.

From August 1990 to April 1991, Wright-Patterson Air Force Base performed its wartime mission as an aerial port of embarkation for the first time. The 2750th Air Base Wing processed and deployed troops and cargo. Thirty-one Wright-Patterson units received deployment taskings. The air base wing, which

Air base wing cargo specialists handle Desert Storm equipment on the flightline.

Army vehicles in desert camouflage line up for loading.

A B-52H delivering 750-pound M-117 bombs during Operation Desert Storm. B-52s delivered 40 percent of all munitions dropped by coalition forces. (*U.S. Air Force photo by Master Sergeant Ralph Hallmon*)

RQ-4 Global Hawk high-altitude, long-endurance, unmanned air vehicle

OHIO TASK FORCE ONE—THE WRIGHT-PATTERSON CONNECTION

In the immediate aftermath of the 9-11 attacks on New York City's World Trade Center (WTC) and the Pentagon in Washington, D.C., the Federal Emergency Management Agency (FEMA) deployed urban search and rescue (US&R) teams to search for survivors. Ohio Task Force 1 (Miami Valley Urban Search and Rescue) responded to the WTC disaster, putting its training into action on a scale of which it never dreamed. The task force's ability to effectively respond to a disaster of this magnitude resulted from the combination of skills, training, and experience possessed by its members—and additional capabilities learned and practiced with Wright-Patterson Air Force Base—capabilities which impacted the entire FEMA US&R system.

On April 19, 1995, domestic terrorists bombed the Alfred P. Murrah Federal Building in Oklahoma City, Oklahoma, leaving the facility in ruins and killing 169 people. In the wake of this disaster, FEMA realized the need to improve its ability to respond to large-scale disasters. The agency decided to increase the number of US&R teams available for nationwide disaster response.

Consequently, the Miami Valley Urban Search and Rescue team competed with 22 other organizations before its selection, along with Boone County, Missouri, as the 26th and 27th teams on March 28, 1997. The teams went before a technical review panel composed of current US&R task force leaders and other technical experts. Evaluation teams then visited the leading candidate sites to determine final selections. The FEMA evaluation teams concluded that the two new organizations were capable of fielding fully operational task forces within two years.

Political leaders and military officials at Wright Patterson Air Force Base developed a support agreement to assist the task force, designated Ohio Task Force 1 (OH-TF-1). The Honorable David L. Hobson, U.S. Representative, Ohio, 7th District, and Lieutenant General Kenneth E. Eickmann, Aeronautical Systems Center commander, became instrumental in arranging for OH-TF-1 to become a tenant organization at Wright Patterson. Thus, OH-TF-1 became the only task force in the United States located on a military installation. This proved very beneficial to the team because of the close personal contacts made with base officials who included OH-TF-1 in their disaster-response plans and exercises.

In March 1999, the team participated in STRONGWIND 99-1, a no-notice recall and deployment in response to a simulated disaster on the West Coast. This exercise represented the first of its kind for any of the FEMA US&R teams. In June 2000, SABER SHIELD 00-2 evaluated OH-TF-1's capabilities and Wright Patterson's ability to support the task force's deployment. This exercise demonstrated improvement in the deployment process over STRONGWIND 99-1 as OH-TF-1 and base personnel developed a smoother rapport. Personnel from the 88th Air Base Wing, Aeronautical Systems Center, and the 445th Airlift Wing supported all aspects of the deployment to ensure a "real world" exercise scenario.

OH-TF-1 team members and base personnel gained considerable experience in the deployment process as participants went through processing lines, accomplished aircraft load planning, complied with FEMA checklists, ensured command post notifications, and accomplished a host of other activities designed to airlift the team to a disaster site within six hours. Even the four rescue dogs completed their annual flying requirement without incident.

Training with the 445th Airlift Wing proved especially beneficial in familiarizing OH-TF-1 personnel with specific tasks involved in preparing cargo for aerial deployment, as well as preparing themselves for rapid recall and departure. The 87th Aerial Port Squadron Cargo

Deploying Function (CDF) worked closely with the emergency responders by preparing their equipment pallets for aircraft loading. At the time of STRONGWIND 99-1, OH-TF-1 had the responsibility to mobilize 62 personnel, 19 short tons of cargo, and several search dogs.

SABER SHIELD 00-2 enhanced and improved the working relationship between OH-TF-1 and a number of Air Force units (active and reserve) at Wright-Patterson. The exercise involved the coordinated actions of many of the base's units. OH-TF-1 members throughout Ohio responded to FEMA's activation order and, shortly thereafter, Wright-Patterson activated its Battle Staff. To handle the deployment, the base also assembled its Transportation Readiness Control Center, CDF, and its Passenger Processing Function. These exercises "opened doors" for the emergency responders as they worked closely with Air Force personnel, learned and practiced minute details involved in aerial deployment, became accustomed to Air Force jargon, and learned the intricacies of mobilizing and deploying from the best in the business—the Air Force. Little did the team know how important this training would become and how quickly they would be called upon to use it in a real world situation!

Just over a year after SABER SHIELD 00-2, the unthinkable happened on September 11, 2001—9-11 struck the American homeland unlike anything else in the nation's history. OH-TF-1 officials

Part of the Ohio Task Force 1 team, which deployed to the World Trade Center, two blocks from ground zero. *(Ohio Task Force 1)*

quickly received a verbal activation order from FEMA but did not know if the team would deploy to the WTC or to the Pentagon. It is significant that FEMA picked OH-TF-1, one of the newest US&R organizations, as one of eight teams responding to the disaster.

As the team assembled at Wright-Patterson, where its vehicles and equipment were stored, base officials immediately prepared to support their deployment. The team had to be self-sufficient once deployed so their equipment included generators, tents, electronic detection gear, and various emergency supplies to aid in the search for survivors. The team departed Wright-Patterson with 62 firefighters and emergency medical technicians, four K-9 search dogs, and

60,000 pounds of equipment. The team included Randall Hawkins, a Wright-Patterson Fire Department employee who was also a member of OH-TF-1. They proceeded to McGuire Air Force Base, New Jersey, arriving on September 12 via ground transportation, and established a base of operations to conduct search activities at the WTC.

Because of its training and experience working with the military system at Wright-Patterson, the Miami Valley Urban Search and Rescue team gained immediate acceptance at McGuire. J. Robert McKee, a task force leader, attributed this to the fact that team members had learned Air Force operations and could "talk their lingo." McKee even briefed the McGuire Air Force Base commander twice daily

about the status of the operations at the WTC.

Sadly, the entire search and rescue effort at the WTC failed to locate any survivors. However, the lessons learned by OH-TF-1 prior to 9-11, at the World Trade Center, and during subsequent emergencies (such as hurricanes and the Space Shuttle *Columbia* disaster in February 2003) have prepared the team to better handle future emergencies. Because of the skills they have developed and because of their "phenomenal relationship" with the personnel at Wright-Patterson, the Miami Valley Urban Search and Rescue team remains the envy of all other FEMA teams throughout the United States.

Sources: J. Robert McKee, former OH-TF-1 task force leader, personal communication with James R. Ciborski, ASC staff historian, June 2003; Report, ASC/IG, "Exercise Evaluation Team Report: STRONGWIND 99-1," March 8-11, 1999, (material used is not privileged information); Report, ASC/IG, "Exercise Evaluation Team Report: SABER SHIELD 00-2" June 26-30, 2000, (material used is not privileged information); *Skywrighter*, ASC/PA, [STRONGWIND 99-1 photo captions] March 19, 1999, p 15; *Skywrighter*, ASC/PA, [9-11 response articles and photos], September 14, 2001, p 1; Bruce Hess, personal communication with James R. Ciborski.

With the support of Wright-Patterson Air Force Base, the Ohio Task Force 1 team departed for the World Trade Center almost immediately following the disaster, arriving on September 12, 2001, via ground transportation. (*Ohio Task Force 1*)

Despite intensive and lengthy efforts, search and rescue workers failed to locate any survivors in the disaster. (*Ohio Task Force 1*)

Rescue workers at the World Trade Center included eight Urban Search and Rescue teams from around the United States. (*Ohio Task Force 1*)

OHIO TASK FORCE 1 MEMBERS

Joyce Bachmann, Technical Information Specialist

James Barrow, Rescue Specialist

Edwin Beacom, Rescue Specialist

Todd Beery, Rescue Specialist

Michael Benedic, Technical Search Specialist

Scott Benjamin, Logistics Specialist

Joshua Blum, Rescue Specialist

Thomas Bourquin, Heavy Rigging Specialist

Terri Boyette, Technical Information Specialist

James Cannell, Rescue Specialist

Mike Cayse, Rescue Specialist

Kelly Clark, K9 Search Specialist

Robert CocKayne, Rescue Specialist

Kevin Coffee, Rescue Specialist

Douglas Cope, Search Team Manager

Dale Dittrick, Safety Officer

Scott Donegia, Rescue Specialist

Gary Flynn, K9 Search Specialist

James Gruenberg, Task Force Leader - IST

Gary Hamilton, Rescue Specialist

Randall Hawkins, Logistics Specialist

Robert Hessinger, Logistics Team Manager

Christopher Hilberg, Rescue Specialist

Stanley Irvin, Heavy Rigging Specialist

Michael Kenney, Rescue Specialist

Kevin King, Rescue Specialist

Danny Kochensparger, Haz-Mat Specialist

David Lambert, Medical Specialist

Trace Lawless, Rescue Specialist

Grant Light, Rescue Squad Officer

Timothy Lombardi, Rescue Specialist

Michael Lotz, II, Rescue Specialist

Douglas Luneke, Logistics Specialist

Joseph Lykins, Technical Search Specialist

Anthony Malone, Medical Specialist

Timothy Manuel, Medical Team Manager

Dr. William Marriott, Medical Team Manager

Dave Martin, Medical Specialist

J. Robert McKee, Task Force Leader - IST

Kevin McMullen, Logistics Specialist

Joseph McNeil, Planning Team Manager

Craig Mignogno, Rescue Specialist

Matthew Modlich, Rescue Specialist

Greg Morris, Rescue Specialist

Michael Muhl, Task Force Leader

David Norrod, Rescue Specialist

Stephen Ober, Rescue Specialist

Michael Palumbo, K9 Search Specialist

Scott Perkins, Medical Specialist

David Pickering, Rescue Specialist

Jack Reall, Rescue Team Manager

Doug Riedel, Rescue Specialist

Thomas Riemar, Rescue Squad Officer

Mitchell Ross, Rescue Team Manager

Michael Rusin, Rescue Specialist

Brian Seabold, Rescue Specialist

Heidi Seiler, Medical Specialist

Alexander Shartle, Rescue Specialist

Donald Sherlock, Rescue Specialist

Steven Shupert, Rescue Team Manager

Al Smith, II, Rescue Specialist

Reid Spaulding, Structural Engineer

Dr. Steven Stephanides, Medical Specialist

David Stitzel, Search Team Manager

Eugene Thomas, Technical Search Specialist

Edward Thomas, Rescue Specialist

Terry Trepanier, K9 Search Specialist

David Tritch, Communications Specialist

Brian Wagner, Structural Engineer

David Young, Rescue Squad Officer

Jeffrey Young, Rescue Specialist

Robert Zickler, Planning Team Manager

Mike Zimmerman, Rescue Specialist

---

mobilized all base personnel, deployed 606 Wright-Patterson personnel in 84 separate deployments. Its initial foray into this wartime mission came on September 22, when the 64th Ordnance Detachment from Fort Benjamin Harrison, Indiana, deployed 14 soldiers and 34.5 short tons of cargo through the base. More than 2,300 soldiers from 21 Army units, representing 33 separate taskings, eventually passed through Wright-Patterson. Air base wing personnel also handled 2,267 short tons of cargo for the Army, Air Force, and Marines. At the conclusion of Desert Storm, the wing redeployed 331 Air Force and 1,916 Army personnel and handled an additional 54 short tons of Air Force cargo. In all, 153 aircraft arrived and departed the base in

support of contingency operations. Another wing activity included processing and shipping more than 10,500 tires by the wing's Aircraft Tire and Storage Distribution Point.

The U.S. Air Force Medical Center Wright-Patterson supplied the largest contingent of base individuals deployed during the war. It sent 207 medical staff members to the Persian Gulf and the European theater of operations, and another 17 to support medical operations at four stateside bases. The medical center received 265 replacements, primarily reservists called to active duty, to continue delivering medical services at Wright-Patterson. The 111 Individual Ready Reservists (IRR) assigned to the Medical

Center were part of the nation's first IRR mobilization since the 1961 Berlin Wall crisis (during which the United States and Soviet Union were on the brink of war over reunification of Germany and Soviet construction of the wall between the two halves of the city). The Air Force also selected the medical center as a primary, casualty-flow location. This required the center to activate its war plans and prepare an initial, casualty-flow triage site. Fortunately, the conflict produced few U.S. casualties. The Gulf War's first casualty, a soldier wounded on March 13 during a Scud missile attack on American barracks in Dhahran, Saudi Arabia, received care at the medical center. Finally, the center's blood-donor facility collected and processed more than 1,000 units of blood for the area of operations.

Other tenant organizations actively participated in the operations as well. All members of the Army's 71st Ordnance Detachment (EOD) at Wright-Patterson deployed. The 401st Combat Logistics Support Squadron, a reserve unit assigned to the 906th Tactical Fighter Group, sent volunteers to Saudi Arabia on August 20, 1990, where they formed a C-130 battle-damage repair team. Although the 906th Tactical Fighter Group did not get the call, its F-16s were ready to deploy within 72 hours of notification. Upon initiation of the air war, the 4950th Test Wing volunteered its fleet of test transport planes and pilots to supply and restock air bases whose supplies had been sent to the Persian Gulf. From January 17 to May 6, 1991, the test wing flew 181 sorties, transported 1,400 tons of cargo, and logged 768 flying hours in this effort.

Air Force Logistics Command and Aeronautical Systems Division played major roles in the Gulf War. Headquarters AFLC directed the Air Force's logistics effort. It procured supplies and spare parts, distributed them to units, and supervised repair and modification work. ASD's Wright Laboratory and system program offices had developed many of the front-line aircraft flown in the war, among them the F-15, F-16, A-10, and F-117. They also had expanded the capabilities of those and other aircraft. Advanced technologies developed by Wright-Patterson's laboratories and program offices allowed Air Force pilots to operate day or night, in high-threat environments, and to precisely place munitions on target. Among the technologies that proved both successful and crucial during Operation Desert Storm were the Low Altitude Navigation and Targeting Infrared system for Night

Ground zero: Smoke clears from the ruins of the World Trade Center, New York City, following the 9-11 terrorist attacks *(Ohio Task Force 1)*

(LANTIRN), the APG-68 and APG-70 attack radar with advanced cockpit displays, digital flight control technologies, improved fuels and engines, and the stealth technology embodied in the F-117 fighter. Pilots later claimed the APG-70 attack radar in the F-15E and the APG-68 in the F-16 offered "phenomenal" range and resolution. In interviews, they proclaimed "if it had metal in it, we could find it," and "with the APG-70, you could tell from 30 miles away whether a MiG-sized target had weapons or fuel tanks on it." These radar systems were an outgrowth of the Forward Looking Advanced Multi-mode Radar program overseen by Wright Laboratory's Avionics Directorate in the 1970s.

Other ASD innovations added to the U.S. Air Force's stunning technological superiority, including advanced cockpit displays with digital processing technology. A-10 tank killers were able to take severe punishment and keep flying thanks to protective materials like titanium armor cockpits. Even the JP-8A fuel that powered the aircraft of Desert Storm was less explosive and more combat-safe.

While the advanced aircraft and avionics developed at Wright-Patterson received the most media attention, other less exotic technologies played vital supporting roles. For example, ASD procured a variety of concealment and decoy products, lightweight camouflage nets large enough to cover F-15s, and the Protective Integrated Hood/Masks for improved protection against chemical warfare. Advanced technology gave the United States and its allies a decisive advantage in Desert Storm, an edge honed over several decades by scientists and engineers working at Wright-Patterson Air Force Base.

Desert Storm marked a beginning rather than an end to military operations in the 1990s. Contingency operations continued throughout the decade as the United States responded to a continuing Iraqi threat, as well as new dangers in the Balkans and other areas of the world. Wright-Patterson organizations sustained these operations with personnel, supplies, equipment, weapons, technology, and airlift flights. They supported ongoing operations

enforcing the United Nations' ban on Iraqi military flights north of the 36th parallel and below the 32nd parallel. In December 1998, Wright-Patterson units aided Operation Desert Fox air strikes against military and strategic targets in Iraq intended to counter that nation's continued buildup of weapons of mass destruction.

United States-led NATO air strikes in 1995 helped end fighting in the Balkans and bring about the Balkan Proximity Peace Talks later that year (see below). Following the talks, units at Wright-Patterson Air Force Base facilitated U.S. and NATO operations that implemented the peace accords. Wright-Patterson's involvement in the Balkans escalated in 1999 when NATO launched an air campaign against the Republic of Yugoslavia in defense of the Province of Kosovo. During Operation Allied Force from March 24 to June 20, 1999, the 88th Air Base Wing deployed personnel through the base mobility center, shipped 225 short tons of munitions to the combat theater, and provided critical weather products to strike-mission planners. Air Force Materiel Command supplied

F-16s from the 178th Fighter Wing, Ohio Air National Guard, stand alert on the base flightline during Operation Noble Eagle. The fighters were stationed temporarily at Wright-Patterson due to construction at their home base in Springfield, Ohio. (*U.S. Air Force photo by Henry Narducci*)

The Wright-Patterson Honor Guard lowers the flag during a retreat ceremony honoring the victims of the September 11, 2001, terrorist attacks. (*U.S. Air Force photo by Henry Narducci*)

munitions and logistics support. During the short conflict, AFMC's air logistics centers surged their depot production to meet immediate and post-operation replenishment requirements. Aeronautical Systems Center managed the aircraft and associated systems programs that made the operation a success. Among its contributions was a new generation of aircraft led by the RQ-1 Predator unmanned aerial reconnaissance vehicle, which made its operational debut over Bosnia in 1996, and the B-2 bomber, which flew its first combat sortie during Operation Allied Force.

As the second year of the new millennium entered fall, the United States suddenly found itself engulfed in a new war. On September 11, 2001, a surprise terrorist attack against the Pentagon and the twin towers of the World Trade Center in New York City mobilized the nation to combat worldwide terrorism. Operation Enduring Freedom directed military operations abroad, while Operation Noble Eagle protected the homeland from further attack. By early 2002, the United States and its allies had successfully disrupted the Al Qaeda terrorist network operating out of Afghanistan and dislodged the Taliban government supporting it. As this phase of the war wound down, U.S. forces began locating and moving against terrorist forces around the world.

On January 29, 2002, President George W. Bush identified Iraq as a threat to world peace. Following Operation Desert Storm in 1991, Iraq agreed to destroy all weapons of mass destruction. In the 12 intervening years, the United Nations could find no

evidence that Iraq had complied with the resolutions. By late 2002, the United Nations declared Iraq in breach of previous resolutions banning its manufacturing of weapons of mass destruction and allowing security inspectors to visit selected sites to monitor Iraq's cooperation. Iraq was given five months to comply with the United Nations' resolutions. When it failed to do so, the United States, along with support from 49 Coalition countries, including Great Britain, Australia, and Canada, invaded the Middle Eastern country on March 19, 2003. Nearly one month later, the United States declared the end of major military operations in Operation Iraqi Freedom and the beginning of a free, democratic Iraq. In December 2003, United States forces captured deposed Iraqi President Saddam Hussein, who had fled the capital city of Baghdad at the outset of the war.

As it had so many times before, during Operations Enduring Freedom and Iraqi Freedom, Wright-Patterson Air Force Base mobilized to ensure that the nation's combat forces had the supplies, weapons, and technology to wage a successful campaign. The 88th Air Base Wing shipped personnel and supplies overseas. AFMC's logisticians expanded the Air Force's supply system and kept war materiel flowing. The 445th Airlift Wing activated 630 reservists and established a C-141 staging facility at Wright-Patterson on February 25, 2003. The airfreight terminal processed more than 850 short tons of cargo, and the 88th ABW increased its support to the base's 445th Airlift Wing's aerial-refueling operations.

A C-141 Starlifter from the 445th Airlift Wing, Wright-Patterson Air Force Base, completes an aerial refueling mission over the North Carolina coastline. (*U.S. Air Force photo by Staff Sergeant Jerry Morrison*)

AFRL's Eagle Sector program made new technologies available to warfighters. The laboratories worked with ASC's Combat Electronics Division to develop the Battlefield Air Operations Kit for close air support controllers. The kit, which included laser range finders/designators and tough book computers, increased speed and targeting accuracy with reduced weight over previous units. Other technologies developed for the warfighter included panoramic night-vision goggles; laser eye protection; a switch that allowed controllers to change rapidly from ground radio to aircraft and satellite communications; and through its Munitions Directorate at Eglin Air Force Base, the Massive Ordnance Air Blast, nicknamed "Mother of All Bombs" by Coalition warfighters.

ASC surged its existing acquisition programs and worked feverishly to meet new requirements levied by combat commanders. Among its early accomplishments was introduction of the RQ-4 Global Hawk unmanned reconnaissance vehicle to the battlefield. Global Hawk flew 15 combat missions totaling 350 flight hours in Iraqi Freedom, captured 1,290 infrared and 2,246 synthetic aperture radar images, and located 13 full surface-to-air missile batteries, 50 surface-to-air missile launchers, 300 tanks, and 70 missile transporters. ASC program offices also delivered Advanced Strategic Tactical Expendable infrared countermeasure flares, Litening Target Pods for the F-15 fighter (also used on the B-52 bomber), "bunker-buster" munitions, portable shelters for B-2 bomber maintenance, "dust eater" valves to protect F-16 onboard systems from desert conditions, and C-40B Communications Command aircraft, among other systems. The C-17 System Program Office modified the airlifter's high-altitude airdrop capability and increased its capacity from 102 to 189 troops for transport into the theater. Aircraft developed by Aeronautical Systems Center flew more than 30,000 sorties during Iraqi Freedom, dropped 31 million psychological leaflets, released 19,000 guided and 9,000 unguided

munitions, and moved more than 55,000 passengers and 40,000 tons of supplies.

The 74th Medical Group, Wright-Patterson Medical Center, deployed more than 150 medics and treated more than 5,000 patients. The National Air Intelligence Center (now known as the National Air and Space Intelligence Center, NAIC) also contributed its specialized analytical skills to the war effort. The 445th Airlift Wing flew missions around the globe, including several medical evacuations of wounded soldiers and former prisoners of war from the combat theater. The reserve unit also transported captured Al Qaeda prisoners from Afghanistan to Camp X-Ray at the U.S. Naval Base, Guantanamo Bay, Cuba. Finally, the 178th Fighter Wing (an Air National Guard unit temporarily operating from Wright-Patterson) flew combat air patrols over the United States as part of Operation Noble Eagle.[23]

Balkan Proximity Peace Talks

Wright-Patterson took center stage as a forum for world peace on October 18, 1995, when Secretary of State Warren Christopher announced that the installation would host peace talks between the combatants in the Balkans conflict. Wright-Patterson was selected due to its excellent airfield, convenient air connections, privacy, security, and logistical support capabilities. Lodging facilities provided the decisive advantage, however. The close proximity of the Hope Hotel and conference facilities

(Building 824, Area A) to visiting officers' quarters (VOQ) in Buildings 832-836 presented the State Department with an excellent diplomatic compound, including separate, but identical, facilities for each delegation.

General Henry Viccellio, AFMC commander, assigned operational support of the peace talks to his deputy, Lieutenant General Lawrence P. Farrell, Jr. General Farrell then selected a base management team drawn primarily from the 88th Air Base Wing. The Balkan Peace Talks Support Team took up residence in Headquarters AFMC's Sarris Auditorium (Building 262, Area A), which became a full-fledged operations center.

The Support Team faced a series of challenges, beginning with the State Department's 51-page list of logistics requirements, and only 10 days in which to accomplish them. Military and civilian personnel across the base, reservists, and the Dayton community immediately pitched in to make the talks a success. Although Headquarters Air Force remained executive agent for Air Force support to the negotiations, the base handled the job so completely that little action was required from Air Staff.

The 88th Air Base Wing's civil engineers constructed a "base within the base" to house the peace talks complex. It ensured the security of the delegates and minimized the impact on normal base operations. Engineering teams prepared the quarters and converted 24 VOQ rooms into 12 presidential suites that met State Department specifications for size, decor,

The Wright-Patterson Honor Guard greeted arriving delegates for the Balkan Proximity Peace Talks.

Secretary of State Warren Christopher arrives to open the Peace Talks.

Secretary of State Warren Christopher and his negotiation team discuss border issues.

and amenities. They installed backup heating and electrical systems with remote controls for emergencies. Engineers installed 1,000 feet of curbing and sidewalk, a major part comprising the "Peace Walk," a lighted sidewalk that meandered through a park-like area between the VOQ and the Hope Hotel. It allowed delegates to walk casually and safely from their quarters to the meeting rooms and dining area in the Hope Hotel.

The air base wing's airfield operators handled responsibility for expanded airspace control, diplomatic arrivals and departures, and support of presidential aircraft. At the direction of the AFMC commander, the 47th Airlift Flight established an Alpha Alert commitment, in anticipation of a possible need for airlift support, with its C-21 aircraft from November 1 to 26. During this period, alert crews launched six sorties to transport delegates to necessary meetings or other engagements. On the ground, logisticians filled supply requests, moved equipment, handled diplomatic cargo, and operated a dedicated fleet of vehicles with a force of volunteer drivers.

Communications technicians installed telephone and computer networks, set up databases, managed air frequencies, and supplied public address systems for numerous events. The Multimedia Center generated graphics, publications, printing, and even designed the peace talks logo.

Base officials dealt with many other tasks as well. They arranged for food service, attended to the religious needs of the delegates, and provided security. Finance and contracting specialists worked in unison to acquire and fund the goods and services. ASC's public affairs office operated a media center and accredited 571 journalists and 470 press representatives from 166 organizations and 20 nations during peak operations. Protocol duties ranged from preparing welcome packages to hosting diplomatic events and setting up bilateral and trilateral meetings. The Wright-Patterson Medical Center handled the medical needs of the delegates. The Dayton community also lent its support; the "Peace Wall" erected in the Hope Hotel became one of its most visible contributions. The wall was filled with letters and drawings from Dayton-area school children encouraging the delegates

The presidents sign the Dayton Peace Agreement.

to bring peace to their troubled lands.

The Balkan Proximity Peace Talks officially began on November 1, 1995. The nine delegations were led by Richard C. Holbrooke, Ambassador, United States of America; Alija Izetbegovic, President, Republic of Bosnia-Herzegovina; Slobodan Milosevic, President, Federal Republic of Yugoslavia; Franjo Tudjman, President, Republic of Croatia; Carl Bildt, Ambassador, European Union; Jacques Blot, Ambassador, Republic of France; Wolfgang Ischinger, Ambassador, Federal Republic of Germany; Igor S. Ivanov, Ambassador, Russia; and Pauline Neville Jones, Ambassador, United Kingdom. Their goal was a comprehensive regional settlement that preserved Bosnia as a single state containing the Muslim-Croat Federation and a Bosnian Serb entity; resolved boundary issues between the Bosnian-Croat Federation and the Bosnian Serb entity; settled the status of Sarajevo; and set forth steps to separate the forces, end hostilities, and return refugees to their homes.

The delegates made progress on the first day when the presidents of Croatia and Yugoslavia agreed to work toward full normalization of relations, recognize human rights for citizens of both countries, and

honor the rights of refugees to return home. Another ray of optimism surfaced on November 10 with the signing of an agreement that bolstered the Muslim-Croat Federation and reunited the city of Mostar. Early optimism soon faded as disputes over slivers of land and boundary lines bogged down negotiations. The atmosphere in the complex alternated between hope and despair as agreements seemingly at hand dissolved into deadlock. Finally, Ambassador Holbrooke suspended the talks. President Tudjman's aircraft was loaded and sitting on the ramp with engines running as the crew prepared to take him home. At the last moment, diplomats salvaged the situation and the delegations reassembled in the Hope Hotel. At 3:00 p.m. on Tuesday, November 21, they signed the Dayton Peace Agreement, officially designated *General Framework Agreement for Peace in Bosnia and Herzegovina.*

The peace talks were a milestone in Wright-Patterson's history. The large base with its organizational complexity and diverse missions delivered quality service in rapid time without disrupting its normal business. High morale, dedication, and cooperation combined with profession-alism to forge the teamwork and can-do

attitude that made the operation an unqualified success. Even the furloughing of "non-essential" base civilians from November 14 to 20, caused by Congress' failure to pass a continuing budget resolution authority, did not interrupt Wright-Patterson's service, support, or commitment to the peace talks. Team Wright-Patt had garnished its greatest achievement. More than 2,000 people received formal recognition for their contributions and the peace talks were a major factor in the awarding of an Air Force Organizational Excellence Award to Aeronautical Systems Center and an Air Force Outstanding Unit Award to the 88th Air Base Wing. The highest honor, however, came from Secretary of State Warren Christopher, who wrote the 88th Air Base Wing commander:

The hard work, superior performance and cheerful enthusiasm which you and your colleagues brought to this effort were key elements in the successful conclusion of the Talks.[24]

AIR BASE SECURITY—A LONGSTANDING, DAUNTLESS TASK

In the twenty-first century, air base security has become more important than ever before in protecting the vital resources required to complete the Air Force mission. Just as society and the world situation have changed over the years, Air Force Security Forces have kept pace. At Wright-Patterson, the 88th Security Forces Squadron (SFS) plays an important role in the base's daily operation. Wright-Patterson has over 1,500 facilities valued in excess of $17 billion spread over more than 8,000 acres. The base is home to the Air Force's research, development, acquisition, and logistic nerve centers, as well as a daytime population of approximately 24,000 personnel. Providing complete security for these valuable resources is no small job, and the 88th SFS's duty performance exceeds that of any small city police department.

Since the Air Force became a separate service in 1947, its security police organizations have undergone a number of changes. In 1948, as the Air Force's identity began to take shape, its military police took on the name Air Police. In 1966, the name changed again to Security Police. Personnel demographics also changed as women finally gained a prominent role in the career field in 1971, when the first six female students graduated from the Law Enforcement Training Course at Lackland Air Force Base. In 1975, Brigadier General Thomas Sadler was appointed the first security police director, within a headquarters-level position designated AF/SP. The following year, realizing the importance of a distinctive, recognizable appearance for security police, Sadler gained approval to authorize the wearing of the blue beret and other uniform changes for his troops. Gone were the white service caps that had been worn since the early years. Still more changes came in 1997, when Security Police became Security Forces, in recognition of the overall role that security police play in keeping the world's premier Air Force safe and sound.

At Wright-Patterson Air Force Base, the 88th SFS uses the latest law-enforcement equipment and modern, crime-fighting techniques. Community-based policing, bicycle patrols, working-dog teams, advanced bomb-sniffing equipment, closed-circuit television, and advanced electronic alarm systems complement the SFS's mission. The 88th SFS's investigators have recovered more than $40,000 in missing property and $1 million in street-value narcotics. The unit also developed a state-of-the-art combat arms training facility that teaches security forces personnel using realistic scenarios, and its antiterrorism programs received outstanding ratings from the Air Force Materiel Command Inspector General. Despite the constant changing threat, the 88th SFS continues to develop and implement new, innovative ways to deter and neutralize any threat or vulnerability that may try to interrupt business at Wright-Patterson Air Force Base.

Wright-Patterson Air Force Base went on high alert following the September 11, 2001, terrorist attacks on the Pentagon and World Trade Center. (*U.S. Air Force photo by Henry Narducci*)

Exercises keep security forces ready for any contingency. (*88th Security Forces Squadron*)

Deployed security forces (*88th Security Forces Squadron*)

Base Activities

The end of the twentieth century was not just an era of work and accomplishment; it was also a time of celebration and remembrance. Wright-Patterson celebrated the 50th anniversary of World War II. Four years of events culminated in the summer of 1995 with Freedom Flight America. A fleet of World War II-era aircraft en route from California to New York stopped at Wright-Patterson Air Force Base. The base opened so the public could see the aircraft. An F-117 on static display and flights by the Wright B Flyer and B-2 Spirit thrilled spectators.

In 1997, Wright-Patterson Air Force Base joined in the United States Air Force's 50th anniversary celebration—"Golden Legacy, Boundless Future." The base was an official celebration site and organizations were encouraged to use the anniversary as a theme in all special activities. Celebrations on base officially began on January 10, when General Ronald R. Fogleman, Air Force chief of staff, presented a special, 50th anniversary flag to the base. The United States Air Force Museum prepared special exhibits and placed the *Sacred Cow* presidential aircraft (in which President Truman signed the National Security Act of 1947 creating the Air Force) on permanent display. On July 18, Dr. Sheila E. Widnall, secretary of the Air Force, and a delegation of congressional and Ohio leaders gathered at the base flightline to christen B-2 aircraft AV-5 *Spirit of Ohio*. The celebration reached a crescendo when the United States Air Force Museum hosted "Salute to Freedom" on the September 18 anniversary date. The salute included a reenactment of the signing of the National Security Act, followed by a military tattoo. A tribute to Orville and Wilbur Wright on December 17 at the Wright Memorial concluded the Air Force's anniversary celebrations.

The year 1997 also marked the 80th anniversary of the establishment of McCook Field, the Engineering Division, and military aeronautical research and engineering in the Miami Valley. On September 6, Aeronautical Systems Center unveiled a new commemorative marker, erected by The Dayton Foundation and the Montgomery County Historical Society, at the former site of McCook Field. One side of the aluminum marker documented the location of McCook as the "Cradle of Aviation" and listed the numerous contributions made to early aviation by

A military tattoo at the United States Air Force Museum concluded Wright-Patterson's celebration of the Air Force's 50th anniversary.

The Wright Memorial was rededicated on August 19, 1998, following the completion of extensive repairs. The renovation prepared the monument for the Centennial of Powered Flight in 2003. (*U.S. Air Force photo by Henry Narducci*)

personnel at the field. The opposite side of the marker included a map of the area.

The nation opened its Korean War 50th Anniversary Commemoration on June 25, 2000, with wreath-laying ceremonies in Washington, D.C. Wright-Patterson was designated a Korean War Commemorative Community and celebrated the anniversary in partnership with the Korean War Veterans Memorial Association.

The most visible and lasting commemoration at Wright-Patterson centered on Huffman Prairie, where the first practical airplane was developed. It became a National Historic Landmark in June 1990 and, the following October, the Wright brothers' original 84.42-acre flying field

was officially dedicated as Huffman Prairie Flying Field. National Park Service Director James Ridenour presented AFLC Commander General Charles C. McDonald with a plaque listing the site as a National Historic Landmark. The adjacent 109 acres, which formed the largest, natural, tall-grass prairie remnant in Ohio, was named Huffman Prairie. A replica of the Wright brothers' 1905 hangar was constructed, and the two sites were opened to the general public for the first time since 1917.

On October 6, 1992, Congress passed legislation (Public Law 102-419) creating Dayton Aviation Heritage National Historical Park, which included Huffman Prairie Flying Field, the Wright Cycle Shop,

U.S. AIR FORCE MARATHON—"THE RACE WITH A DIFFERENT ATTITUDE"

Andrew Herr captured the gold during the first Air Force marathon. (*U.S. Air Force photo by Spencer P. Lane*)

Until 1997, the Air Force was the only military service without a marathon. Wright-Patterson changed this by holding the first Air Force Marathon on September 20, 1997, in conjunction with the Air Force's 50th anniversary celebration. Beginning at the United States Air Force Museum, the 26.2-mile course traversed historic Wright Field, passed through the Wright State University campus, looped around the Patterson Field runway, passed by the historic Huffman Prairie Flying Field, and returned to Wright Field before ending at the finish line at the museum. The first marathon's 2,751 runners represented all military services, 48 states, and 7 foreign countries. A force of 2,000 volunteers provided water, nutritional supplements, and medical aid to runners along the course. More than 10,000 spectators lined the route to provide moral support. All 1,660 marathon finishers and the 908 relay team members received commemorative medallions portraying the Wright Flyer.

The marathon proved so successful that it became an annual event. By 2003, race entrants totaled 3,300, representing 49 states and 8 countries, an unofficial participation record. Participants ranged in age from 10 to 80 years. In addition to the 26.2-mile marathon, the race included a wheelchair competition, a 13.1-mile half-marathon, a four-person relay race, and a 5-kilometer (3.1-mile) race.

Air Force Marathon Winners

Year	Winner	Time
1997	Andrew Herr, USA	2:28:34
1998	Andrew Herr, USA	2:27:41
1999	Tony Meyers, Belgium	2:29:50
2000	John Agnew, USA	2:30:50
2001	Marathon cancelled due to the events of September 11	
2002	Jeffery Gibson*, USA	2:42:15
2003	Hendrick Vanloon, Belgium	2:37:44

* 2002 winner Staff Sergeant Jeffery Gibson was the first Air Force member to win the marathon.

Source: Brett Turner, "Air Force Marathon a success," *Skywrighter*, September 26, 2003, p 1A.

Bang! The first Air Force marathon is underway. (*U.S. Air Force photo by Spencer P. Lane*)

ASC Command Chief Master Sergeant Thomas A. Edwards holds medals for those who persevere and complete the grueling race, 1997.

Volunteers constructed a replica of the Wright brothers' 1905 hangar on Huffman Prairie Flying Field in 1990.

Archaeological excavations pinpointed the location of the Wright brothers' 1910 hangar. (*U.S. Air Force photo by Henry Narducci*)

The Huffman Prairie Flying Field Interpretive Center anchored the east end of the Dayton Aviation Heritage National Historical Park. (*U.S. Air Force photo by Henry Narducci*)

Paul Laurence Dunbar State Memorial, and the Wright Flyer III at Carillon Historical Park. Wright-Patterson worked with the National Park Service to develop and manage the park, while the Ohio Chapter of the Nature Conservancy helped maintain flora and fauna on the adjacent prairie.

Wright-Patterson and the National Park Service continued to rejuvenate monuments honoring the Wright brothers and their accomplishments. In view of the upcoming 2003 Centennial of Flight celebrations, the installation undertook an extensive $600,000 rehabilitation of the 60-year old Wright Memorial plaza in the late 1990s. Cultural landscape studies helped base officials restore and maintain the plaza and the area around it. The 88th Air Base Wing's Office of Environment Management awarded a contract to Kelchner Environmental of Centerville, Ohio, and its subcontractor Parks Masonry, to replace the bluestone at the plaza area; repair the steps, walls and entry gates; and restore some of the landscaping around the plaza. A second subcontractor restored the bronze tablets at the memorial. On August 19, 1998, a rededication ceremony marked the work's completion.

Lieutenant General Robert F. Raggio, ASC commander, officiated at the May 26, 2000, groundbreaking ceremony for the Huffman Prairie Flying Field Interpretive Center. The 5,400-square-foot facility, financed with $1 million from the state of Ohio, was located on Wright Brothers Hill, directly across from the Wright Memorial plaza. It anchored the eastern end of Dayton Aviation Heritage National Historical Park. The center's exhibits, funded by the National Park Service, interpreted the Wright brothers' work at Huffman Prairie and continuation of their legacy at Wright-Patterson Air Force Base.

The Air Force, National Park Service, and state of Ohio funded preparations at Huffman Prairie Flying Field. Analyses from remote-sensor flights and archeological excavations confirmed the site of the Wright brothers' 1910 hangar and other landmarks on the field. Cultural landscape studies established 1904-1905 as the most historically significant period for interpretation and defined how the field should be restored. Wayside exhibits were designed; original roads re-established; and a mock-up of the Simms Station interurban rail stop constructed. Wright-Patterson and the state of Ohio constructed a special base entrance to make the park more accessible to the public. The Flying Field and Wright Memorial were also connected to a regional bicycle path.

In July 2000, Colonel Michael W. Hazen, the 88th Air Base Wing commander, hosted a ceremony on Huffman Prairie Flying Field with Eric Metzler from the Ohio Lepidopterist Society and Dr. David Adamski from the Smithsonian Institution. Six years earlier, Mr. Metzler had identified

Gnorimoschema huffmanellum, a new species of moth found on Huffman Prairie. This moth, along with another new species found on the prairie (Glyphidocera wrightorum), were given names honoring the Wright brothers and their use of Huffman Prairie. (Smithsonian Institution)

a new species of moth on the adjacent Huffman Prairie. At the ceremony, the moth was officially named Glyphidocera wrightorum in honor of the Wright brothers.[25] The following month, Wright-Patterson received a White House plaque honoring Huffman Prairie Flying Field as one of 13 local sites on the National Millennium Trail, an honorary designation conferred by the then White House's Millennium Council, which focused on connecting people with their land, history, and culture. All 13 local sites already had been designated as public sites by Aviation Trail, Inc., a nonprofit group that highlights aviation landmarks in the Dayton/Miami Valley region.

In 2002-2003, Wright-Patterson supported the U.S. Air Force's Centennial of Flight Office in celebrating 100 years of powered flight. The base actually hosted the nation's first centennial event: a December 17, 2002, First Flight Ceremony at the newly rehabilitated Wright Memorial. The Huffman Prairie Flying Field Interpretive Center was officially opened by Secretary of the Interior Gail Norton and Director of the National Park Service Fran Mainella in ceremonies preceding the traditional wreath laying at the Wright Memorial. Wright-Patterson continued the celebration with its Air Power 2003 open house in May. Reminiscent of Air Force Day celebrations earlier in the century, Wright-Patterson proudly displayed modern and historic aircraft for public viewing. In July, in conjunction with the Dayton Inventing Flight events, the United States Air Force Museum held a hot air balloon festival, a blimp meet, and a dedication of its newest facility, the Eugene Kettering Gallery. The celebratory year ended with the annual wreath laying at Wright Memorial on December 17, 2003 (see Chapter 5: Wright-Patterson and the Miami Valley).

The Peace Walk was dedicated November 21, 1997, in memory of the Dayton Peace Agreement. The walkway led from the Hope Hotel to the Five-Plex, where the delegates were housed for the Balkan Proximity Peace Talks. (U.S. Air Force photo by Henry Narducci)

ASSOCIATE ORGANIZATIONS

As Wright-Patterson's host organization, the 88th Air Base Wing (and its predecessors) provided support services for a wide range of associate, or tenant, organizations. The largest of these on base were the various commands headquartered in Building 262, Area A: Air Materiel Command (1946-1961), Air Force Logistics Command (1961-1992), and Air Force Materiel Command (1992-present). Second largest were the acquisition program offices of Aeronautical Systems Center and its predecessors (Wright Air Development Center, Wright Air Development Division, and Aeronautical Systems Division) and the laboratories, which in 2003 were collectively referred to as the Air Force Research Laboratory. Other major on-base units included Air Force Institute of Technology (AFIT); National Air and Space Intelligence Center (NAIC); Orientation Group, U.S. Air Force (AFOG); United States Air Force Museum; U.S. Air Force Medical Center Wright-Patterson Air Force Base; and a variety of other units.

In 1950, the 2750th Air Base Wing provided support services to a wide variety of tenant organizations. By 1969, on- and off-base associate organizations numbered 166. To support these organizations, the wing managed real property resources amounting to nearly $281 million. These fixed capital assets consisted of real estate, supply, and other facilities, and utilities and ground improvements at 20 locations in

Ohio, Indiana, Kentucky, Maryland, and West Virginia.[26]

Restructuring of the nation's military forces at the end of the Cold War affected many of Wright-Patterson's tenant organizations. New units arrived as others departed. By 2003, the 88th Air Base Wing supported 74 tenant organizations (see Appendix 5). Some of the more prominent organizations, many of which are still located on base, are discussed below.

15th Weather Squadron

Effective July 8, 1961, Detachment 15, 15th Weather Squadron, became an on-base tenant. Fifteen years later, its host organization, the 15th Weather Squadron (WS) of Military Airlift Command (MAC), moved without personnel or equipment from Scott Air Force Base, Illinois, to Wright-Patterson, where it was reassigned from the 5th Weather Wing to the 7th Weather Wing. In addition to Detachment 15, the 15th WS was assigned the following detachments: Detachment 1, Tinker Air Force Base, Oklahoma; Detachment 6, Hill Air Force Base, Utah; Detachment 7, Kelly Air Force Base, Texas; Detachment 8, McClellan Air Force Base, California; and Detachment 13, Robins Air Force Base, Georgia. On June 1, 1980, the 15th WS moved, again without personnel or equipment, from Wright-Patterson to McGuire Air Force Base, New Jersey. Detachment 15, however, remained at

WHEN DISASTER STRIKES—WRIGHT-PATTERSON AND THE MIAMI VALLEY ARE PREPARED

After the disastrous 9-11 terrorist attacks, President George W. Bush combined all homeland defense missions performed by Department of Defense organizations and placed them under the direction of a single military command—U.S. Northern Command, or NORTHCOM (not to be confused with the Department of Homeland Security). NORTHCOM provides "one-stop shopping for military support" while performing its dual mission of homeland defense and military assistance to civil authorities for disaster relief and other emergencies. A variety of organizations at Wright-Patterson Air Force Base exist to respond to such situations.

The Wright-Patterson Disaster Preparedness (DP) office manages the full spectrum threat response (FSTR) program to ensure that base personnel can respond to accidents, natural disasters, terrorist use of weapons of mass destruction, and also to perform wartime passive defense actions. DP develops and updates contingency plans for command and control, medical, security, firefighting, bioenvironmental, public affairs, and other responders. Wright-Patterson's FSTR plans must also comply with the Posse Comitatus Act that prohibits direct military involvement in enforcement of state laws or local ordinances. For this reason, NORTHCOM, as represented by facilities such as Wright-Patterson, is limited to military homeland defense and civil support to lead federal agencies. However, the Wright-Patterson Fire Department provides fire protection and hazardous-materials incident response under a formal, mutual-aid agreement with the Greater Dayton Area Fire Departments.

The military can also assist, for a maximum of 10 days, local communities overwhelmed by disaster. Some examples of such off-base support include Wright-Patterson's assistance during the 1974 tornado in Xenia, a severe blizzard in 1978, and following several aircraft crashes on and near Wright-Patterson over the years.

Primarily, the Disaster Preparedness and Inspector General (IG) offices ensure that Wright-Patterson organizations can respond to emergency situations as they occur. In most actual or exercise scenarios, the base sends its initial response element consisting of firefighting, security, and medical personnel. The follow-on element includes the disaster control group (DCG) consisting of the commanders/managers of the 18 organizations at Wright-Patterson tasked under FSTR plans. An on-scene commander directs activities inside the cordon established around the disaster or exercise area and coordinates actions with the Battle Staff, which handles operations outside the cordon and ensures the base performs its normal missions.

The IG and DP conduct ongoing response exercises, including exercises to handle attacks; mass casualty; aircraft crashes; tornados; nuclear, biological, chemical weapons of mass destruction; hazardous materials; and major accidents. The DP office possesses a state-of-the-art mobile command post fully equipped with weather-analysis equipment (for temperature readings, wind speed, etc.), a generator, a map of the base, "go-lights", radiation-detection equipment, and a winch. A response support vehicle, command trailer, and support trailer provide responders with an "office on wheels," sustainment and decontamination equipment, a tent for the DCG, backup radio communication equipment, radios for communication with local emergency organizations, a locker room, briefing room, and equipment storage.

Wright Patterson's 74th Medical Group serves as one of eight Federal Coordinating Centers (FCC) in the Air Force for the National Disaster Medical System (NDMS). The 74th manages memorandums of understanding with civilian medical facilities volunteering to provide inpatient care in the event of a natural disaster, catastrophic event, or military contingency operations. The 74th's area of responsibility encompasses the four major metropolitan areas of Cincinnati, Columbus, Dayton, and Toledo, Ohio.

The close ties existing between disaster-response organizations at Wright-Patterson Air Force Base and the communities in the Miami Valley provide the best available resources for effectively handling practically any disaster.

Sources: Air Force Instruction 10-205-1, "Full Spectrum Threat Response (FSTR) Planning and Operations," December 24, 2002; Stephen M. VanDegrift, disaster preparedness specialist, interview with James R. Ciborski, Aeronautical Systems Center History Office, August 26, 2003; R. Michael Rives, chief, Medical Readiness Flight, personal communication with James R. Ciborski, October 7, 2003; Agreement, "Mutual Aid Agreement for Fire Protection," January 24, 2001; U.S. Northern Command website, viewed online July 29, 2003, at http://www.northcom.mil/index.cfm?fuseaction=s.whoweare.

Wright-Patterson Air Force Base personnel regularly participate in disaster preparedness exercises. This exercise required participants to wear protective chemical suits and gas masks. *(Disaster Preparedness Office)*

The Disaster Preparedness Office modified its Mobile Command Post to create a state-of-the-art vehicle for controlling disaster response forces. *(Disaster Preparedness Office)*

Detachment 15 meteorologists from the 15th Weather Squadron prepare a base weather forecast.

Wright-Patterson to operate the base weather station. With the 1992 restructuring of the Air Force, Detachment 15 and the 15th WS inactivated. On September 24, 1992, the 645th Weather Squadron activated at Wright-Patterson. It went through a number of subsequent redesignations, including 645th Weather Flight (1993), 88th Weather Flight (1994), and 88th Weather Squadron (1996). The 88th Weather Squadron continues to operate the weather station for Wright-Patterson.[27]

17th Bombardment Wing

In early 1958, Air Materiel Command signed a joint tenancy agreement with Strategic Air Command (SAC) for support of SAC units at Wright-Patterson. Basing of a B-52 strategic wing at the base was part of the SAC dispersal program initiated in 1958 to decentralize large concentrations of bomber aircraft, making them less vulnerable to enemy attack. Under the Air Force-mandated program, not more than 15 aircraft were to be stationed at each of 33 locations around the nation. A $549 million construction bill passed by the U.S. House of Representatives authorized construction at 29 locations, including $22.6 million for facilities at Wright-Patterson.

Due to the existing layout of Wright-Patterson and tight security requirements associated with nuclear operations, Air Force officials decided to construct facilities for the new strategic wing as a

separate complex in Area C. The base acquired 465 acres on its northeast boundary in 1954 for construction of the West Ramp. Construction began in August 1958. To accommodate the complex, the Fairborn, Ohio, sewage and waste treatment plant and a base recreational area known as Hadden Park were relocated. The land was leveled and subgraded, and the Mad River was widened. In 1960, the extensive $25 million complex was ready to support the 4043rd Strategic Wing (SAC) and its B-52 missions.

The 4043rd activated at Wright Patterson on April 1, 1959. Its operational units included the 42nd Bombardment Squadron and the 922nd Air Refueling Squadron (assigned effective December 1959).[28] This

assignment represented the 42nd's second tour at Wright-Patterson. The organization's earliest antecedent, the 42nd Aero Squadron, had transferred from Camp Kelly, Texas, to Wilbur Wright Field on August 25, 1917. On October 1, 1918, the squadron had been redesignated Squadron I. It demobilized at Wilbur Wright Field on February 21, 1919.

The 4043rd gained several additional squadrons in late 1959. These included the 66th Aviation Depot Squadron on September 15, 1959; and the 4043rd Armament and Electronics Maintenance Squadron, 4043rd Organizational Maintenance Squadron, 4043rd Field Maintenance Squadron, and 4043rd Support Squadron, all on October 1, 1959. The first KC-135 refueling aircraft arrived on February 29, 1960. On June 15, 1960, the 42nd Bombardment Squadron moved its B-52E's in combat-ready status from the 11th Bomb Wing at Altus Air Force Base, Oklahoma, to Wright-Patterson Air Force Base. The SAC wing was then fully operational and, for the next 15 years, Wright-Patterson was home to a strategic, nuclear alert mission.

On July 1, 1963, the 4043rd Strategic Wing (SAC) reorganized as the 17th Bombardment Wing (Heavy). The 4043rd's redesignation was part of an Air Force-wide program to retain units with rich historical lineages. The 17th had distinguished itself in World War II, most notably for the 1942 air raid on Tokyo. The Black Knights also earned two Distinguished Unit Citations for exceptional service in Europe, and a third for interdiction and close support missions during the Korean War. One of the 17th Bomb Wing's components, the

B-52s of the 17th Bombardment Wing on the West Ramp at Patterson Field, 1960s

Facilities constructed for the 4043rd Strategic Wing at the north end of Area C (known as the West Ramp). The first KC-135 aerial tankers assigned to the wing arrived in February 1960, followed by B-52s in June.

42nd Bombardment Squadron (Heavy), inactivated and was replaced by the 34th Bombardment Squadron (Heavy), a unit that had been prominently connected with the old 17th Bomb Group. The 17th Bomb Wing converted to B-52H bombers in 1968 and, from 1968 to 1973, deployed 70 B-52 strategic bomber crews to Southeast Asia in support of the Arc Light bombing campaign. More than 125 KC-135 aerial tankers and crews also participated in Young Tiger or Combat Lightning operations in Southeast Asia.

An official "Buckeye Farewell" was extended to the 17th Bombardment Wing (Heavy) in 1975, when its two remaining KC-135s and one B-52H left Wright-Patterson's Very Heavy Bomber (VHB) runway for the last time. The nuclear alert mission ended at Wright-Patterson on September 30, 1975, when the 17th Bombardment Wing (Heavy) moved to Beale Air Force Base, California. Strategic Air Command reassigned the bomb wing's 14 B-52H bombers to bases in Michigan and North Dakota. The 15 KC-135 tankers went to the Ohio Air National Guard at Rickenbacker Air Force Base near Columbus, Ohio. Bomb wing personnel were reassigned to bases throughout the United States and overseas, with a large segment stationed at SAC bomber bases in the upper Midwest. Approximately one-third of the wing's 1,200 military members remained at Wright-Patterson, where they were assigned to other base units. Facilities on the West Ramp in Area C vacated by

the 17th Bomb Wing were reassigned to the 4950th Test Wing.

58th Air Division

While Wright-Patterson received its first fighter squadron in 1951 (the 97th Fighter-Interceptor Squadron, see below), it became further involved in air defense with establishment in September 1955 of the 58th Air Division (Air Defense Command). Air defense consisted of four functions: detection, identification, interception, and destruction. The 58th Air Division, upon receipt of ground observation and radar information, transmitted "scramble" messages to appropriate fighter-

interceptor squadrons, Air National Guard components, and Navy units, as well as target information to Army anti-aircraft artillery organizations. Furthermore, the 58th served as control center for air defense forces in 11 states of the Eastern Air Defense Region: Alabama, Georgia, Illinois, Indiana, Kentucky, Mississippi, North Carolina, Ohio, Tennessee, Virginia, and West Virginia.

The 58th Air Division activated at Wright-Patterson on September 8, 1955, with an authorized personnel strength of 75 officers, 24 airmen, and 14 civilians. Buildings 1419, 1420, and 1421 in the Sherwood Forest barracks section of Area A were renovated to serve as headquarters for the new division. The 58th remained at Wright-Patterson for three years. On August 11, 1958, both the 58th Air Division and its associated 4717th Ground Observer Squadron inactivated.

97th Fighter-Interceptor Squadron

The 97th Fighter-Interceptor Squadron (FIS) was a component of the Eastern Air Defense Force, headquartered at Stewart Air Force Base, New York. It arrived at Wright-Patterson on January 8, 1951, as part of Air Defense Command's program to provide aerial defense for all industrial areas of the United States. The squadron's specific mission was to provide air defense for heavily populated and industrial areas of the greater Miami Valley, including Ohio, Indiana, Kentucky, and Tennessee. Its arrival marked the first time in the long history of Wright-Patterson that an Air Force fighter squadron had been based there.

The 97th moved its aircraft into new alert operations facilities (Building 153)

Lockheed F-104 Starfighters were flown by the 56th Fighter-Interceptor Squadron, assigned to Wright-Patterson from 1955 to 1960.

Alert hangars (Building 153) were constructed on the north end of the Area C flightline in 1952 to house an air defense unit.

The 906th Tactical Fighter Group transitioned from F-4 to F-16 fighters in 1989. Here, F-16 Fighting Falcons sit on the 906th ramp during an Operational Readiness Exercise for the 9th Air Force, January 1992.

constructed for them at the north end of Area C in December 1952. The squadron, equipped with F-86D Sabre all-weather jet aircraft, operated under control of the 56th Fighter Wing at Selfridge Air Force Base, Michigan. The unit maintained 24-hour alert readiness. Four aircraft, loaded with live ammunition, and their crews were stationed near the end of the Patterson Field flightline, ready to scramble within five minutes. The squadron also occupied Buildings 1445 thru 1451 in the Sherwood Forest housing area, as well as portions of Building 206 (base operations) and Building 146 (the airfreight terminal).

In August 1955, the 97th FIS was redesignated the 56th Fighter-Interceptor Squadron and later transitioned to F-104

Starfighters in the summer of 1958. On March 1, 1960, the 56th FIS, with a strength of 300 assigned individuals and 25 F-104s, inactivated at Wright-Patterson, one of four such squadrons inactivated by Air Defense Command.

Wright-Patterson did not host another tactical combat unit until July 1982 when the 906th Tactical Fighter Group (Air Force Reserve) with its 89th Tactical Fighter Squadron (Buckeye Phantoms) activated. The 906th was one of five U.S. Air Force Reserve F-4 fighter units. This group, however, had the distinction of being the first Reserve fighter unit assigned to Wright-Patterson. Dedication of an F-4D Phantom II fighter as the *City of Fairborn* highlighted the activation ceremony. The

906th Tactical Fighter Group transitioned to F-16s in 1989. At that time, two of its retiring F-4s, the *City of Fairborn* and the *City of Dayton*, were placed on static display in front of Headquarters Air Force Logistics Command, now Air Force Materiel Command (Building 262, Area A). The 906th Tactical Fighter Group inactivated in October 1994.

445th Airlift Wing

In the 1990s, Wright-Patterson welcomed many new members to its community, including three flying units. One of these, the 907th Airlift Group (Air Force Reserve Command), arrived at Wright-Patterson in April 1993 on reassignment from Rickenbacker Air Force Base, Columbus, Ohio. On October 1, 1994, the 907th, along with the 906th Tactical Fighter Group (established at Wright-Patterson in 1982, see above), inactivated. The 445th Airlift Wing activated at Wright-Patterson on the same date. All personnel previously assigned to the 907th and 906th transferred to the 445th.

The 445th Fighter-Bomber Wing originally activated in the Air Force Reserve in July 1952, at Buffalo, New York. After several moves, inactivations, and reactivations, the wing reactivated as the 445th Airlift Wing at Wright-Patterson on October 1, 1994. In 2003, the wing remained a viable team player for local base functions, maintaining its commitment to transporting troops and cargo, via the C-141C Starlifter, around the world whenever and wherever needed.

The 445th Airlift Wing had a distinguished career before its activation at Wright-Patterson in 1994. It participated in missions such as Operation New Life, evacuation of babies and other refugees from Vietnam in 1975; Operation Just Cause in Panama in 1989; Operation Provide Hope, bringing emergency food and medical supplies to the former Soviet Union in 1992; relief flights to supply earthquake and hurricane victims in Mexico City, Florida, and Hawaii; and return of the remains of America's missing-in-action from Vietnam in 1985 and again in 1993. During Operations Desert Shield and Desert Storm, the wing provided strategic airlift, medical, maintenance, and other support for allied forces in the Persian Gulf.

For Operations Enduring Freedom and Iraqi Freedom, numerous personnel of the 445th Airlift Wing activated and played major roles in transporting troops and cargo

A 445th Airlift Wing crew readies its C-141 for another mission.

Air Force Reserve Major General Edward Mechenbier stands in front of the "Hanoi Taxi," December 2002. In 1973, Mechenbier, then a captain, was one of the first POWs to receive a ride on the C-141 out of Hanoi.

Retired Brigadier General James E. Sehorn points to signatures of fellow prisoners of war on display in the "Hanoi Taxi." Sehorn was one of the POWs who returned to the United States on the C-141 Starlifter. (*U.S. Air Force Photo by Dan Doherty*)

and providing aeromedical evacuation missions. The wing hosted contingency aircraft operations supporting all Reserve-owned C-141s during Iraqi Freedom. Additionally, the 445th participated in numerous deployments supporting the Air Expeditionary Forces throughout the world.

The 445th Airlift Wing has the honor of being home to "Hanoi Taxi," a C-141 that played a major role in Operation Homecoming, repatriation of Americans held as prisoners of war in North Vietnam. On February 12, 1973, this aircraft, tail number 66-0177, transported 40 freed POWs from Hanoi to Clark Air Base in the Philippines. Several days later, they returned to the United States. Thirty years later, the aircraft remained an active and reliable component in the airlift business, but was particularly known as a "flying museum." Plaques, photographs, and etchings of the names of those listed as missing in action, taken from the Vietnam Wall in Washington, D.C., were mounted in the aircraft.

On September 8, 1996, the 445th Airlift Wing dedicated the aircraft to those first 40 POWs with a plaque bearing their names. At that dedication, retired Navy Captain Robert Doremus, a former POW who flew on the C-141, reported that when the North Vietnamese drove them to the aircraft, every prisoner vowed to remain totally silent. "We kept straight faces until takeoff, but as soon as the wheels were in

EVOLUTION OF THE DAYTON AIR SHOW

The premier air show in the United States is held in Dayton, Ohio, every July, and it can trace its roots back to a variety of local aviation events. It began with a small gathering of barnstormers at the Montgomery County Airport, now Dayton General Airport South, and has blossomed into a huge event attended by tens of thousands of spectators. It can trace its lineage along the lines of the 1924 International Air Races, the National Air Olympics, the National Aircraft Show, and Wright-Patterson's annual base "open house" that was usually held in conjunction with Armed Forces Day.

In 1941, the National Air Olympics were held at Dayton Municipal Airport, and one of Wright Field's many contributions to the event was its nine-member "Beauty Squadron," a group of young female typists, secretaries, stenographers, and clerks who hosted the air show in specially made blue uniforms in order to promote the Army Air Corps. (One noticeable outcome of their appearance was an almost immediate surge in female job applicants at Wright Field.) The main event of the show was the National Aerobatic Championship, which drew entrants from all over the nation and abroad. Air show attendees were treated to spectacular stunts and maneuvers.

In honor of the 50th anniversary of powered flight, the National Aircraft Show ("aviation's greatest spectacle") was held at Dayton Airport from September 5-7, 1953. (Earlier National Aircraft Shows had been held in Cleveland and Detroit.) It was sponsored by the Dayton Chamber of Commerce and the Air Foundation of Cleveland and was viewed as the successor to the National Air Races, which were now considered too risky to be held in heavily populated areas. Major General Albert Boyd and Lieutenant General Edwin W. Rawlings of Wright-Patterson Air Force Base served on the executive committee of the Aircraft Show, which 195,000 people attended. There were ground exhibits and flight demonstrations by the U.S. Air Force, Army, Navy and Marines, indoor commercial exhibits, a National Parachute Jumping Contest, appearances by famous pilots and the cast of the new Hollywood film *Sabre Jet*, and several trophy contests, including a competition to break the world's aviation speed record. The Karl Taylor Orchestra provided a musical background to the air show and television station WLW-D (Channel 2) broadcasted the events.

A second National Aircraft Show was held in Dayton the following year, attracting even more spectators. Wright-Patterson Air Force Base was the staging base for most of the participating aircraft and provided logistical support to the show, just as it had the year before. Unfortunately, the show was marred by the fatal crash of a F-86H Sabre jet one mile east of Tipp City; the pilot, Major John L. Armstrong, was trying to recreate his record-breaking General Electric trophy run from two days before, when his plane disintegrated for unknown reasons.

Wright Field "Beauty Squadron" members (left to right) Marion Martin, Wanda Smith, and Betty Stoddard pose for a publicity shot at the National Air Olympics in 1941. *(Robert Durrum)*

The popularity of the National Aircraft Shows made Dayton officials yearn for a permanent, annual Dayton air show, and aviation leaders such as Eddie Rickenbacker thought the birthplace of aviation would be the ideal site for the nation's most spectacular air show. In 1975, the first Dayton Air Fair was held and 65,000 people attended. Wright-Patterson Air Force Base initially served as the staging area for aircraft in the show. It provided logistical support, including housing, mess facilities, and aviation fuels for participating armed services organizations. Its maintenance facilities and services also remained on call for emergencies. In 1982, the Dayton Air Fair became the Dayton International Air Show and Trade Exposition, and six years later, it was renamed the Dayton Air and Trade Show. After a temporary period as the United States Air and Trade Show in 1990 to reflect its official sanctioning by the U.S. Department of Commerce, it returned to its roots as the Dayton Air Show in 2001. Today, the event attracts more than 100,000 people, and Wright-Patterson Air Force Base continues its support.

Sources: "The Story of the Dayton Air Show," viewed online July 21, 2003, at http://www.usats.org/html/history.htm; Bob Batz, "Beauty Squadron: Women served their country with pride at 1941 air show," *Dayton Daily News*, August 3, 2003, p E3; "Flying Field To Be Divided For National Air Olympics," *Dayton Herald*, August 25, 1941, n.p.; Mr. Robert Durrum, interview with Robin Smith, Aeronautical Systems Center History Office, September 24, 2003; "National Aircraft Show Official Program and Log, Dayton, Ohio, 1953;" "Report on Armed Service Participation in the National Aircraft Show, Dayton, Ohio, 4-6 September 1954," by General E. W. Rawlings to Director, Office of Public Information; "'Huffman Prairie' First 'Airfield,'" *Dayton Daily News*, December 13, 1953, n.p., ASC History Office Archive, Folder: *Event: Dayton Air Show 1953*.

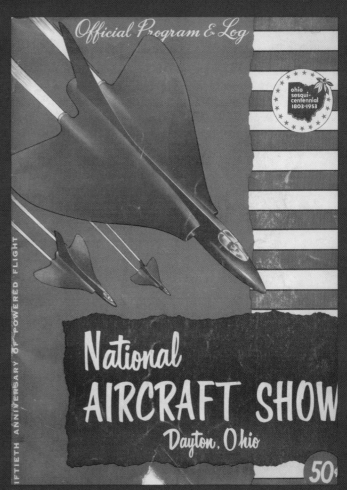

Program of the 1953 National Aircraft Show, which commemorated the 50th anniversary of powered flight. Nearly 200,000 people attended the three-day event.

Colonel Lester T. Miller, left of center, commanding officer at Wright Field, stands by while an unidentified officer awards a trophy to participants in the National Air Olympics at Dayton Airport in 1941.

the well, we let go. That cheer will be with me forever. I'll never forget my ride on 177." On July 19, 1997, at a second dedication ceremony, the aircraft was officially christened "Hanoi Taxi." The aircraft was painted with its original colors of white and gray. When it retires, it will take its place among other notable aircraft at the United States Air Force Museum. In 2003, "Hanoi Taxi" was one of four C-141s that were involved in Operation Homecoming assigned to the 445th Airlift Wing. The tail numbers of the other three C-141s are 65-0258, 66-0132, and 67-0031.

The 445th Airlift Wing received four Air Force Outstanding Unit Awards. The first award recognized the unit's support during the Pueblo Incident of 1968-1969. A second was given for outstanding support to the Air Force and the community during the year 1983. The third award resulted after superior performance in competitions and unit inspections during 1986 and 1987. The wing received its fourth award for its excellence in supporting contingency operations around the world during the years 2000 and 2001.[29]

3500th U.S. Air Force Recruiting Wing

Prior to the mid-1950s, recruiting for the Air Force had been a joint responsibility of the Air Force and the Army and, in 1951, Wright-Patterson opened a processing center for Air Force recruits. On March 6, 1954, however, the secretary of defense ordered the Air Force to begin recruiting its own military personnel. The 3500th U.S. Air Force Recruiting Wing (Air Training Command) activated on April 10, 1954, and representatives of the new wing arrived at Wright-Patterson later in the month to arrange for establishment and location of their headquarters. They established their headquarters on May 31, 1954, and began full operations in Buildings 287 and 288 (Area A) on July 1, 1954. With an authorized strength of 422 officers, 2,723 airmen, and 53 civilians, the recruiting wing consisted of a headquarters and six recruiting groups located in or near principal population centers in the United States. The unit mission was to select a sufficient number of qualified men and women from civilian sources to meet requirements of the Air Force. The wing also monitored the important career retention program. The unit remained at Wright-Patterson until 1965, when it moved to Randolph Air Force Base, Texas.

4950th Test Wing

In 1970, Aeronautical Systems Division's Directorate of Flight Test transferred its all-weather flight test mission, which it had conducted for nearly two and a half decades, to Edwards Air Force Base, California. The next year, the 4950th Test Wing (Technical) activated at Wright-Patterson Air Force Base to conduct ASD's remaining flight test functions. As originally constituted, the test wing had 10 organizational elements: Headquarters Squadron Section, Administrative Security Office, Computer Center, and Plans and Programs Office; and six divisions, for Test Engineering, Test Operations, Engineering Standards, Civil Engineering, Research and Development Procurement, and Logistics.

In 1975, Air Force Systems Command substantially reorganized the 4950th Test Wing. In anticipation of Project Realign, the test wing transferred its Administrative Security Office, Computer Center, and R&D Civil Engineering and R&D Procurement divisions to other ASD organizations in late 1974. At the same time, the wing reorganized its remaining sub-elements, creating three new deputates for Operations, Aircraft Modification, and Maintenance. This action was significant because, for the first time, it clearly separated aircraft modification from maintenance. The reorganized wing also included a Headquarters Squadron Section, Safety Office, Administrative Office, Directorate of Flight Test Engineering, and

Directorate of Support. In 1975, the 4950th Test Wing absorbed the Precision Measurement Equipment Laboratory (PMEL) and the Wright-Patterson base-level aircraft maintenance and allied support functions that had previously been the responsibility of the 2750th Air Base Wing.

In addition to receiving added responsibilities, the 4950th Test Wing also obtained new resources. The wing received 20 additional aircraft: 10 C-135s from Patrick Air Force Base; two C-135s from Edwards; one T-39 from Eglin Air Force Base; and two C-135s and five C-131s from Griffiss Air Force Base. Eight of the C-135s comprised the Advanced Range Instrumentation Aircraft (ARIA) fleet. ARIA aircraft served as tracking stations for Apollo space launches beginning in 1968. The Apollo mission sent them around the world to receive and transmit communications with the astronauts and to track and record information from the spacecraft. The 4950th used the ARIA fleet to receive, record, and retransmit telemetry data on orbital, re-entry, and cruise missile missions. In 1982, the test wing acquired four retired Boeing 707 aircraft, which it converted to the EC-18B configuration. With the EC-18B's, which had greater range and capabilities than the C-135s, the wing's ARIA flight test mission expanded.

The 4950th Test Wing handled other Air Force programs, such as Rough Rider in 1960. This combined effort between the Air Force and the U.S. Weather Bureau tested the effects of lightning on aircraft weapon systems and gathered information

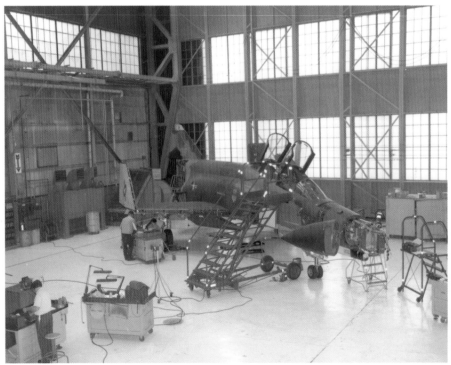

4950th Test Wing technicians modify an F-4 for flight testing.

Air Force Institute of Technology

The 4950th Test Wing assumed occupancy of the West Ramp facilities vacated by the 17th Bomb Wing in 1975. In the center of the photograph, an EC-135N ARIA aircraft awaits its next mission.

A seven-foot-diameter antenna for tracking missile and space systems telemetry accounted for the unusual bulbous nose in the 4950th Test Wing's EC-135N ARIA aircraft.

about cloud formation and thunderstorm electricity. The Zero-G program, started in 1962, used the KC-135 and the C-131 to simulate zero gravity. By 1972, more than 48,000 zero-g flights had been flown. Also developed and used in the 1960s was the ARD-21 Air Rescue Hovering set, an electronic device that sent a signal through jungle terrain into the air to announce the location of a downed pilot. ARD-21 made it possible to rescue pilots without establishing visual contact, greatly enhancing rescue operations in Southeast Asia. Through the years, the 4950th Test Wing also evaluated the capabilities of a

variety of radar, satellite communications, the Navstar Global Positioning System, the microwave landing system, electronic countermeasures systems, identification friend or foe programs, and the early stages of the Airborne Laser Laboratory program.

On April 21, 1994, the ASC commander, Lieutenant General James A. Fain, Jr., officiated at a "Mission Out" ceremony marking the transfer of the 4950th Test Wing and its assets from Wright-Patterson to the 412th Test Wing at Edwards Air Force Base, California. The wing officially inactivated on June 29, 1994.[30]

The Air Force Institute of Technology (AFIT), located at Wright-Patterson Air Force Base, Ohio, is the Air Force's primary graduate school and its premier, professional, continuing education institution. As a component of Air University, the institute provides responsive, graduate and professional continuing education, research, and consulting programs to keep the Air Force and the Department of Defense on the leading edge of science, technology, logistics, and management.

AFIT traces its roots to the end of World War I when Colonel Thurman H. Bane, commanding officer of McCook Field, advocated establishment of an aeronautical school to assure a solid corps of technical experts to manage Air Service programs. The Air Service School of Application was established in 1919 at McCook Field in Dayton, Ohio, to provide "proper technical training" in aeronautical science and allied technical fields for military officers assigned to the Air Service. It was renamed the Air Service Engineering School the next year and became the Air Corps Engineering School in 1926. In 1927, the school moved to newly opened Wright Field where it flourished until World War II. During that time, it graduated more than 200 officers.

Classes were suspended shortly after the Japanese attack on Pearl Harbor and the school remained closed until 1944, when it reopened as the Army Air Forces Engineering School to meet emergency wartime training requirements. After World War II, the Army Air Forces Institute of Technology was established as part of Air Materiel Command. The institute consisted of two colleges: Engineering and Maintenance, and Logistics and Procurement. These colleges were later redesignated the College of Engineering Sciences and the College of Industrial Administration.

When the Air Force became a separate service in 1947, the institute received its current designation as the Air Force Institute of Technology. The School of Civil Engineering Special Staff Officer's Course began the same year and civilian institution programs were transferred to AFIT the following year.

In 1950, command jurisdiction of AFIT shifted from Air Materiel Command to Air University, headquartered at Maxwell Air Force Base, Alabama. In 1951, the two AFIT colleges combined to form the Resident

FLIGHT TEST AT WRIGHT-PATTERSON

The Wright brothers' successful flights on Huffman Prairie in 1904 and 1905 established their claim as the "fathers of aviation" and Dayton, Ohio, as the "birthplace of aviation." The Dayton bicycle makers also made Dayton the birthplace of flight testing.

Dayton became firmly established as a flight test center during World War I when the Army opened a temporary installation for aeronautical research, development, and flight testing. McCook Field began operations in 1917 and, in the next decade, was home to the pioneers of flight testing, including Harold Harris, Rudolph "Shorty" Schroeder, John A. Macready, Eugene "Hoy" Barksdale, and Eddie Allen. Most famous of all was Lieutenant Jimmy Doolittle, who captured the imaginations of many with his record-setting flights and air-racing feats.

McCook pilots kept the skies above Montgomery and neighboring counties humming. In 1919 alone, more than 1,275 test flights and 3,550 incidental flights were recorded by McCook's Flight Test Section. The section pioneered work on turbosuperchargers, high-altitude flight, controllable and reversible pitch propellers, bulletproof and leak-proof gas tanks, radio beam navigation, and air-cooled and liquid-cooled radial engines.

The aircraft for the Airborne Laser Laboratory was a highly instrumented NKC-135 (serial number 55-3123). The Air Force investigated the integration and operation of high-energy laser components in a dynamic airborne environment and the propagation of laser light from an airborne vehicle to an airborne target.

McCook's small size could not accommodate the larger aircraft developed after World War I and the Air Service began a search for a new site. It accepted a gift of land from the Dayton Air Service Committee, and in 1927 opened Wright Field, thus ensuring that Dayton would remain the home of Army aeronautical research, development, and test flying. Test pilots at Wright Field during this period included Stanley M. Umstead, Donald Putt, Benjamin Kelsey, Ann Baumgartner, Albert Boyd, and Fred Bordosi, among others.

As aircraft became more sophisticated during World War II, the Materiel Command established the Flight Test Training Unit at Patterson Field to provide test pilots with advanced engineering expertise and flying skills. In 1945, Colonel Albert Boyd became chief of the flight test school. The "father of modern flight test" set and enforced exacting standards for experimental test pilots, examining their skills, intelligence, temperament, and interest in the job. College-level engineering training became a prerequisite for the flight test program. In 1949, the school became known as the Air Materiel Command Experimental Test Pilot School.

Once at war, the Army had no time to test and modify prototype aircraft. Instead, Wright Field tested early production models, while the combat versions were in production. Modifications suggested by Wright Field were incorporated into production models—an early example of spiral development. Wright Field's Accelerated Service Test Branch also conducted accelerated flight testing at the Dayton Army Air Field, Vandalia, Ohio, while the Glider Branch of the Aircraft Laboratory conducted troop-carrying glider tests at Clinton County Army Air Field near Wilmington, Ohio.

After World War II, the urban location of Wright and Patterson fields presented safety concerns, and the flight test mission began moving to Rogers Dry Lake, Muroc, California, later renamed the Air Force Flight Test Center at Edwards Air Force Base.

Remnants of the flight test mission remaining in Dayton were small but critical to flight research. The All-Weather Flying Group, established in 1945, attempted to solve problems of flight control under adverse weather and light conditions using radar in air traffic control. Results of those tests led to the development of equipment and techniques as prototypes for future commercial service. The All-Weather Flying Group also operated the "On-Time Every-Time Air Line," a regularly scheduled service between Clinton County Air Force Base and Andrews Air Force Base between 1946 and 1948. No scheduled flight was ever canceled, even for severe weather.

Wright Air Development Center test pilots also conducted flights to determine the effects of icing, turbulence, and thunderstorms on aircraft. They tested windshield rain repellants, combat traction systems for landing on wet runways, electronic location finders for downed airmen, tow wires for the accurate aerial delivery of cargo, frangible canopies for pilot ejection, and air cushion landing systems. The United States' space program astronauts were trained and tested in zero-gravity environments as was the lunar rover vehicle used in the Apollo missions.

In 1960, the Air Force partnered with the U.S. Weather Bureau, and eventually the Air Force Cambridge Research Laboratory and the Sandia Laboratories, on Project Rough Rider. Their mission was to gather data about thunderstorms, electricity, cloud structure dynamics, and weapons effects vulnerability. Results of that project led to improved handling of aircraft and weaponry in adverse weather conditions.

In 1971, the Aeronautical Systems Division's (ASD) flight testing components reorganized into what became the 4950th Test Wing. During the next two decades, the 4950th tested advanced radars and other avionic systems, infrared missile guidance systems, lasers, and satellite systems. Its most visible program was the Advanced Range Instrumentation Aircraft (ARIA, known as the Apollo Range Instrumentation Aircraft until completion of the Apollo program). ARIA aircraft deployed worldwide as mobile tracking stations that received and transmitted astronaut voices, and recorded telemetry information from spacecraft and other NASA (National Aeronautics and Space Administration) and Department of Defense space vehicles.

In 1977, the 4950th took delivery of a testbed for a program that became one of the Air Force's most critical future technologies—the Airborne Laser Laboratory (ALL). A highly instrumented NKC-135 (serial number 55-3123) was delivered to Wright-Patterson for installation and test of a canopy and weapons-quality laser. After preliminary testing at Wright-Patterson, the testbed returned to Edwards Air Force Base for continued tests. That test vehicle was eventually retired to the U.S. Air Force Museum at Wright-Patterson.

In 1994, the 4950th's remaining test mission moved to Edwards Air Force Base, marking the end of an era. For nearly 75 years, flight testing in the Miami Valley had made significant contributions to the winning of two world wars and breaking the nuclear stalemate of the Cold War, as well as innumerable contributions to civil aviation in the areas of all-weather flying, air traffic control, and tracking technologies. For the first time since 1917, the skies above Dayton remained silent to the sound of flight test aircraft.

Source: *Against the Wind: 90 Years of Flight Test in the Miami Valley* (Aeronautical Systems Center History Office, Wright-Patterson Air Force Base, Ohio, 1994).

F-86A of the All-Weather Flying Center undergoing major inspection, about 1951

Wright Field test pilots, 1936 (left to right): Captain Stanley M. Umstead, Captain J. S. Griffith, Captain Harold R. Harris, Lieutenant Eppright, Lieutenant Donald Putt, Major McClellan, and Captain Frank G. Irvin

Test Pilot School Class of 1949 at Patterson Field. Front row, left to right: Captain J. R. Amann, Major J. C. Wise, Major G. V. Lane, Major P. P. Haug, Lieutenant R. J. Harer, Lieutenant R. D. Hippert, Captain S. P. Parsons. Back row: Major K. O. Chilstrom, J. Krug, Captain G. B. Quisenberry, Major D. A. Johnson, Captain L. K. Nesselbush, Captain R. M. Roth, Captain R. L. Stephens, Captain R. M. Howe. (*United States Air Force Flight Test Center History Office*)

WADC Commander Major General Albert Boyd, right, viewing a piece of test equipment used during flight test of a Russian MiG-15 at Kadena Air Force Base, Okinawa. Also shown are Major Chuck Yeager, left, and Captain Earl Ashworth, center.

College. The institute established a logistics education program with Ohio State University conducting the first courses on a contract basis in 1955. AFIT began a series of short logistics courses in 1958 as part of the Air Force Logistics Command Education Center. Later that year, the School of Logistics became a permanent part of AFIT.

In 1954, the 83rd Congress authorized the commander of Air University to confer degrees upon students in the AFIT Resident College. The college was subsequently divided into the School of Engineering, the School of Logistics, and the School of Business. The first undergraduate engineering degrees were granted in 1956, and the first graduate degrees in business in 1958. The School of Business programs transferred to civilian universities in 1960.

In 1963, the School of Logistics was redesignated the School of Systems and Logistics and the Civil Engineering Center became the Civil Engineering School. The next year, the Civil Engineering School moved into Building 640, a brand new facility. General Curtis E. LeMay, Air Force chief of staff, presided at the December 1962 groundbreaking for the building. This facility marked the first step in AFIT's conversion from temporary, makeshift quarters to a modern campus. The School of Systems and Logistics received its own home on October 4, 1977, when the Honorable Hans M. Mark, under secretary of the Air Force, dedicated Building 641.

In 1967, AFIT became a member of the Dayton-Miami Valley Consortium, which later became the Southwestern Ohio

Council for Higher Education. This association of colleges, universities, and industrial organizations in the Dayton area promoted educational advancement. AFIT traditionally has been active in both the council and in other community and joint institutional programs.

AFIT's operations expanded in 1976, when it activated the Defense Institute of Security Assistance Management (DISAM), an element of the Defense Security Assistant Management Education program. DISAM held its first classes in Building 288, Area B, on January 18, 1977, later relocating to Building 125.

The AFIT campus continued to grow in the 1990s. The institute accepted the Science, Engineering, and Support Facility (Building 642) in 1990 and opened the School of Civil Engineer and Services (Building 643) in 1994. In 2003, AFIT accomplished its mission through three resident schools: the Graduate School of Engineering and Management; the School of Systems and Logistics; and the Civil Engineer and Services School. AFIT also managed educational programs of Air Force officers enrolled in civilian universities, research centers, hospitals, and industrial organizations through its Civilian Institution Programs.

By the close of the twentieth century, effects of the institute's educational programs upon the Air Force and the Department of Defense had been profound. More than 266,000 Department of Defense personnel, including 30 United States astronauts, had attended AFIT programs. Since resident degrees were first granted in 1956, AFIT had awarded more than 900

bachelor of science, 13,000 master of science, and 300 doctor of philosophy degrees. Furthermore, more than 12,000 Air Force students attending civilian institutions had attained undergraduate and graduate degrees. In 2003, AFIT accepted a group of senior noncommissioned officers into its master's degree programs.

Air Force Research Laboratory

The Air Force laboratories at Wright Field have undergone many changes over the past eight decades. Their origins date back to World War I when the Army decided to establish a center for experimental engineering and, in December 1917, opened McCook Field, in Dayton, Ohio. By the end of war, McCook had organized its experimental work into five technical sections: Airplanes, Power Plants, Equipment, Armament, and Materials. In the early 1920s, it added organizations for Lighter-than-Air and Aircraft Radio development. The staff was small and made little attempt to conduct advanced research on a large scale. Instead, the technical sections concentrated on test and evaluation of industry-produced airplanes and equipment, especially engines. Results of these tests were disseminated to industry via technical reports and information circulars. The technical sections did undertake several major in-house projects, such as design, fabrication, and tests of a 55-foot, all-metal, stressed-skin wing in the late 1920s and design,

A student arrives at the Air Force Institute of Technology.

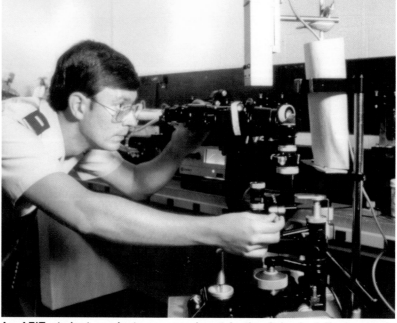

An AFIT student conducts an experiment in the School of Engineering's laser laboratory.

fabrication, and flight test of a pressurized cockpit/cabin for high-altitude flight in the mid-1930s.

By 1939, the Engineering Division at Wright Field had seven laboratories: Aircraft, Propellers, Power Plants, Materials, Equipment, Photography, and Armament. During World War II, the laboratories grew in size and number. The engineering staff increased more than eightfold and the number of laboratories grew from seven to 12, with new laboratories for Radar, Aero Medical Research, Communication and Navigation, Personal Equipment, and Electronic Components. The number of experimental facilities also increased to include at least four new wind tunnels, a new structures-testing building, and expansion of the power-plant complex for jet engine and rocket motor testing. The experimental budget increased nearly 100 percent over its prewar total, and the number of experimental projects surpassed a thousand by war's end.

In 1951, the Air Force set up a separate command for R&D, Air Research and Development Command (ARDC). At Wright Field, ARDC's field unit was Wright Air Development Center (WADC). WADC essentially took over the 12 laboratories of the old Engineering Division. Several minor reorganizations during the 1950s did not materially affect the overall mission of the laboratories: support of weapon systems development and advanced research.

In 1959, ARDC reorganized WADC when it established Wright Air Development Division (WADD). More than half of the old organization transferred to the Directorate for Systems Engineering, while the remainder realigned under the Directorate for Advanced Technology Development. The laboratories were placed in the latter directorate. In 1961, ARDC gave way to Air Force Systems Command (AFSC), and WADD became Aeronautical Systems Division (ASD). Shortly thereafter, the number of laboratories decreased to six: Materials, Aero Propulsion, Avionics, Flight Dynamics, Aerospace Medicine, and Aerospace Research. The last two laboratories reported to the Aero Medical Division and the Office of Aerospace Research, respectively. The other four labs were placed under a new Research and Technology Division (RTD). When RTD was disestablished in 1967, the laboratories transferred to AFSC's Director of Laboratories. Throughout all these reorganizations, laboratory personnel and facilities continued to reside at Wright Field.

Throughout the 1960s, laboratory functions at Wright Field expanded, manpower increased, and new facilities were built. The laboratories were semi-autonomous organizations and established their own science and technology agendas, largely independent of the day-to-day needs of weapon system development, which was now the province of systems engineering. The laboratories embarked on several programs to "grow" technology, such as Aero Propulsion Laboratory's Advanced Turbine Engine Gas Generator (ATEGG) program for engine cores; Avionics Laboratory's pursuit of solid-state, phased-array radar development; and Materials Laboratory's programs in advanced composites and carbon-carbon development. During the Vietnam War, the laboratories worked closely with ASD's system program offices (SPOs) to transition advanced technology to the battlefield. The laboratories advocated technology insertion into weapon systems, such as Flight Dynamics Laboratory's championing of fly-by-wire technology that revolutionized aircraft flight controls and battlefield capabilities.

The end of the war, however, resulted in a downsizing of the defense establishment,

The Radar Test Facility, Building 821, Area B, constructed in 1947, was made entirely of wood, including the use of wooden pegs rather than metal nails.

When completed in 1965, the Nuclear Engineering Test Facility, Building 470, Area B, was the seventh largest nuclear reactor in the United States.

Radar reflectivity and cross-section testing performed in Building 821 formed the foundation for stealth technology.

and in 1975, the Air Force decided to consolidate its laboratories nationwide. At Wright Field, this resulted in consolidation of the four remaining laboratories for Materials, Aero Propulsion, Avionics, and Flight Dynamics under a new corporate headquarters and staff—Air Force Wright Aeronautical Laboratories (AFWAL). In 1982, formalizing the spirit of cooperation that existed during the Vietnam War, AFWAL was reassigned from AFSC to ASD to promote technology transition from the laboratories to weapon systems. Six years later, AFWAL was reorganized and redesignated Wright Research and Development Center (WRDC) and added an Electronic Technology Laboratory and directorates for Manufacturing Technology, Cockpit Integration, and Signature Technology. In 1990, WRDC was reorganized once again as Wright Laboratory, one of four Air Force "super" laboratories nationwide.

In the early 1990s, the Department of Defense reduced its R&D facilities and manpower infrastructure. This decision resulted in consolidation of all Air Force laboratories under a single Air Force Research Laboratory (AFRL) in 1997. Whereas each of the four "super" laboratories had reported directly to a product center (for example, in the case of Wright Laboratory to Aeronautical Systems Division/Center), AFRL was aligned under and reported directly to Air Force Materiel Command. In 1998, AFRL's headquarters moved from Headquarters AFMC in Building 262, Area A, to Building 15 in Area B.

Air Force Security Assistance Center

In July 1976, Air Force Logistics Command (AFLC) created Air Force Acquisition Logistics Division (AFALD) from existing AFLC sources, primarily its Deputy Chief of Staff for Acquisition Logistics and the 2732nd Acquisition Logistics Operations Squadron. AFALD's mission was to expand and strengthen the interface between AFLC and AFSC, thereby improving operational utility, field availability, and supportability of new systems, while reducing operating and support costs. AFALD acted as a catalyst for stimulating and improving the interchange of knowledge between AFLC, AFSC, and the combat commands.[31]

On May 1, 1978, AFLC activated the International Logistics Center (ILC), which it identified as a "major field organization."

ILC merged elements from the Headquarters AFLC Office of the Assistant for International Logistics and most of the international logistics functions from AFALD, which closed at this time. The new ILC had three principal offices: Plans and Procedures, Programs and Resources, and Operations. It established and implemented an international logistics program for development, negotiation, and management of AFLC's security assistance programs. These programs included foreign military sales, grant aid, and international military education and training. In 1993, ILC became the Air Force Security Assistance Center (AFSAC) in Air Force Materiel Command at Wright-Patterson Air Force Base. In 2003, AFSAC supported more than 9,000 aircraft, ranging from World War II-era C-47 transport aircraft to the latest versions of the F-16 fighter, in 90 different countries around the world.

Medical Center Wright-Patterson Air Force Base

Major Alfred G. Farmer, the first post surgeon, established Wilbur Wright Field's Medical Department with the aid of a lieutenant and three enlisted men in July 1917. A converted barracks served as their infirmary until the hospital, a temporary wooden structure, opened the next month. Within a year, the hospital expanded to 89 beds and a staff of 14 physicians and dentists, 11 commissioned nurses, and 63 enlisted men.

Shortly after the start of World War II, the installation opened its first permanent medical facility. The Patterson Field Post Hospital in Building 219 served military personnel from both Patterson and Wright fields. A staff of 15 doctors, six civilians, and 550 enlisted men supported the 50-bed hospital. Rapid wartime expansion, however, soon overwhelmed the facility. By December 1942, a temporary cantonment-type hospital (Building 1113) opened in Wood City. It remained the main base hospital until 1956.

After the war, Building 219 converted to non-medical functions, including bachelor officers' quarters, Women Airforce Service Pilots (WASP) barracks, and offices for the Foreign Technology Division. The old hospital returned to medical use in 1977 when the medical center moved several clinics to the facility. It transformed into a pediatrics clinic in 1989. The World War II structure was also one of several buildings on base that employees claimed to be haunted.

In July 1944, health officials opened a station dispensary on Wright Field. Building 40 accommodated a host of medical services through the years, including a flight surgeon's clinic added in 1963. However, occupational medicine became its main function. Building 40 remained in medical service until 2002, when a new Occupational Health Clinic, Building 675, opened in Area B.

The era of temporary hospitals ended in September 1956 with dedication of the 2750th U.S. Air Force Hospital, Building 830 in Area A. The $5 million facility had a patient capacity of 348 beds. A dental clinic was added in 1969 and a new wing built in 1970 increased the hospital's capacity by 75 beds.

Until January 1, 1961, the base hospital was assigned to the 2750th Air Base Wing.

The 348-bed base hospital was completed in June 1956.

On that date, it was reassigned to Headquarters Air Materiel Command (later AFLC). On July 1, 1969, the U.S. Air Force Hospital Wright-Patterson became the U.S. Air Force Medical Center Wright-Patterson. In its new capacity, the hospital served as a major referral and consulting center for Department of Defense Health Service Region 5, a seven-state area containing more than 675,000 eligible beneficiaries. A $123 million expansion completed in May 1989 established the Wright-Patterson Medical Center as the second largest hospital in the Air Force.

A major Air Force restructuring resulted in yet another redesignation for the facility in 1993. On July 1 of that year it became the Medical Center Wright-Patterson Air Force Base. The hospital also gained a new supporting unit, the 645th Medical Group, which was redesignated the 74th Medical Group the following year.

Besides supporting the base military population and Region 5, the medical center was one of at least two Air Force centers to offer hyperbaric medicine therapy. The medical center's residencies in obstetrics/gynecology, pediatrics, general surgery, emergency medicine, internal medicine, and psychiatry, which were fully integrated with Wright State University's School of Medicine, were unique within the Department of Defense. They allowed Air Force physician trainees and their civilian colleagues to share clinical experiences and faculties. Clinical rotations were shared using the medical center and several major teaching hospitals in Dayton and the surrounding area. Additionally, many of the medical center's clinical staff held appointments as faculty members at Wright State University as well as the Uniformed Services University of the Health Sciences in Bethesda, Maryland, and actively shared teaching responsibilities with civilian clinical faculty members in the private health sector.[32] The medical center and the Veterans Affairs Medical Center (VAMC) in Dayton also shared treatment capabilities that were highly specialized but in relatively low demand. For example, gynecological services were offered at Wright-Patterson, while patients at the base that needed inpatient psychiatric or sleep-disorder treatments were referred to the VAMC.

National Air and Space Intelligence Center

In 2003, the National Air and Space Intelligence Center (NAIC), headquartered at Wright-Patterson Air Force Base, was the Air Force's single, integrated, intelligence production center and the primary Department of Defense producer of foreign air and space intelligence. NAIC assessed current and projected foreign forces, threats, and weapon system capabilities and employment, and evaluated evolving technologies of potential adversaries. The center traced its heritage from the Foreign Data Section of the Army Signal Corps' Airplane Engineering Department established at McCook Field in 1917. The section was charged with evaluating foreign scientific and technical programs related to aircraft. This mission passed through many organizations before being assigned to the National Air Intelligence Center (NAIC) in 1993.

During the interwar years, NAIC's predecessors gained responsibility for disseminating aviation-related technical information to businesses and military organizations, operating the Army's Aeronautical Museum, and producing motion picture studies of engineering experiments. With the advent of World War II, the impact of the enemy's radical and advanced weapon-design concepts forced the Allies to a new appreciation of technical intelligence. As the war neared its end, the scientific and technical intelligence mission shifted emphasis from air operations support to technical exploitation of enemy technology. Front-line troops captured

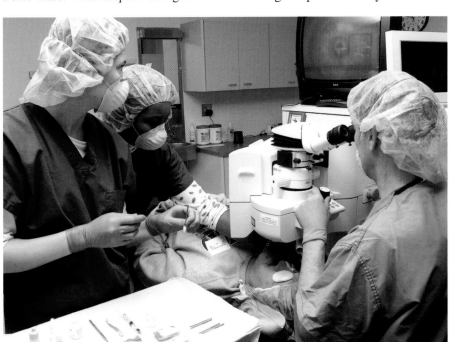

Dr. (Colonel) Leo Hurley performs laser eye surgey on a patient in the Laser Surgery Center at the medical center. (*U.S. Air Force photo by Robert Memering*)

Medical Center Wright-Patterson Air Force Base in the twenty-first century

National Air and Space Intelligence Center in Area A (*National Air and Space Intelligence Center*)

NAIC replaced its MiG-21UM display with this MiG-29UB Fulcrum B training aircraft in 2000. Built in 1988, the Fulcrum served both the Soviet and Moldovan Air Forces. The U.S. government purchased it and NAIC extracted it, along with 21 other MiG-29s and 507 air-to-air missiles, in 1997. The project was the largest extraction of foreign aircraft to the United States since World War II. (*National Air and Space Intelligence Center*)

enemy equipment and sent it back to Wright Field for assessment, with the first German and Japanese aircraft arriving in 1943. Officers and civilian scientists, mostly from Wright Field, followed the armies to exploit captured enemy materiel and documents.

Under Project Lusty, Wright Field's Colonel Harold E. Watson and "Watson's Whizzers," a group of handpicked pilots, gathered German aircraft from the battlefield and sent or flew them back to Wright Field and Freeman Field, Indiana, for study. The Messerschmitt 262 jet fighter was the best known of these aircraft. Operation Paperclip, a follow-on project, brought more than 200 German scientists and technicians to Wright Field to work with their American counterparts. Initially assigned to the intelligence branch, many scientists eventually worked in Wright Field laboratories.

Captured documents also flowed into Wright Field. By the end of 1947, Wright Field personnel had processed more than 1,500 tons of documents, adding 100,000 new technical terms to the English language. The technical knowledge gained revolutionized American industry. Besides aviation-related advances, new designs for vacuum tubes, development of magnetic tapes, night-vision devices, improvements in liquid and solid fuels, and advances in textiles, drugs, and food preservation were made available to American manufacturers.

Establishment of T-2 Intelligence at Wright Field in 1945 began the move toward a balanced integration of engineering and intelligence. T-2 became responsible for creating air intelligence; identifying foreign aircraft and related equipment needed for study; receiving, translating, and distributing foreign language documents; and distributing air intelligence products. By decade's end, T-2's efforts turned increasingly toward the emerging Russian technological threat. T-2 also opened an office in 1947 for the study of unidentified flying objects (UFOs).

With establishment of the Foreign Technology Division (FTD) under the new Air Force Systems Command in 1961, the Air Force gained a formal organization that handled its scientific and technical intelligence mission for the next 30 years. FTD's investigation of new foreign technology provided a yardstick against which American research and development could be measured. During the Vietnam War, FTD contributed significant data regarding countermeasures against surface-to-air attacks.

Over the years, the intelligence-gathering mission grew increasingly complex. In the late 1940s, NAIC's predecessors had just begun to develop a scientific and technical database through exploitation of published foreign literature. By the mid-1950s, they had created an impressive file of retrievable information. The addition of raw intelligence gained from the Korean conflict, however, established a need for more modern techniques to reduce this vast amount of documentation to useable information. As a result, intelligence analysts began pioneering the use of computers. Radar intelligence, electronic intelligence, and machine translation capabilities were also added to the intelligence arsenal.

By 1961, FTD automated the photo analysis process. It added the capability to provide invaluable information on foreign aerodynamic, ballistic missile, and space vehicle systems in 1963. That same year, the database became automated and a computerized library of scientific and technical information from many sources was available for instant recall.

In the 1970s, FTD acquired capabilities in human intelligence targeting and laser signal analysis. It consolidated all scientific and technical databases into a single, comprehensive database. The use of automated microfilm storage, retrieval, and display equipment improved accessibility to document processing and dissemination.

As the intelligence organization's capabilities expanded, so too did its facilities. The opening of Building 828, a 100,000-square-foot complex in Area A, in the late 1950s marked the first phase in the development of a state-of-the-art working complex for the Air Force's scientific and technical intelligence workforce. This structure was specifically designed to house Readix, FTD's first computer. Building 829 joined 828 in 1967 to accommodate FTD's growing sensor data processing and analysis mission. Building 856 was added to the complex in 1976, and the Photo Laboratory was modernized in 1982. A Foreign Materiel Exploitation Facility (Building 4023) was erected in 1994, and a Special Operations Intelligence facility, Building 858, opened in 1997. In June 1992, the National Air Intelligence Center placed a MiG-21 on display in front of its headquarters (Building 856). The center replaced this monument with a MiG-29 Fulcrum B purchased from the Republic of Moldova in August 2000.

The name change to National Air Intelligence Center (NAIC) resulted from the Department of Defense's 1993 trend to merge scientific and technical intelligence centers with general military intelligence organizations. In 2002, NAIC's name changed slightly to the National Air and Space Intelligence Center to more accurately reflect its mission. NAIC's mission continued to included assessing current and projected foreign aerospace capabilities, developing targeting and mission planning intelligence materials, and evaluating evolving technologies of potential adversaries.[33]

Orientation Group, U.S. Air Force

The Orientation Group grew out of the hugely successful 1945 Army Air Forces (AAF) Fair held at Wright Field at the end of World War II. The fair displayed technological advances made in aviation during the war, showed off captured German and Japanese weapons, and presented the AAF story to the American people. More than a million visitors from across the United States and 26 foreign countries attended the event. Although originally intended as a local weekend show, public response was so overwhelming that Air Technical Service Command (ATSC) quickly decided to extend the fair for an additional week.

General Henry H. Arnold, commanding general of the AAF, recognized the public relations coup and ordered ATSC to develop a traveling version of the fair. ATSC assembled a selection of exhibits into a touring exhibition and formed the 4140th Army Air Forces Base Unit (Research and Development Exhibition) to organize and handle the road show. The new unit toured most of the northeastern United States during 1946 and also presented aerial demonstrations of the P-51 Mustang.

In 1956, the traveling exhibit unit was renamed the Orientation Group, United States Air Force. However, the group was more commonly referred to as the Air Force Orientation Group (AFOG). Its mission was to create and display exhibits that informed

AFOG teams often took their traveling displays to high schools.

Master Sergeant Dave Menard briefed President Gerald Ford and Secretary of Defense Donald Rumsfeld at an AFOG display in Washington, D.C. As part of the Armed Forces Bicentennial Caravan Project, the group traveled around the nation with displays from each of the military services.

The AFOG team with key exhibits assembled in front of their facilities at Gentile Air Force Station, Ohio, 1990.

the American public about Air Force people, equipment, and contributions to the nation. These efforts played a major role in Air Force public affairs and recruiting efforts.

The Orientation Group produced both outdoor and indoor exhibits. Its outdoor exhibits featured full-size aircraft and display vans showing audiovisual programs on aviation history and the Air Force's technological contributions to the nation. Indoor exhibits depicted a variety of Air Force stories through large color transparencies, graphics, models, artwork, and multi-image productions. Displays also featured selections from the Air Force Art Collection. Besides traveling exhibitions, AFOG maintained permanent displays at the Pentagon and at the Chicago Museum of Science and Industry.

Originally headquartered in Wood City, AFOG experienced a devastating fire on November 25, 1961. The fire claimed three buildings and damaged four others. Most of the equipment in the destroyed buildings was also lost. AFOG then moved to Area B. The organization moved again in 1981, this time to the Defense Electronics Supply Center at Gentile Air Force Station in Dayton, Ohio. It remained there until its inactivation in 1992, a victim of the post-Cold War drawdown. Although the unit closed its doors, it left behind a rich heritage evidenced by its numerous Air Force Organizational Excellence Awards for "increasing public understanding and awareness concerning the Air Force."[34]

United States Air Force Band of Flight

The Air Force Band of Flight originated in 1942 to help new recruits at Patterson Field adjust to marching at the rate of 120 paces a minute. The commander of the recruits asked Paul Shartle, himself a recruit and music major in college, to organize a marching band. (Patterson Field already had a dance and show band.) Shartle convinced 28 of his fellow recruits to volunteer. They sent home for their instruments and a local music store donated some used instruments.

The band was organized as the 361st Army Air Forces Band with a table of organization for 28 members, a captain, and a warrant officer. Shartle played trumpet and served as first sergeant. Although the band formed to play marching music for drills along the flightline, it also performed for savings bond drives, Red Cross events, and award presentations to companies

Paul Shartle, a trumpet player and music major in college, organized the 361st Army Air Forces Band in 1942 with 28 volunteers and donated instruments. After redesignation as the 661st Army Band in 1944 and the 661st Air Forces Band in 1947, it became the Air Force Band of Flight in 1991.

The Air Force Band of Flight, which originated at Wright-Patterson during World War II, performs for both official ceremonies and public entertainment. (*U.S. Air Force photo by Henry Narducci*)

producing products for the war effort.

After being redesignated the 661st Army Air Forces Band in 1944 and the 661st Air Force Band in 1947, it became the Air Force Band of Flight in 1991. At the turn of the century, the 60-member Band of Flight averaged 450 performances a year for official, recruiting, and community relations events, including appearances at the Ohio bicentennial celebration, Armed Forces Day, parades and concerts for Memorial Day and Veterans Day, and various local monument dedications. Although one of 12 Air Force bands, in 2003, the Band of Flight had the distinction of being the oldest in the service and one of only two that remained at its station of origination. It was comprised of six ensemble components that played everything from Broadway show tunes to rock-and-roll. In

1997, the 45-member Concert Band, the largest ensemble within the Band of Flight, received the Colonel George S. Howard Citation of Musical Excellence for Military Concert Bands.[35]

United States Air Force Museum

The United States Air Force Museum located at Wright-Patterson Air Force Base is internationally recognized as the largest and oldest aviation museum in the world. Established in 1923 at McCook Field, the museum collection first consisted of an informal exhibition of World War I airplanes and equipment. In 1927, the museum relocated to Wright Field, where it occupied 8,000 square feet in a laboratory building.

In 1935, it moved into newly constructed Building 12, designed and built by the Works Progress Administration to house and display artifacts. By this time, the museum collection had increased to more than 2,000 items. Due to the urgent need for administrative space and manpower in 1940 to support the burgeoning war effort, the museum closed its doors to the public and placed its collection in storage.

Near the end of World War II, a Museum Office was created to collect, preserve, and safeguard both foreign and American aeronautical equipment for historical purposes. With establishment of the National Air Museum at the Smithsonian Institution, Washington, D.C., the Air Force Technical Museum at Wright-Patterson was tasked with supporting the National Air Museum's collection and preparing public displays and study collections of aeronautical items. The Technical Museum's aircraft and reference collections also were used extensively for research and patent litigation.

Museum officials at Wright-Patterson continued to plan for an eventual public re-opening. These efforts reached fruition in April 1954, when the museum opened to the public in a temporary World War II structure, Building 89, Area C. Its first year attendance figures totaled 10,000 visitors. In 1956, the facility officially became the United States Air Force Museum. It soon began converting from a huge, open-storage display format to one featuring a unique "maze" floor plan. This more modern display technique directed visitors along a controlled walkway presenting the unfolding story of military aviation in chronological sequence. The United States Air Force Museum was reassigned from Air University, Maxwell Air Force Base,

Alabama, to Headquarters AFLC in 1965. Soon thereafter, the museum was attached to the 2750th Air Base Wing for administrative and logistical support.

By the early 1960s, the museum outgrew its accommodations in Building 89. Furthermore, the structure itself also made it unsuitable for properly protecting and displaying the growing collection; the temporary wooden building was not fireproof and lacked air-conditioning. Local interest in aviation dating back to the Wright brothers sparked creation of the Air Force Museum Foundation, Inc., whose goal was to finance larger and more modern facilities for the museum. Eugene W. Kettering, son of the Dayton inventor and president of the Charles F. Kettering Foundation, was chairman of the Museum Foundation board. Frank G. Anger, president of Winters National Bank and Trust Company, served as the foundation's president. In 1962, the foundation launched a nationwide fundraising campaign for a new facility.

In November 1964, Secretary of the Air Force Eugene Zuckert presented the deed to 225 acres of Air Force land along the west edge of Springfield Pike to the Air Force Museum Foundation. Public-spirited citizens in Dayton and across the nation contributed more than $6 million to the foundation for construction of Building 489. On September 3, 1971, President Richard M. Nixon, joined by members of the Wright family, opened the new United States Air Force Museum.[36]

Designed especially to display aircraft, the building provided environmentally controlled indoor protection for exhibits, a 500-seat theater, administrative offices, a gift shop, a bookstore, and reception services. Five years later, the foundation donated another $1 million to construct a visitor center. The Modern Flight Hangar, a $10.8 million expansion funded equally by the foundation and the federal government, opened in 1988. This expansion was followed in 1991 by a $7.3 million gift from the foundation that added the IMAX Theatre and atrium. The Eugene Kettering Gallery, opened in 2003, added a third major building, granting the museum additional space to bring more aircraft indoors, revamp the flow of its exhibits into a more chronological order, and create a Cold War Gallery. In addition, construction of a Hall of Missiles and a new Space Gallery reflected the U.S. Air Force's expanding commitment to space.

By 2003, the museum collection had grown to more than 300 aircraft and missiles on display, plus thousands of personal artifacts, documents, photographs, and mementos of Air Force history. More than one million visitors toured the museum annually. It represented Ohio's top, noncommercial, tourist attraction. The museum also supported other Air Force, Department of Defense, and civilian organizations in the United States and abroad with technical advice and loans of artifacts and aircraft.

In 1999, the museum welcomed the National Aviation Hall of Fame into its complex. The Hall of Fame was established in 1962 to recognize the nation's leading air and space pioneers. Since 1962, 178 men and women, including the Wright brothers, Charles Lindbergh, Amelia

The United States Air Force Museum resided in Building 89, Area C, from 1954 to 1971.

Virginia Kettering is greeted at the dedication ceremonies for the new museum by General Jack G. Merrell, commander of Air Force Logistics Command.

United States Air Force Museum during construction of its third display hangar

Earhart, Jimmy Doolittle, Chuck Yeager, John Glenn, and Neil Armstrong, have been enshrined.

A HALF-CENTURY OF GROWTH

Transitions between peace and war over the last half of the twentieth century significantly altered the physical appearance of Wright-Patterson Air Force Base. Wartime structures received modifications and new facilities were constructed to meet the needs of Wright-Patterson's military and civilian personnel and residents. Additional housing and community facilities were major facets of the construction program. These facilities enhanced the quality of life—morale, welfare, and recreational opportunities— of base employees and supported the Air Force's goal of retaining quality airmen. New laboratories and workspaces for the base's many tenant organizations also became frequent sights on the base.

Housing Complexes

Adequate quarters for military and civilian workers at the base remained a persistent problem throughout World War II and into the early postwar years. With the swift upturn in defense requirements accompanying the Korean War, the housing shortage became critical. In 1950, base housing accommodated only a small percentage of military families assigned to Wright-Patterson. These families resided primarily in Wood City across from Patterson Field and a few smaller housing complexes near the base gates. Several large barrack complexes scattered throughout the base housed single airmen and the Brick Officers' Quarters, constructed between 1933 and 1935, offered space for 92 officers. With the exception of the permanently constructed brick quarters, many of the barracks were of temporary wartime construction, like those in Wood City, which had been slated for demolition until the onset of the Korean War gave them new life. After civil engineers installed new siding, base officials declared the rehabilitated barracks suitable for many more years of productive use. In fact, some of these barracks remained in service into the 1990s.

With the influx of new personnel in the early 1950s, hundreds of officers and enlisted personnel legally entitled to government-furnished quarters had to commute from areas around the base, some from as far away as Cincinnati. Many airmen found off-base housing hastily built and substandard. Some military families even found themselves residing in shacks and cabins without running water or indoor plumbing. These deplorable conditions precipitated an in-depth study of the problem, which in turn led to a coordinated military/civilian housing drive. The base commander initiated conferences between civil, government, and military officials to explore proposals for more housing for the Wright-Patterson area. An extensive publicity campaign urged home and apartment owners to make units available to base personnel. Builders were encouraged to begin construction programs to alleviate critical housing shortages, and area colleges were asked to provide additional space for unmarried personnel.[37] In 1949, the base also opened a trailer camp in Wood City with space for 40 to 60 residences.

Real relief for the housing crisis finally arrived in 1953, when the 2,000-unit Page Manor Housing Development (named in honor of Brigadier General Edwin Randolph Page) was completed on the south side of Airway Road in Mad River Township. Construction plans for Page Manor had

The Wood City trailer park opened in 1949 to help alleviate a critical housing shortage.

Construction of the first 1,000 units of the Page Manor housing development began in July 1951. An additional 1,000 units were constructed during 1953.

The Cottage, originally built as a kindergarten for children in the Brick Quarters, became a residence for visiting dignitaries.

Building 826, Area A, opened as visiting officers' quarters in 1959.

Bible school was held in Skyway Park for children of Wright-Patterson's service members. (*Alan W. Rault*)

Woodland Hills housing complex under construction in 1971. The Radar Test Facility is visible in the lower half of the photograph.

been drawn up in 1949-1950 under authorization of the Wherry-Spence Amendment to the National Housing Act (August 1949), which permitted the Federal Housing Administration to insure privately financed housing on or near military installations. On July 12, 1951, a groundbreaking ceremony was held for the new complex, and on October 29, 1952, Brigadier General C. Pratt Brown, base commander, cut the ribbon during opening ceremonies for the first 1,000 units. Page Manor's second 1,000 units were completed in 1953.

The influx of Wright-Patterson families into the new housing project precipitated yet another concern for base planners—providing education for children of the postwar "baby boom." In October 1953, base officials requested permission from the U.S. Department of Health, Education, and Welfare, through the Ohio Department of Education, to construct an elementary school on Wright-Patterson to accommodate children living in Page Manor. Simultaneously, the superintendent of the Mad River Township School District, which adjoined the base, applied for federal assistance to construct additional school facilities if the base request was denied. While they waited, Wright-Patterson provided the school district with space in Buildings 1445 and 1448 for emergency schoolrooms. In 1954, the Mad River Township School District received a federal grant to construct a new elementary school on Spinning Road in lieu of a base school for Wright-Patterson.[38] Page Manor Elementary School continued to accommodate children of military families until 1983, when the 2750th Air Base Wing leased it from the Mad River Township Board of Education to house an overflow of base functions. The base later purchased the school from the Mad River Township School District in 1991.

The Page Manor housing complex remained in private ownership until August 15, 1960, when the Air Force purchased the 103 acres at a cost of $18,876,154. The 2750th Air Base Wing acquired jurisdictional and operational responsibility for the property later that year. Page Manor continued to improve over the rest of the century. In the 1970s, 904 apartments in the housing area were converted into 640 larger, more modern quarters. In the following decade, a phased upgrade of the housing units began and continued into the next century.

In addition to family housing, new on-base quarters for single military personnel were also constructed in the 1950s. For example, in 1954, Building 825 in Area A was erected as bachelor officers' quarters (BOQ) with space for 223 officers. Two years later, three new dormitories and a dining hall for airmen (Buildings 1212, 1213, 1214, and 1215) were built and, in 1959, 194 units of visiting officers' quarters (VOQ) opened in Building 826.

Construction of these new housing units

in the 1950s allowed base officials to remove some of the temporary World War II billeting. The most noteworthy demolition occurred at Skyway Park, at the intersection of Colonel Glenn Highway and Kauffman Avenue. By the end of 1957, all 536 family units constructed in 1944 had been removed. Many of the homes were sold for scrap lumber; however, a number of units were purchased by area residents and transported to new locations where they were installed as private housing. Skyway Park, known as Area D, later became part of Wright State University.

Family-housing shortages continued to plague the base throughout the 1960s. In 1970, for example, approximately 4,900 families assigned to Wright-Patterson had to compete for only 1,900 on-base accommodations. Fortunately, the base's housing capacity increased again, when work began May 21, 1971, on the 300-unit Woodland Hills project in Area B. This package, awarded to National Homes Construction Corporation, Inc., of Lafayette, Indiana, represented the first, permanent, military family-housing complex built on base since the brick quarters were erected during the mid-1930s. On July 9, 1973, National Homes broke ground for an additional 500 military family-housing units. Seventy of the quarters were located off Zink Road, near Woodland Hills. The remaining 430 units were adjacent to the U.S. Air Force Medical Center, forming the Pine Estates and Green Acres housing complexes.

Shortly after the turn of the twenty-first century, the Air Force chose to privatize much of its military housing, including that at Wright-Patterson Air Force Base. The Wright Field Development Company (including MV Communities, Woolpert LLP, and Hunt Building Corporation) was selected to assume ownership and management of Page Manor and Woodland Hills, which were, subsequently, renamed The Prairies at Wright Field and The Woods at Wright Field, respectively. Construction of nearly 800 new homes and townhouses in the complexes was slated for completion in 2006.

Community Services and Recreation Facilities

New facilities devoted to community services and recreation became frequent additions to the base landscape in the second half of the twentieth century. The

Warehouse 209 in Area C served as the base exchange from 1957 to 1980.

The original Wood City NCO Club, opened in 1943, was destroyed by fire in 1953. A new facility was ready for business by 1956.

The NCO Club was renamed "The Wright Place" in 1996.

1950s witnessed the addition of a new Noncommissioned Officers (NCO) Club in Wood City (1956); a 348-patient Air Force hospital in Building 830, Area A (1956); a base exchange store in remodeled warehouse 209, Area C; and Chapel 2 in Building 1220 of Wood City (1959). The NCO Club replaced the original club, established in 1943 and destroyed by fire on November 8, 1953. Opening of the new medical facility, later to become the Medical Center Wright-Patterson Air Force Base (see above), allowed base officials to close the wooden cantonment hospital constructed in Wood City during World War II.

Further construction in the 1960s continued to improve the quality of life for military personnel on base. The NCO Club received major renovations in 1962, and at the same time, a Grand Ballroom was added to the Officers' Club. In 1968, Major General Edwin R. Chess, chief of Air Force Chaplains, dedicated a new chapel, Chapel 3, in the Page Manor housing area. Completion of the Twin Base Golf Club in 1963 added another 18 holes to the base, complementing the original Wilbur Wright Field Officer's Golf Club built in 1922. Club membership was opened to base civilians for the first time since World War II. In 1961, when the first nine holes of the William Diddel-designed course were completed, Arnold Palmer played the dedication round.

Structures were also lost in the 1960s. In late 1961, a major fire broke out in the Wood City area (now Kittyhawk Center, see below). It destroyed three buildings (1118, 1045, and 1046) and damaged four others (1044, 1084, 1087, and 1089). Base firefighters used "Hi-Reach," the first Air Force water-tower vehicle, to combat the fire. The vehicle was a modified aircraft maintenance cherry-picker that held firefighters above the burning structure so they could direct a wall of water onto the fire. Except for a small amount of evacuated equipment, the buildings and their contents were ruined. Losses were valued at $693,920. AFOG, the main occupant of these structures, moved its operations to Area B as a result of the fire.

A second major fire in Wood City destroyed the Airmen's Service Club on January 28, 1963. The spectacular blaze was particularly difficult to control in the sub-zero January weather, and 14 persons had to be treated for frostbite at the base hospital. During the ensuing summer, a $110,000 two-story, brick Airmen's Service Club was built in Wood City to replace the destroyed facility.

The Airmen's Service Club in Wood City was a popular gathering place for nearly 20 years before it was destroyed by fire on January 28, 1963. A new Airmen's Service Club opened on the same site in 1964.

The base's enlisted dormitories in Kittyhawk Center were dedicated June 22, 1979, in honor of Ohio airmen who died from enemy action in South Vietnam. The dining hall was dedicated to Medal of Honor recipient Airman First Class William H. Pitsenbarger. (*U.S. Air Force photo by Henry Narducci*)

In the 1970s and 1980s, the World War II appearance of Wright-Patterson Air Force Base continued to fade as old facilities were torn down or modernized and new ones erected. Wood City underwent the most dynamic change as it converted from a quiet, World War II-era, residential neighborhood into a bustling commercial and recreational community. Christened Kittyhawk Center in 1972, the area was redesigned and developed to be "people oriented," with the products, services, and accommodations needed to support Wright-Patterson's service members and their dependents. Opening of a 1,000-seat motion picture theater started the transition. Shortly thereafter, the dormitories and dining hall were renovated. The Noncommissioned Officers' Open Mess was modernized in 1978, and an Airmen's Club was added to the recreation center in 1984. A childcare center (Building 1235) opened in August 1979, replacing three temporary, wood-frame buildings erected during World War II. A new athletic complex sporting softball diamonds, a track, and a football and soccer field was installed in 1982 to replace sites lost to the construction of other facilities. Jarvis Gymnasium and an Olympic-size swimming

Ribbon cutting ceremonies for the new base exchange and commissary complex, August 26, 1980. The $7 million community shopping complex formed the centerpiece of a revitalized Kittyhawk Center.

The Hope Hotel and Conference Center became the first privately owned and operated hotel in the Air Force. (*U.S. Air Force photo by Henry Narducci*)

pool opened for business the next year. Finally, the base library moved into newly remodeled Building 1044 in 1984.

The focal point of Kittyhawk Center, however, was its four-acre, $7 million Community Shopping Center Complex (Building 1250). Opened in 1980, the complex brought together in a modern structure an assortment of base enterprises that had been previously scattered about the installation. It housed the commissary, as well as the main base exchange, a restaurant, concession shops, and a commercial bank branch. A credit union office also opened in Building 1224.

While Kittyhawk Center prospered, the base's decision to close the Civilian Club in Building 274 meant the end of an era. The club, a World War II legacy, had opened on December 1, 1944, to all Wright-Patterson civilian employees, military members, and their families and friends. The club hosted nationally famous dance orchestras, square dance groups, countless dances, wedding receptions, and other popular events. Although it held memories for several generations of base employees, television, the demise of nightclubs, and the changing entertainment tastes of the "baby boomer" generation robbed it of the

patronage needed to survive. The club closed in 1979 and the building was converted into a conference center before being transformed into the base civilian personnel center.

Additions to Wright-Patterson's community services continued in the 1990s. The Hope Hotel and Conference Center (Building 824) at Chidlaw Road and Spruceway opened for business in July 1990, as the first, privately owned and operated hotel in the Air Force. Faced with a serious lack of visitor quarters, funding shortages, and stiff competition for military construction dollars, base officials turned to the private sector for help. They leased the land at no cost for 40 years to a private firm in exchange for funding, constructing, and operating the hotel for exclusive use of the Air Force. The Air Force retained the right to assume ownership and maintenance of the facility when the lease expired. The 266-bedroom hotel and 850-seat conference center were named in honor of actor and comedian Bob Hope. The facility also had a full-service restaurant and sports lounge capable of catering for up to 1,000 guests.

Also in 1990, Nightingale House, the first of several compassionate care facilities, opened near the base hospital. Located at 304 Red Bud Lane, it was the first of its kind in the Department of Defense. A brainchild of Colonel Dennis P. Tewell, 2750th Air Base Wing commander, the house offered a temporary home for families accompanying patients under treatment at the Wright-Patterson Medical Center. A nonprofit support association provided volunteers and ongoing material support.

In 1994, Fisher House (Building 831) opened its doors at the corner of Chidlaw Road and Schlatter Drive. Fisher House was the gift of Zachery and Elizabeth M. Fisher, New York philanthropists who believed that family support was crucial during medical crises. They wanted to make it possible for families of service members to be together during those difficult times. Through their Fisher House Foundation, they funded similar lodgings around the world. Fisher House at Wright-Patterson accommodated individuals whose family members were hospitalized with complicated or critical medical problems at the base medical center. The foundation completely furnished and equipped the home, which was staffed by volunteers. The Wright-Patterson facility was the seventeenth donated by the Fishers to military installations.

In its ongoing mission to provide much-

Nightingale House was the first compassionate-care facility of its kind in the Department of Defense.

The Turtle Pond reflecting pool has been enjoyed by generations of Wright-Patterson families.

Fisher House was a gift from the Zachery and Elizabeth Fisher Foundation.

A basewide team-building project in October 1994 produced the Super Playground.

Bass Lake, a popular recreational area for the Wright-Patterson community

needed services to its employees, Wright-Patterson Air Force Base opened a child development center in Area C (Building 1404) in August 1993. In September 2000, a second childcare center opened in Building 630. This 46,000-square-foot complex in Area B represented the largest such center in the Department of Defense. The lives of Wright-Patterson's children were further enriched in October 1994 with the addition of a Super Playground adjacent to the Hope Hotel in Area A.

Architects held brainstorming sessions with the children of base residents to determine the best features to include in their designs. Military personnel and civilians from across the base erected the playground in a weeklong, self-help, team-building venture.

In 1988, the base's recreational offerings were enhanced by construction of a log cabin and beach facilities at Bass Lake, located east of the West Ramp in Area C. In 1998, Wright-Patterson also opened the

135,000-square-foot Wright Field Fitness Center in renovated Hangar 22 on the historic Wright Field flightline.

Area A Command Headquarters

On the evening before Thanksgiving in 1961, fire totally destroyed Building 262A, the annex to Air Force Logistics Command headquarters in Area A. Two base firemen,

WRIGHT-PATTERSON GOLF COURSES

Wright-Patterson's golf courses are a historic part of the installation. Major Augustine Warner Robins, the commanding officer of the Fairfield Air Depot from 1921 to 1928, established the first base golf course in 1922. Robins had a golf course constructed in an "unsightly weed patch" on the facility. Volunteers on their off-duty hours used the facility's horse and mule-drawn equipment to construct and maintain a six-hole course with greens. Area businesses donated grass seed, materials, and technical advice. The links expanded to nine holes in 1924, and 18 the following year. The first tee was conveniently located adjacent to the commander's residence, today's Arnold House Heritage Center, Building 8.

The course proved popular, but maintenance required manpower, so in 1927 Robins invited the military and civilian communities of Wright Field and the Fairfield Air Depot to become members of the Wilbur Wright Golf Club and share in the sport and maintenance. The new club, 86 members strong, took up residence at the original Officers' Club across the street from the first tee. A family membership to the club in 1932 cost $5 per month.

Augustine Warner Robins played golf extensively in the 1930s with a 10 handicap. He loved the game so much he frequently closed his office and adjourned to the golf course. A tree on the eighth hole proved Robins' particular nemesis, until he ordered the post's civil engineer to cut the tree down. Robins also hated slow play and ordered slow groups to report to his office the next morning, where he banished them from the course.

In the 1930s, the course had major hazards: the rough was not mowed; fairways looked like tunnels with six-foot walls of prairie grass lining their edges; and trees and bunkers were added in the mid-1930s when the Brick Quarters were built. Deadly rattlesnakes infesting the course presented a true hazard and balls hit into the tall rough were automatically abandoned as unplayable. Golfers carried razors so they could treat snakebites, and women and children were not permitted to play the course. The snake problem was solved in the late 1930s when hogs were allowed to freely roam the area until they cleaned out the snakes.

A 1947 rulebook for the Wilbur Wright Field Officers' Golf Club boldly proclaimed, "Golf is a game for ladies and gentlemen." Proper attire, courtesy, speeding up play, and protecting the course were major themes. Practice swings on tees and greens were prohibited because "we are trying to grow grass." Similarly, "high heels (ladies and cowboys)" were forbidden. Green fees were 50 cents on weekdays and $1 on weekends. Caddies, usually children of military personnel, worked for $1 per 18 holes. "This fee is low according to present standards," the rulebook advised players. "These caddies are young. Please help instruct them." Private golf lessons by the course professional were $2 per hour; group lessons ran 50 cents per person. Tony Pena, Jimmy Demaret, Gene Sarazen, and Byron Nelson were among the notable golfers who played exhibition rounds at the base course in the 1940s and 1950s.

As the base military population grew during World War II, membership was gradually closed to non-military individuals. In the late 1950s, Wright-Patterson civilian employees led by Thomas Z. Jones, a former base restaurant manager, initiated efforts to construct a golf course for civilian employees. Noted golf course architect William Diddel designed the Twin Base Course, which opened as a nine-hole facility in 1961 and became an 18-hole course two years later. Arnold Palmer played the course's dedication round and, in 1967, shot a course record 65 during an exhibition. Although built for civilian employees, all members of the Wright-Patterson community were welcome to play the course.

Wright-Patterson built one other golf course in the 1960s. To serve the recreational needs of Strategic Air Command's alert crews on the West Ramp, a one-hole "course" was constructed adjacent to Chambersburg Road.

On July 8, 1997, the Wright-Patterson Golf Club became the Prairie Trace Golf Club. "Prairie" represented the historic flying field, the cradle of aviation. "Trace" was the remnant from the past, the natural Huffman Prairie that was nearly lost.

At the turn of the twenty-first century, Wright-Patterson offered the base community a variety of golfing challenges. The East Course was a nine-hole, 2,732-yard, executive course. The rolling 18-hole championship West Course spanned 6,799 yards, and the 18-hole Twin Base Course tested players over its 6,843 level yards. Today only the eighth green on the East Course and the fourth on the West Course remain in their original locations. Old timers claim that the modern course was much easier than it had been 60 years earlier.

Twin Base Golf Course opened in 1961.

East Course

Prairie Trace Golf Club (*U.S. Air Force photo by Henry Narducci*)

Building 262A ablaze, Thanksgiving eve, 1961. Two base firemen, Station Chief Dale V. Kelchner and William J. Collins, lost their lives fighting the fire. Approximately 3,200 official personnel files and countless other records smoldered in the ruins. Building 266, built on the same site, opened in July 1964.

Fire Station No. 1 was dedicated July 7,1989, in memory of Dale V. Kelchner and William J. Collins, who lost their lives November 21, 1961, fighting the fire that destroyed the AFLC headquarters annex.

Warrior Hall, Building 271 in Area A, honored AFLC's civilian workforce. It housed the command's logistical system operations center.

A newly designed installation main entrance greeted visitors in the 1990s. The two stars on the F-4D Phantom represent MiG kills in the Vietnam War.

Station Chief Dale V. Kelchner and William J. Collins, lost their lives fighting the blaze. (The Area C fire station, Building 163, was subsequently dedicated to their memory.) The fire also destroyed approximately 3,200 official personnel records housed in the Central Civilian Personnel Office. Row after row of files, heavy office equipment, and safes crashed to the basement of the two-story, wooden building as the first and second floors collapsed. The only documents spared were those stored in classified safes, but identification of individual safes was difficult because dials, numbers, and other markings were burned off. Reconstruction of personnel files was not completed until April 1963. A new $2.7 million structure, Building 266, opened for business on the same site in 1964.

Warrior Hall (Building 271) opened in 1990 as AFLC's logistical system operations center. The name "Warrior Hall" was selected to honor the command's civilian workforce. Two years later, the command opened a modernized $4.3 million command center in Building 266. A new

breezeway entrance connected Building 266 to the command's headquarters in Building 262. Across from the headquarters building stood the newly designed base entrance (Gate 12A) and visitor center.

Area B Acquisition Complex and Laboratories

In Area B, many of the laboratories gained major additions to their facilities in the past 50 years. New construction in the 1950s included a Propeller Control and Fatigue Research Building, a Fuel Systems Components Test Building, a Rocket Test Laboratory, a Compass Testing Building, and Microwave Building. The Gas Dynamics Research Building and the Universal Dynamic Sight and Computer Test Facility were completed at the end of the decade.

The Nuclear Engineering Test Facility, Building 470, in Area B became a major addition to the Wright-Patterson

landscape. This facility was originally conceived as a small, 10-kilowatt, nuclear test reactor for the Materials Laboratory. Strategic Air Command's desire to develop an intercontinental bomber powered by nuclear energy, however, quickly expanded the scope of the project. The facility was eventually scaled up to a 10-megawatt reactor and the test cell expanded to accommodate a full-scale jet engine. In 1956, the project was removed from the Materials Lab and placed under Air Research and Development Command's Directorate of Research. Work on the seventh largest reactor of its kind in the United States began in 1958, but the reactor's internal facilities were not fully completed until late 1965. By the time the government accepted the building in 1960, the project it was built to support—development of a nuclear-powered aircraft engine—had been abandoned in favor of aerial refueling.[39] The facility was then assigned to AFIT to assist the school's nuclear-engineering program. The reactor operated for five years before being

The historic exterior of Building 32, an original Wright Field structure, was restored in the 1990s while the interior was completely modernized. (*U.S. Air Force photo by Henry Narducci*)

Lieutenant General Robert F. Raggio and Mr. Patrick Carroll unveil a plaque naming the last building in the Doolittle Management Complex in honor of Major General Franklin Otis Carroll, June 1, 2001.

The General James H. Doolittle Acquisition Complex gave Wright-Patterson a state-of-the art acquisition management work center.

deactivated in 1970 due to high operational costs. It was entombed in concrete the following year.

Rapid developments in aerospace technology continued to stimulate a construction boom in Area B throughout the 1960s as new facilities were built to house new research and testing equipment. A 1960 addition to the structural test facility supported aerospace operations

simulations for the X-20 Dyna Soar and XB-70 bomber programs. The $7.7 million addition made the test facility the largest and most versatile of its kind in the nation. In 1961, Wright-Patterson also became home to the free world's largest, aerospace and missile sonic test chamber. The chamber was used to measure the effects of sonic fatigue—the weakening and malfunction of flight vehicles and their

components from sound-wave pressures.

The Air Force Avionics Laboratory substantially improved its capabilities with the completion of several facilities in 1967. A $5 million Optics Laboratory for its Reconnaissance Division became home to the world's largest optical collimator used for testing precision photographic lenses of all sizes for accuracy and clarity. The 9,000-pound lens of the gigantic collimator

HIDDEN GEM—BUILDING 12

Building 12 is the most distinctive building at Wright-Patterson Air Force Base and arguably one of the most unique within the U.S. Air Force. It was designed by Dayton architects Albert and Freeman Pretzinger and built in 1934-1935 with funding provided by the Works Progress Administration. It initially housed the holdings of the Administrative Section of the Materiel Division's Technical Data Branch, including a library of technical reports dating as far back as 1917 and a collection of artifacts packed for storage when McCook Field closed in 1927. It also served as the filming location for Air Force propaganda and training films, some later examples starring war veterans Ronald Reagan and James Stewart. The need for space during World War II led to the re-packing (and, unfortunately the eventual loss) of the museum exhibits, as Building 12 was used for a variety of administrative purposes.

Following the war, the Technical Data Branch (renamed T-2 Intelligence) took on the critical task of translating and cataloging the captured German technical documents brought to Building 12. The process took several years and, before its completion, the process and mission had outgrown the facility. As the Cold War set in, the sensitivity of technical information called for a specially designed building to house information and provide a secure work environment for personnel gathering and interpreting intelligence data. By the late 1950s, the Aerospace Technical Intelligence Center, successor to T-2, had moved its headquarters to a larger complex in Building 828, Area A. The museum collections also moved from Building 12 following World War II. The Air Force's museum was first reestablished in Hangar 89 on Patterson Field and, in 1976, moved to its current location on Springfield Street near Area B.

With the departure of all Technical Data units, Building 12 reverted to administrative use. The large rotunda, former home of the museum displays, was divided into office space. In the 1950s, civilian employees of Wright-Patterson knew Building 12 as the Civilian Personnel Office (as well as a Civil Defense Bomb Shelter).

In 1976-1977, the F-16 System Program Office moved into Building 12. The unique architecture was an ideal setting for a program with customers from 19 foreign nations. The grand entrance remained little changed from 1935 with the ornamental stainless steel balcony, decorated with flags of all customer nations, overlooking the marble pavilion, wherein sits a model of an F-16 fighter. Official functions are held in the lobby, and the balcony serves as launching pad for an annual paper airplane contest, during which administrative personnel pit their aeronautical design skills (often successfully) against experienced engineers and design experts. Proceeds go to the Combined Federal Campaign (the federal version of the United Way).

In 1993, the first chief of the F-16 program office, Lieutenant General George L. Monahan, Jr., died unexpectedly. Building 12 and the street on which it is located were memorialized as Monahan Hall and Monahan Way, respectively.

Sources: Bonnie Moutoux, *An Historic Tour of Wright-Patterson Air Force Base* (Aeronautical Systems Center History Office, Wright-Patterson Air Force Base, Ohio, 1995); Emma J. H. Dyson, et. al, *The Engineering of Flight: Aeronautical Engineering Facilities of Area B, Wright-Patterson Air Force Base, Ohio* (National Park Service, 1993).

Historic Building 12 in Area B was dedicated in honor of Lieutenant General George L. Monahan, Jr., in 1993. (*U.S. Air Force photo by Henry Narducci*)

Architectural details of Building 12

was a fused silica mirror 100 inches in diameter and 12 inches thick. It was installed at the bottom of a 155-foot vertical vacuum chamber that extended 85 feet above and 70 feet below ground.[40] In June of that same year, General John P. McConnell, Air Force chief of staff, opened the $2 million Electronic Warfare Research Facility (Building 620). The reinforced concrete structure with its distinctive twin towers represented the first in a three-phase construction program for the Avionics Laboratory. Prior to its completion, the Electromagnetic Warfare Branch and the Electromagnetic Warfare Applications Branch had conducted their research in Building 22, Area B.[41]

In 1969, the Aerospace Medical Research Laboratory completed a six-year program that doubled its capacity to conduct toxicology research. The improved Toxic Hazards Research Facility contained eight "Thomas Dome" long-term exposure chambers, four ambient-pressure laboratory exposure chambers, several laboratories, and animal preconditioning facilities.[42]

Progress, however, was occasionally tinged with bittersweet. Many current and former employees had cause to reminisce in 1979 when Wright Field's original hangars 2, 3, and 10 (constructed in 1928)

were razed to make way for a new Fuels and Lubricants Laboratory.

The 1980s continued to be a period of major laboratory construction in Area B. Among new research facilities added during the decade were an Aircraft Survivability Research Facility, a new wing for the Flight Control Development Laboratory (Building 146), a biotechnology facility for the Aerospace Medical Research Laboratory (Building 248), an expanded biodynamics laboratory (Building 824), a new metals and ceramics center for the Materials Laboratory (Building 655), and a High Power Research Laboratory.

The Area B laboratory complex also expanded with the addition of the Materials Processing Laboratory (Building 655) in 1992 and the Optical System Laboratory, which connected Buildings 33 and 248, in 1993. A multi-phased expansion of the Avionics Laboratory brought reality to the long-held dream of a consolidated avionics research center. The electronic warfare segment, completed in 1965, had been the initial component, with another added in 1972. Additions to Building 620 in 1994 and 1996, coupled with construction begun in 2000, finally enabled Air Force Research Laboratory's Sensors Directorate to consolidate its widely separated

laboratories and engineering workstations. AFRL also broke ground in 2001 for a 36,000-square-foot, consolidated, aerospace-structures research laboratory. The structure was an addition to Building 65, the historic World War II static test facility.

Perhaps Wright-Patterson's most visible construction project at the turn of the twenty-first century was the General James H. Doolittle Acquisition Complex. This multi-phased complex was designed to consolidate ASC's acquisition programs in a single, state-of-the-art work center. The first structure, Building 557, opened in 1994. Buildings 556, 558, and 560 followed in 1997, and Building 553 was added in 2001. When completed, these buildings housed ASC's Directorate of Engineering and program offices for the C-17, B-1, B-2, F/A-22, F-117, and F-16 aircraft. Aeronautical Systems Center also completely restored the exteriors and renovated the interiors of several original Wright Field structures for use by its program managers. A thoroughly rehabilitated Building 32 reopened in 1994 and a revitalized Building 11A was reoccupied in December 2000. ASC's new Simulation and Analysis Facility in Building 145 opened in July 1999.

A new control tower rises on the flightline side of base operations, Building 206, February 2002.

ARNOLD HOUSE HERITAGE CENTER

In 1841, Henry Hebble moved from Pennsylvania to Dayton, Ohio, and built a farmhouse on the Dayton-Springfield Turnpike. The house was within view of Huffman Prairie, and the inhabitants watched the now-famous Wright brothers practice their flying techniques. In 1924, the house became part of the land given to the government by the Dayton Air Service Committee for establishing Wright Field. Between 1929 and 1931, Major Henry "Hap" Arnold lived in the farmhouse while serving as commander of the Fairfield Air Depot. The house was inhabited by installation commanders until World War II, when it temporarily served as headquarters for the 478th Air Base Squadron.

After the war, Building 8, Area C, reverted to officers' quarters and continued in that capacity until 1980, when the farmhouse, the oldest building on Wright-Patterson Air Force Base, awaited the wrecking ball. Through the concerted efforts of Norma Woods, the base's protocol officer, and Lois Walker, the base historian, however, the landmark was saved. They proposed preserving it as a base heritage center and protocol reception area. With the aid of the Huffman Prairie League, a private nonprofit organization created to support the heritage center, the 1841 house was completely restored and converted for its new function.

A ceremony on May 16, 1986, formally dedicated the building as the Arnold House Heritage Center in honor of General Hap Arnold, the home's most distinguished former resident and the Air Force's only five-star commander. Four months later, on September 19, 1986, the Miami Valley Military Affairs Association dedicated a memorial in honor of former prisoners of war and those still missing in action. The black marble stone was placed next to the new heritage center.

Building 8, the oldest building on Wright-Patterson Air Force Base, was dedicated as the Arnold House Heritage Center in 1986. (*U.S. Air Force photo by Henry Narducci*)

The Arnold House serves as the base heritage center and reception area for honored guests.

FOULOIS HOUSE

The house now known as the Foulois House on Wright-Patterson Air Force Base was constructed around 1874. At that time, it sat along the Dayton-Springfield Turnpike near the small towns of Fairfield and Osborn in Greene County, Ohio. Originally occupied by farmers, by 1890 it had come to be owned by Dr. Duff W. Greene, an oculist and aurist who pioneered new treatments for blind and hearing-impaired men at the Dayton Soldiers' Home. In 1897, he sold the house to Dr. F. S. Patton, head of the Medical Department at the Soldiers' Home from 1884 until his death in 1898. Dr. Patton died shortly after purchasing the farmhouse and his widow, Eliza, lived there with her two children until they married. In later years, her grandchildren, Tom Carter and Alice, along with brother Fred Wolf, watched from her porch as Orville and Wilbur Wright learned to turn and bank their airplane during the summers of 1904 and 1905.

Eliza Patton was still living on the property when the entire Miami Valley flooded in 1913 and the Miami Conservancy District was formed to develop a comprehensive flood control system for the area. The conservancy purchased the land from Mrs. Patton and, almost immediately, leased the land to the Army. It then became part of Wilbur Wright Field and the farmhouse served as the home of the field commander. Brigadier General Benjamin D. Foulois, chief of the Materiel Division in 1929 and 1930, lived in the house, as did Brigadier General Augustine Warner Robins. With the dramatic increase in workload and manpower during World War II, the residence was often used for administrative offices. Following the war, however, it reverted to military family housing, generally assigned to the officer at Wright or Patterson Field with the largest number of children.

A major renovation of the farmhouse was completed in 1990, and it has since served as the residence of the Aeronautical Systems Center (ASC) commander, who also serves as the installation commander. Its large living and dining rooms, spacious wrap-around sunroom, tree-covered lawn and gardens, and large patio with a beautiful view overlooking Huffman Prairie and the Patterson Field runway, make it an excellent site for social events hosted by the ASC commander.

In 1989, the house was memorialized to honor Major General Benjamin Foulois, who retired in 1935 as chief of the U.S. Army Air Corps. He made his mark at Wright Field in 1931, as commander of a provisional air division during Air Corps maneuvers. The maneuvers involved 670 airplanes that flew nearly 38,000 hours, often in large numbers and in close formation. This highly coordinated series of aerial demonstrations occurred over major cities in the eastern United States and was, at the time, the largest exercise staged in the nation. It occurred with coordination and precision thought to be impossible and with no major mishap. For this feat of organizational skill, General Foulois received the National Aeronautic Association's Mackay Trophy "…for the most meritorious flight of the year."

Sources: "National Military Home, Dayton, Ohio: Virtual Tour and Exhibits," viewed online August 4, 2003, at http://www.dayton.med.va.gov/museum/homehosp.html; Diana G. Cornelisse, *The Foulois House: Its Place in the History of the Miami Valley and American Aviation* (Aeronautical Systems Center History Office, Wright-Patterson Air Force Base, Ohio, 1991).

Captain Benjamin D. Foulois in 1914

Building 88 was officially named the Foulois House in 1989 and has since served as the residence of Aeronautical Systems Center's commander.

In June 1963, a new control tower on the West Ramp became operational. It replaced the old tower atop the north hangar on Building 206, Area C, across the airfield. (*U.S. Air Force photo by Henry Narducci*)

Area C Base Operations

The skyline of the Area C flightline changed in 1963 with the addition of a new airfield control tower (Building 4041). The old tower atop Building 206, although modified and upgraded several times since the 1940s, had become too expensive to maintain and could no longer accommodate modern safety standards. The new nine-story tower constructed on the West Ramp near the 17th Bomb Wing (SAC) area served Wright-Patterson into the next century.

Other construction on the West Ramp included hot-water transmission mains, a liquid-oxygen generating plant, missile-fuel storage, a missile research test shop, and a propulsion research test facility. Purchase

of 1.32 acres of land along State Route 235 in September 1959 prevented private ownership adjacent to this sensitive area and provided right-of-way for highway acceleration and deceleration lanes. Easements for 269 acres in Clark and Greene counties extended the Area C airfield approach area in 1961. Easement deeds granted the United States the right to remove all aerial obstructions from the land. However, Universal Atlas Cement Division of the United States Steel Corporation retained ownership of the properties and continued quarry operations.

In the mid-1980s, the air base wing altered Building 110 to create a consolidated readiness facility. It centralized the base's mobility functions by converting Building 142 into a base mobility-processing center. In 1985, the wing opened a new logistics

airfreight facility, Building 143, which replaced the soon-to-be demolished Building 146. A new air-passenger terminal opened in Building 206 the following year. Building 1, the original depot building and oldest military structure on base, finally received much needed attention at the end of the decade. A complete makeover of its exterior façade modernized the building's appearance, but at the cost of its historic architecture, save for a small segment of its distinctive trainway entrance.

Following a brief moratorium on military construction at the start of the decade, the 1990s became a boom period for new construction. New facilities rejuvenated old Patterson Field. In 1995, the base added a hazardous-materials warehouse (Building 247). The next year, the 88th Air Base Wing commander renamed the portion of Pearson Avenue running between Wright Avenue and Breene Drive as "Patterson Parkway" so the air base wing headquarters resided at the intersection of "Wright" and "Patterson." Several major construction projects improved operations on the base flightline. Taxiway B, which was added in 1994, incorporated hazardous-material cargo pads, the result of lessons learned during Operation Desert Storm. Although ramp and runway repairs remained a never-ending endeavor, replacement of the 50-year-old West Ramp in 2001-2002 was a major milestone. At the same time, a new aircraft control tower was installed in Building 206. It replaced one built on the West Ramp in 1963 and returned the tower function to Building 206, where it had resided prior to 1963.

WRIGHT-PATTERSON INTO THE TWENTY-FIRST CENTURY

At the dawn of the twenty-first century, Wright-Patterson Air Force Base remained one of the nation's most important military installations. It was the Air Force's largest, most diverse, and organizationally complex base. It served as headquarters for a vast, worldwide logistics and acquisition network, housed a major research and development center, was a hub of professional education, and sustained a national center for aerospace intelligence. A major medical center and the United States Air Force Museum also called Wright-Patterson their home. In fact, more than 100 organizations, representing numerous Air Force, Department of Defense, as well as other federal, state, and

Today, Wright-Patterson Air Force Base encompasses 8,145 acres of land, with approximately 450 buildings on base and over 1,000 family-housing units.

local organizations were located at Wright-Patterson Air Force Base. Visitors often compared the base to a large industrial park with city-like characteristics.

From any perspective, Wright-Patterson's vital statistics were impressive. It encompassed 8,145 acres of land in Montgomery and Greene counties, Ohio, with approximately 450 buildings on base plus more than 1,000 family-housing units. It was the fifth largest employer in Ohio and the state's largest employer at a single location. The installation was a major economic force in the region, a leader in technology, and a partner in the community. Wright-Patterson represented a tremendous investment, not only in its physical plant, but also in technical skill and knowledge. These qualities, coupled with a dedication to excellence, earned Wright-Patterson its reputation as a significant force in the nation's defense.

Each year on December 17, the Dayton and Wright-Patterson communities gathered at the Wright Memorial to celebrate the courage, perseverance, and achievements that led Wilbur and Orville Wright to conquer the skies. The ceremony also honored Dayton and Wright-

Dayton in the twenty-first century *(Jim Aldridge)*

Patterson's efforts to build upon that legacy. In a community steeped in tradition, the Wright brothers' legacy remained a crucial part of daily life. The employees of Wright-Patterson Air Force Base considered that tradition a proud part of their heritage. For them, it remained an enduring foundation for the role their work

played in the U.S. Air Force and the life of the Miami Valley. As the first century of flight drew to a close, the two communities looked back with pride at the role they jointly played in the growth of aviation. As the second century of flight dawned, they stood ready to continue their commitment to master the skies.

Chapter Five

WRIGHT-PATTERSON AND THE MIAMI VALLEY

YESTERDAY, TODAY, AND TOMORROW

In the half-century following World War II, Dayton and Wright-Patterson Air Force Base have grown closer together than at any time in their history. This has been particularly marked in the past several decades as many "smokestack" industries in the Miami Valley have downsized or closed and the emphasis has shifted to "high-tech" products and services. Wright-Patterson has played a key role in attracting both high-tech ventures to the greater Dayton area and in promoting curricula and institutes at local colleges and universities with the same high-tech focus. This has extended beyond products of a purely military nature following the end of the Cold War and the Department of Defense's (DOD's) emphasis on "dual use" technologies and products. Wright-Patterson's laboratory and engineering corps, moreover, have infused the local population with citizens knowledgeable about and appreciative of science, technology, and engineering enterprises. This final chapter of *Home Field Advantage* surveys the growing synergy between greater Dayton and the Air Force. It is a synergy that pays ultimate tribute to the two Daytonians who launched mankind on the enterprise of flight a mere century ago.

DAYTON DURING WORLD WAR II

World War II restored Dayton to a position of industrial preeminence and affluence that it had not enjoyed since the 1920s. During the war, most Dayton industries, like National Cash Register (NCR) Company, Frigidaire, and Delco, turned to producing war materiel. Many specialized in aircraft development and parts production (see Appendix 2). NCR, Frigidaire, and Delco won the Navy's prestigious E (for excellence) flag. Typically, NCR was the first Dayton company so recognized.[1] General Motor's (GM's) Inland Division turned from the production of auto parts to production of

Downtown Dayton, 2003 *(Helen Kavanaugh-Jones)*

During World War II, NCR switched from cash register production to the manufacturing of magazines for 20-mm anti-aircraft guns. *(Library of Congress, Farm Security Administration, Office of War Information Photograph Collection)*

America's standard ground weapon, the M-30 carbine. By war's end, Inland had produced more than 2.6 million of them.[2]

NCR also turned some facilities over to the military for direct use. Every day, 600 Navy WAVES (Women Appointed for Voluntary Emergency Service) trooped in to the company's Sugar Camp in the Far Hills, where they worked on code-breaking machines.[3]

Some buildings with no previous military or industrial use were also turned over to the government. At a warehouse on West First Street, once the home of Dayton's Bonebrake Theological Seminary, scientists of the Monsanto Chemical Company studied the characteristics of polonium for the manufacture of atomic bombs.[4]

Overall, Dayton's wartime contracts exceeded $1.645 *billion.*[5]

For a time, a real labor shortage existed in Dayton, leading to the government's identification of the city as a "No. 1 critical labor area." Indeed, there was a danger for a while that the government would shift work from Dayton to areas of labor surplus.[6]

The war swelled Dayton's population and industrial workforce.[7] Between 75,000 and 125,000 newcomers poured into the city during the war.[8] Most were employed in local industry. A substantial portion, however, worked for Wright and Patterson fields. At NCR alone, employment jumped from 8,000 just before the war to 20,000 by

1945.[9] In 1943, employment was up 123 percent from 1933.[10]

The housing shortage was also acute. The Corwin home in Lower Dayton View, a commodious but by no means vast house, for example, accommodated up to 15 servicemen at any one time during the war.[11] Frank Wattendorf's provident investment also assured base personnel got good accommodations at reasonable rents[12] (see Chapter 1: Birthplace of Aviation).

The war also transformed Wright and Patterson fields to a mainstay of the national and local economy. In 1943, 30,926 civilians and 14,821 military personnel were assigned to Wright and Patterson fields.[13] By the end of the war, in 1945, the total payroll for Wright and Patterson fields was $131.5 million, or one third of Dayton's total industrial payroll.[14]

THE RISE OF GREATER DAYTON

In the decade following World War II, certain key trends became apparent that characterized the development of Dayton to the end of the twentieth century.

One of the most important trends was the trek to the suburbs, first by many of Dayton's white, middle and upper class citizens and, beginning in the 1960s, an increasing number of Dayton's black citizens as well.

One of Dayton's earliest suburbs was Lower Dayton View, located across the Great Miami River from downtown. The area was promoted for residential building in the early 1800s. At the time, however, no bridge linked Dayton proper to Dayton View and the project failed. However, by the end of the nineteenth century, Lower Dayton View blossomed with homes of the wealthy and near-wealthy, including the Stoddards, Lowes, Hawes,[15] and Canfields. Most of the grandest mansions had given way to larger, public buildings, like the Art Institute and Masonic Temple, during the 1920s. Some, like the Canfield house, reminded Dayton Viewers of their grand and glorious past until the late 1980s, when it, too, fell to the wrecker's ball. However, even more modest homes in Dayton View were built in the grand tradition, and it was not unusual for foursquare brick houses, built around 1910 on Stoddard Circle, to have a bathroom for every bedroom. Upper Dayton View, which centered on the "Dayton Triangle," began to fill out during the 1920s along Salem Avenue.

Both Lower and Upper Dayton View were significant for remaining within the political, administrative—and taxation— jurisdiction of the city of Dayton. However, beginning in the early twentieth century a more ominous development occurred as wealthy Daytonians, egged on by John H. Patterson, among others, began to settle to the south of the NCR complex. (Indeed, Wilbur, Orville, and Katharine Wright decided to place their mansion, Hawthorne Hill, in this area, after earlier considering and deciding against Upper Dayton View along Salem Avenue.)[16] The 1913 flood also infused a growing number of well-heeled downtown Daytonians with longing for the increased altitude of the Far Hills.[17] Shunning Dayton's repeated attempts of annexation, this community incorporated as the city of Oakwood in 1931.[18]

The Depression put an end to any further suburbanization, which did not take off again until after World War II—and then it took off with a boom.

Fueling the postwar move to the outlying areas north and south of Dayton were two main factors: the automobile and the rise of relatively inexpensive "ranch-style" homes, often built without a basement, on a concrete "slab." Within 10 years of the end of World War II, the remainder of Van Buren Township had incorporated as the city of Kettering.[19] About the same time, in the mid-1950s, real estate developer Charles H. Huber began building what, in 1981, became the city of Huber Heights.[20] Another suburb that incorporated the same

CHUCK TAYLOR AND HIS "CHUCKS"

A U.S. Navy lieutenant coaching an Army Air Forces basketball team? A famous basketball personality to boot! How was that possible? The full story of Lieutenant (senior grade) Charles "Chuck" Taylor's coaching exploits at Wright Field in 1944-1945 are not fully known, but he clearly had a big impact on local basketball fans, the Air Technical Service Command (ATSC) sports program, and thousands of military and non-military fans around the United States.

Like many Americans during the dark days of World War II, Chuck Taylor answered the call to duty, reportedly receiving a commission in the Naval Reserve in or about 1942. Military service eventually took him to Wright Field in Dayton, Ohio. A December 1944 Dayton newspaper announced the formation of a new, command-level basketball team for ATSC—one to be coached by Chuck Taylor, "one of the pioneers in modern professional basketball while a member of the legendary New York Celtics."

What a legend he was! After his playing days, Chuck became an "ambassador" on the road teaching basketball around the world. An innovator from the start, he organized the first basketball clinic at North Carolina State University in 1922. His real fame, however, came from his development of basketball equipment, especially shoes designed for the sport. In 1921, he complained to officials of the Converse shoe company about his sore feet from playing basketball. Converse took the hint and created a shoe for playing the game—and they hired Taylor to promote the new product. In 1932, Converse added Taylor's signature to its five-pronged star shoe and "Chucks" were born.

Clearly, by 1944, Chuck Taylor had made a name for himself, and people at Wright Field and in the Dayton area had a star coach in Navy lieutenant's clothing. Taylor assembled the Wright Field "Air-Tecs" from a pool of talented players around ATSC. The team, as a whole, averaged 6 feet 6 inches in height, making them a formidable rebounding and scoring threat. The Air-Tecs boasted some talented players with plenty of college experience. Technical Sergeant Johnny Schick played on the 1939-1940 Ohio State Big Ten championship team. Other standouts included former Bradley Tech star Lieutenant Chris Hanson and Corporal John Mahnken, who had played for Georgetown University.

The Air-Tecs quickly made a name for themselves traveling around the United States playing a variety of amateur, military, and professional teams, at one point in the season even winning 15 of 17 road games. One sportswriter stated: "As for the Air-Techs, it is easy to see why the soldier team has been running roughshod over opponents around the country. They have size and they can move and they play spirited basketball." They attracted the fans, too—as many as 3,000 at one game. On one occasion, 2,425 Dayton fans watched them beat the city's own professional team, the Acme Aviators.

At many games, Chuck Taylor took the opportunity to "put on one of his famous exhibitions…explaining the fundamentals of passing, shooting and team play…." The season was rolling along and, in March 1945, they received an invitation to play in the National Service Invitation Basketball Tournament. Unfortunately, Army Headquarters in Washington had other ideas when it ordered all teams to quit playing games away from their home installations.

During his remaining time at Wright Field, Chuck Taylor remained involved with basketball in some capacity as many of his Air-Tec players joined the Acme team. The Aviators, featuring mostly former Air-Tec players, made it all the way to the finals of the National Professional Basketball Tournament, losing to the Fort Wayne Zollners (who later became the Detroit Pistons). At about this time, Chuck Taylor "shifted" from the Navy to the Army (presumably the Army Air Forces) and was promoted to captain. It is unknown where military service took him next.

One thing is known. For one basketball season, Chuck Taylor guided a military powerhouse team, creating plenty of excitement at Wright Field, in Dayton, and at all of the Air- Tecs' stops on hardwood courts across America at a time when people needed something to take their minds off the world situation.

After the war, Chuck's legend, and that of his Chucks, continued to grow. In 1958, the Sporting Goods Industry Hall of Fame recognized his contributions to the development of basketball equipment by enshrining him, only the second living person to be so honored. In 1969, he received the ultimate basketball honor—election to the Naismith Memorial Basketball Hall of Fame. To date, over 600 million Chucks have been sold—making them a footwear legend! Chuck Taylor passed away on June 23, 1969, following a heart attack. His impact on the game of basketball, and on the feet of those who play the game, continues to this day.

Chuck Taylor first served as a U.S. Navy lieutenant during World War II. *(Abe Aamidor)*

Current version of the Chuck Taylor All Star "Chucks" *(The Converse Company)*

Sources: Abe Aamidor, currently writing Chuck Taylor's biography, personal communication with James Ciborski, ASC/HO; "Taylor […] Schedule […]," Dayton *Journal-Herald*, December 3, 1944, n.p.; "The Air Techs," *Wright Field Take-Off*, December 30, 1944, p 10; "Employee '5' Books Fast Air Tecs," *Postings*, January 26, 1945, p 3; various articles, Dayton Daily News, February 9, 1945 – April 10, 1945; Chris Doyle, Converse Public Relations, personal communication with James Ciborski, October 14, 2003; Naismith Memorial Basketball Hall of Fame website, viewed online September 22, 2003, at http://www.hoophall.com/halloffamers/TaylorC.htm.

Grafton Hill, one of a half-dozen "historic districts" ringing the city of Dayton. Pictured here are some of the elegant residences of Grafton Hill and two of its fine public buildings, the Masonic temple and the Greek Orthodox Church. (*Jim Aldridge*)

Dayton, Ohio, 1945

year was the city of Beavercreek,[21] which served primarily as a "bedroom suburb" of Wright-Patterson Air Force Base. Meanwhile, several outlying, older towns around Dayton became the focus of further suburban growth. Among these were Fairborn, Vandalia, Miamisburg, and Centerville.[22]

URBAN RENEWAL AND DOWNTOWN REVITALIZATION

The rise of suburbia after World War II posed several problems—and opportunities—for the city of Dayton.

Dayton still remained the geographic hub of the region and lent nearly all the outlying communities its "identity." Letters to residents of Oakwood, Kettering, Beavercreek, or Huber Heights as often as not bore addresses (and even return addresses) to "Dayton, Ohio." For many years, Dayton's tax bureau computed the

income taxes for suburban residents, and not just those who worked within the city. Until the late 1970s, downtown Dayton was the retail and commercial center of the Miami Valley, which, given its central

location and placement of streets, superhighways, and public transportation grid, made (and still makes) eminent good sense. Dayton also remained, into the twenty-first century, the cultural and (to a

Gone but not forgotten: Dayton's memorial to its principal founder, Daniel Cooper, is Cooper Park, shown here in 2003 with downtown Dayton's business district in the background. *(Jim Aldridge)*

CLINTON COUNTY AIR FORCE BASE

In the early 1940s, Clinton County Army Air Field, located near Wilmington, Ohio, conducted all types of glider tests and reported the results to the Glider Branch at Wright Field. In 1945, the Army Air Forces sought a base near Wright Field that could be utilized exclusively for Air Materiel Command's All-Weather Flying Division. Clinton County was initially chosen but, soon afterwards, the decision was made to move the all-weather facilities to Lockbourne Army Air Base in Columbus instead. Before long, however, Lockbourne became involved in a different mission, and the All-Weather Flying Division's activities returned to Clinton County.

The All-Weather Flying Division at Clinton County Army Air Field supported the Berlin Airlift (Operation Vittles) by developing, installing, and maintaining an AN/CPS-5 radar that helped Berlin air controllers manage heavy air traffic. The division's personnel also designed and built the rhombic array antennas used for receiving and transmitting VHF communications along the air corridors used in the airlift.

In the late 1940s, Clinton County Air Force Base (CCAFB) participated in the Office of Naval Research's famous "Skyhook" project, in which huge, polyethylene balloons were sent into the stratosphere for testing instruments at extreme altitudes. In the early 1960s, CCAFB became a Strategic Air Command bomber alert facility and, later in the decade, as a Special Operations Squadron Field Training Detachment, it provided gunship crew training for personnel headed to war in Vietnam.

CCAFB also became the home of the 302nd Troop Carrier Wing, which flew missions of mercy to disaster areas and transported heavy cargo to places as far away as Puerto Rico, Cuba, and the west coast. Later, as the 302nd Special Operations Wing, it added to its list of duties the provision of C-119 aircrew training for U.S. Air Force pilots, navigators, mechanics, and friendly foreign forces.

As a result of Department of Defense cuts, CCAFB closed in 1971. The economic impact would have been devastating to the area if not for the resourcefulness of the citizens and business leaders of Wilmington. A college, a vocational school, and more than 40 corporations sprang up on land once occupied by CCAFB. In 1980, Airborne Express bought the old air base and made it the main hub in its central package-sort and maintenance operation, bringing thousands of jobs to the Wilmington area. In 2003, it was the nation's largest, privately owned airport.

Sources: "Wilmington Square," viewed online October 15, 2003, at http://www.rgprop.com/Wilmington_square_overview.htm; "Clinton County Air Force Base, Ohio," viewed online October 15, 2003, at http://www.globalsecurity.org/wmd/facility/clinton_county.htm; "Appendix 17: The Navy in Space," viewed online October 15, 2003, at http://www.history.navy.mil/avh-1910/APP17.PDF; ASC History Office Archives: "Clinton County AFB Guides," "History of 302D Special Operations Wing 1 January - 31 March 1971," and "Clinton Co. Army Air Field, Wilmington, Ohio" folders.

South Vietnamese Air Force officers received training in electronics at Clinton County Air Force Base in the late 1960s.

C-119s prepare for takeoff from the flower-bordered runway of Clinton County Air Force Base.

Final Retreat: Clinton County Air Force Base closed on June 30, 1971.

lesser extent) business and financial center of the area *par excellence*.

However, from the first postwar decade, Dayton confronted a number of challenges that, unless surmounted, threatened its preeminence. A decade and a half of Depression and war had led to neglect of streets and buildings. Slums had grown up in formerly prestigious residential areas around downtown.

Dayton emerged from World War II flush with money that only burgeoning local industry and full employment could provide. The city proceeded to think and act proactively in spending this windfall. Dayton became, for instance, the third city in the United States to install directional traffic lights for both cars and pedestrians.[23]

Dayton also tackled problems of downtown traffic congestion produced by rising car ownership. Dayton had been fortunate to have broad streets, which it owed to the foresight of Daniel Cooper, who laid them out in the eighteenth century.[24] However, there were still obstacles to contend with. In 1948, Private Fair was unceremoniously removed—pedestal and all—from the junction of Main Street and Monument Avenue and placed across the river in Sunrise Park, on the west bank of the Great Miami:[25] one traffic circle was one too many for downtown drivers! (When Private Fair returned in triumph four decades later, he was placed in the middle of Main Street, near but not on the original site.)[26] Dayton's city fathers also reacted with enthusiasm to the opportunities that the Defense Act of 1956 afforded and made sure that one of the new superhighways,

provided for in the act,[27] passed right by downtown within view, for example, of billboards advertising Rike's department store. Interstate (I) 75, whose downtown corridor replaced old Route 25 (Dayton's first superhighway), was the result.[28] At the same time, work began on state route (SR) 35 connecting Dayton with Xenia, and SR-4 was extended to intersect I-75 at a point soon dubbed by startled Daytonians and out-of-towners alike "malfunction junction."[29]

During the 1950s and 1960s, Dayton also "cleaned house" by removing a number of structures—and even whole neighborhoods—under the rubric "urban renewal." Steele High School was perhaps the most prominent of the grand buildings of turn-of-the-century Dayton to fall to the wrecker's ball in the 1950s. It was replaced, first by a parking lot, then by Dayton's first high-rise parking garage[30]—an accoutrement of the new automobile culture and necessary (it was thought) to keep the driving public coming downtown. In the early 1960s, the demolition of Steele High was joined by the old public library (downtown Dayton's last superb example of Romanesque-revival architecture),[31] which was replaced by a modern structure reminiscent of a cereal box lying on one side. A more successful example of "modern architecture" was the Sinclair Community College complex that rose from the debris of Dayton's west-end downtown (and run-down) residential neighborhood. The multi-building Sinclair complex was a tribute to the versatility—and even aesthetics—of preformed concrete construction and soon teamed with life. By

the end of the century, nearly 20,000 students enrolled in Sinclair's classes and practica.[32]

Dayton's business and commercial district was also subject to several makeovers and continued architectural evolution. Aside from *de rigueur* parking garages, some of truly dizzying aspect, the city also witnessed older structures give way to such elegant "skyscrapers" as the Kettering Tower (originally called the Winters Tower)[33] and the Mead Tower,[34] both built in the 1970s, and the Citizens Federal Centre and the Arcade Centre towers, both built in the 1980s.[35] Indeed, Dayton continued to attract world-class architects, such as I. M. Pei and Partners, who designed the Gem Savings Building and Plaza, built in the 1980s,[36] and Cesar Pelli, who designed the new Schuster Performing Arts Center that debuted in 2003.[37]

All this happened as commercial businesses—particularly retail—deserted downtown for suburban shopping centers and malls. The first major retailer to abandon downtown was J.C. Penney, in the mid-1950s.[38] The last was Elder-Beerman, which closed its downtown store in 2002.

To keep suburbanites coming downtown and spending money, Dayton tried a number of expedients over the years, mostly under the slogan "downtown revitalization." Two ambitious projects revolved around the courthouses, on the one hand, and the arcade, on the other. In 1979, the 1888 "new" courthouse was demolished. No new building replaced it. Instead, the site was made into a grand, open space in the center of downtown,

To beautify the city and lure suburbanites back downtown, Dayton and Montgomery County MetroParks laid out a new recreation center on the site of old Van Cleve Park on the bank of the Great Miami River. Called "RiverScape," the new park featured pavilions recalling Dayton inventions, such as the airplane, cash register, ice-cube tray, and pop-top can. (*Jim Aldridge*)

DAYTON'S AFRICAN-AMERICAN HERITAGE

Dayton's African-American community has made major contributions to the social, economic, political, religious, cultural, and technological accomplishments of the Miami Valley over the past 200 years.

The first recorded mention of blacks in Dayton dates to 1798, two years after the settlement was established. By 1820, about 141 African Americans lived in Dayton, which had a total population of 1,139. They settled primarily in two areas: a collection of cabins and houses called "Africa" near the Miami-Erie Canal in the 1820s, and across the Great Miami River on the west end of town from the 1840s.

During the first half of the nineteenth century, African Americans, in Dayton and around the United States, lived under the shadow of slavery. Even in nominally "free" states like Ohio, laws and regulations severely restricted blacks' political and economic activities. There was also, however, considerable sympathy for the plight of black slaves. Both whites and free blacks supported the activities of the so-called "Underground Railroad" that assisted slaves in escaping to freedom in Canada. In Dayton, several homes served as "stations" of the Underground Railroad, and concerned citizens established the Dayton Abolition Society in 1832. However, Dayton also had many citizens hostile to the abolitionists, and the two groups often came into conflict.

Following the Civil War, Dayton's black population increased modestly due in part to freed slaves coming north seeking economic opportunity and (after the end of Reconstruction in 1876) to escape the pernicious regime of Jim Crow laws. Joshua and Matilda Dunbar, both former slaves, were among the many who settled in Dayton, where their son, Paul Laurence, was born in 1872.

Paul Laurence Dunbar represented the acme of African-American achievement in Dayton and the nation at the turn of the century. Dunbar attended Dayton public schools and graduated from Central High School in 1890. He published his first book of poetry in 1892 and sold copies while operating the up and down switch of the Callahan Building's elevator—the only job available to a black high-school graduate. But his career was definitely on the way *up*. In 1893, he was invited to recite his poetry at the Chicago World's Fair. His second book of poetry caught the attention of national literary arbiter, William Dean Howells, who opened to Dunbar the international world of *belles lettres*. Dunbar spent the next decade traveling widely, before returning to Dayton in 1904. He died there two years later, succumbing to tuberculosis. He left a literary corpus comprising a dozen books of poetry, four books of short stories, five novels, and a play. He was 33 years old when death stilled his pen.

The fortunes of Dayton's African-American community began to change in the years immediately preceding World War I, the beginning of the Great Migration, when blacks left the South in large numbers for better job opportunities in the industrial north. By the end of World War I, 9,052 blacks lived in Dayton, up from just short of 1,000 in 1900. However improved blacks' economic lot became, dangers and petty humiliations still existed throughout the general prosperity of the "roaring twenties." The new Roosevelt High School welcomed both white and black students; however, it also had segregated swimming pools. During the 1920s, Dayton was one of the centers of Midwest Ku Klux Klan activity. The Promised Land it was not. Then came the Depression years of the 1930s, when many blacks—still at the bottom of the economic pyramid and frequently the first to be fired—lost their homes to foreclosure.

However, the 1920s and 1930s were also a time of black cultural renaissance. This was the golden age of West Dayton's black community, particularly the businesses and centers of recreation and entertainment on West Fifth and West Third streets.

Paul Laurence Dunbar *(Dayton Metro Library)*

William Carter, a skilled machinist with the 4950th Test Wing's Fabrication and Modification Division, began working at Wright-Patterson in 1942.

World War II brought another influx of African-American immigrants from the south to Dayton, more than doubling Dayton's black community by war's end. The early postwar years also saw fall the first barriers of the "separate but equal" regime between blacks and whites in Dayton and across the nation. In 1958, NCR admitted African Americans to its apprentice program for the first time. One local center of advancement for blacks, increasingly from the 1950s onwards, was Wright-Patterson Air Force Base, where many black officers, airmen, engineers, and clerical workers got their start and laid the foundations of professional advancement and major contributions to national security. By the beginning of the twenty-first century, the commander of the Air Force Material Command was an African American, General Lester L. Lyles.

The 1960s began with marches, protests, and sit-ins by black Daytonians fed up with being "separate and *unequal*" in their own hometown. In 1961, Dayton elected its first black city commissioner, Don L. Crawford. In 1963, protests opened up Rike's and other downtown retail establishments to fair-hiring practices for blacks. Increased racial tensions also produced tragedy. In 1966, rioting broke out in West Dayton when a group of white men shot and killed a black man. The National Guard was called into Dayton, and later in the year President Lyndon Johnson spoke to all Daytonians at the Montgomery County Fairgrounds. Blacks likewise protested discrimination in the housing market, and before the decade ended, the U.S. Supreme Court struck down all local ordinances that sanctioned neighborhood housing discrimination.

The 1970s opened with James H. McGee as Dayton's first black mayor—one of the first black mayors of a major American city. During the 1970s, Dayton's schools were desegregated, in part by busing students across town, to achieve racial balance. Meanwhile, an increase in crime due to illegal drug trafficking, among other causes, resulted in an accelerated exodus of both white and black Daytonians to the suburbs.

During the 1980s and 1990s, Dayton blacks made more strides in achieving political and cultural, if not always economic, parity with their white fellow citizens. In 1982, Dayton's Black Leadership Program was established. The next year Tyree Broomfield became Dayton's first black chief of police. In 1984, Richard Hunter became the first African-American superintendent of Dayton's public schools. In 1986, Richard Clay Dixon became Dayton's second black mayor. The following year, Isaiah Jackson became the first black music director of the Dayton Philharmonic Orchestra. In 1988, when long-serving state representative C. J. McLin, Jr., died, he was succeeded by his daughter Rhine McLin, who went on, in 1994, to the Ohio State Senate, the first black woman to hold that position. In 1994, Dr. Harvey Flack became the first black president of Wright State University. And in 2002, Ms. McLin was elected mayor of Dayton, the third black and the first woman to be so honored.

Secretary of Defense Donald Rumsfeld visits Wright-Patterson in October 1976. Greeting him are Aeronautical Systems Division Commander Lieutenant General George H. Sylvester and 2750th Air Base Wing Commander Colonel Titus C. Hall. The career of Colonel Hall (later brigadier general) included assignments at Wright-Patterson as chief avionics engineer for the B-1A bomber, systems program director of the Avionics Program Office, assistant deputy for the Reconnaissance/Strike/Electronic Warfare SPO, and deputy for Systems, all in the Aeronautical Systems Division.

Sources: Margaret E. Peters, *Dayton's African American Heritage* (Virginia Beach, 1995); *For the Love of Dayton: Life in the Miami Valley, 1796-1996* (Dayton, Ohio, 1996); Virginia and Bruce Ronald, *The Lands between the Miamis: A Bicentennial Celebration of the Dayton Area* (Dayton, Ohio, 1996); Herbert Woodward Martin and Ronald Primeau, *In His Own Voice: the Dramatic and Other Uncollected Works of Paul Laurence Dunbar* (Athens, Ohio, 2002); Alan L. Gropman, *The Air Force Integrates, 1945-1964* (Washington, D.C., 1978).

Air Force Chief of Staff General George S. Brown and Aeronautical Systems Division Commander Lieutenant General James T. Stewart converse with Dayton Mayor James McGee at a Dining-In, January 1971. McGee was Dayton's first black mayor and reached out to all elements of the greater Dayton community, including Wright-Patterson Air Force Base.

General Lester Lyles, commander of Air Force Materiel Command, receives red-carpet treatment at Wright-Patterson Air Force Base. *(U.S. Air Force photo by Al Bright)*

While urban renewal swept away many monuments of Dayton's architectural heritage, several worthy representatives remained into the twenty-first century, including (clockwise from top left) the Kuhns building (1880s), the *Dayton Daily News* building (1900s), the Dayton arcade (1900s), the old post office building (1900s), Sacred Heart church (1880s), and the old courthouse (1850s). (*Jim Aldridge*)

Beginning in the 1980s, Dayton began encouraging the development of residential apartments and condominiums in and near downtown. These included construction of the Cooper townhouses (shown here) and later the Cooper Lofts, redevelopment of the Dayton YMCA building into condos, and the construction of an 18-story tower housing offices and condominiums as part of the new Schuster Performing Arts Center. (*Jim Aldridge*)

bordered by the old courthouse to the south, the new Mead Tower to the north, and the (then) new Elder-Beerman store on the west.[39] Courthouse Square became a respite largely for business people taking in lunch and sunshine; itinerant know-it-alls, megaphones in hand; transients; and pigeons. In the early 1980s, the city also underwrote a complete rehabilitation of the old arcade complex, particularly the magnificent, domed rotunda area, the site of Dayton's first indoor "mall" built in 1904.[40] The arcade reopened to great fanfare in 1980.[41] However, it failed to pay its own way and finally closed in 1991.[42] The "reopening" of the arcade remained an open issue into the new century.

More successful than these efforts were four entertainment and recreation center projects. The first revolved around the refurbishing of the old Victory Theater, which reopened as the Victoria Theater, in 1990.[43] The second was the construction of a new baseball stadium in the late 1990s, which opened as Fifth Third Field, the "Home of the Dayton Dragons," in time for the 2000 baseball season.[44] The third was the overhaul of old Van Cleve Park, on the

banks of the Great Miami, with a larger park and recreation area called "RiverScape," in 2002.[45] RiverScape featured Dayton's contributions to science and invention memorialized with a number of pavilions; one, for instance, featured the invention of the pop-top can by Ermal C. Fraze,[46] another the invention of the cash register by James Ritty.[47] The fourth project was the construction of the Benjamin and Marian Schuster Performing Arts Center and condo complex on the site of the old Rike-Lazarus department store.[48] The Schuster was an instant success with both the tuxedo and tennis shoe crowds and promised to bring to Dayton many off-broadway and operatic performances formerly staged in Cleveland, Cincinnati, and Chicago. A fifth project (actually designed and built in conjunction with RiverScape), a giant water fountain in the middle of the Great Miami near Deeds Point, nearly fizzled when the original design did not perform as advertised. The addition of a central fountain in 2002 cast aspersion—literally!—on the project's doubters and naysayers.

Finally, beginning in the 1960s, the city encouraged the revitalization of downtown

neighborhoods. The first so-called "historic district" was the Oregon District east of downtown. (Charlotte Reeve Conover recalled that, in the 1860s, "Fifth Street crossed the canal and reached a semi-suburb called Oregon, of which there was nothing much, except the Darst mills, a fire engine house, some small stores and then the...'commons' full of gypsum weed, dandelions and rubbish heaps.")[49] The Oregon District was soon joined by St. Anne's Hill, McPhersontown, South Park, Old North Dayton, and the Grafton Hill historic districts. The latter was the new name for part of Lower Dayton View. The most recently revitalized neighborhood was the Wright-Dunbar historic district located near Third Street and the bicycle shop of Wilbur and Orville Wright and the home of Paul Laurence Dunbar.[50] Beginning in the early 1980s, the city also began to encourage the building of condo and apartment complexes in or near downtown. The first was the Cooper townhouse complex, built across from Cooper Park and the downtown Dayton Public Library.[51] However, it was not until the mid to late 1990s that the "condo craze" took hold,

Dayton was the Miami Valley's focal point for the arts and entertainment. Pictured here (clockwise from top left): the Schuster Performing Arts Center, the Dayton Art Institute, the Victoria Theater, and Fifth Third Field, home of the Dayton Dragons baseball team   (*Jim Aldridge*)

and a number of old industrial and warehouse buildings were converted to condos and "loft" apartments, many selling or renting for premium prices.

BUSINESS AND INDUSTRY IN TRANSITION

Dayton in the early twentieth century was one of the nation's leading centers of manufacturing—everything from agricultural equipment and train cars to bicycles, automobiles, and airplanes were produced by several dozen, small- and medium-size firms. Several dozen more companies produced parts, supplies, and services for the larger manufacturers, both locally and nationally. Dayton was also a national center of publishing. Then—lest we forget!—there was the cash register, the accountant for the god mammon that, almost literally, touched the lives, one way or another, of every man, woman, and child in town. Dayton's leading industrialists— Patterson, Deeds, Kettering—were not only local but national figures (see Chapter

1: Birthplace of Aviation). Dayton's leading publisher, James M. Cox, served as congressman, governor, and presidential candidate—and then went on to create one of America's first "media empires" of print and broadcast journalism. What other medium-size city in the United States could boast as much?

Dayton's business fortunes dipped during the Depression decade of the 1930s. However, the extent and diversity of Dayton's manufacturing enterprises helped ward off the utter devastation wreaked by economic hard times on other, less well endowed communities. When world war turned full on the government's fiscal spigot, Dayton's manufacturers were poised to soak up the federal largesse.

Dayton's industries also transitioned successfully from wartime to peacetime commercial production. Indeed, Dayton was aided by pent-up consumer demand for durable goods, such as automobiles, that 15 years of Depression and world war had produced. From the beginning, some companies, like NCR, began the transition from smokestack to high-tech operations, while others, like GM, stuck to a more traditional industrial model.

As early as 1952, NCR purchased a California-based computer company and, 11 years later, established Dayton's first microelectronics research laboratory.[52] By the mid-1970s, NCR's product line included sophisticated retail terminals, automated teller machines, and computers[53]—a far cry from innovative "electric" cash registers that won fame and fortune for Edward Deeds and Charles Kettering. Within another decade, NCR had become the top-ranked international manufacturer of information processing equipment.[54] John H. Patterson would have been proud.

Success was not without its ups and downs, however. NCR had started the postwar period carrying some 20,000 employees locally on its books.[55] It continued to do so until the late 1960s.[56] However, in the economic hard times ushered in by the 1970s, this number fell to 5,000 by decade's end[57]—a lower figure by almost half than during the Great Depression.

The makeup of the local NCR workforce also changed. As much of the company's manufacturing activity moved out of Dayton, and some of it overseas, the remaining local employees increasingly

In the mid-1970s, the National Cash Register Company changed its name to the NCR Corporation and built a new world headquarters building (shown here), located in Old River Park. (*NCR Corporation*)

During the 1970s and 1980s, Dayton's business district was transformed with the construction of a handful of sleek, modern buildings. Pictured here (clockwise from top left): the Kettering Tower, the Citizens Federal Centre tower, the Arcade Centre tower, and the Mead Tower (*Jim Aldridge*)

THE BIRTHPLACE OF AVIATION HAS NO AIRPORT—
BELIEVE IT OR NOT!

The Dayton International Airport traces its roots back to 1924, when Dayton construction engineer Frank Hill Smith built a 311-acre airport for local aviation entrepreneur E. A. "Al" Johnson. It was located just north of Vandalia, near the "Crossroads of America"— National Road (U.S. Highway 40) and North Dixie Highway. Johnson moved his flight operations from Wilmington Pike to the new airport, which consisted of two 80-by-120-foot hangars, heating plants, a water tower, and macadam runways. Johnson Airplane and Supply Company failed to thrive, however, and in 1928, a group of businessmen led by Frederick B. Patterson bought the airfield and named it Dayton Airport, Inc. They developed the airport with the construction of three paved runways and administrative buildings, but the business did not succeed. Winters National Bank, holder of the airport's deed, dissolved Dayton Airport, Inc. The airfield went unused for several years, inspiring Robert Ripley, of Ripley's "Believe It or Not" fame, to publicize, in 1932, the ironic fact that Dayton, the birthplace of aviation, had no municipal airport.

In early 1936, former Ohio governor and *Dayton Daily News* editor-publisher James M. Cox took action to remedy the embarrassing situation. He first appealed to President Franklin D. Roosevelt for federal funds, but when his request was denied, he turned to local business leaders for help. Cox told them that, if the city of Dayton owned the airport, a half million dollars for badly needed airport improvements would be made available through the Works Progress Administration (WPA). Business leaders raised $65,000 and donated it to the city of Dayton so it, in turn, could purchase the airport property. On December 17, 1936, the 33rd anniversary of powered flight, the new Dayton Municipal Airport was dedicated and opened for business. Trans Western, later known as TWA, was the first airline to operate out of Dayton Airport.

During World War II, the U.S. government took control of the airport, renaming it Dayton Army Air Field and assigning it to Wright Field to test all new production aircraft, engines, and accessories. The airfield nearly tripled in size; its runways were lengthened; and a new hangar and terminal area were built. In June 1946, the government sold the airport back to the city of Dayton for $1.

In 1947, the 162nd Fighter Squadron of the Ohio Air National Guard was assigned to the airport. On October 8, 1952, the airport was renamed the James M. Cox Municipal Airport in recognition of Cox's contributions to its establishment. In 1959, work began on a new $5.5 million terminal building and, a year later, Dayton had jet service for the first time. In 1975, the airport added "International" to its name and, in 1986, as it celebrated its 50th anniversary, Dayton International Airport serviced 4.5 million passengers. In 1989, a $50 million renovation was completed.

By 2003, James M. Cox Dayton International Airport had three runways and employed more than 5,000 people. It was the North American hub for Emery Forwarding, serving more than 400 freight terminals. It had long been the site of various spectacular air shows, including the National Air Olympics, the National Aircraft Show, and the Vectren Dayton Air Show. It had the nation's largest, 90-minute air travel market, extending to almost 55 percent of the United States and half of the population of Canada.

Sources: "Huffman Prairie First Airfield," *Dayton Daily News*, December 13, 1953, n.p.; "About Dayton International Airport," viewed online October 15, 2003, at http://www.flydayton.com/airport_info/airport_info_frmst.htm; "Dayton International Airport History: Not Always Smooth Flight," *Bowling-Moorman Newspapers, Inc.*, 1990; ASC History Office Archive, "Dayton Army Air Field, Vandalia, Ohio" folder.

During World War II, the U.S. government took over Dayton Airport and transformed it into Dayton Army Air Field, shown here.

Al Johnson's 311-acre flying field is now part of the 4,000 acres owned by Dayton International Airport. (*Timothy Boone, Dayton International Airport*)

An era in Dayton history closed with the deaths of James Middleton Cox in 1956 and Edward A. Deeds in 1960. Pictured here: the Cox cenotaph and Deeds Mausoleum, both on the same hillock in Woodland Cemetery (*Jim Aldridge*)

wore a white collar. Indeed, when the last of NCR's local manufacturing activity closed down in the mid-1990s, only 125 employees were affected.[58]

Success also attracted the attention of predatory firms bent on acquiring other companies at almost any cost. In 1990, AT&T opened a full frontal attack to acquire NCR, which it succeeded in doing the following year for $110 a share.[59] John H. Patterson's nightmare had come to pass. AT&T's stewardship was less than happy. In 1995, faced with mounting operating losses, AT&T cut 20 percent of NCR's workforce, including 1,300 local jobs and, the following year, split up the company into three publicly traded companies—the largest corporate breakup in U.S. history.[60] In 1996, NCR won its independence of AT&T and regained its local, national, and international identity.[61]

Coming full circle: the Dayton Bicycle Club in 2003 (*Jim Aldridge*)

While NCR went white-collar, General Motors largely retained its blue-collar image in the Miami Valley. GM's several local divisions, among them Inland, Delco, and Frigidaire, generally prospered throughout the 1950s, 1960s, and 1970s. However, in 1978, GM shocked Dayton when it sold its Frigidaire Division to White Consolidated Industries, which thereupon moved the operation to Cleveland—along with 8,000 jobs.[62] Three years later, GM opened plants in Moraine to manufacture Chevrolet's new compact truck and diesel engines, thereupon returning 3,000 jobs to the greater Dayton area.[63] The following year, in 1984, GM added a second shift to the Moraine truck and assembly plant— and an additional 1,000 jobs.[64] The Moraine plant continued to be busy; in 1990, it was producing four-door Chevy Blazers and was slated to begin production of Chevrolet's next generation sport utility vehicles and small pickup trucks.[65]

Dayton became the headquarters of Mead's paper-manufacturing business

when the company moved to Dayton from Chillicothe in 1944.[66] By the mid-1980s, the company had become the nation's second largest paper products distributor when it acquired the Zellerbrach Distribution Group from the James River Corporation.[67] The company continued to make acquisitions throughout the 1990s. However, it also downsized its operations in several areas. In 1994, the company sold Mead Data Central (MDC) to the Anglo-Dutch publisher Reed Elsevier. MDC was thereupon renamed LexisNexis, which continued to have a presence in the Miami Valley as it marketed electronic information retrieval databases.[68] The most momentous change of all occurred, however, in 2002, when Mead merged with the Westvaco Corporation to become MeadWestvaco. The merger included the transfer of corporate headquarters to Westvaco's base of operations in Stamford, Connecticut.[69] Although the new corporate entity continued to have a division in Dayton, *MEAD* no longer lit up the Dayton

skyline atop its namesake building downtown.

There were other changes in Dayton's corporate landscape in the decades following World War II. The Chemineer Company was established in the early 1950s to manufacture mixers and agitators.[70] Also in the 1950s, the IAMS pet food company was founded by Paul Iams, who thought that if dogs could not have mink coats, they should at least have coats as glossy as minks'.[71] Dayton also lost several companies of long-standing. In 1970, a Canadian company bought Dayton-based Buckeye Iron and Brass and transferred operations elsewhere.[72] Two years later, the Joyce-Cridland Company, manufacturer of hydraulic lifts, transferred from Dayton to Indiana and, the same year, the Harris-Seybold Company, a manufacturer of printing machines, moved to Cleveland.[73] In 1980, the Firestone Tire and Rubber Company closed its subsidiary, the Dayton Tire and Rubber Corporation.[74]

One of the most dramatic changes in the

corporate and industrial picture was the virtual disappearance of the publishing industry from Dayton. In the decade after World War II, the McCall Corporation printed more than 52 national publications in its local plant.[75] In 1962, McCall's output averaged more than a *billion* magazines, including issues of *Newsweek, Reader's Digest,* and *U.S. News & World Report.*[76] However, two decades (and a corporate name-change) later, Dayton Press Inc. closed its doors with the loss of more than 2,500 jobs.[77] When, in 1989, United Color Press left the area, Dayton was no longer a center of national publishing—a status it had enjoyed since the turn of the twentieth century.[78]

If Dayton were to continue to remain competitive on the national and world stage, it needed an educated and technically literate corporate leadership and workforce. Providentially, it laid the foundations for this, beginning in the 1960s, when it began to expand and diversify its educational institutions. In 1961, Ohio State and Miami universities agreed to jointly operate a division of Ohio's state university system near Wright-Patterson Air Force Base.[79] Four years later, this division was christened Wright State University.[80] By the end of the century, Wright State was a leader, both locally and statewide, in such high-tech areas as engineering and medicine. Meanwhile, Dayton's Sinclair College, which had begun as a YMCA school in the 1890s, went public in 1966.[81] By all measures, Sinclair Community College was a genuine "success story" both in providing additional educational opportunity for residents of the Miami Valley and in revitalizing a hitherto "bad" section of downtown real estate.[82] Finally, the venerable University of Dayton, founded in the mid-nineteenth century by the

Marianist Order of the Roman Catholic Church, continued to expand its menu of educational opportunities to both Daytonians and out-of-towners, while upholding a national reputation for excellence in academics.[83] In association with the University of Dayton was the University of Dayton Research Institute that got its start primarily in response to Air Force needs for a local center of high-tech expertise (see below).[84]

Local business and industry leaders also looked beyond downtown and other regional manufacturing areas to establish new high-tech centers or "research parks." One of the most notable was the Miami Valley Research Park that, in 1985, opened its first factory, operated by Diconix, a subsidiary of the Eastern Kodak Company.[85] Other companies from around the country also considered locating at this site. In addition, office and industrial complexes of a variety of firms began sprouting up in the corridors along I-75 and I-675.[86] Many of the latter had direct connections with the laboratories and program offices at Wright-Patterson Air Force Base.

THE WRIGHT-PATTERSON CONNECTION

When Edward Deeds arranged to have the Army Air Service's principal research and development and test and evaluation airfield located just north of downtown, he "did good" by Dayton—and perhaps better than even he knew at the time.

The value of McCook Field's work was, however, already evident to the aged John H. Patterson in the early 1920s, when the Army bruited its intention of transferring

its R&D operations elsewhere. Deeds had already left town for "greener" pastures on Wall Street, but Patterson and his son and heir, Frederick, spearheaded the effort to keep the Army's experimental work in Dayton. The establishment of Wright Field rendered permanent Deed's good deed to his hometown (see Chapter 2: Military Aviation Comes to Dayton).

The modern-day story of Wright-Patterson and its interface—economically, technically, educationally, and socially—with the greater Dayton area owes, at least, a nod to the earliest residents of the Great Miami River valley. People have lived on land in and around Wright-Patterson Air Force Base for several millennia. The original Paleoindian and Adena period peoples were followed by Woodland peoples. From about 400 B.C. to A.D. 400, Early Woodland Native Americans, sometimes known as "mound builders," thrived in southwestern Ohio and other areas east of the Mississippi River. In fact, vestiges of their presence—burial mounds—have been documented on Wright Brothers Hill, a memorial to man's remarkable discovery of powered flight, located on Area B of the base. Totally undisturbed and, in fact, cared for by the same modern "tribe" that has crafted flying machines of astonishing technical prowess for the past 100 years, the six mounds doubly mark the site as significant and of great worth.

A bronze plaque located on one side of the circular wall defining the memorial's plaza pays tribute to those long-ago people. It reads:

Centuries ago this hill was a center of life and culture of the mound builders. Two large burial mounds and four small ones have been identified here by the Ohio State Archeological and

Dayton skyline from Deeds Point alongside the Great Miami River *(Helen Kavanaugh-Jones)*

Indian mounds, evidence of Native American presence atop Wright Brothers Hill—linking the prehistoric past to sophisticated research for future aerospace systems at Wright-Patterson

WRIGHT FLYING REPLICAS

Several replicas of early Wright brothers' aircraft have been constructed in the Dayton area. Perhaps the most public are the flyable, full-scale, 1911 Wright "B" Flyer, a reproduction of the world's first military production aircraft, and the Centennial Flyer mounted in the Dayton International Airport.

The idea for the Wright "B" was conceived by the Dayton Area Chamber of Commerce in 1974 and begun the following year with base employee Charles A. Dempsey (now deceased) as program manger. The Flyer project brought together a true volunteer army—as many as 500 base employees, civilian aviation enthusiasts, and drafting students from Fairmont West and East high schools. Ninety-eight Air Force and private organizations, businesses, and individuals contributed funds, materials, or services to the Wright "B," a nonprofit project. The Flyer and a civilian version of the plane, used primarily for static display, were constructed in Hangar 145 in Area C.

Wright "B" Flyer *(Jim Sandegren)*

The Wright "B" was constructed to the Federal Aviation Administration's regulations for experimental aircraft: it has an aluminum-and-steel airframe the same size as the original (28 feet long with a 39-foot wingspan); dimensions of its engine also match the original, but it is heavier. After assembly at Wright-Patterson, the Flyer flew its flight tests in summer 1982, then made a 22-mile "official" first flight to a new destination, then Dayton General Airport South, on May 5, 1983, with John H. Warlick, retired U.S. Navy lieutenant commander, as pilot, and W. A. Sloan, Jr., retired U.S. Air Force colonel, as copilot. A crew of five volunteers, including retirees from Wright-Patterson, maintained and serviced the Wright "B."

Since its maiden flight, the military Wright "B" replica has logged impressive flight data at local, national, and international events. More than 1,725 passengers have taken rides on the Wright "B," and it has flown every year at the Dayton air show. Other special-occasion flights included the celebration of the 50th anniversary of the Berlin Airlift at Templehoff Air Base, Germany, in 1990; the arrival of the first B-2 stealth bomber at Whiteman Air Force Base,

Base retiree Harold Edinger, a volunteer with the Wright "B" Flyer, at Dayton Wright Brothers Airport. Edinger first offered his services for the Flyer in 1983 and today is the Flyer's only mechanic certified by the Federal Aviation Administration. *(U.S. Air Force photo by Spencer P. Lane)*

Missouri; installation of a Wright A-B sculpture at Maxwell Air Force Base, Montgomery, Alabama; and the 70th anniversary celebration of Brooks Air Force Base, Texas. On January 1, 2003, the Wright "B" flew over the Rose Bowl stadium in Pasadena, California, and on May 25, 2003, flew around the Statue of Liberty in New York Harbor. The latter flight was a special centennial-of-flight event following Wilbur Wright's route on September 29, 1909, as part of the Hudson-Fulton Celebration.

The Centennial Flyer, an exact replica of the 1903 Wright Flyer, hangs in the lobby of Dayton International Airport, visible to the 1.5 million travelers who pass through the airport annually. Under the leadership of Nick Engler of West Milton, Ohio, and his Wright Brothers Aeroplane Company (WBAC), the replica was assembled at Dayton's Boonshoft Museum of Discovery and Wilkie's, a bookstore formerly located in downtown Dayton. The Centennial Flyer is one of six replicas of Wright brothers' inventions (one kite, three gliders and two Flyers) built by Engler's company for the 100th anniversary of powered flight. A seventh replica, a 1905 Flyer, is scheduled for completion in early 2004.

Engler received assistance from numerous sources, such as Lieutenant Colonel Scott Martin and Major Andrew Thurling, Air Force flight test officers from the F/A-22 System Program Office of Aeronautical Systems Center at Wright-Patterson. In April 2003, Martin and Thurling flew reproductions of the 1900, 1901, and 1902 gliders at Warren Dunes State Park, Michigan, to gain a better understanding of the Wrights' engineering methodology and to give feedback to Engler regarding the crafts' aerodynamics. (The officers' expedition was considered "new archeology," or the process of learning about historic events by recreating them.)

Nearly 300 children across the United States and in five countries participated in a "rib-workshop" program mailed to schools, scout troops, and other youth organizations upon request. The workshop allowed young people to build a rib while learning about the Wright brothers and the invention of the airplane. At the Flyer's dedication on April 16, 2003, (Wilbur Wright's birthday in 1867), officiated by Dayton Mayor Rhine McLin, children from Russia School in Shelby County, Ohio, and the choir from Wilbur Wright Middle School participated. Two months later, during the Vectren Dayton Air Show, the WBAC premiered the "Birth of Aviation" pavilion and displayed their one kite, three gliders and a second 1903 Wright Flyer replica in a comprehensive, early-flight exhibition.

In October 2003, the second 1903 Flyer (named Friendship Flyer), five military officers, and two civilian commercial pilots traveled to Currituck County Airport (40 miles from Kitty Hawk, North Carolina) for flight testing. The participants were Lieutenant Colonel Martin; Major Thurling; Lieutenant Colonel Dawn Dunlop (also from the F/A-22 office); U.S. Army Captain Tanya Markow, an Apache helicopter instructor; U.S. Navy Lieutenant Commander Klas Ohman, flight test officer; Dudley Mead, a pilot for DHL and hang-gliding champion; and Connie Tobias, a pilot of the U.S. Airways Airbus 330 and the Bleriot, a contemporary craft of the Wright Flyer. At Currituck, the group performed numerous tests of the Flyer—including acceleration, velocity, and launching—with overall good results. Plans called for sustained, powered flight before the close of 2003.

Sources: "Wright 'B' Flyer Circles again," *Dayton Daily News*, May 26, 2003, p A1; "Jumper takes to skies in Wright 'B' Flyer," *Skywrighter*, June 27, 2003, p B1; 1911 Wright "B" Flyer pamphlet, Wright "B" Inc.; "Centennial Flyer at the Dayton International Airport" pamphlet, City of Dayton, Department of Aviation, and Wright Brothers Aeroplane Company of Dayton, Ohio, n.d.; "Test pilots evaluate Wright brothers' gliders," *Skywrighter*, May 2, 2003, p 3A; Wright Brothers Aeroplane Company website, viewed online October 15, 2003, at www.wright-brothers.org; Marion Schniegenberg, WBAC public relations, personal communication with Helen Kavanaugh-Jones, October 15, 2003.

Exact replica of the 1903 Wright Flyer, built as a community project spearheaded by Nick Engler, hangs in the lobby of the Dayton International Airport. *(Helen Kavanaugh-Jones)*

Base camp at Warren Dunes, Michigan, for Wright glider replicas built by the Wright Brothers Aeroplane Company. Left to right: the 1900, 1902, and 1901 gliders *(Major Andy Thurling)*

Major Andy Thurling aloft in a reproduction of the 1902 Wright glider at Warren Dunes, Michigan. Thurling was a U.S. Air Force flight test officer assigned to the F/A-22 System Program Office at Aeronautical Systems Center.

Historical Society. They were once part of a mound community which from all indications was a place of importance in the mound builders era. Thus, from a site with considerable prehistoric interest rises a shaft commemorating one of the greatest advances in the history of mankind.

In the twenty-first century, Dayton-area communities remained as diverse as the "mound builders" were tribal: large and small, urban and rural, industrial and agricultural. In 2003, a population of about 950,000 resided within an area encompassed by four southwestern Ohio counties: Clark, Greene, Miami and Montgomery. The greatest number lived in Dayton, host to the county seat of Montgomery County.

Thus, Wright-Patterson in the early twenty-first century potentially had the eyes and ears of nearly one-tenth the population of Ohio, in addition to the national and international aerospace communities and scientific minds that closely monitored its programs. Anyone newly acquainted with the base must be struck by its energy, sophistication, and economic value to the area and its military importance to national defense.

Those interested in glimpsing the many-faceted contributions of McCook Field, Wright Field, and Wright-Patterson Air Force Base to Dayton and the nation during the past century need only read *Splendid Vision*, the first volume of this two-volume history of the contributions of the Air Force's science, engineering, and logistics operations from Dayton, Ohio. What follows herein, constituting the heart of this chapter, are "snapshot" stories from the recent past detailing a number of collaborations, in technology and education, between Wright-Patterson and greater Dayton. The chapter culminates with the Air Force's and Dayton's celebration of the 100th anniversary of powered flight during the centennial year, 2003.

WRIGHT-PATTERSON'S ECONOMIC IMPACT ON SURROUNDING COMMUNITIES

As the twenty-first century began, few denied that Wright-Patterson was a powerhouse in the economic life of southwestern Ohio. In 2003, approximately

24,000 military and civilian employees and government contractors worked at Wright-Patterson in more than 60 organizations, making it the largest, single-site employer in the state and one of the largest employers among U.S. Air Force bases worldwide. Civilian workers comprised more than half the workforce, and the base's replacement value approximated $8.5 billion. The following 13 major organizations located on base provided the most significant impact throughout the Air Force and the Department of Defense (DOD).

Aeronautical Systems Center (ASC), Wright-Patterson's host unit, carried an average annual budget of $12 billion (12 percent of the annual Air Force budget), employed a workforce of 12,000-plus civilians and military members, and maintained 37 units worldwide. ASC has been accorded high visibility, not only within DOD but the nation as a whole, due to its mission of developing, acquiring,

modernizing, and sustaining Air Force aerospace systems, such as the F-15, F-16, F-117A, and F/A-22 fighters; Global Hawk and Predator unmanned aerial vehicles; B-2 and B-1B bombers; C-17 transports; MC-130H special operations aircraft; AC-130U gunships; and T-6A and T-1A trainers. Acquisition personnel also contributed to development of the F-35 Joint Strike Fighter and began managing the Airborne Laser weapon system.

System program office managers worked in the largest, most advanced complex of research and management facilities in the country: 1.7 million square feet of floor space neared completion along the closed Area B flightline as 2003 drew to an end. ASC also maintained the Major Shared Resource Center, a DOD asset comprising one of four high-performance, computing locations in the United States; the eighth-largest super computer in the world; and the Simulation and Analysis Facility that

Headquarters of Aeronautical Systems Center, Building 14, Area B, Wright-Patterson Air Force Base *(Diana Cornelisse)*

enabled engineers to visualize, build, test, and operationally evaluate systems in virtual reality before committing work and money to a project.

ASC's *88th Air Base Wing* served as landlord and caretaker to all organizations on Wright-Patterson and serviced many other defense organizations in Ohio, Michigan, Indiana, and Kentucky. In 2003, its domain encompassed 8,000 acres, 1,600 buildings, and two runways that supported an average of 40,000 aircraft operations annually. Wing employees also handled the payroll, kept records on the base workforce, maintained and provided security to buildings, and managed communications, transportation, weather forecasting, and legal and chaplain services.

United States Air Force Medical Center Wright-Patterson, organizationally in DOD Health Service Region 5, served as one of six regional Air Force medical centers and provided care to more than 54,000 beneficiaries in a 40-mile radius. It employed more than 2,000 medical professionals and maintained a mission emphasizing referrals, community care, and military readiness.

Air Force Materiel Command (AFMC), headquartered at the base, provided oversight for the entire life cycle of weapon systems, from the drawing board to final disposition. Worldwide, its 87,000 military and civilian employees comprised the Air Force's largest command and managed one-third of the total Air Force budget.

Air Force Research Laboratory (AFRL), with an annual budget of approximately $2.3 billion, planned and executed the Air Force's science and technology programs. Its partners in academia and industry worked with engineers and scientists in AFRL's five on-site directorates: Sensors, Materials and Manufacturing, Propulsion, Human Effectiveness, and Air Vehicles. AFRL also had directorates at other locations, including the Information Directorate at Rome, New York; Munitions at Eglin Air Force Base, Florida; Space Vehicles at Kirtland Air Force Base, New Mexico; and the Air Force Office of Scientific Research in Arlington, Virginia.

Air Force Security Assistance Center (AFSAC) represented AFMC's administrative center for foreign military sales, the Air Force program that provided materiel and services to support more than 9,000 aircraft (170 different models) in more than 80 countries and international organizations. The center also served as AFMC's international liaison to DOD, the secretary of the Air Force, the North Atlantic Treaty Organization (NATO), United States/foreign embassies, and foreign defense ministries.

Materiel Systems Group, organizationally a part of the Electronic Systems Center at Hanscom Air Force Base, Massachusetts, maintained employees in the Dayton area and at Ogden and Oklahoma City Air Logistics Centers. The group supported the Air Force's goals for information dominance through development, acquisition, maintenance, reengineering, and technical support of information solutions and systems.

The *55th Wing E-4B National Airborne Operations Center* (NAOC), a detachment from Offutt Air Force Base, Nebraska, maintained a 24-hour alert facility in support

Main entrance to the Air Force Institute of Technology, Area B, Wright-Patterson Air Force Base *(Helen Kavanaugh-Jones)*

Communities surrounding Wright-Patterson Air Force Base benefit economically from the base's employment and research contracts. *(Helen Kavanaugh-Jones)*

of the E-4B aircraft, a Boeing 707-200. A highly survivable command, control, and communication platform, the E-4B provided the National Command Authorities essential connectivity with civil authorities and force commanders, allowing direction of United States forces, execution of war orders, and coordination of actions taken by civil authorities. The wing's personnel were comprised of 12 enlisted airmen.

Air Force Institute of Technology (AFIT) provided graduate (masters and doctoral) degrees and continuing education in science, engineering, and advanced technologies to officers, enlisted personnel, and civilians. Its three resident schools included the Graduate School of Engineering and Management, School of Systems and Logistics, and Civil Engineer and Services School. AFIT also managed

the educational programs of Air Force officers, enlisted personnel, and civilians enrolled in civilian universities, research centers, hospitals, and industrial organizations around the world.

The *445th Airlift Wing*, a unit of the U.S. Air Force Reserve Command, provided strategic, worldwide transport, including aeromedical evacuations, via the C-141C transport. Approximately 2,000 reservists

WRIGHT-PATTERSON'S ECONOMIC IMPACT ON THE DAYTON COMMUNITY IN THE TWENTY-FIRST CENTURY

Expenditures (in millions per fiscal year)

	2000	2001	2002
Annual payroll (military and civilian)	$804	$896	$917
Indirect jobs created	$595	$656	$629
Retirees in area (not factored into total expenditures)	$594	$608	$615
Construction, services, materials, equipment, and supplies	$768	$770	$901
Total (in billions)	$2.17	$2.45	$2.31

Sources: "Economic Impact Analysis," September 30, 2002, Wright-Patterson Air Force Base, 88 ABW/FMA, p 6; "Economic Impact Analysis," September 30, 2001, Wright-Patterson Air Force Base, 88 ABW/FMA; "Economic Impact Analysis," September 30, 2000, Wright-Patterson Air Force Base, 88 ABW/FMA.

and trainees comprised the wing, which flew humanitarian missions to Somalia and Rwanda, provided ground support for Operation Provide Promise in Bosnia and Operation Allied Force in Kosovo, and transported military firefighters to the western United States to fight uncontrolled forest fires. In 2001-2002, the 445th supported Operation Enduring Freedom by transporting Taliban and Al Qaeda detainees to Guantanamo Bay, Cuba (see Chapter 4: Fulcrum of Base Support).

National Air and Space Intelligence Center (NAIC), with a personnel strength of 1,600 people, was a unit of the Air Intelligence Agency, the primary DOD collector of foreign aerospace intelligence. It analyzed all available data on foreign aerospace weapon systems to determine performance characteristics, capabilities, and vulnerabilities. The center played a key role in assuring that American forces could counter the foreign aerospace threat and avoid technological surprise.

The *United States Air Force Band of Flight* provided professional music and entertainment for a wide variety of official military, recruiting, and community-relations events. With a continuous record of service dating back to 1942, the band had 60 active-duty members who performed primarily within a seven-state area of the Midwest that included almost one-quarter

of the nation's population.

The *United States Air Force Museum* in Area B of Wright-Patterson Air Force Base represented the largest and oldest military aviation museum in the world and the national museum for the U.S. Air Force. As the Midwest's largest, free tourist attraction with nearly 1.2 million visitors

yearly, the museum displays illustrated the exciting story of military aviation development, beginning with the days of Wilbur and Orville Wright. Exhibits included more than 300 aircraft and missiles and thousands of artifacts, including personal memorabilia, uniforms, and photographs.

The Air Force Band of Flight performed a musical celebration of the Ohio Bicentennial by performing in all 18 of Ohio's congressional districts.

Geographically, these 13 organizations headquartered at Wright-Patterson Air Force Base financially impacted, most keenly, Clark, Greene, Miami, and Montgomery counties, which surrounded the base. However, the base's contract dollars for research, production of aircraft systems, and logistical supplies stretched across the country. Locally, research contract dollars flowed to large and small businesses and academic institutions, including historically black colleges such as nearby Central State University, which, in 2003, was conducting materials research for the Air Force Research Laboratory. In addition, many local, small businesses provided support services, as well as research expertise, to units on base, contributing to the area's reputation for sophisticated research and innovation.

Front entrance of the National Air and Space Intelligence Center, Area A, Wright-Patterson Air Force Base, 2003 (*National Air and Space Intelligence Center*)

The financial impact "footprint," called the Dayton-Springfield Metropolitan Statistical Area, encompassed 29 cities, villages, and suburban communities that counted Wright-Patterson employees as residents. These communities included Beavercreek, Bellbrook, Centerville and Washington Township, Dayton, Englewood, Enon, Fairborn, Huber Heights, Jamestown, Kettering, Riverside, Miamisburg, Moraine, New Carlisle, Oakwood, Piqua, Sidney, Spring Valley, Springboro, Springfield, Tipp City, Trotwood, Troy, Vandalia, Waynesville, West Carrollton, Xenia, and Yellow Springs.

Wright-Patterson's employees formed a strong economic base, the envy of many other areas of similar size. Thus, for the first years of the twenty-first century, Wright-Patterson's report to "stockholders" continued to list an impressive array of financial data. For example, the annual payroll of military and

United States Air Force Museum, Area B, Wright-Patterson Air Force Base (*United States Air Force Museum*)

civilian personnel at Wright-Patterson at the close of fiscal year 2002 (October 1, 2001 – September 30, 2002) topped $917 million, an average of $2.5 million per day. Fifty-seven percent of that payroll was earned by employees living in Montgomery County, with Greene County second at 31 percent. Between 1998 and 2003, in fact, the base payroll averaged more than $895 million annually.[87]

In addition to employing its own workforce, Wright-Patterson indirectly created jobs for numerous local residents. Using the "average annual pay" defined by the Bureau of Labor Statistics, U.S. Department of Labor, in 2002, Wright-Patterson created 18,723 indirect jobs with a total value of more than $629 million. Those included, for example, jobs in

restaurants, hotels, and retail establishments. More than 27,000 retirees from all military services resided in the area, providing an additionally significant economic impact on the community. Their annual retirement salaries for 2002 totaled about $615 million.[88]

In 2002, military disbursements for construction, services (including use of civilian, medical-care facilities), and materials, equipment, and supplies in the local economic area totaled more than $767 million. These expenditures combined—$2.3 billion—represented a figure that any community—indeed any state—would covet.

The financial impact of Wright-Patterson extended, as well, to primary educational institutions. Based on Public Law 81-874,

passed in 1951, school districts received funds from the federal government when a certain percentage of their enrollment included children from families employed by the federal government. The government annually distributed these funds, known as "impact aid," on the basis of formulae set in the legislation. In 2002, local school districts received $2.7 million in educational impact aid for 3,386 students whose families were federally employed. In previous years, the figures were $2.6 million for 4,964 students (2001) and $2.5 million for 5,873 (2000).[89]

Finally, Wright-Patterson employees, both military and civilian, contributed annually to the base's Combined Federal Campaign, which raised money for qualified charitable organizations. Over the past 20 years, employees' contributions totaled more than $41 million. In 2002, contributions reached nearly $2.78 million; in 2001, $2.85 million; and in 2000, $2.2 million. These generous contributions represented Wright-Patterson employees' commitment to the broader community outside the base.[90]

MUTUAL BENEFITS OF RESEARCH AND DEVELOPMENT

The transfusion of military technology into business, academic, engineering, and scientific life represented another of Wright-Patterson's impacts on the local community. The use of military technologies in commercial aviation, as well as other industries, extended back to the days when Orville and Wilbur developed their Military Flyers for the United States government. Nearly all major advances in aviation since that time had their roots at Wright-Patterson Air Force Base or one of its predecessors, and base personnel often shared their expertise and laboratory results with the general public. This interchange of ideas significantly impacted development of technology.

Annually, several thousand scientists, engineers, business leaders, media representatives, foreign nationals, and specialized groups visited Wright-Patterson and, in turn, laboratory and program office personnel traveled to symposia and meetings worldwide to present technical papers on Air Force-sponsored programs. For example, for more than 50 years, technical personnel from the base supported the National Aerospace and Electronics Conference (NAECON) held annually in Dayton. The annual National Aeronautical Systems and Technology Conference (NASTC), begun in 2000 and also held locally, attracted representatives from the Air Force, Army, Navy, and private industry. In particular, at NASTC, senior management from the base updated attendees on military operations, requirements, challenges, and solutions, including the role of technology in accomplishing goals. Many other organizations continued to choose the greater Miami Valley for technical meetings (large and small) to take advantage of the contributions base personnel made to their scientific and engineering fields.

In addition to sharing technology,

Lieutenant General Kenneth E. Eickmann, commander of Aeronautical Systems Center, holds aloft the Thomas Edison Award presented by Ohio Governor George Voinovich (at podium) for Wright-Patterson's achievements in technology and community contributions.

Wright-Patterson played a mentor role for numerous academic institutions. Through the use of Education Partnership Agreements (EPAs), federal laboratory personnel loaned or donated equipment declared surplus; taught or assisted in developing courses and course materials; cooperated with academic institutions so students could earn credits for laboratory research work; and provided academic and career advice and assistance to students. Since the early 1990s, the Air Force Research Laboratory signed EPAs with academic institutions ranging from local elementary and high schools to universities and nonprofit organizations dedicated to improving science, mathematics, and engineering education.[91]

Whether through technology transfer or academics, Wright-Patterson's impact on technology has affected nearly every resident in the Dayton area. On September 17, 1997—the eve of the Air Force's 50th anniversary—Ohio's Thomas Edison Program recognized the "technology and economic development assets" at Wright-Patterson when Ohio Governor George Voinovich presented the Thomas Edison Award to the base in a ceremony at the United States Air Force Museum. Lieutenant General Kenneth E. Eickmann, commander of Aeronautical Systems Center, accepted the award, along with General George Babbitt, commander of Air Force Materiel Command. Governor Voinovich commented at the presentation, "It's my hope that the partnership that has been established between the state of Ohio and this base will continue to expand as Ohio builds upon its reputation for technological excellence well into the twenty-first century."[92]

Technology Transfer

Beginning in 1917, Wright-Patterson and its predecessors have been preeminent for the research and development occurring within the field of military aviation. Since it was organized in 1997, the Air Force Research Laboratory (AFRL), with headquarters and five of 10 directorates located on base, has had management responsibility for technology transfer, ensuring that technology developed as part of its mission transferred in a timely manner to private and public sectors. AFRL's Technology Transfer (T^2) Program promoted the transfer and/or exchange of technology with state and local governments, academia, and industry to create jobs, improve productivity, and increase competitiveness while supporting the Air Force mission.

For example, in the mid-1960s, Air Force materials experts began studying and developing lightweight but strong and stiff composite materials as an alternative to traditional, structural metals for airframes, aerospace engines, space vehicles, and satellites. By the twenty-first century, a variety of industries manufactured composites for commercial products ranging from sporting goods equipment and firemen's helmets to automobile parts

A BRIDGE IN BUTLER COUNTY—TECH 21

In the summer of 1997, a collaborative effort by a four-member military and civilian team brought a full-size, vehicular bridge made of fiberglass, a composite material, to southwestern Ohio and had it installed—at no cost. Air Force materials research engineer John Mistretta, of then Wright Laboratory, worked for three years with the Butler County Engineer's Office in Hamilton, Ohio; Lockwood, Jones and Beals, Inc., of Dayton and Cincinnati, Ohio; and Martin Marietta Materials, Inc., of Raleigh, North Carolina, to promote the use of advanced aerospace materials to reinforce or replace existing concrete and steel highway bridges in Ohio. Composite bridges, with a life-expectancy of 150 years were ideal replacements for the typical concrete-and-steel bridges, which tended to last only 50 years due to Ohio's severe winters, spring freeze/thaw cycles, blazing summers, and the corrosive effects of road-salt.

Through perseverance, some serendipity, and lots of cooperation, on July 25, 1997, officials gathered to dedicate Tech 21, the first composite bridge in the state. Representatives from the U.S. Air Force, the Ohio Governor's Office, state departments of Development and Transportation, Butler County Engineer's Office, and 250 additional distinguished guests were present at Smith Road Bridge #03.730, located just two-tenths of a mile west of State Route 128 in Hamilton. The structure is a 33-foot-long, 24-foot-wide, single-span, two-lane vehicular bridge load-rated by industry standards to carry vehicles weighing more than 80,000 pounds. It carries about 1,000 cars and trucks per day and can withstand, engineers estimate, an Army M-1 tank; the asphalt approaching the bridge would collapse, but the bridge itself would withstand the weight.

In 2003, the Butler County bridge was still in place with no observable deterioration. Sophisticated, embedded sensors recorded the bridge's long-term environmental and life cycle data for more than three-and-one-half years, and that data proved useful to the team members. The materials technology used in fabricating the bridge demonstrated the use of advanced composite materials for the repair and/or replacement of some of America's 230,000 structurally deficient, functionally obsolete bridges—close to half the 575,000 bridges that the Federal Highway Administration monitors.

Source: ASC/PAM news release 97-124, 'Tech 21 Team Promotes First All-Composite Bridge in Ohio," August. 6, 1997.

Off-loading a 22,000-pound fiberglass bridge beam built by Martin Marietta Materials, Inc., of Raleigh, North Carolina, during installation of Ohio's first, all-composite bridge on Smith Road in Butler County, July 8, 1997. Wright Laboratory at Wright-Patterson was one of several partners in the bridge program, exemplifying technology transfer in practice. *(U.S. Air Force photo by Al Bright)*

OSCAR BOONSHOFT

Oscar Boonshoft obtained his Mechanical Engineer degree from Stevens Institute of Technology in Hoboken, New Jersey. In 1939, he started as project engineer in the Armament Laboratory in the Engineering Division of the Army Air Corps Materiel Command. After a 30-year career as a project engineer, U.S. Air Force contracting officer, and supervisory production engineer at Wright -Patterson Air Force Base, he retired in January 1970. Since retiring, he has remained active as a successful, speculative trader in commodity contract futures.

Boonshoft has supported science and medicine in a number of capacities. He helped to establish and continues to actively support the Boonshoft Museum of Discovery, focusing it on science education for children by means of interactive hands-on exhibits.

At the Wright State University School of Medicine, he established a chair that evolved into the Division of Health Systems Management. This division is presently establishing a Physician Leadership Development Program, which will provide a fully integrated five-year dual MD/MBA degree to make available appropriate education for future management and administrative leaders in the health-care system. Boonshoft is a member of the Wright State University Foundation Board of Trustees. He also is a supporter of the Wallace-Kettering Neuroscience Institute at Kettering Memorial Hospital and is an active member of its Governance Board.

Boonshoft continues to support the arts and charitable organizations in the greater Dayton area and has been honored by the National Conference for Community and Justice (NCCJ) and Big Brothers/Big Sisters organizations. He has been identified as one of the influential persons in the greater Dayton area in the category of philanthropy.

Oscar Boonshoft, a retired engineer from Wright-Patterson Air Force Base, donated millions of dollars to several worthy causes in the Dayton area, including the Museum of Discovery, which was created through the joint efforts of the Dayton Society of Natural History and the Children's Museum Board. (*Oscar Boonshoft*)

Boonshoft Museum of Discovery (*Helen Kavanaugh-Jones*)

Children learn about air, water, and motion at an interactive water table, part of the "Oscar Boonshoft Science Central" exhibit area at the Boonshoft Museum of Discovery. (*Lori Tagg*)

and bridge structures (see below). Scientists and engineers in AFRL's Materials and Manufacturing Directorate also played an instrumental role in the 1998 establishment of the National Composite Center (NCC) at Kettering Business Park in Kettering, Ohio (see below).

In 1980, Congress enacted a series of laws to promote T^2 and to provide mechanisms and incentives for its implementation. These laws and related executive orders encouraged the pooling of resources during development of commercial technologies. The Federal Technology Transfer Act of 1986 made T^2 a responsibility of all federal laboratories. Several T^2 mechanisms used at Wright-Patterson to accomplish this federal mandate included collaborating on research, testing innovations or products, providing excess equipment to schools, or licensing Air Force technologies.[93]

Wright-Patterson's laboratories used several mechanisms in their technology transfer program. One such instrument was the Cooperative Research and Development Agreement (CRADA), which had been authorized by the 1986 Federal Technology Transfer Act. Through written CRADAs, government laboratories provided personnel, facilities, equipment, or other resources (but not funds), with or without reimbursement, to non-federal parties. In turn, the non-federal parties provided funds, people, services, facilities, equipment, or other resources to conduct

specific research and development efforts consistent with AFRL's mission. While the Air Force retained the license to use a technology and share it with other government agencies, the company or university benefited from commercial application of the product.[94] AFRL's first CRADA was a joint effort with AdTech Systems Research, Inc., located in Dayton, for sophisticated computer software to use in the analysis of structural composites. More recently, an executed agreement between AFRL's Human Effectiveness Directorate, Aural Display and Bioacoustics Branch, and the Kettering Memorial Hospital's Wallace-Kettering Neuroscience Institute focused on reducing the exposure of patients to noise during Magnetic Resonance Imaging (MRI) scans. By 2003, AFRL signed more than 850 CRADAs, ranging from materials processing to solid-state electronics, along with potentially beneficial products for health care.[95]

Other mechanisms for technology transfer included the Small Business Innovative Research (SBIR) program and Independent Research and Development. Projects funded under the SBIR program became a mainstay of Air Force T^2 in 1982, when the program was created to focus, in particular, on high-risk research. These projects served an Air Force need and the resulting products had potential for commercialization in private and military markets. To qualify for a SBIR grant,

businesses had to meet certain qualifications, such as employing no more than 500 employees and being 51 percent owned by United States citizens. Since 1982, the laboratory complex at Wright-Patterson entered into 14,306 SBIR agreements with a total value of more than $2 billion.[96]

Independent Research and Development (IR&D), a term for company- or corporate-funded research and development, represented another component of T^2. While a company planned, initiated, and controlled its own IR&D projects (spanning all economic and technical areas necessary in the development of a product or service), it also submitted annually its IR&D project descriptions to the Defense Technical Information Center. That data, in turn, facilitated government strategic planning and the use of resources. The research, engineering, and management offices at Wright-Patterson became vitally interested in IR&D programs in the early 1970s and considered them an important element of its T^2 efforts.[97]

During the 1980s and 1990s, the predecessors of AFRL created ties with several organizations, especially at the state level, to promote and enhance its

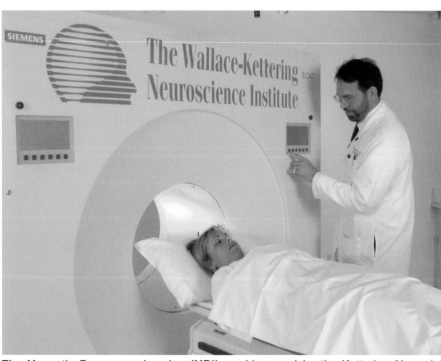

The Magnetic Resonance Imaging (MRI) machine used by the Kettering Memorial Hospital's Wallace-Kettering Neuroscience Institute. A study to test a prototype, active-noise cancellation system for the machine was a collaborative effort between the institute and Air Force Research Laboratory's Human Effectiveness Directorate, Aural Display and Bioacoustics Branch. *(Kettering Medical Center)*

technology transfer efforts. These groups included the Ohio Technology Transfer Organization, which worked with AFRL's Office of Research and Technology Applications; the Ohio Advanced Technology Center (precursor to the Wright Technology Network); the Ohio Aerospace Institute; the Edison Materials Technology Center; and the National Composite Center (see below). AFRL's headquarters also maintained a "single point of entry," called TECH CONNECT, for individuals seeking information about programs under the T² umbrella.[98]

Edison Materials Technology Center

In 1987, a nonprofit, materials resource center, named the Edison Materials Technology Center (EMTEC), opened in the Miami Valley Research Park, Kettering, Ohio. The new facility represented one of several centers in the state created through Ohio Governor Richard F. Celeste's Thomas Edison Program. Each center focused on specific technologies with established expertise in their respective areas. In Dayton, with the assistance of specialists in the Materials Laboratory of the Air Force Wright Aeronautical Laboratories (AFWAL), EMTEC emphasized materials and manufacturing processes. Like all Edison program centers, EMTEC offered its members—federal laboratories, industry, and academic institutions—problem-solving in relevant technologies. The center's long-range goals included accelerating the development, deployment, and commercialization of materials technologies to strengthen Ohio and United States industry, thus creating new jobs. EMTEC covered its operating costs through membership fees, the Ohio Department of Development's Thomas Edison Program, a variety of government contracts, and direct project investment.

Planning for EMTEC had strong support from officials at Wright-Patterson, both in Air Force Logistics Command and Air Force Systems Command, represented by Aeronautical Systems Division (ASD). On August 24, 1987, Colonel James M. Walton, commander of AFWAL (then an element of ASD), and Dr. Frank Moore, EMTEC director, signed a Memorandum of Understanding (MOU) permitting cooperation between the two organizations in materials and manufacturing technology. The agreement, a response to presidential directives and congressional legislation mandating accelerated transfer of federally sponsored technology to state and local governments and the private sector, was one of the first in the nation to link an Air Force research laboratory with a state-funded technology center. Such coordinated efforts expedited the movement of federal technologies from the laboratory into commercial applications. Top officials at the base also played an instrumental role in EMTEC's initial organizational structure. Lieutenant General William E. Thurman, commander of ASD, served on the center's first board of governors; Dr. Gary L. Denman, director of the Materials Laboratory, served on EMTEC's regional advisory council; and Dr. Harris M. Burte, chief scientist of the Materials Laboratory, sat on the technical steering committee.

In 2003, EMTEC, led by President and Chief Executive Officer Frank Svet, functioned as an information conduit among member industries, including more than a dozen universities and multiple government laboratories staffed by thousands of scientists and researchers. In 16 years, the center carried out more than 1,000 projects to accelerate technology utilization by participating companies. Specifically, if a member encountered a problem with a product or process, EMTEC offered assistance. Additionally, EMTEC members maintained access to expert advisors on virtually any process that transformed a material into a product (e.g., casting, forging, finishing, rapid prototyping, and nondestructive testing). The center's materials expertise spanned metals, composites, ceramics, plastics, lubricants, adhesives, and coatings.

National Composite Center

The products of the National Composite Center (NCC) in Kettering, Ohio, offer a viable example of a technology—composite materials—that originated at Wright-Patterson and moved, literally and technically, into the civilian community. The center represents a unique genesis and mode of operation—a success story of the Air Force at Wright-Patterson and communities of several types working together to demonstrate and commercialize a technology pioneered in laboratories on base in the 1960s.

Beginning in the 1960s, the development of aerospace composites, from test-tube samples to aerospace manufacturing processes, was centered at Wright-Patterson. Composites, a matrix reinforced with fibers (not unlike ancient "bricks" made of straw embedded in mud), proved to be lighter, stiffer, and stronger than most known alternative materials, and the Air Force found them useful in many applications, such as radar-evading structures. The high costs of composites, however, limited their use in airframes, engines, and space vehicles. Lower cost through wider use in the commercial marketplace, especially in the automotive industry, would make them more affordable for the Air Force.

In the mid-1990s, an Air Force initiative, led by senior managers and engineers at

National Composite Center in Kettering Business Park, Kettering, Ohio *(Helen Kavanaugh-Jones)*

Wright-Patterson, sought to reduce the cost of manufacturing composite materials. Air Force officials from the base, the major United States aircraft manufacturers, and representatives from USCAR (United States Consortium for Automotive Research)—composed of the country's "big three" auto manufacturers: Chrysler, Ford, and General Motors—met in Detroit to discuss the potential of composites in industry. USCAR had previously begun conducting innovative manufacturing research with fiberglass composites in a program called P4 (Programmable Powdered Preform Process), developed by Owens Corning Company, and Air Force officials saw promise in the work, especially if graphite—a stiffer, stronger material favored in military-use composites—could be substituted for fiberglass. Encouraged by what they saw and heard at the meeting, four officials from the Miami Valley returned to Detroit shortly thereafter: Dr. Vincent J. Russo, director of the Materials Laboratory at Wright-Patterson; Joseph Rowe, director of the University of Dayton Research Institute (UDRI); Jon Miller, vice president of Research, General Motors; and Dr. Gordon Sargent of UDRI. In Detroit, the group saw small-scale demonstrations of composite manufacturing technology for the flat beds of pickup trucks. To scale up the truck piece to its production size, 5 feet by 7 feet, the experimenters needed a "big press"—technically, a 1,000-ton Structural Reaction Injection Molding (SRIM) press. Dr. Russo struck an agreement with the automakers that the Dayton contingent would help find financing for the SRIM press if it would be located in the Dayton area.

Assisted by Richard Mollen, chief executive officer of Huffy Corporation, and Dr. Sargent, the group received financial support from several sources, including the state of Ohio, the Department of Energy, the Department of Commerce, (through the National Institute of Standards and Technology), and the automakers. They realized a $2.5 million capital investment and another $11.5 million for a facility. The center officially opened for business on September 25, 1998, in one of several buildings formerly occupied by the Defense Electronics Supply Center (DESC). (During a Department of Defense downsizing, DESC had moved to Columbus, Ohio, leaving vacant 2,000,000 square feet of real estate, now known collectively as the Kettering Business Park. NCC occupied 200,000 square feet of that space.) Once the SRIM press was installed in Building 1 at the Kettering Business Park, the National Composite Center—then called the National Center for Composite Systems Technology (NCCST)—was truly under way. The SRIM's first demonstration project was fabrication of fiberglass composite truck beds.[99]

By 2003, NCC was an established, national source in promoting, developing, and applying advanced composite technology to non-aerospace markets. Its more than 50 members included national industries, educational institutions, and local, small businesses. President and Chief Executive Officer Lou Luedtke and Vice President of Technical Operations Scott Reeve employed 31 full- and part-time personnel. NCC tailored its manufacturing processes for very specific uses: it operated a pilot production plant that drew companies of diverse types—large automakers to smaller, niche businesses—to refine composite manufacturing methods and develop standards for their industries. The center was not only a resource for pilot runs and production quantities of composite parts for such items as automobiles, bridges, and safety helmets, but also an information center or "clearinghouse" of data on the technology.

NCC also remained a viable company for aerospace research. In 2002, Boeing Phantom Works donated to the center a robotic fiber placement machine, valued at $3.9 million, which could produce large-scale parts up to 14 feet by 43 feet in size. The center planned to use the machine on projects for aerospace, automotive, marine, and other industries. The center also announced plans to create, in the Dayton area, an aerospace preform composite parts company, predicting the availability of up to 250 new aerospace jobs. That company, Vector Composites, Inc., began operations September 1, 2003, in the Kettering Business Park.[100]

Louis A. Luedtke, left, president and chief executive officer of the National Composite Center, Kettering, Ohio, and Dr. Vincent J. Russo, executive director of Aeronautical Systems Center at Wright-Patterson, discuss NCC's P4A machine (in background). *(U.S. Air Force photo by Al Bright)*

U.S. Senator George Voinovich of Ohio; Dr. Vincent J. Russo, executive director of Aeronautical Systems Center at Wright-Patterson; and Louis A. Luedtke, president and chief executive officer of the National Composite Center (NCC), in front of the 1,000-ton molding press at NCC, Kettering, Ohio. *(U.S. Air Force photo by Chase Simon)*

GENTILE AIR FORCE STATION

Gentile Air Force Station, a 165-acre site on Wilmington Pike in Kettering, was the home of the Defense Electronics Supply Center (DESC). The facility first opened in 1944 as a centralized storage facility for the Army Signal Corps and was known as Dayton Air Force Depot. Prior to that time, it served as a commercial flying field for the Johnson Flying Service. On August 14, 1943, the chief signal officer granted approval for construction of a $3 million depot on Wilmington Pike. Ground was broken October 5, 1943, and the project was completed in less than a year. Formal dedication ceremonies were held in October 1944.

In 1945, Signal Corps functions were integrated into the Army Air Forces and the installation became known as the 862nd Army Air Forces Specialized Depot. In 1951, it was renamed Gentile Air Force Depot in honor of World War II flying ace Major Don S. Gentile of Piqua, Ohio. In 1955, under the jurisdiction of the Air Force Logistics Command (AFLC), separate titles were given to the organization and the installation. The organization was designated Dayton Air Force Depot and the installation became Gentile Air Force Station. AFLC transferred the Dayton Air Force Depot to the Department of Defense Supply Agency in 1962. The agency then established the Defense Electronics Supply Center (DESC) at Gentile. The Air Force retained ownership of the air station and leased it to DESC. The 2750th Air Base Wing served as the station's real property manager.

In June 1993, the Commission on Base Realignments and Closures recommended the closure of Gentile Air Force Station as part of the post-Cold War drawdown. Gentile Air Force Station, Ohio, was closed in 1996. The site was then turned over to the city of Kettering, Ohio.

Major Don S. Gentile, a Piqua native, for whom Gentile Air Force Depot was named. He was killed in the crash of a T-33 jet trainer in 1951.

Gentile Air Force Station

The Dayton Air Force Depot at Gentile Air Force Station transferred from Headquarters AFLC to the Department of Defense Supply Agency in January 1962 for the establishment of the Defense Electronics Supply Center.

Butterscotch Jones test drives "Car-Vet," one of the one-tenth-scale composite cars manufactured by NCC for the "Start Your Engines" art program celebrating the 50th anniversary of Kettering, Ohio. Area artists decorated the cars, which were sold at a charity auction. This car was sponsored and purchased by Dr. and Mrs. E. Eugene Snyder of Bigger Road Veterinary Clinic, where Butterscotch often travels for tune-ups. *(Bigger Road Veterinary Clinic)*

At the end of 2002, NCC listed assets of more than $16.3 million, with capital equipment (including the SRIM press, robotic fiber placement machine, and molding equipment) accounting for $4.8 million. NCC's financial stability resulted, at least in part, from its business with Wright-Patterson organizations, as well as the state of Ohio. On June 6, 2003, Ohio Governor Bob Taft ceremonially presented the National Composite Center with a check for $2 million for purchase of equipment for manufacturing very large composite products, promising to add as many as 100 jobs in Ohio. (Another $2 million was awarded simultaneously to the University of Dayton Research Institute for non-related technical projects.) On October 16, 2003, the governor returned to NCC with $1 million for further work in reducing the manufacturing costs of composites. The University of Akron, the University of Alabama at Birmingham, and Fiber Form of Newark, Ohio, were collaborators on the effort, which also promised to create new jobs locally. The monies were part of Governor Taft's Third Frontier Project, which sought to stimulate Ohio's advanced technology projects, their resulting companies, and jobs.[101]

In addition to its manufacturing initiatives (some of which are discussed below), the National Composite Center also advanced the flow of information about composites and their fabrication to its members and prospective members. Beginning in 2000, it hosted an annual Member Day, at which member companies gave technical presentations on new composite technologies and products. The center's technical oversight committee also awarded grants, funded by membership dues, for research programs. In its seven-year history, the center exhibited displays at more than 20 meetings and symposia around the country and its member companies often presented technical papers. The center also maintained on site a fully equipped Education and Conference Center to facilitate education in composites-related fields, and it leased that center for technical meetings and presentations on a variety of subjects, including Town Hall meetings hosted by the commander of Aeronautical Systems Center (ASC) at Wright-Patterson.

Composite Airframe and Automotive Parts

In 1997, Wright Laboratory at Wright-Patterson awarded a two-year, $7.2 million contract to NCC to adapt USCAR's P4 process to aerospace components (P4A) using graphite fiber as reinforcements in composites. Ceremonially, on July 9, Dr. Russo presented an oversized check to Jon Miller, center president and chairman of the board. The "big four" military aircraft manufacturers—Boeing, McDonnell Douglas, Lockheed Martin, and Northrop Grumman—each contributed funds to the project. The University of Dayton Research Institute served as administrative and fiscal agent for the P4A endeavor.[102]

Once P4A preform equipment was installed, a dozen aerospace components were considered for fabrication before the C-17 tail cone and the F-18 access hatch were selected for design, development, and manufacture. Several of each part were fabricated and tested for accuracy and strength, while careful cost-analysis models were developed. Promising savings potential of 40 to 80 percent resulted in continuation of the developments and additional demonstration parts for the C-17 and other aircraft.

Other milestones followed, each lending credibility to the innovative, composite manufacturing methods developed by the center for commercial and aerospace uses. In 1998, NCC demonstrated the P4 preform process for Focal Project II, with a goal of fabricating an automotive component of composites. Focal Project II, an initiative of the Automotive Composites Consortium (a division of USCAR), sought innovative uses of composites on mass produced cars and trucks. The single component was a pickup truck box that replaced 17 individual steel parts and saved weight as well. In 1999, the NCC delivered, for test purposes, nearly 700 pickup truck boxes to partners in the Automotive Composites Consortium. Successful completion of this part of Focal Project II contributed to the technology for composite pickup boxes on the 2000 model of the GM Silverado truck.

Using Focal Project II as an example, NCC demonstrated to the Air Force that fabricating complex aerospace shapes, such as the tail cone and the access hatch, and realizing cost reductions were possible by using the P4's follow-on process, P4A. In 2001, following the successful completion of Focal Project II, the center began work on Focal Project III, which explored graphite fiber used for the support piece between the front and rear doors on a four-door sedan. By 2003, the part remained under development at NCC with involvement by the "big three" automakers.

In 2001, NCC also entered the market for truck springs and suspension links made of composites with its Liteflex™and Litecast™products. It purchased Dayton-based Delphi Automotive Systems' spring business and established Liteflex LLC (Limited Liability Company) composite springs as a replacement for steel leaf springs. With Litecast, NCC debuted composites for suspension links in the medium-to-heavy-duty truck market. These composite products have been used in American- and European-made trucks, giving them springs lighter in weight and more durable then steel springs, with uniquely engineered ride properties. The Liteflex springs were used in the Corvette and the newly introduced Cadillac XLR sports car, as well.[103]

Composite Bridge Decks

Perhaps the most visible contribution of NCC's composite research and manufacturing has been in the development of fiber-reinforced composites for bridge decks. In 2000, NCC launched Phase I of Project 100, a state-backed initiative designed to capitalize on the anticipated growth of these fiber-reinforced composites. The center teamed with Hardcore Composite LLC and WebCore Technologies, Inc., to develop lower-cost composite preforms, resulting in the cost of preforms dropping from $85 per square foot to $75 per square foot. Eight bridge decks installed under this project remained in use in 2003.

The following year, NCC partnered with Wright-Patterson and WebCore to install the Hebble Creek Bridge, the first composite vehicle bridge deck on federal property, complete with substantial performance-monitoring equipment. The bridge was located on the new access road into Huffman Prairie Flying Field on Wright-Patterson. The center also partnered with Martin Marietta Composites (MMC, Inc., a subsidiary of Martin Marietta Materials, Inc.) to launch the new Composites for Infrastructure (C4I) initiative that sought to apply advanced composites to the bridge construction industry. By 2002, NCC and MMC installed in Ohio, under C4I, the largest, vehicular composite bridge deck to date. Located on Fairgrounds Road in Greene County, Ohio, the bridge was officially opened during a July 1, 2002, ceremonial ribbon cutting attended by representatives of NCC and MMC. The fiber-reinforced, polymer composite bridge deck, one of six composite structures installed in 2002, boasted three spans totaling 7,074 square feet. To accomplish

the project, MMC, in December 2001, relocated its bridge assembly operation from Pennsylvania to Ohio, and in the process, created six jobs for the local area.

The Fairgrounds Road bridge deck took less than three days to install using MMC's patented DuraSpan system. DuraSpan, a fiber-reinforced composite material with continuous glass fibers and fabrics in a polymeric matrix, made the deck lightweight and highly resistant to corrosion, yet strong enough to carry heavy loads. Weighing less than one-fifth the weight of a comparable concrete slab, the new bridge deck was predicted to have greater durability with lower fatigue maintenance. Monitoring equipment on the bridge deck continually provided a database of standardized information for local and state transportation engineers.

The Greene County bridge deck installation was the first produced by the NCC-MMC partnership and the tenth such deck under NCC's programs. Martin Marietta successfully completed 15 other deck installations throughout the United States, including bridges in Oregon, California, Idaho, Maryland, New York, Iowa, Illinois, North Carolina, and South Carolina. The other five bridge decks installed in Ohio under the NCC initiative were constructed in Summit, Geauga, Washington, Clinton, and Defiance counties.[104]

ACADEMICS

Academic training at Wright-Patterson extends back more than 75 years and has always involved the community. In late 1917, for example, the Aviation Section of the U.S. Army Signal Corps, facing a severe

Group of college representatives at Wright-Patterson Air Force Base for the Professors' Tour, August 2003. Mr. Chase Simon (third from right) of the Aeronautical Systems Center History Office served as tour guide. *(Diana Cornelisse)*

UNIVERSITY OF DAYTON RESEARCH INSTITUTE

"…initiate and perform research programs which are responsive to the needs of the community and which contribute to the broad educational goals of the University."

In this simply stated goal of the University of Dayton Research Institute (UDRI), "the needs of the community" often included those of the technical community at Wright-Patterson Air Force Base, located only 12 miles from UDRI's parent organization, the University of Dayton (UD). In fact, while the institute grew from the scientific and engineering interests of its faculty, its first government-funded program was a response to Air Force needs at the base. The project, awarded under a $10,000 contract in 1949, sought reduction of raw, aircraft flight-loads data. A UD mathematics professor, Dr. Kenneth Schraut, and his 10 top students undertook the work with slide rules and calculators. That initial effort grew into the multidisciplinary UDRI, a nonprofit organization supported by contracts.

In the early 1950s, the Air Force-sponsored work at UDRI, then called the University of Dayton Research Center, included a contract from the Structures Branch of the Aircraft Laboratory at Wright-Patterson to determine the effects of a nuclear detonation on aircraft materials and structures. Beginning in 1951, the center conducted eight weapons-effects programs, which became known at the institute as Project Delta. The work involved data reduction, design and construction of experiments, on-site test monitoring during nuclear detonations above the atmosphere, and post-test analyses. The Air Force B-36, F-80, and F-86 aircraft were part of the projects code-named by the government: Greenhouse (1950), Tumbler-Snapper (1952), Upshot-Knothole (1953), Castle (1954), Teapot (1955), Redwing (1956), Plumbob (1957), and Hardtack (1958).

On September 1, 1956, two years before the nuclear-effects programs concluded, Dr. John Westerheide, a research engineer at UD, became the institute's first director. Dr. Westerheide created a more formal structure for the institute, which had approximately 20 projects under contract. Under Dr. Westerheide's management, UDRI grew from a support function to a research organization with faculty involvement.

When the U.S. government ended atmospheric testing of nuclear weapons in 1958, the institute focused on scientific and technical areas in which it already had expertise—particularly data processing, electrical engineering, and psychology—and placed emphasis on its strength in interdisciplinary cooperation. Its teaching and research staff grew, thanks to "dual appointments": qualified researchers received the same academic status as faculty members. By 1970, UD rated among the top 100 schools in the country based on UDRI's level of federal funding. In the early 1970s, the institute received its first contract for off-site work: materials research at Wright-Patterson. Its first contract to set up a remote research location—laser-window research at Kirtland Air Force Base, New Mexico—also was Air Force-funded. Between 1969 and 1977, the institute's professional staff entered 16 new research areas, including aircraft transparencies, holography, solar heating and cooling, technological forecasting, and wind-shear analysis. Its reputation in aircraft propulsion was enhanced, in 1979, when jet-engine inventor Dr. Hans von Ohain joined UDRI as a senior scientist after his retirement as chief scientist of the Aero Propulsion Laboratory at Wright-Patterson. Until his retirement from UDRI 14 years later, Dr. von Ohain focused not only on unique aerothermodynamic machines for government and commercial uses, but also on the students in his engineering classes who would make the scientific advances of the twenty-first century.

By the 1980s, when the U.S. government encouraged the transfer of federally funded technology into the commercial marketplace, UDRI had products primed for market—several of which had been developed under Air Force contract. In fact, its first license agreement signed in 1982 with Unisys Company, Overland Park, Kansas, was for MAGNA, a software program that modeled the high-speed impact of large birds on aircraft canopies. Other aviation and non-aviation-related commercial products followed, including a test to identify hydroperoxides in jet fuels; a fluorocarbon cleaning process for parts in the hot section of jet engines; a "smart dipstick" for analysis of oils; and small, rare-earth "super magnets" for a variety of applications, including high-performance motors and supercomputers. By fiscal year 1988, UDRI had been awarded 50 patents, entered three license agreements, and tallied $429,000 in royalties. By the mid-1990s, UDRI submitted an average of five patent applications on 20 to 30 invention disclosures annually, indicating a significant diversity from its earlier, heavily federally funded work. In the spring of 2003, the Air Force Research Laboratory at Wright-Patterson awarded the institute its largest multi-year contract ever: $31.5 million for development of more efficient and less expensive fuels for aerospace systems.

In 2003, UDRI specialized in aerospace mechanics, electrical and computer engineering, structural integrity, composite materials engineering, and experimental and applied mechanics. Its 160,000 square feet of research facilities and offices filled the Kettering Laboratories building, the Caldwell Street Center, and the Shroyer Park Center, in addition to sections of buildings on campus. Looking into the twenty-first century, the institute planned continued growth—thanks to the multi-faceted talents of its employees, the power base of its success. Its 400-plus employees, of which 157 worked at Wright-Patterson, worked on more than 1,000 projects annually and its yearly revenues in 2003 topped $50 million under the leadership of its current director, Dr. Mickey McCabe. UDRI's customers numbered more than 200 government agencies and 800 industrial businesses. In 2002, UD, along with the research institute, was the biggest Department of Defense-sponsored research university in the state of Ohio. UDRI's presence in the Dayton-area community remained solid; its support to the U.S. Air Force, including Wright-Patterson, encompassed 75 percent of its workload; and its commitment to federally funded research remained foremost in its mission.

Sources: UDRI, *The University of Dayton Research Institute Sponsored Research: 40 Years of Progress* (1998); "UD researchers get $31.5M contract," *Dayton Daily News*, May 12, 2003, p D1; "Research Institute Breaks Record," *University of Dayton Quarterly*, Autumn 2003, p 3.

UDRI researchers in front of the B-36 they instrumented in 1954 for Operation Castle at the Pacific Proving Grounds—part of a nuclear weapons-effects research project funded, in part, by the U.S. Air Force. Left to right (standing): John Moreau, Dwayne Page, Jerry Busch, Fred Pestian, Ed Schlei, Chuck Hutchins, and John Feeney. Left to right (seated): Ron Gabriel, Bob (C.R.) Andrews, Bill Hovey, Herb Mildrum, John Westerheide, Gordon Mills, and Bernie Mahle. *(University of Dayton Research Institute)*

UDRI began research in 3-D virtual environment systems for pilots in the early 1990s, much of it for the U.S. Air Force at Wright-Patterson. *(University of Dayton Research Institute)*

Sam Liu, PhD, manager of UDRI's magnetics laboratory, holds a sample of a hot-pressed/hot-deformed nanocomposite magnet, which rests atop a larger magnet. Lighter and tougher with much more magnetic energy than traditional magnets, nanocomposite magnets are just one area of research explored by scientists at UD's Wright Brothers Institute Endowed Chair in Nanomaterials, announced in fall 2003. *(University of Dayton Research Institute)*

shortage of aviation mechanics, began two training programs: the first with local, private industry and the second on military installations. In the former, the Signal Corps asked local aircraft and engine factories and garages to train groups of 25 soldier-students in the intricacies of their machines. Twenty local companies responded; it was a co-op program in reverse. The second effort saw schools opened at five locations, including Wilbur Wright Field, where an aviation mechanics' class opened on December 17, 1917, with Major W. R. Weaver in command. Before it closed on April 7, 1918, it had trained 1,181 graduates (see Chapter 2: Military Aviation Comes to Dayton).

The flow of students (military and civilian) for further training and degrees—off base into schools near and far and onto the base from other installations—continued over the years and remained vital into the twenty-first century. Southwestern Ohio offered academic opportunities at a variety of colleges and universities, including the Air Force's premier institution for advanced degrees, the Air Force Institute of Technology at Wright-Patterson (see below). The base pursued several initiatives to encourage young people to consider engineering and science degrees. For example, the base's Educational Outreach Program reached out to students in the community from kindergarten through twelfth grades (see below). To educate college advisors on the opportunities available in the Air Force, the Civilian Personnel Division resurrected the College Professors' Tours, originally started in the 1950s. Officials from top-ranking university engineering departments witnessed first-hand the research, development, and systems acquisitions programs available at the base. The fully funded tours helped the officials properly advise their undergraduate and graduate-level students of potential employment and developmental opportunities in the Air Force, and particularly at Wright-Patterson. In addition to representatives from several Ohio universities and colleges, engineering professors came from as far away as Texas and Puerto Rico, illustrating the base's efforts to recruit from Hispanic and other minority populations.

Another initiative pursued by the Air Force Research Laboratory's Materials and Manufacturing Directorate included partnerships with institutions of higher learning. The partnerships provided students and faculty with increased access to the directorate's state-of-the-art instruments, especially computational tools, for research projects, while giving directorate personnel the opportunity to advise and shape future career paths for the students, and encourage possible new laboratory employees. As of November 2003, the directorate had entered into five mutually beneficial partnerships with Fisk University, Tennessee; Tuskegee University, Alabama; and the University of Dayton, Wilberforce University, and Wright State University, all in Ohio. Additional partnerships were planned for the future.

Wright-Patterson also developed programs to assist high-school students financially with their college educations. For example, since 1985, the dependents of both military and civilian employees at Wright-Patterson have had available to them the Wright-Patterson Educational Fund. The fund was begun as a private, not-for-profit scholarship by retired U.S. Air Force Colonel Nathaniel R. Rosengarten and his wife, Irma, to enable the children—and grandchildren—of Wright-Patterson's employees to pursue higher education. Initially, the Rosengartens raised monies and managed the fund out of their home but, by 2000, the program had grown considerably and responsibility for the program was turned over to the Western Ohio Senior Executive Association, with administrative support by the Educational Outreach office at Wright-Patterson. As of 2002, the program had distributed 618 scholarships, 154 no-interest loans, and 267 loan renewals totaling $536,500, rewarding the excellence of local students.

Air Force Institute of Technology

The Air Force Institute of Technology (AFIT) traces its roots to the early days of powered flight, when the progress of military aviation depended on special education in a variety of new technical disciplines. In 1919, the Air Service School of Application was established at McCook Field in Dayton. When the Air Force became a separate service in 1947, the institute—then located at Wright-Patterson—acquired its current name, and by 1955, Ohio State University conducted the first courses on a contract basis (see Chapter 4: Fulcrum of Base Support). During an 80-year span, more than 266,000 Department of Defense (DOD) personnel,

including 30 United States astronauts, attended AFIT programs, illustrating that its presence extended beyond the boundaries of Wright-Patterson and even the Air Force.[105]

While AFIT historically existed for the graduate-level education of military personnel, the school also interacted with the local community and educational institutions in the greater Miami Valley. In 1967, AFIT became a member of the Dayton-Miami Valley Consortium, which later changed its name to the Southwestern Ohio Council for Higher Education (SOCHE), an association of colleges, universities, and industrial organizations in the Dayton area united to promote educational advancement. The Civilian Institution Program, AFIT's most visible program on other academic campuses around the country, managed the graduate degrees of 2,300 Air Force personnel (military and civilian) in non-military universities, research centers, hospitals, and industries. AFIT also managed its Continuing Education program for 3,600 military personnel annually, with students attending schools throughout the United States.[106]

AFIT's partnership with the Dayton Area Graduate Studies Institute (DAGSI) represented its most prominent tie to the local community. Through DAGSI, AFIT's Graduate School of Engineering and Management partnered with Wright State University and the University of Dayton. This agreement, started in 1995, offered an exceptional opportunity for cross-school sharing, allowing non-DOD students to attend AFIT classes and AFIT students to take courses at its partner schools. According to DAGSI, more than 100 AFIT students took advantage of the opportunity, and just as many non-DOD students did the same. Beyond that, 184 non-DOD students received full scholarships from DAGSI to attend AFIT full-time.[107]

The Graduate School of Engineering and Management also focused on undergraduates in a summer engineering program. Beginning in 1997, students hired on through the Student Temporary Employment Program—a process conducted by the civilian personnel office—and were matched with AFIT professors to conduct research or work in the laboratories. The program opened to all United States citizens pursuing appropriate higher education. While Reserve Officer Training Corps (ROTC) members were the primary targets, the mix

TOM D. CROUCH, PH.D.

Tom D. Crouch, a preeminent aviation historian and senior curator of aeronautics at the National Air and Space Museum is—like the Wright brothers he has chronicled—a product of the Miami Valley. Raised in Medway, Ohio, Crouch says that his father Harold Crouch, a now-retired propeller test engineer at Wright-Patterson, took him early-on to the United States Air Force Museum at Wright-Patterson, where he learned about airplanes. Later, he would ride his bike to the museum to stroll among the exhibits.

Today a prolific writer of aerospace-related persons and events, Crouch began his career—while still himself a student—in South Charleston, Ohio, teaching social studies to grades 7-12. He earned three history degrees at Ohio institutions: a bachelor of arts from Ohio University (1966), a master of arts (1968) from Miami (Ohio) University (both with honors) and a doctoral degree from Ohio State University (1976). In 2001, Wright State University conferred on him the honorary degree of Doctor of Humane Letters.

In 1969, Crouch joined the Ohio Historical Society, first as director in its education department and then as director of the Ohio American Revolution Bicentennial Advisory Commission. An employee of the Smithsonian Institution since 1974, he has served both the National Air and Space Museum (NASM) and National Museum of American History in a variety of curatorial and administrative posts.

Crouch's published writings are voluminous. Among his 16 books is a 1989 seminal work on the Wrights entitled *The Bishop's Boys: A Life of Wilbur and Orville Wright*—the title referring to the Wrights' father who was a bishop in the Church of the United Brethren in Christ—and *Wings: A History of Aviation from Kites to the Space Age* published in the fall of 2003. The subject matter of his books varies from Ohio aerospace events and personalities, American rocketry, ballooning, the Bleriot XI aircraft, Apollo moon missions, NASA history, and Charles A. Lindbergh. Crouch's magazine and journal articles, even more extensive and equally varied in aeronautical subjects, number some 80 narratives in American, French, and German publications.

A native son who remembers his roots, Crouch also lectures frequently at professional conferences, including the 1996 and 2003 Wright Brothers symposia hosted by Wright State University. He has also assisted in the development of museum exhibitions in the Neil Armstrong Museum, Wapakoneta, Ohio; the Ohio Historical Center, Columbus, Ohio; the NASM; and the National Museum of American History. In fall 2000, Crouch was appointed by President Bill Clinton to chair the First Flight Centennial Federal Advisory Board, an organization created to advise the Centennial of Flight Commission on activities planned to commemorate the 100th anniversary of powered flight.

Crouch has received numerous honors including, in 2001, the Trailblazer Award presented annually by the Dayton-based Aviation Trail, Inc., which recognizes individuals and organizations for contributions to aviation in the Miami Valley. His history book prizes are from both the American Institute of Aeronautics and Astronautics and the Aviation/Space Writers Association. For *The Bishop's Boys*, he was awarded the 1989 Christopher Award literary prize recognizing "significant artistic achievement in support of the highest values of the human spirit" and, in 2002, he earned the Smithsonian Distinguished Lecturer Award.

Tom D. Crouch

(Air Force Institute of Technology)

of students usually split 50-50; half were ROTC students and half were non-ROTC, regular AFIT students.[108]

AFIT's Graduate School of Engineering and Management was not alone in its community involvement. The School of Systems and Logistics, along with the Defense Acquisition University Midwest Region, made its presence equally prominent outside the Wright-Patterson gates. In fall 2003, the school moved from the AFIT campus on Wright-Patterson to a three-story building at the Miami Valley Research Park in Kettering, Ohio. The new, temporary location for 2,000 students put AFIT in closer contact with local industry leaders during renovation of its on-base building.[109]

While AFIT's presence in the community continually manifested itself in new ways, some of its partnerships with the community existed for many years. In 2003, AFIT's involvement with Lincoln Elementary School in Dayton, for example, began its nineteenth year. While its largest role in the elementary school was, quite naturally, with the science department, AFIT staff members also served as mentors and tutors to some of the young students, and faculty served as volunteer judges at science fairs.[110]

AFIT's involvement did not end in the local community; it also maintained a global presence. For example, the Air Force Office of Scientific Research selected Dr. Yung Kee Yeo, professor of physics at AFIT, for a Window on Asia assignment. The Window programs, designed when the Air Force "identified a need to increase the participation of top quality Air Force scientists and engineers in global foreign research communities," provided great opportunities for collaborative research. Scientists performed research in leading foreign labs and assessed foreign technology programs, strengthening global ties throughout the process.[111]

Advanced Degree Programs Offered on Base by Regional Universities

Two universities in the greater Dayton area initiated academic programs on Wright-Patterson to offer master's degrees for the base's military and civilian personnel. The University of Dayton's School of Engineering offered degrees in either engineering management or management science; the University of Cincinnati's College of Business Administration provided an opportunity to earn a master of business administration (MBA). Both programs advanced the higher education of the base's employees, allowed the Air Force to retain a qualified workforce, and provided for efficient, timesaving delivery of instruction.

The Aeronautical Systems Center's Directorate of Engineering initiated the University of Dayton (UD) program in February 1999, after recognizing a need to make advanced degrees more accessible to its technical workforce. After several cooperative meetings among UD officials and engineering and personnel specialists from Wright-Patterson, a program was created to offer engineers and other technical professionals the analytical tools they needed to plan, design, optimize, and direct complex programs, processes, and systems, and to manage the teams that made them work. The stated goal for the UD engineering management degree was to prepare practicing engineers for the management of engineering activities in industry, government, and academia. The management science degree prepared professionals from a variety of backgrounds for positions requiring competence in the quantitative methodologies of operations research, system analysis, and simulation.

Students in the inaugural academic year (1999-2000) numbered 49 and attended classes in Building 560, Area B, during lunch hours. Thirty-three students enrolled for the 2003-2004 academic year. In four years, the UD master's program on base awarded 52 graduate degrees in engineering management or management science. Five full-time faculty and former professors from the Air Force Institute of Technology, all of whom were published authors holding doctoral degrees in engineering, taught the classes. The Air Force covered the tuition for all classes.[112]

The on-base Mid-Day MBA program with the University of Cincinnati (UC) began the fall quarter 2002, with an initial enrollment of about 30 military and civilian students. Classes met on Wright-Patterson with instruction accomplished both on site and electronically—video teleconferencing, streaming video, and Internet-based instruction. Grants from the Cleveland Foundation and the Cincinnati Bell Foundation financed the school's video teleconferencing capability, which reached worldwide. The Mid-Day MBA program provided a highly structured curriculum for the first five quarters, but allowed for additional electives taken on base, at UC, at UC's Blue Ash facility, or on UC's Clifton Campus.[113]

Co-op Program

Since 1960, Wright-Patterson managed an aggressive, college students' co-op program called the Student Career Experience Program. By 2003, participating institutions numbered more than 55 schools of engineering, business, and two- and four-year colleges and universities. The program aimed to integrate college-level academic study with full-time work experience in cooperating employer

DAYTON CHRISTIAN-JEWISH DIALOGUE

In the early 1970s, a small group of Jews and Christians in Dayton, Ohio, began to meet in private homes in an effort to dispel old myths and stereotypes about each other and to better understand one another's faith. These "living room" dialogues eventually evolved into the Dayton Christian-Jewish Dialogue (DCJD). The founding members of the DCJD were Paul and Shirley Flacks; Eileen Moorman; Dr. Eric Friedland; Dr. Louis Ryterband; Fathers John (Jack) Kelley, Bert Buby, and Phillip Hoelle of the Society of Mary; and Air Force Institute of Technology (AFIT) logistics professor Harold Rubenstein and his wife Sophie. In ensuing years, the group was joined by Wright-Patterson Air Force Base personnel Arthur and Judith Auster, Harry and Eleanor Koenigsberg, and Dieter and Suzie Walk, as well as other members of the community. One of the unique features of the DCJD is the high percentage of lay people who have been its leaders as well as its members.

In November 1973, the first National Workshop on Christian-Jewish Relations was held under the auspices of the Dayton Christian-Jewish Dialogue at Mount St. John's Bergamo retreat center in Dayton. Since that time, Dayton has been acknowledged as the birthplace of the Christian-Jewish dialogue movement, a movement that has contributed greatly to understanding, cooperation, and friendship between Christians and Jews.

Sources: Shirley Flacks, personal communication with Robin Smith, June 26, 2003; Dr. Eugene Fisher, "National Workshop on Christian-Jewish Relations," viewed online October 15, 2003, at http://www.ibiblio.org/bgreek/archives/96-08/1004.html.

AFIT professor Dr. Harold Rubenstein, one of the founding members of the groundbreaking Dayton Christian-Jewish Dialogue

---

agencies. This arrangement allowed students to enhance their academic knowledge, develop personally, receive financial assistance, and prepare professionally.

Students in the program—166 in 2003—came from a wide range of geographical locations in the eastern United States, including Florida, New York, Minnesota, as well as the Miami Valley. Schools ranged from small institutions, such as Miami-Jacobs College (Dayton) and Evangel University (Springfield, Missouri), to educational giants, like Michigan, Ohio State and Purdue universities.

Students enrolled in the program were required to take at least a half-time course load in an accredited school, with employment arrangements accommodating either part-time (16-32 hours a week) or full-time (40 hours per week) schedules. Federal co-op workers enjoyed benefits similar to those of a career employee, including annual and sick leave, enrollment in the Federal Employees Retirement System, and eligibility for health benefits and regular and optional life-insurance coverage. Following completion of the course requirements, students were converted into the federal workforce.[114]

Educational Outreach Program

In 1999, a partnership between ASC and AFRL created the Wright-Patterson Air Force Base Educational Outreach Program. By targeting students in the primary and secondary grades in the surrounding communities, the program enticed students to pursue future scientific and technical careers at Wright-Patterson. Although Wright-Patterson had *ad hoc* educational programs around the base, the Educational Outreach Program paved the way for activity at a corporate level. Its mission was to facilitate partnerships with local schools, kindergarten through twelfth grade, and increase awareness and excitement in all fields of math and science. The program's goal was to become a science information clearinghouse and to encourage and enhance the study of science throughout the United States. Over four years, the program reached more than 53,000 students through a wide variety of activities and contributed more than $1.4 million in free programs to local schools. Base personnel worked 1,500 volunteer hours each month in support of the local educational programs.[115]

In addition to coordinating base volunteers needed within the schools, the Educational Outreach office developed several programs to connect with local students and educators.

• The Scanning Electron Microscope Educators (SEMEDS) after-school program brought local students to the base to explore the Air Force's laboratories. Students used a $500,000 scanning electron microscope in a real-world laboratory setting under the guidance of practiced researchers. The mobile version of SEMEDS, the Tech Trek Mobile Research Laboratory, provided a reverse field trip by taking a portable microscope to local schools and businesses. Tech Trek consisted of a 40-foot, customized bus emblazoned with paintings of scientific themes on its exterior. Since 2001, the mobile lab reached more than 13,000 students.

• Many base scientists, engineers, and technicians volunteered for the "Wizards of Wright!" (WOW!) program, taking science and math demonstrations directly to the classroom. Each class project, requiring thinking, planning, teamwork, and construction skills, taught students

SINCLAIR COMMUNITY COLLEGE

Sinclair Community College is named after David Ainslie Sinclair, who was born in Scotland on May 27, 1850. Sinclair's family immigrated to Canada and, as a young man, Sinclair became an enthusiastic member of the Young Men's Christian Association (YMCA). In 1874, his local YMCA sent him as a delegate to a YMCA convention in Dayton, and the Dayton YMCA (which at that time operated out of rented rooms over a saloon at 65 North Main Street) offered him a job as secretary general. Sinclair accepted the position and later became director of education for the Dayton YMCA.

Sinclair noticed many healthy young men loitering around the city of Dayton, apparently with nothing to do. When he asked them why they were not working, they replied no one would hire them because they had no job skills. Although the YMCA had moved into new quarters on East Fourth Street, it did not have space to hold vocational classes for jobless men. Sinclair's motto was "Find the need and endeavor to meet it," and he was determined to find a way to meet the vocational needs of the young men he saw wandering the streets.

In 1883, Sinclair Community College founder David A. Sinclair disguised himself as a homeless person to see what it felt to be treated like a "tramp." His experience inspired him to design a vocational curriculum that would prepare young men for employment. (*Sinclair Community College Archives*)

In 1886, E. A. Daniels, the director of the Dayton YMCA, and Edwin Shuey of the United Brethren Publishing House, joined David Sinclair to form the Educational Committee of the Dayton YMCA and to plan a curriculum. The old YMCA quarters on East Fourth Street (now Dave Hall Plaza) was torn down to make room for a new facility spacious enough to hold vocational classes. In 1887, 55 men enrolled in bookkeeping and mechanical drawing courses at the new YMCA and, with each succeeding year, the number of students and courses grew. Sinclair worked hard to raise funds for another new building, which is now the City of Dayton Building on Third and Ludlow streets. Unfortunately, Sinclair died in 1902, six years before the new building opened.

On April 15, 1948, the YMCA College was renamed Sinclair College in honor of David A. Sinclair. With the passage of a 10-year, 1-mil levy in 1966, Sinclair became a public college and, in 1972, it moved from its YMCA home onto a new, seven-building campus. Today, 19 buildings make up the campus of Sinclair Community College, and over 20,000 students are enrolled in nearly 100 different programs of study, such as aviation technology, which prepares students for employment in high-tech aviation careers, including hard-to-fill positions at Wright-Patterson Air Force Base. Sinclair also offers many of its general education courses at Wright-Patterson Air Force Base through a satellite program.

Source: Julie Orenstein, "David A. Sinclair: A Biography," Sinclair Community College Archives, 2003.

Dr. Wade Adams was chief scientist of the Materials and Manufacturing Directorate, Air Force Research Laboratory, when he and Dr. Vincent J. Russo created Wright-Patterson's Educational Outreach Program. Ohio's First Lady Hope Taft called the program "the best in Ohio."

Before his retirement as ASC executive director in August 2003, Educational Outreach cofounder Dr. Vincent J. Russo was made an honorary "Wizard of Wright." One of Dr. Russo's last acts before retiring was to sign an order for the creation of a STARBASE unit at Wright-Patterson. STARBASE (Science and Technology Academies Reinforcing Basic Aviation and Space Exploration) is a Department of Defense program designed to teach math, science, and technology skills to at-risk students.

how to think and act in the worlds of math, science, and engineering. Since its inception in 1995, WOW! volunteers performed an average of 30 demonstrations per month, reaching more than 23,000 students.[116]

• Project MISSE (Materials on the International Space Station Experiments), made possible by AFRL's Materials and Manufacturing Directorate, allowed students to develop piggyback experiments for the International Space Station. Students were challenged to come up with experiments to solve problems involved with long-duration space flight. Selected experiments were launched in August 2001. (Upon returning to Earth, the experiment results will be analyzed by teams of students and experts.)

• In 2001, a grant from the American Society of Mechanical Engineers provided funding to the Educational Outreach office for conducting introductory robotics workshops for teachers using LEGO Mindstorms Robotics Kits. As the project evolved, Wright-Patterson hosted the State of Ohio FIRST (For Inspiration and Recognition of Science and Technology) LEGO League "Championship" in December 2001. The competition attracted 20 teams and more than 300 students/ mentors from across the state. In 2002, the base continued as the lead, coordinating activity in Ohio and providing workshops and guidance for activities in Cleveland, Toledo, Cincinnati, and Columbus. Wright-Patterson, in partnership with Sinclair Community College, also hosted the 2002 State of Ohio Championship, which involved 78 teams and almost 800 students.

• Through partnership with the Miami Valley Interactive Distance Learning Consortium, the Educational Outreach office received $45,000 worth of distance-learning teleconferencing equipment in 2001 to provide educational programming for and interaction with high-school students across Ohio.

• The Educational Outreach office at Wright-Patterson developed an organized support program for local school science fairs, providing volunteer judges and experts to assist schools with their annual programs. In addition, the "Scopes for Students" program allowed high-school students free use of computer-based, digital-storage, oscilloscope systems that provided a variety of scientific measurements.

• The Math Resource Program posted on the Internet basic math formulae and their proofs, conversion tables, and other study aids for high-school and college students.

Tech Trek bus used by the Educational Outreach office at Wright-Patterson to interest students in science and engineering during frequent visits to schools and other venues in the community

Children from Kent Elementary School, Columbus, Ohio, work at a computer aboard the Tech Trek Mobile Research Laboratory, May 2003. The lab, a project of the Educational Outreach office at Wright-Patterson, is outfitted with a portable microscope and support equipment.

The personnel of the Educational Outreach Program, while not alone in their work, remained proud of their connections to many local organizations, including the Miami Valley Interactive Distance Learning Consortium, Wright STEPP, various school improvement task forces, Junior Achievement, the American Society of Mechanical Engineers, the American Institute of Aeronautics and Astronautics, Inventing Flight, the Boonshoft Museum of Discovery, Southwest Ohio Instructional Technology Association/ Greater Miami Valley Educational Technology Council, Armed Forces Communications and Electronics Association, Air Force Association, National Teachers Training Institute, FIRST LEGO League, Ohio Space Grant Consortium, and Miami Valley READS. In 2001, the READS program awarded Wright-Patterson, as an "Outstanding Friend of Literacy in the Miami Valley," a "Grimmy" award for providing more than 2,300 volunteer hours in support of the program.

Wright Scholar Research Assistant Program/Wright STEPP

The Air Force Research Laboratory supported two additional programs complementary to those of the Educational Outreach office: the Wright Scholar Research Assistant Program and Wright STEPP. Both aimed to interest students in science and engineering and depended heavily on the volunteers who generously donated their time and technical expertise.

The Wright Scholar Research Assistant Program, begun in summer 2002, encouraged top-notch high-school juniors

Matt Mullins, an 18-year-old Wright Scholar, spent the summer of 2003 at the Air Force Research Laboratory helping run tests for the pulsed-detonation engine program and acquiring, analyzing, and storing test data for future reference. *(Bill McCuddy)*

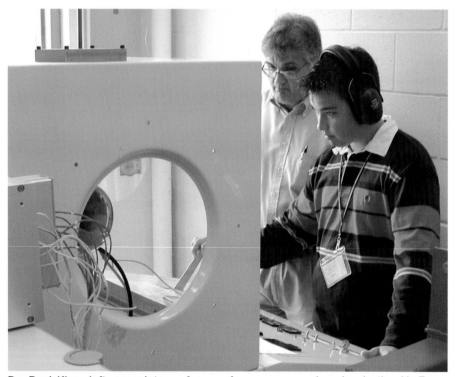

Dr. Paul King, left, associate professor of aerospace engineering in the Air Force Institute of Technology's Graduate School of Engineering and Management, watches 17-year-old, Wright Scholar Casey Holycross as he "runs up" a turbojet engine in a propulsion test cell at the Air Force Research Laboratory, summer 2003. *(Steven Poland)*

and seniors to pursue those science- and engineering-related fields predicted to be beneficial to the future Air Force. The Propulsion Directorate of the Air Force Research Laboratory managed the program, which hired under the Student Temporary Employment Program authority, and provided positions in most AFRL directorates, the Air Force Institute of Technology, and Aeronautical Systems Center's Engineering Directorate. Students in the program worked 30 to 40 hours per week for 10 to 12 weeks under the guidance of a mentor. At the conclusion of their assistantship, the students summarized in writing their summer employment accomplishments, and made short presentations to their peers and mentors. While applicants were solicited from high schools in the greater Dayton area, the Wright Scholar program was opened to any student who met the criteria, with preference given to those most in need of financial assistance.

The Air Force Institute of Technology and the University of Dayton (UD) also participated in the Wright Scholar program. Throughout the summer months, AFIT offered short tutorials in science and engineering, and the students took field trips to UD to participate in the university's Summer Science and Engineering Enrichment Program, specifically structured for Wright Scholar. UD also offered college pre-admission to those juniors who, at the time of application, met all Wright Scholar eligibility requirements and indicated on their application a desire to take advantage of the offer. The scholars also attended weekly lectures by guest speakers on topics ranging from rocket science to preparation of a 30-second "info-commercial" for use when meeting someone the first time. Twenty-nine students participated in the Wright Scholar program's inaugural year, and in 2003, 48 students (13 girls and 35 boys) joined the program.[117]

The Wright STEPP program began in 1988 as an initiative targeted to bring low-income and minority Dayton public school students into careers related to science, technology, engineering, and mathematics (STEM). Wright State University, in conjunction with Dayton Public Schools, sponsored Wright STEPP (Science, Technology, and Engineering Preparatory Program), and Wright-Patterson provided volunteer instructors during the three-week, engineering preparatory program. Each summer the program prepared 160 pre-eighth grade through pre-eleventh grade students from the Dayton public school system for STEM-related careers. The program aimed to increase the overall graduation rate of Dayton students and provide a thorough background for the students to achieve a bachelor's degree in STEM disciplines. In 1999, 26 of the 40 students who participated in Wright STEPP and graduated from high school gained admittance to Wright State University, the largest number ever.

Initially, funding for the program came from Wright State University. Additional benefactors later included Standard Register, General Motors, the U.S. Department of Energy, Mead Data Central, the Ohio Board of Regents, the American Association of Blacks in Energy, and Corning Glass. In 2000, the program received its first federal support from the National Science Foundation: $214,000 for scholarships over a period of two years.

WRIGHT STATE UNIVERSITY

For many years, Daytonians sought to expand higher education opportunities in the Miami Valley and, in 1963, Dayton business and civic leaders Stanley C. Allyn and Robert S. Oelman of NCR took steps to create a new state institution that eventually became Wright State University. Wright-Patterson Air Force Base officials recognized the benefits of having a college nearby and were closely involved in the development of the new campus, which was built on land adjacent to the base. Acting on behalf of the Air Force, General Mark E. Bradley, Jr., commander of Air Force Logistics Command, formally transferred 190 acres of vacant, Wright-Patterson property to the new branch campus in April 1963. This gift represented a significant portion of the 613 acres eventually acquired by the campus. The donation involved two tracts along Kauffman Avenue, including the old Skyway Park housing area and a former section of the Miami Conservancy District, both of which had been declared excess by the base. General Bradley presented title to the lands to Dr. John W. Millett, president of Miami University, and Dr. Novice G. Fawcett, president of Ohio State University, at a luncheon at the Wright-Patterson Air Force Base Officers' Club. Former General Motors administrator Frederick A. White became the campus business manager.

Since the new institution was a commuter campus, its traffic control and master land-use plan required close coordination between Wright-Patterson and campus officials. On September 18, 1964, formal dedication ceremonies were held in Allyn Hall, with representatives from Wright-Patterson and the local community in attendance. The institution opened as a joint branch of Ohio State University (Columbus) and Miami University (Oxford), but on October 1, 1967, Senate Bill 212 recreated it as Wright State University (WSU), named in honor of Wilbur and Orville Wright. Wright-Patterson's relationship with WSU has continued over the years: many base employees have taken WSU courses, and WSU medical students have completed internships at Wright-Patterson's medical center. In addition, the Aeronautical Systems Center History Office at Wright-Patterson has donated many items to WSU's aviation archives.

Sources: Lois E. Walker and Shelby E. Wickam, *From Huffman Prairie to the Moon: The History of Wright-Patterson Air Force Base* (Wright-Patterson Air Force Base, Office of History, 2750th Air Base Wing, 1986); Toni Jeske, Wright State University Libraries Special Collections and Archives, email to Robin Smith, dated June 27, 2003; Bessie Shina, Wright State University Records, email to Robin Smith dated July 1, 2003; Charles W. Ingler, *Founding and Fulfillment: 1964-1984* (Wright State University, Dayton, Ohio, 1987).

The WSU campus encompasses 190 acres of former Air Force property, including the site of Skyway Park at the intersection of Colonel Glenn Highway and Kauffman Avenue.

Wright State University's first building, Allyn Hall, stands alone in the middle of a pasture in 1964. (*Wright State University*)

Ivonette Wright Miller, niece of the Wright brothers, speaks with WSU campus officials John Jeffrey and Fred White in 1973 at the Library Dedication of the Wright Brothers Collection, which is the crown jewel of Wright State University's Special Collections and Archives. (*Wright State University*)

WRIGHT-PATTERSON SUPPORT OF CENTENNIAL OF FLIGHT EVENTS

Besides impacting the economics, technology, and academics of the greater Miami Valley, Wright-Patterson interfaced with the community socially. Many of these activities relied on volunteers for their success and, over the years, Wright-Patterson personnel, military and civilians, gave countless hours to a multitude of civic, fraternal and special-interest groups. Most area organizations that accepted volunteers had people connected in some way to Wright-Patterson. This applied to active-duty military, civilian employees, retirees, and spouses who, by their tireless service, enabled the functioning of hospitals to fraternal clubs to community

parades. Often base units joined in, thereby enriching the event for residents and tourists by their presence, whether the support was substantial (e.g., the Vectren Dayton Air Show) or limited.

The area's celebrations in 2003 to commemorate the first powered, controlled, heavier-than-air flight of Dayton's sons, Wilbur and Orville Wright, offered the best example of the cooperative, community spirit held by the base and its personnel. Wright-Patterson upheld its unique reputation as the bearer of the Wright brothers' name during community celebrations in 2003. Besides supporting events sponsored by the Air Force Centennial of Flight Office in Washington, D.C., base senior officials ensured that several Air Force events held on base focused on the centennial and that many, locally organized centennial activities received strong base participation. These

venues ranged from full-scale air shows to individual endeavors, such as the 930-mile Dayton Cycling Club tour of 15 persons, including three base employees and one base retiree, from Dayton to Kitty Hawk, North Carolina. Event attendees comprised a wide cross-section of government officials, base personnel, their families, friends, and the public.

An effort in 2003 of a singular, enduring type was publication of the new, two-volume history of Wright-Patterson by the Aeronautical Systems Center's History Office led by Diana Cornelisse. With graphic design by Curtis Alley of the National Air and Space Intelligence Center, the book outlined the story of military aircraft developed since 1903 for U.S. aerial forces (*Splendid Vision, Unswerving Purpose*, Volume I) and the continuing interface between the base and its surrounding communities (*Home Field Advantage*, Volume II).

Splendid Vision, Unswerving Purpose:
Developing Air Power for the United States Air Force during the First Century of Powered Flight

Cover illustration, *Splendid Vision, Unswerving Purpose (Curtis Alley)*

Air Power 2003

Air Power 2003, the most ambitious of Wright-Patterson's centennial celebrations, opened to the public on May 10 and 11 near Huffman Prairie Flying Field, where the Wrights learned to fly, perfected their flying skills, and opened the first aviation school. More than 54 aircraft from as far away as California, Mississippi, and Massachusetts—as well as the Predator and Global Hawk unmanned aerial vehicles—lined up along the Area C flightline. In Base Operations, Building 206, the base displayed a variety of static and interactive exhibits, and in the base's oldest hangar, Building 145 (built in 1928), Lieutenant General Richard V. Reynolds, commander of ASC, hosted opening ceremonies. Despite heavy rains on May 10 and high winds on May 11, approximately 15,000 individuals attended the static displays and viewed aerial demonstrations focusing on the missions of the Air Force and base organizations. The Air Force Band of Flight provided musical entertainment and notable speakers gave presentations both days.

On May 10, three former U.S. Air Force officers spoke about aviation legends and the early days of the National Aeronautics and Space Administration (NASA). ABC-TV host David Hartman moderated the discussion with retired Major General Joe Engle, Air Force test pilot and former NASA space shuttle commander, and retired

900-PLUS-MILE BIKE TREK

In August 2003, three employees and one retiree from Wright-Patterson undertook a 13-day, 930-mile bike trek through four states from Dayton to Kitty Hawk, North Carolina, in commemoration of 100 years of powered flight. The employees were Mark Minardi, who rode the route in advance to research it; Lynn Brucker; and Chuck Smith. The retiree was Tom Purkey. (Brucker's husband Roger and Smith's 18-year-old son Casey also were part of the group.) The trip, sponsored by the 700-member Dayton Cycling Club, was an official part of the Inventing Flight celebration as well as an Ohio bicentennial event.

The 15 bikers (11 men and four women) began their trek August 2, in Beavercreek, a suburb of Dayton, and rode through southern Ohio, Kentucky, the Appalachian Mountains of Virginia, and the rolling hills and piedmont flatlands of North Carolina. On August 14, all 15 arrived at Kill Devil Hills, near Kitty Hawk, to be welcomed by members of the local bike club, Wheels of Dare. Along the way, the bikers alternated between camping out and staying in motels, and they had their share of technical "bike" problems (17 flat tires and one wrecked bike).

The North Carolina Highway Patrol, Kitty Hawk police, and Kill Devil Hills police escorted them in parade fashion, then handed them off to officials from the Wright Memorial National Park, who led them around the Wright Brothers National Memorial, while tourists and sightseers watched. Bill Harris, mayor of Kitty Hawk; Sherry Rollason, mayor of Kill Devil Hills; and Dare County Commissioners Renee Cahoon and Warren Judge welcomed them. Despite rain most days, tired muscles, and high temperatures, the group achieved a fitting tribute to Wilbur and Orville Wright, the bicycle makers and repairmen who also invented the airplane.

Sources: "Dayton Cycling Club pedals more than 900 miles to the Outer Banks for the flight Centennial," viewed online August 25, 2003, at http://kittyhawkfreepress.com/Daytonbicycle.htm; "Cyclists weather rain to coast into Kitty Hawk," *Dayton Daily News*, August 15, 2003.

Three Wright-Patterson employees and one retiree were among a group of 15 who rode 930 miles from the Dayton area to Kitty Hawk, North Carolina, in celebration of 100 years of powered flight: retiree Tom Purkey (wearing a white headband), Mark Minardi (wearing wide-brim straw hat), Lynn Brucker (in yellow helmet), and Chuck Smith (standing in the far rear right). *(Joanne Clodfelter)*

During Air Power 2003, crowds waited in line under a B-1B Lancer from Ellsworth Air Force Base, South Dakota, to get a glimpse inside the bomber. *(U.S. Air Force photo by Spencer P. Lane)*

During Air Power 2003, retired Colonel Gail Halvorsen plays with Zach Luttrell and a mini-parachute at the Wright-Patterson Educational Outreach exhibit. Halvorsen, known as the "candy bomber," dropped candy to the children of Berlin on his missions for the Berlin airlift. *(U.S. Air Force photo by Spencer P. Lane)*

A van DeGraff generator at the Wright-Patterson Educational Outreach exhibit during Air Power 2003 gives Sarah Munroe's hair a charge from static electricity. *(U.S. Air Force photo by Spencer P. Lane)*

Brigadier General Chuck Yeager, once a test pilot at Wright Field and the first man to break the sound barrier. Retired Colonel Gail Halvorsen, the "candy bomber" who, during flights for the Berlin Airlift, dropped treats by mini-parachutes for children, spoke during the opening ceremonies and manned a booth with the Civil Air Patrol. Halvorsen, who was assigned to Wright-Patterson after his missions over Berlin, also spent hours showing children how he made the small parachutes—then gave them away, with treats attached.

Women in aviation dominated the program on May 11. Betty Stagg Turner, World War II member of the Women Airforce Service Pilots (WASPs), spoke on the same forum as Captain Kim Black, a B-1B bomber pilot who flew combat missions in Afghanistan. Dr. Peggy Chabrian, founder and president of Women in Aviation International, also was one of the speakers.[118]

Music Power 2003

While airplanes and static displays pleased the crowd at Wright-Patterson for Air Power 2003, an Air Force collaboration of a different kind took place in downtown Dayton. May 11 marked a moment in musical history, when the Air Force Band of Flight and the Dayton Philharmonic Orchestra came together in concert to perform a specially commissioned piece in honor of the centennial of flight. It

EYES OF THE HOME SKIES

The Civil Air Patrol (CAP) is the official civilian auxiliary of the United States Air Force. Aviation pioneer Gill Robb Wilson and New York Mayor Fiorello LaGuardia founded the organization on December 1, 1941. The CAP's initial tasks included liaison and reconnaissance flying, but after the United States entered World War II, many CAP planes were transformed into bombers, which attacked German submarines threatening America's eastern shore.

For over 60 years, the Civil Air Patrol's three primary missions have included aerospace education, emergency services, and cadet programs. As part of its education mission, the CAP created aerospace education materials and provided them at no cost to all grade levels nationwide; it also provided training to teachers and speakers to schools and colleges. Within its sphere of emergency services, the CAP performed aerial reconnaissance, provided search-and-rescue operations and disaster-relief support, and made urgent medical deliveries, saving an average of 100 lives per year.

In 2003, the Civil Air Patrol offered a variety of activities and opportunities to young people through its cadet program, including aerospace education, orientation flights, an International Air Cadet Exchange program, military drills, encampments, and scholarships. Locally, CAP squadrons existed in Dayton, Wright-Patterson Air Force Base, Piqua, Xenia, Middletown, and Springfield, all of which come under the jurisdiction of the Great Lakes Region CAP. The national headquarters of the CAP was located at Maxwell Air Force Base.

Sources: Civil Air Patrol website, viewed online October 15, 2003, at http://www.cap.gov/; "Civil Air Patrol Fact Sheet," viewed online October 15, 2003, at http://www.cap.gov/data/FactSheet.pdf; "Our History: Volunteers Serving in Times of Need," viewed online October 15, 2003 at http://www.cap.gov/about/history.html.

During World War II, the Civil Air Patrol discovered 173 German submarines off the eastern coast of the United States. Of the 57 submarines the CAP attacked, 10 were hit and two were sunk. (*Library of Congress*)

Young people between the ages of 12 and 21 learn discipline and team work when they join the Civil Air Patrol. The cadets shown here are participating in the 2003 CAP Cadet Competition at Wright-Patterson Air Force Base. (*U.S. Air Force photo by Brittany Fisher*)

THE WRIGHT ART

Several sculptures and replicas of Wilbur and Orville Wrights' aircraft and memorials to the famous brothers are located on Wright-Patterson Air Force Base and around the greater Miami Valley as artistic evidence that the community cherishes its aviation heritage. Wright-Patterson personnel volunteered hundreds of hours on many of these projects, a mix of public and private art, which speaks well for the community involvement of base personnel. The sculptures range from the grand (e.g., the Wright Memorial) to more humble markers and plaques.

A sculpture of the 1909 Wright Flyer, the first aircraft purchased by the U.S. government, was installed at Wright-Patterson's Gate 1B in the summer of 2003. Its emplacement topped off the redesign of the approach roadways and relocation of the original 1927 gatehouse structures, a project initiated by Jon Ogg, deputy for engineering at Aeronautical Systems Center (ASC). The stainless-steel sculpture was made possible with the assistance of the Wright Memorial Chapter 212 of the Air Force Association (AFA), and was executed by sculptor Larry Godwin and Brown & Bills Architects of Dayton. Under the Wright Flyer, a pavilion called *Field of Dreams* is paved with commemorative bricks purchased by past and present Wright-Patterson employees and contractors. On August 2, 2003, Lieutenant General Richard V. Reynolds, commander of ASC, dedicated the monument to all men and women who have contributed and who continue to contribute to the technological superiority of the Air Force.

West of Gate 1B in the atrium of the United States Air Force Museum stands a 20-foot-high statue of Icarus, a tribute to all Air Force Institute of Technology (AFIT) graduates who have given their lives for their country. Commissioned by the AFIT Foundation and produced by the Eleftherios Karkadoulias Bronze Company of Cincinnati, Ohio, the statue took three years to complete and was donated to the museum. Its dedication took place on May 29, 1998.

A mosaic-tile mural, 60 feet long and 20 feet high, replicating the famous photo of the Wright brothers' first flight at Kitty Hawk, North Carolina, on December 17, 1903, is mounted on a wall overlooking Kettering Hall in the museum. The work was the result of several interests in and around the community: the Rike Family Foundation of Dayton provided funding; Read Viemeister of Yellow Springs, Ohio, conceived and developed the mural; Ivonette Wright Miller, niece of the Wright brothers, donated the photo made from the original glass photographic negative; and computer technicians from Mead Technology Laboratories analyzed the photo and broke it down into varying densities of color represented by shades of gray. The 63,000 separate ceramic tiles were made in Florence, Italy, and shipped to Dayton in 1,134 one-foot squares. The mural, originally installed at the Dayton Convention Center in the 1970s, was placed in storage in 1986. In 1998, the mural was donated to the United States Air Force Museum by the National Aviation Hall of Fame and mounted in its new location.

In downtown Dayton, an abstract, stainless steel, fixed sculpture, *Flyover*, soars over the southern end of Main Street between Fourth and Fifth streets. A tribute to the Wrights and their first flight at Kitty Hawk, the sculpture measures 120 feet in length, the distance of the first successful flight on December 17, 1903, and features a stepped series of white, curved forms resembling wings. *Flyover* was sponsored by Dayton's Public Arts Commission and funded with federal and state grants in the early 1990s. Its creator, David Black from Columbus, Ohio, was chosen from a field of 88 artists.

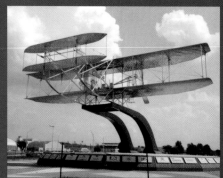

Sculpture of the 1909 Military Flyer, the first aircraft purchased by the U.S. government, graces the new approach to Gate 1B at Wright-Patterson. *(Helen Kavanaugh-Jones)*

Icarus statue in the lobby of the United States Air Force Museum *(Helen Kavanaugh-Jones)*

Flyover on Dayton's Main Street between Fourth and Fifth streets is a 120-foot-long abstract sculpture in tribute to the Wright brothers' first flight. Funded with federal and state grants in the early 1990s, its sculptor was David Black. *(Helen Kavanaugh-Jones)*

On Monument Avenue north of *Flyover*, a more recent, literal tribute to the Wrights is represented by a full-scale, stainless steel interpretation of the Wright's 1905 Flyer in a climbing turn. A bronze figure of Wilbur is aboard the aircraft and Orville watches from the ground. This artistic work was the idea of Chuck Curran, a Montgomery County commissioner, and its sculptor was Larry Godwin of Brundidge, Alabama. Marilyn and Bill Crotty funded the project, and Bob Siebenthaler and his family donated the juniper trees behind the sculpture. Brown & Bills Architects, Inc., designed the plaza and base of the Wright Flyer. The work was unveiled May 7, 2001, and is clearly visible from, and artistically integrated with, RiverScape, a public gathering and recreation area across the street.

Another 1903 Wright Flyer replica hangs in the atrium of the Paul Laurence Dunbar Library at Wright State University. Made of basswood, muslin, and metal, its specifications match those of the original, except that the replica is 55 pounds lighter than the original. Beginning in 1999, using the original drawings for the Wright Flyer, 16 volunteers (three military veterans, one Wright-Patterson employee, and 12 area residents) under the direction of Howard F. DuFour logged 4,000 hours of labor in its construction. The National Composite Center donated workspace at the Kettering Business Park, and the Wright State University Foundation provided funding. Propellers were donated by Hartzell Propeller, Inc., which has been making propellers in Piqua, Ohio, since 1917. The group completed the project in September 2001, after which they also constructed a replica of a 1902 Wright glider, on display in the lobby of the National Composite Center. The group also plans to construct identical Flyers for Wright State University's Student Union and Carillon Historical Park in Kettering, Ohio.

Back in Dayton, in the Wright-Dunbar neighborhood, a bas-relief executed in brick depicts Wilbur and Orville Wright and the poet Paul Laurence Dunbar. The work graces the Wright-Dunbar Plaza connecting the Aviation Trail Visitor Center, the Hoover Block, and the Wright Cycle Company on Williams Street off West Third Street. Jack Curran of Bella Vista, Arkansas, executed the 8-by-10-foot curved sculpture, focal point of the public plaza. Two historic markers beneath the bas-relief describe the work of the Wrights, and a paving pattern of concentric circles radiates from the wall, highlighting the artwork and entrance.

Across the Great Miami River from the Wright-Dunbar Village, *Pathway*, a 9,000-pound, polished steel sculpture towers 70 feet high on the lawn of the Dayton Art Institute. Its sculptor, John Safer, donated the work to the institute as a permanent installation. The work, valued at $3 million, is dedicated to the spirit of Orville and Wilbur Wright and suggests man's pathway into space. The institute's celebration of the 100th anniversary of powered flight also included a special exhibit of photographs and lithographs from the first two decades of flight; a showing, entitled "SOARING," of 45 contemporary sculptures, also by Safer, that evoked images of flight in the mind's eye; and artwork by children in the area who drew their interpretations of "flight."

Sources: "Memo: Columbus sculptor has Wright design," *Dayton Daily News*, October 10, 1994, p B1; Fact Sheet, Brown & Bills Architects, "Addendum to Award Submittal," n.d.; Wright State University website, viewed online April 2003, at http://ww.libraries.wright.edu/special/1902/; "1903 Wright Flyer Project at WSU Special Collections & Archives"; Fact Sheet: Wright-Dunbar Plaza, Brown & Bills Architects, n.d.; "Soaring: The Sculpture of John Safer," *The Dayton Art Institute Member Quarterly*, July 2003, p 9; *The Dayton Public Art Catalogue* (City of Dayton Public Arts Commission, 2001); Karen Pittman, personal communication with Helen Kavanaugh-Jones.

Sculpture of the 1905 Wright Flyer III with bronze figures of Wilbur and Orville Wright on Monument Avenue, Dayton *(Helen Kavanaugh-Jones)*

Howard DuFour, at the National Composite Center, spearheaded a group of volunteers, including Wright-Patterson personnel, in construction of the Wright replica that hangs in the atrium of the Paul Laurence Dunbar Library, Wright State University. *(Chase Simon)*

Replica of the Wright brothers' 1903 Flyer in the atrium of the Paul Laurence Dunbar Library, Wright State University. *(Helen Kavanaugh-Jones)*

Outdoor *Pathway* sculpture at the Dayton Art Institute *(Helen Kavanaugh-Jones)*

represented the first time the two organizations had played together and the first time that an Air Force band had joined an orchestra in equal partnership to perform.

Lieutenant Colonel Alan Sierichs, band commander, led the combined orchestra, and General Lester Lyles, commander of Air Force Material Command, provided a special introduction. Neal Gittleman, Dayton Philharmonic conductor, dubbed the concert Music Power 2003 and noted that the "…composition is proof how music can move us figuratively as flight moves us literally."

Retired Air Force Master Sergeant Ronald Foster, known to both conductors, took more than a year to complete the commissioned composition, titled "On the Wings of Angels." The combined band and orchestra rehearsed only twice before the concert, and during the second rehearsal, Foster fine-tuned his composition and some new percussion instruments—aircraft parts such as brake drums and chimes made from bleed tubing salvaged from aircraft wings. Two propellers were mounted and played, "…not only bringing the piece an interesting sound, but a wonderful visual effect."[119] A capacity audience of 2,300 persons attended the concert held in the Schuster Performing Arts Center in downtown Dayton.

In addition to its performance with the Dayton Philharmonic Orchestra, the Air Force Band of Flight supported other centennial-related events throughout 2003, including the Vectren Dayton Air Show and premiers of five special exhibits by aviation artists at the United State Air Force Museum. Ken Miller, a retired Air Force

Academy band member who had written music for the academy band for 20 years, wrote the music for the museum exhibit debuts. Other composers commissioned for centennial pieces included Julie Giroux, a Hollywood composer and daughter of an Air Force member; Lisa DeSpain, whose compositions were a concert fanfare and another entitled *The Bicycle Shop*; and Ed Hureau, also a retired Air Force band member. The Band of Flight's own composer, Master Sergeant Al Wittig, wrote music scheduled for performances at more than 400 events throughout 2003.[120]

Events at the United States Air Force Museum

As the world's largest and oldest military aviation museum and the national museum for the Air Force, the United States Air Force Museum at Wright-Patterson played a prominent role in celebrating the 100th anniversary of powered flight and served as the location for many events. Its "headliner" event was a 4th of July visit by President George W. Bush, who addressed a huge crowd, many of whose members had stood in long lines for tickets to see and hear their commander-in-chief.

The museum's "Centennial Celebration of Aviation Art" was a five-part, yearlong display of more than 250 original paintings by the world's most renowned aviation artists. Each group of paintings debuted at a free public reception. The five exhibitions, on view sequentially throughout the year, included: "A Century

of Flight," "Air Power," "Aviation Art Worldwide," "Those Magnificent Flyers," and "Fly Me to the Future."

On May 22, the museum hosted the first-day issue ceremony for "First Flight," a 37-cent postage stamp commemorating the Wright brothers' first flight at Kitty Hawk, North Carolina. Retired U.S. Air Force Major General Charles D. Metcalf, director of the museum, welcomed the crowd of several hundred museum visitors, stamp collectors, and the general public in the Modern Flight Gallery. Greg Gamble of the Cincinnati District of the U.S. Postal Service served as master of ceremonies, and Amanda Wright Lane, great grandniece of Wilbur and Orville Wright, spoke about the

George W. Bush, president of the United States, speaks to a huge audience at the United States Air Force Museum on July 4, 2003.

United States Air Force Museum, showing its new third building, 2003 *(James Ferguson, United States Air Force Museum)*

Debbie Dorst of Selma, Indiana, was one of many volunteers who re-enacted military life during the United States Air Force Museum's 2003 Dawn Patrol Rendezvous World War I Fly-In. Here, she explains a U.S. Army Nurse Corps uniform to Barbara Bauer of Geneseo, Illinois. *(U.S. Air Force photo by Spencer P. Lane)*

In July 2003, despite bad weather, some hot air balloons were inflated during the United States Air Force Museum's Balloon Celebration, a joint effort with Inventing Flight and RE/MAX. *(United States Air Force Museum)*

On May 22, 2003, the United States Air Force Museum, the U.S. Postal Service, and Inventing Flight hosted the unveiling of the Postal Service's 37-cent, "First Flight" stamp. Here, Louis Reif of Grove City, Ohio, watches as Rita Schemmel, postmistress of Kettlersville, Ohio, cancels his postcard with the Wright brothers' stamp. *(U.S. Air Force photo by Spencer P. Lane)*

NATIONAL AVIATION HALL OF FAME

For more than 40 years, the National Aviation Hall of Fame (NAHF) has preserved and promoted the legacies of America's outstanding air and space pioneers. Founded in Dayton in 1962 and chartered by U.S. Congress in 1964, the NAHF is a nonprofit organization supported primarily through membership dues and contributions, with on-site support from volunteers, including retired Air Force members. While some aviation-oriented organizations emphasize aircraft, events, or technology, the Hall of Fame focuses strictly on individuals.

Since its founding, the organization has hosted an annual formal banquet and induction ceremony enshrining (by 2003) 178 men and women who achieved prominence in aerospace endeavors. In the past, this event, called the "Oscars of Aviation," brought as many as 2,000 attendees to the Miami Valley. The Hall of Fame includes such notables as Orville and Wilbur Wright, Charles Lindbergh, Amelia Earhart, Jimmy Doolittle, Chuck Yeager, John Glenn, Neil Armstrong, and more than 70 members affiliated with the Air Force, Army Air Corps, and Army Signal Corps.

The NAHF's three-pronged mission of education, inspiration, and motivation dominate its Learning Center, located in a building adjacent to the United States Air Force Museum. The center is part of, and integrated into, six distinct exhibit galleries: the Early Years, World War I, The Golden Age, World War II, the Jet Age, and Into Space, which trace the history of aviation. A variety of interactive displays also offer hands-on introduction to air and space activities, and in the Harry B. Combs Resource Laboratory—a branch of the Harry B. Combs Research Center—visitors can access the NAHF's archives of enshrinee-related material.

In July 2003, the NAHF marked the centennial celebration of powered flight with a salute to all 178 enshrinees and a "reunion" attended by nearly two dozen past enshrinees. During the event, at which actor and pilot Harrison Ford served as master of ceremonies, Bob Taft, governor of Ohio, received the 2003 Milton Caniff "Spirit of Flight Award" in recognition of the state's contributions to the advancement of aviation over the past 100 years.

Source: National Aviation Hall of Fame website, viewed online May 2003, at www.nationalaviation.org.

Even a very young visitor to the National Aviation Hall of Fame can "try on" an astronaut's space suit. *(National Aviation Hall of Fame)*

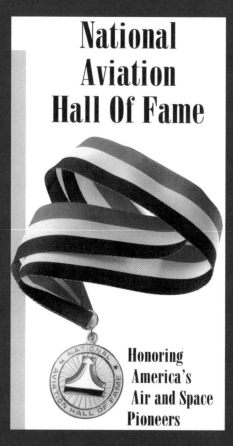

National Aviation Hall Of Fame

Honoring America's Air and Space Pioneers

brothers' use of the mail to seek information about early flight endeavors and correspond with one another, their family, and friends. The first-day issue ceremony was co-hosted by Inventing Flight, the Dayton-based organization that spearheaded the area's 100th anniversary celebration; the museum; and the U.S. Postal Service.[121]

The museum again partnered with Inventing Flight and RE/MAX in early July for a planned, hot air balloon celebration. Poor weather cancelled balloon glows, a gas balloon race to Kitty Hawk, North Carolina, and an outdoor concert by the U.S. Air Force Band of Flight. Less than a week later, however, the museum and Inventing Flight brought together a record number of blimps in one location. The Great Blimp Meet, held at the museum, attracted four participants: Goodyear, Met Life, Saturn Ion, and Fujifilm.

Finally, on September 12-14, the museum and the Great War Aeroplanes Association co-hosted the Dawn Patrol Rendezvous World War I Fly-In with flying aircraft, radio models, re-enactors, an auto show, and a collectors' show. James Dietz, whose paintings were on display as part of the museum's centennial art show, made a special appearance.[122]

Celebration Central at Deeds Point

Celebration Central, the main venue for the greater Dayton community's celebration held July 4-20, was located on a plot of land called Deeds Point at the convergence of the Miami and Mad rivers north of downtown Dayton. The event's organizer, Inventing Flight, staged a variety of performances on a two-acre area named FlightScape Plaza, the entrance to which was called Deeds Legacy Plaza and was accessed by a newly constructed pedestrian bridge over the Miami River. Plaques commemorating people, events, and inventions that advanced man's pursuit of powered flight throughout the ages encircled Deeds Legacy Plaza. The entrance to the plaza was fronted with permanent, full-size sculptures of Orville and Wilbur Wright by Mark Allen Henn, a former Dayton resident. The figures captured the Wrights' elegant discussion of the phenomena of aeronautical wing warping. The statue of Orville held a bicycle inner tube box which, when twisted, aided the brothers' understanding of the concept of flexing an aircraft's wings to accomplish turns.

Opening ceremonies at Celebration Central featured two guest speakers from Wright-Patterson: Brigadier General Rusty Moen, commander of the 445th Airlift Wing, discussed the unit's mission and support to operations Enduring Freedom and Iraqi Freedom; and Captain Chad Clementz, program manager for the Global Hawk unmanned aerial vehicle, talked about his experiences while deployed to support Global Hawk.

Four temporary pavilions housed exhibits, including three Air Force-sponsored displays: the Air Force Centennial of Flight display from Washington, D.C., focused on the activities of the nine Air Force commands; an exhibit sponsored by Air Force Materiel Command (AFMC) at Wright-Patterson; and an Educational Outreach booth hosted by Aeronautical Systems Center and the Air Force Research Laboratory. The latter included "Wizards of Wright," an educational display targeted at the younger crowd. Personnel from Wright-Patterson tended the displays, answered questions, and showed the Air Force presence to the patrons. Other on-site attractions included a space shuttle mock-up, an outdoor stage, and a youth-activity area. The Air Force Band of Flight also participated during the 17-day celebration.

Full-size bronze sculptures of Wilbur and Orville Wright welcomed visitors to Celebration Central at Deeds Point north of downtown Dayton during the centennial of flight. *(Helen Kavanaugh-Jones)*

THE OHIO BICENTENNIAL

The year 2003 was not only the centennial of powered flight; it also was the Bicentennial of Ohio, which became a state on March 1, 1803. Ohio commemorated its rich history through a variety of projects, including the Barns Project, honoring Ohio's agrarian heritage with the painting of the Ohio Bicentennial logo on at least one barn in each of Ohio's 88 counties; the Bells Project, paying tribute to the state's manufacturing and industrial heritage with the casting and dedication of a bicentennial bell in every county; and the Historical Markers Project.

The Ohio Memory Project celebrated Ohio's history via the Internet. The online scrapbook gave Ohio's archival institutions, historical societies, and museums the opportunity to display in digitized format some of their most treasured photographs and documents. The contribution of the Aeronautical Systems Center History Office at Wright-Patterson Air Force Base to this "Bicentennial Scrapbook" was this photo showing the link between the work of Wright-Patterson Air Force Base civilian engineer William E. Lamar and the creation of the first space shuttle, the *Enterprise*. Officials and engineers of the National Aeronautics and Space Administration signed the photograph.

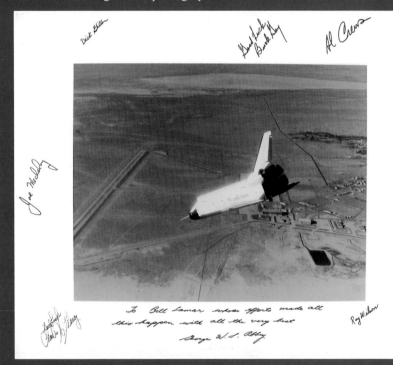

Wright-Patterson Air Force Base commemorated the bicentennial in a variety of ways. The Air Force Band of Flight played musical tributes to Ohio's aerospace heroes, the state's eight United States presidents, its land and rivers, literary artists, famous entertainers, sports figures, and the Rock and Roll Hall of Fame. On March 15, 2003, the United States Air Force Museum hosted a "Family Day," at which children and their parents learned about significant events and persons in Ohio's history, with an emphasis on aviation history. On July 16, 2003, the Wright B Flyer circled over Huffman Prairie Flying Field as part of a special ceremony in which a representative of the Ohio Bicentennial Commission dedicated an Ohio Historical Marker at the historic site. In the fall of 2003, an Ohio Historical Marker also was dedicated at Gate 1B, after Wright Field's original stone gates were moved and put into use once again.

Sources: *Skywrighter* articles: Susan Ferns, "Huffman Prairie Flying Field to be Honored Next Week," July 11, 2003; Deborah S. Csutoras, "Family Day at the Air Force Museum," March 14, 2003; Chris McGee, "Band of Flight in Concert at Museum," March 21, 2003; Technical Sergeant Greg Sell, Band of Flight Director of Operations, personal communication with Robin Smith, July 25, 2003.

Time Flies at Huffman Prairie Flying Field

From July 5-20, 2003, the Office of Environmental Management (EM) at Wright-Patterson supported *Time Flies*, a living history experience staged by Carillon Historical Park, a major program partner of Inventing Flight. The EM office arranged for siting of the event tents and rest facilities and ensured minimal conflicts with other activities in that area. Act Two of the three-act "experience" focused on Huffman Prairie Flying Field—now part of Wright-Patterson—where the Wright brothers learned to fly in 1904 and 1905. *Time Flies* activities included a variety of programs: staged plays, interactive vignettes between actors and the audience, and hands-on activities for children and families. During an interactive play, "A Matter of Balance," visitors "experienced" aircraft pitch, roll, yaw, and wing warping. A replica of the 1911 Wright "B" Flyer was on display, as well as a model based on the 1908 Wright plane, designed and built by the National Composite Center, Kettering, Ohio, and Utah State University. The NASA-Glenn Research Center, Cleveland, Ohio, displayed a prototype fuel-cell plane.[123]

2003 Vectren Dayton Air Show

During the four-day Vectren Dayton Air Show (July 17-20) in 2003, Wright-Patterson provided its usual strong support to the aircraft on static display and in the air, and to the Air Force-related exhibits on the ground. For the special 100th anniversary year of powered flight, the show featured the three premier, aerobatic military flight teams in North America: the U.S. Air Force Thunderbirds (celebrating its 50th anniversary), the U.S. Navy Blue Angels, and the Canadian Forces Snowbirds. (The Thunderbirds' first team show took place on June 8, 1953, at Luke Air Force Base, Arizona.)

More than 250 military personnel, civilian employees, and contractors from the base teamed with Air Show personnel, and units from the Air Force Band of Flight performed all four days. Base volunteers staffed the Air Force exhibits. To service all military aircraft, the Military Flight Service Station, also staffed by Wright-Patterson personnel, operated within the Dayton Automated Flight Service Station at Dayton International Airport from July 14-21. Six military and civilian controllers on

Program for the 2003 Vectren Dayton Air Show

A unit of the Air Force Band of Flight performed at the 2003 Vectren Dayton Air Show. *(Helen Kavanaugh-Jones)*

At the 2003 Vectren Dayton Air Show, Captain Mike Myers of Aeronautical Systems Center's Global Hawk System Program Office talks with a visitor about the unmanned aerial reconnaissance aircraft in the background. *(U.S. Air Force photo by Spencer P. Lane)*

The Air Force Thunderbirds aerial demonstration team, celebrating its 50th anniversary, flew in the 2003 Vectren Dayton Air Show. *(U.S. Air Force photo by Spencer P. Lane)*

site assisted the movement of military aircraft, including at least one of almost every aircraft in the U.S. Air Force inventory, either on static display or part of a flyby, as well as U.S. Army and U.S. Navy aircraft. Specifically, the Military Flight Service Station provided support to more than 30 government-owned aircraft, four single and two-ship demonstration teams, and the three North American military jet teams. Airfield management operators with the service station provided flight-plan input, aircrew briefings, severe weather advisories, and liaison with the Federal Aviation Administration and U.S. Air Force throughout the arrival phase and until the last military aircraft departed. Additionally, base air traffic controllers

coordinated aircraft arrivals and departures, and transient alert crews safely and expeditiously parked all participating aircraft.

General Lester Lyles, commander of Air Force Materiel Command, spoke during the show's opening ceremonies on July 17 and commented, after the F-117 and other military aircraft had flown overhead, "I couldn't help but think about all the significance of this. Look at what the Wright brothers started. Look at where we are today."[124]

During another feature of the opening ceremonies, Major General Paul Nielsen, commander of AFRL at Wright-Patterson, presented Centennial of Flight Awards to five high-school students whose science

projects in aerospace research had been judged top in the nation. Each winner received a cash award, with an overall total of $17,000, and a tour of the base. The $6,000 grand prize went to Ray He of Hempfield High School in Landisville, Pennsylvania, for his paper entitled "Magnetoplasmadynamics: Ionization and Magnetic Field-Improving the Efficiency of Ion Propulsion." Second-place winner was Laura Wong ($4,000); third place went to Stephen Kennedy ($3,000); and fourth- and fifth-place winners Eric Mueller and Lauren Hansen each won $2,000. The one-time contest, part of the centennial-of-flight events, provided a way to involve students in Dayton's anniversary activities. After reviewing 175 10-page papers written by

During the 2003 Vectren Dayton Air Show, a smoke-trail loop in the sky—formed by the Aeroshell Aerobatic Team of four T-6 Texans—is framed by a B-52H from the 5th Bomb Wing, Minot Air Force Base, North Dakota. *(U.S. Air Force photo by Spencer P. Lane)*

The "Hanoi Taxi," a C-141 from the 445th Airlift Wing at Wright-Patterson, dwarfs visitors to the 2003 Vectren Dayton Air Show. Among its feats, the aircraft was the first U.S. military transport to repatriate prisoners of war from Vietnam on February 12, 1973. *(U.S. Air Force photo by Spencer P. Lane)*

Ray He of Landisville, Pennsylvania, grand prize winner in the Centennial of Flight Awards sponsored by the Air Force Research Laboratory, checks out night-vision goggles in the laboratory's Human Effectiveness Directorate. For his award-winning technical paper, He received a tour of the laboratory and a check for $6,000, which was presented as part of the opening ceremonies at the 2003 Vectren Dayton Air Show.

competitors from regional and state science fairs across the United States, AFRL judges selected the winners.[125]

For the annual Air Show parade hosted by the Vandalia-Butler chambers of commerce, the Wright-Patterson Honor Guard and the U.S. Air Force Band of Flight marched, while Lieutenant General Richard V. Reynolds, commander of ASC, served as grand marshal.

International Air and Space Symposium and Exposition

The International Air and Space Symposium and Exposition, held July 14-17, 2003, at the Dayton Convention Center, focused on the next century of flight, paid tribute to the 100 years passed, and was heavily attended by personnel and technology from Wright-Patterson.

A pulsed-detonation engine being developed by the Propulsion Directorate of the Air Force Research Laboratory, for example, was a star attraction among the 70 exhibits. The proof-of-concept demonstrator engine—its main feature resembling a pipe organ's four pipes laid horizontally—utilized a series of controlled explosions of fuel and air in the tubes in a way that increased the intensity of the explosions. The system used any type of general aircraft fuel or even regular, unleaded gas used in cars. The Propulsion Directorate expected data from this demonstrator to forge designs for an inexpensive, simply constructed, and efficient engine for future war-fighting planes that will fly in the Mach 2-4 range, and be more maneuverable and lightweight.

While the engine had only been "bench-tested" at Wright-Patterson, the directorate planned to sponsor its flight tests in a Long-EZ home-built airplane from the Mojave Airport in Mojave, California. Scaled Composites LLC, the contractor installing the engine in the Long-EZ, planned to conduct flight tests before the end of 2003. Dr. Fred Schauer of AFRL and Dr. John Hoke of Innovative Scientific Solutions, Inc., directed the team. Joshua Burger of Wright State University and Scott Cruciger from the University of Dayton, engineering students who worked on the Air Force project, explained its innovative design to exposition onlookers. The engine also was displayed at the Experimental Aircraft Association's (EAA) AirVenture, Oshkosh, Wisconsin, in the summer of 2003.[126]

Dayton Aviation Heritage National Historical Park

Eleven years before the centennial of flight, an event occurred at the national level that was to have a more lasting impact on the Dayton-area community and Wright-Patterson than even the 100th anniversary events. On October 16, 1992, President George H. W. Bush signed legislation creating the Dayton Aviation Heritage National Historical Park to commemorate the legacy of three exceptional men and their work in the Miami Valley: Wilbur and Orville Wright and their friend and acclaimed poet, Paul Laurence Dunbar. Through their invention of powered flight, the Wrights advanced technological history, building and flying the first heavier-than-air, practical, powered flying machine. Paul Laurence Dunbar, an African-American writer, achieved national and international acclaim in a literary world that was then almost exclusively reserved for whites.

Dayton's community leaders and legislators nurtured the park's legal creation for 14 months. Tony Hall of Dayton and David Hobson of Springfield sponsored the

A pulsed-detonation engine, shown with its horizontal "pipes," was a highlight among 70 exhibits at the 2003 International Air and Space Symposium and Exposition. A project of the Propulsion Directorate, Air Force Research Laboratory, the engine is shown here at the EAA AirVenture at Oshkosh, Wisconsin. *(Air Force Research Laboratory, Propulsion Directorate)*

High-school seniors Joe Miller, left, and Matt Mullins, participants in the Wright Scholar program, worked closely in summer 2003 with propulsion scientists and engineers in the turbine-engine and combustion-science branches of Air Force Research Laboratory testing a pulsed-detonation engine. The engine was to be flown on the Long-EZ, a popular, home-built aircraft, positioned between them. *(Bill McCuddy)*

ON THE AVIATION TRAIL

Nearly 50 sites around the Miami Valley are marked with blue and white signs bearing the words *Aviation Trail* and a sketch of the 1903 Wright Flyer. Each identifies an important place in aviation history. The "trail" they comprise leads visitors from the elemental, such as Orville and Wilbur Wright's 1904 flying field, to the sophisticated, such as the country's largest and oldest aviation museum at Wright-Patterson Air Force Base. The markers were researched and posted by a group of dedicated aviation professionals and enthusiasts who incorporated Aviation Trail, Inc., in 1981 to preserve and promote the Dayton/Miami Valley area's unique aviation heritage and stimulate economic development.

The discovery that a building at 22 South Williams Street in Dayton that once housed one of Wilbur and Orville Wright's bicycle shops was still at its original location served as the group's impetus. Consequently, Aviation Trail, Inc. bought the building and, subsequently, another—the Hoover Block at the corner of Williams and West Third streets, where the Wrights had a job-printing business. Restoration work began on the Wright Cycle Company building and, in 1992, the two buildings became part of the new Dayton Aviation Heritage National Historical Park, with headquarters in West Dayton. Without the work of Aviation Trail, a national park may never have been established in Dayton.

Four sites on Wright-Patterson are located on the aviation trail: the base itself, the United States Air Force Museum, Huffman Prairie Flying Field, and the Wright Memorial on Wright Brothers Hill. Other sites around the Miami Valley range from the former Waco aircraft factory in Troy, Ohio, to the Carillon Historical Park in Kettering, Ohio, which displays the original 1905 Wright Flyer III, the world's first practical airplane.

Each year since its founding, Aviation Trail has sponsored three public events. A luncheon and wreath laying at the Wright Memorial on Wright-Patterson commemorates the anniversary of the Wright brothers' first flight on December 17, 1903. At an annual banquet held on April 16, Wilbur Wright's birthday, the trail presents its Trailblazer Award to an individual, group, or business for significant contributions to aviation in the greater Miami Valley. Finally, the trail hosts a luncheon on National Aviation Day, August 19, which is the birthday of both Orville Wright and his sister Katharine.

Recently, Aviation Trail began developing a Parachute Museum on the second floor of its restored building adjacent to the Hoover Block. The museum covers the important story of parachute development at the government's McCook Field, Wright Field, and Wright-Patterson Air Force Base. The museum, which first opened in June 2003, the centennial year of the first powered flight by the Wrights, occupies 1,350 square feet and features artifacts and papers from the unique Dave Gold parachute collection donated to the trail in the mid-1980s by Gold's children and heirs. The trail also has an educational program through which it sponsors students for aviation camp at the Waco Learning Center in Troy.

Source: *Aviation Trail* brochure (Aviation Trail, Inc., n.d.)

Hoover Block on the corner of West Third and Williams streets in the Wright-Dunbar Village. In the corner of the second floor, Orville and Wilbur Wright operated a job-printing business. Next door is the original façade of a structure that now houses Aviation Trail, Inc., a local nonprofit organization dedicated to preserving Dayton's aviation heritage. *(Helen Kavanaugh-Jones)*

A bas-relief graces the Wright-Dunbar Plaza, location of the Wright-Dunbar Interpretive Center and Aviation Trail Museum and Visitor Center building at West Third and Williams streets in Dayton, Ohio. *(Helen Kavanaugh-Jones)*

park in the U.S. House of Representatives, and proponents in the U.S. Senate included John Glenn, former astronaut, and Howard Metzenbaum. U.S. District Court Judge Walter Rice led the park initiative as head of the 2003 Committee planning the area's much-anticipated celebration of 100 years of powered flight. The park's creation also acknowledged that a relatively small patch of land, then called Huffman Prairie—today part of Wright-Patterson Air Force Base— was seminal to Wilbur and Orville Wright's early achievements with heavier-than-air flying machines.

The park served also as a tribute to the grassroots group Aviation Trail, Inc., and a partnership formed among federal, state, and local governments and the private sector, which sought to preserve the Miami Valley's aviation heritage. That partnership grew from cooperation between the city of Dayton and the trail, a small, nonprofit organization formed in 1981, whose early successes included saving and restoring the building that housed the Wrights' 1895-1897 Wright Cycle Company and the Hoover Block where the Wrights operated their job-printing business. Trail members also recognized the need for long-term care of the historic buildings and the possibilities for development in the neighborhood—now called Wright-Dunbar Village—where the Wrights had lived and maintained their printing and bicycle enterprises.

The park legislation specified a scattered-site park at four locations in the Miami Valley: Huffman Prairie Flying Field, the 84-acre pasture used by the Wright brothers in 1904-1905 and where they later trained more than 100 military and civilian aviators; the Wright Cycle Company building at 22 South Williams Street, where the Wrights operated their bicycle sales and repair business; Paul Laurence Dunbar's home located in the same neighborhood as the Wrights' cycle shop; and the 1905 Wright Flyer III at Carillon Historical Park south of Dayton. While each site received care by one or another group, the National Park Service recognized that much work remained to be done before the expected influx of tourists during the 17-day celebration of the Wrights' achievement.

William Gibson, first superintendent of the park, arrived in 1993 as a staff of one. His initial tasks included completing a General Management Plan, hiring additional staff, and beginning development of the park. Lawrence Blake replaced Gibson in 1999 and, by 2003, had acquired a staff of seven rangers and five support personnel, opened two interpretive centers (one in the Hoover Block next to the Wright Cycle Company and the other on Wright Brothers Hill in Area B of Wright-Patterson), and debuted a restoration of the Wrights' job-printing business on the second floor of the Hoover Block. Park personnel quickly

The Wright Cycle Company at 22 South Williams Street, Dayton. The building was "found" still on its original location in west Dayton in the early 1980s and became a centerpiece and impetus for establishment of the Dayton Aviation Heritage National Historical Park. *(Helen Kavanaugh-Jones)*

At Carillon Historical Park in Kettering, Ohio, this 1905 Wright Flyer is one of four sites in the Dayton Aviation Heritage National Historical Park. *(Carillon Historical Park)*

Home of Paul Laurence Dunbar, one of four sites in the Dayton Aviation Heritage National Historical Park *(Helen Kavanaugh-Jones)*

Park Ranger Julia Frasure, National Park Service, interprets Huffman Prairie Flying Field for a group of tourists. The field is one of four sites in the Dayton Aviation Heritage National Historical Park. *(Helen Kavanaugh-Jones)*

Wright Brothers Hill, Wright-Patterson Air Force Base, location of the Huffman Prairie Flying Field Interpretive Center.

became an accepted, visible presence in the Miami Valley.[127]

Many improvements around the Huffman Prairie Flying Field made it easier for visitors to access and experience the environment in which the Wright brothers tested the world's first practical airplane and opened the world's first flying school. As early as the 1970s, Wright-Patterson's Civil Engineering and then Environmental Management offices managed Huffman Prairie Flying Field as a historic site. Then, with help from the National Park Service (NPS), the base civil engineers conducted an official survey to better delineate the field's 84-acre boundaries. Since 1990, the base's Office of Environmental Management has maintained funding and management responsibility for the site, tasking other organizations, such as Civil Engineering, as needed for proper maintenance. Civil Engineering also reopened Gate 16A and installed fences, gates, and directional signs for open, public access to the area, and managed the construction of trails, a parking lot, and improvements to an intersection along Marl Road. Adjacent to the flying field, the 109-acre, tall-grass prairie, the largest prairie remnant in Ohio, was designated an Ohio Natural Landmark in 1986. The base

continued to maintain that landscape in a manner consistent with its past.

The base also maintained, in Area B, the grounds of Wright Brothers Hill, location of the Wright Memorial where the park's Huffman Prairie Flying Field Interpretive Center explained, in written, audio, and visual exhibits, the Wrights' aviation accomplishments on the flying field. The base civil engineers provided building and grounds maintenance, utilities, fire protection, and janitorial services. During development of the interpretive center, they provided design and construction management of the building, and the Environmental Management office and the 2750th Air Base Wing History Office accomplished technical design review for the exhibits. Civil Engineering also resurfaced Memorial Road and provided oversight for expansion of the parking lot.[128]

Rededication of Huffman Prairie Flying Field

On July 16, 2003, General Reynolds, commander of ASC, and Lawrence Blake, superintendent of the Dayton Aviation

Heritage National Historical Park, hosted a public, rededication ceremony of Huffman Prairie Flying Field near the newly restored Marl Road entrance to the field.

Lieutenant General Richard V. Reynolds presents Amanda Wright Lane, great grandniece of the Wright brothers, with a memento of the rededication of Huffman Prairie Flying Field, July 16, 2003. Dayton Mayor Rhine McLin, dressed in turn-of-the century apparel, is in the background. *(Helen Kavanaugh-Jones)*

The Ohio Historical Society (OHS), in conjunction with the Ohio Bicentennial Commission and the American Institute of Aeronautics and Astronautics (AIAA), unveiled two new historical markers and the relocation of a third, the National Historic Landmark Plaque, which had been erected in October 1990, removed during site restoration, and then later placed in a more appropriate location. Dr. John C. Blanton, director for Region III of the AIAA designated the flying field as a Historical Aerospace Site, and Phil Ross, representative of the OHS and the Ohio Bicentennial Commission, unveiled the Ohio plaque.

Opening remarks by General Reynolds were followed by the keynote speaker, Jerry Ross, a retired U.S. Air Force colonel and former shuttle astronaut. Ross, a distant relative of the Wright brothers, also served at Wright-Patterson from 1972-1975. During his NASA career, he launched into space seven times and held the record for the most United States space walks (nine). Ross paid tribute to the 116 men and three women who learned to fly on Huffman Prairie at the Wright School of Aviation and called the site the "springboard" from which Ohioans John Glenn and Neil Armstrong became the first to orbit the earth and walk on the moon, respectively.

Special guests Amanda Wright Lane, great grandniece of Orville and Wilbur Wright, and Rhine McLin, mayor of Dayton, also made remarks at the ceremony, whose start was signaled with a flyover by the Wright "B" Flyer piloted by William Sloan, a retired U.S. Air Force colonel, with retired Lieutenant Colonel Don Stroud, U.S. Air Force, as copilot.[129]

December 17 Memorial Ceremony

On December 17 of the 100th year of powered flight, base officials again cooperated with local aviation and Air Force groups to place a wreath at the Wright Memorial on Wright Brothers Hill. The event, which varied in composition from year to year, commemorated the first powered, controlled, heavier-than-air flight of 120 feet by Orville Wright at Kitty Hawk, North Carolina. The Wright Memorial Chapter of the Air Force Association graciously donated the wreath for the annual memorialization, and Aviation Trail, Inc., traditionally hosted a luncheon that day.[130]

Wright Memorial on Wright Brothers Hill, Wright-Patterson, where traditionally a wreath is placed each December 17th in remembrance of the Wrights' first flight at Kitty Hawk, North Carolina, in 1903

* * *

And so, Dayton and Wright-Patterson Air Force Base closed out the first century of human, heavier-than-air, controlled, and powered flight with a flourish.

In his autobiography, *I Could Never Be So Lucky Again*, Jimmy Doolittle tells about the day at McCook Field when he and another pilot were sent aloft in two planes to see which could dive the fastest. As the planes hurtled earthward, the propeller of Doolittle's plane disintegrated. Fortunately Doolittle was close enough to the ground to land safely. However, propeller fragments flew in all directions. "[O]ne blade of the prop, four feet long and weighing 40 pounds, landed among three children playing on the sidewalk a mile and a half away. It actually smashed a hole in the concrete walk. Rebounding, it imbedded itself in a nearby porch. Not one of the children was touched!"[131] Over the years, there were other mishaps, too. During World War II, a plane from Wright Field crashed into the schoolyard of nearby Fairfield High School. Another plane crashed into the Wright Field parking lot, killing a crew member but no commuters. Sometimes the earthbound fought back. Tiring of years of airplane noise, a 60-year-old man, one day in the 1964, began taking potshots at planes taking off and landing from Wright-Patterson Air Force Base. No

one was hurt before the man was apprehended.[132]

Yes, the relationship between Dayton and its Air Force airfields and air bases has had its ups and downs over the years. Generally, that relationship has resulted in much mutual advantage, as the foregoing chapters have illustrated. Generally, too, the community and the base have had far more in common than perhaps each realized at the time. Again, Jimmy Doolittle tells how Colonel Thurman Bane, the "master of McCook" kept a small notebook in his shirt pocket in which he jotted down his plans for the day and the coming week or two. "I was so impressed with that little notebook, which I first started using in 1922 that I am still using one."[133] 1922 was the year John H. Patterson died. Patterson was likewise an inveterate user of the pocket notebook (had been since the 1870s) and— in true Patterson fashion—required its use by all his subordinates.[134] Great minds, as they say, think alike.

Now, the second century of flight beckons. If anything, the twenty-first century will require even greater forethought, planning, and *mutual* cooperation and understanding than the previous hundred years between the citizens of the Miami Valley and the United States Air Force. So, go and take out your pocket notebooks—or electronic notepads. Planning that future is in your hands!

PET THERAPY AT WRIGHT-PATTERSON AIR FORCE BASE

In the aftermath of the 9-11 attacks, there was a great need in the nation for grief counseling in all sectors of society, regardless of whether a personal connection with the victims existed. Americans were hurting emotionally and Pet Therapy teams across the nation provided immeasurable comfort and hope to those in pain. The act of rubbing a face across a satin muzzle, petting curly ears, listening to a beating heart, crying into a silky fur coat, and hugging a loving, living being who carried within no sorrow, fear, or hatred, brought a significant measure of peace to many people.

In the 1970s, the medical profession undertook clinical studies to determine what, if any, impact a controlled use of animals had on patients. The findings were measurable and dramatic. The presence and touch of a gentle animal reduced stress in humans: a child undergoing a CAT scan relaxed when holding a rabbit in his arms; a stroke victim brushing the coat of a cat or throwing a ball for a dog to fetch enjoyed the exercise far more than when performing the same physical activity with a machine and physical therapist. A child with ADHD sat quieter, longer, and listened more intently if he stroked the head of a dog lying in his lap. A person who underwent trauma and would not speak to family or doctors, confided secrets of their abuse to a nonjudgmental dog that brought them only joyous kisses and asked for nothing but hugs. Pet Therapy soon became an accepted form of stress reduction for patients with physical and mental illnesses.

Pet Therapy was not a program that had dogs running loose in hospital wards. All aspects of Pet Therapy had strict training, rules, and standards. Animals were always accompanied by their human companions, as well as a nurse or therapist. Animals were bathed before each visit and required periodic veterinary checks and immunizations while their human companions were likewise trained, tested, and immunized. All teams were retested every two years.

The Wright-Patterson Medical Center was a pioneer in the Air Force by introducing a Pet Therapy program in 1990. Under the direction of Major Janice White, a curriculum was created and, over the next decade, thousands of animals, along with their human companions, were trained to become therapy animals.

Therapy animals come in all sizes and colors with fur, feathers, fins, and wings, and they have many special talents. The main qualification is temperament, although impeccable manners, friendliness, and cleanliness are required.

Teams that switched duty stations spread the Wright-Patterson system throughout the Air Force and established Pet Therapy programs at other bases. Because the base had a large civilian population, many teams from that sector were trained and they introduced the Pet Therapy concept to hospitals throughout the Miami Valley. The program continued to grow and, in 2000, a private, nonprofit organization was established, the Miami Valley Pet Therapy Association (mvPTA), to assume management of training and certification.

Pet Therapy is now an accepted form of treatment in many clinical settings in the nation and as long as humans need medical care and emotional support, the animals will be on duty.

MY DAUGHTER IS A DOCTOR

MY SON IS A LAWYER

But MY DOG IS A THERAPIST

Aeronautical Systems Center employee James Terpenning relaxes with Peggy Sue.

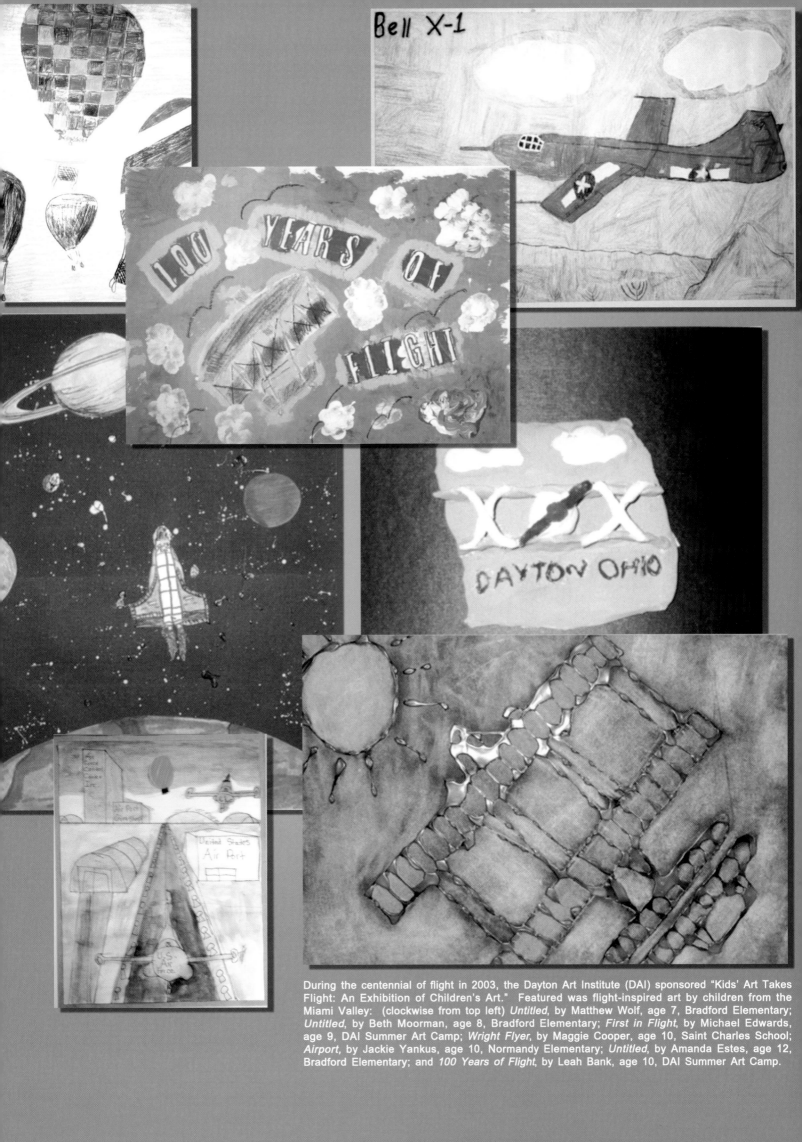

Bell X-1

100 YEARS OF FLIGHT

DAYTON OHIO

United States Air Port

U.S Air Force

During the centennial of flight in 2003, the Dayton Art Institute (DAI) sponsored "Kids' Art Takes Flight: An Exhibition of Children's Art." Featured was flight-inspired art by children from the Miami Valley: (clockwise from top left) *Untitled*, by Matthew Wolf, age 7, Bradford Elementary; *Untitled*, by Beth Moorman, age 8, Bradford Elementary; *First in Flight*, by Michael Edwards, age 9, DAI Summer Art Camp; *Wright Flyer*, by Maggie Cooper, age 10, Saint Charles School; *Airport*, by Jackie Yankus, age 10, Normandy Elementary; *Untitled*, by Amanda Estes, age 12, Bradford Elementary; and *100 Years of Flight*, by Leah Bank, age 10, DAI Summer Art Camp.

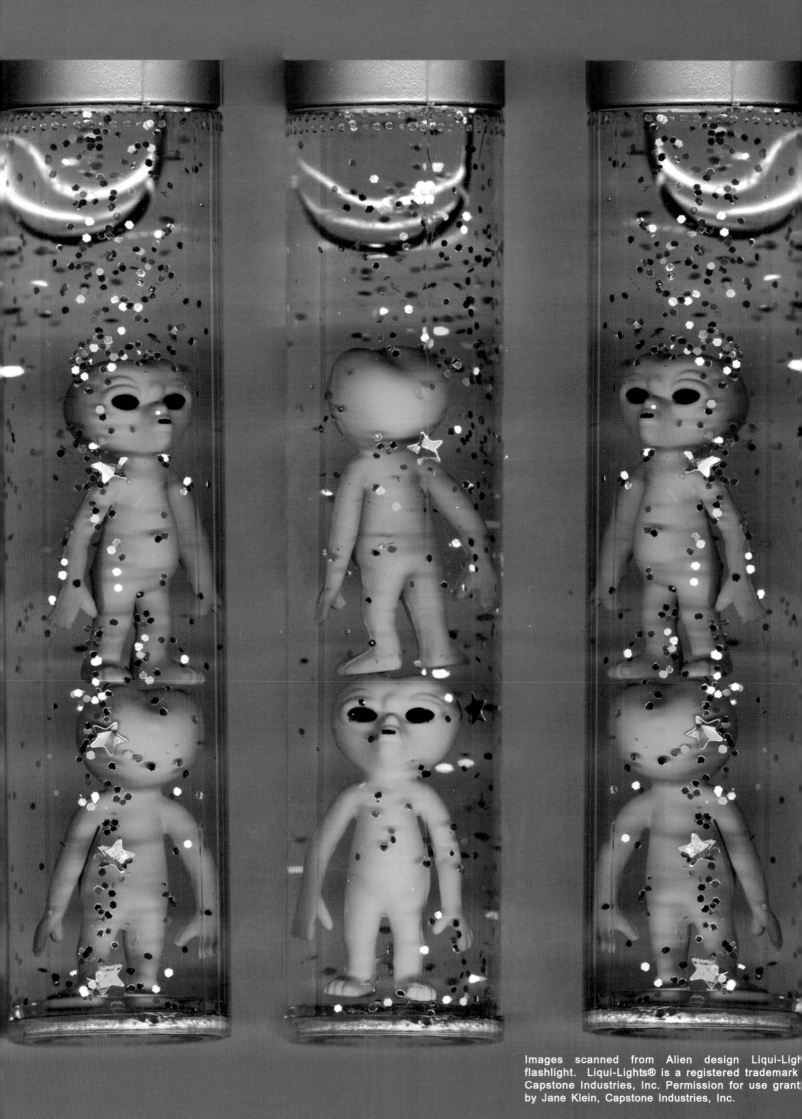

Postscript

THE SAGA CONTINUES

Historians from the Aeronautical Systems Center (ASC) History Office made a remarkable discovery on May 27, 2003, while visiting the abandoned site of old McCook Field, the original aeronautical research and development facility north of Dayton. While looking for evidence of the airfield, which existed between the years 1917 and 1927, the researchers discovered the petrified remains of what can only be described as a "little green person," since scientists have so far been unable to determine the sex of the creature. This finding potentially moves the evidence chain of extraterrestrial presence on planet Earth back nearly 40 years.

Scientists with the Air Force Research Laboratory at Wright-Patterson Air Force Base have made a *preliminary* conclusion that the creature was *not* from planet Earth and—even more remarkable—potentially originated on planet Mars. Researchers from Ohio State University speculated that the small size of the specimen, along with the presence of a compound thought to be plastic, lends credence to speculation that the relic originated on a distant planet, perhaps the green planet Venus or the red planet Mars (since both green and red look identical when photographed with black-and-white film, thus creating great confusion in the astrological world when identifying artifacts from either planet.)

The discovery of the creature has led to speculation regarding a number of occurrences at McCook Field that, quite frankly, have never had plausible explanations. What did Lieutenant Shorty Schroeder see to cause his eyes to freeze open? Why did Frank Stuart Patterson's perfectly fine DH 4 crash? Why did test pilots choose pit bulls as mascots? Why did they name an airplane the Question Mark? What caused one test pilot to jump out of a nice airplane and, by some miracle, make a parachute landing in a grape arbor? Who was that unidentified person wearing a flight cap and goggles in all the pictures taken of the Round-the-World Flight? Why didn't the GeeBee racer kill everyone who tried to fly it? Who actually landed Jimmy Doolittle's plane during his blind-flying tests? These and a thousand other questions have remained unanswered for decades and, finally, ASC historians may have discovered—quite by chance—some hint as to why these, and a thousand other peculiar situations, occurred decades before the world heard of Roswell.

Since its discovery, the SGLPFE (strange green little person from elsewhere), as it has entered the Air Force lexicon, has been secured in Hangar 14 at old Wright Field. The hardened facility is under 24-hour guard, with access by appointment only to recognized scholars and experts in the fields of materials and extraterrestrials. Access remains so restricted because ASC History Office personnel, who made the initial discovery at McCook Field, have since developed what can only be described as "odd" behaviors. The intensity of peculiarities appearing in personnel is proportional with the amount of direct contact they had with the specimen. Abnormal behaviors observed include an increased proclivity to nocturnal AM radio, long hours devoted to web searches, unfocused wandering of steam tunnels and sewers beneath Wright Field, unexplained attacks on computers and related equipment, muttering, and an obsessive desire to write Volume 3 of the U.S. Air Force's Roswell report, first issued in 1997.

As of this writing, the Air Force Research Laboratory is continuing its exhaustive analysis of the remains, and numerous students from the Air Force Institute of Technology have expressed interest in taking on this finding as a research topic. Dr. Vincent Russo, retired civilian director of the Aeronautical Systems Center and past director of the Materials Laboratory, is writing, with the intent to publish, a comprehensive review of documentation and the photographic record of the find that will surely serve as the Air Force's definitive word on the topic. As for the hapless historians, they have embarked on a number of team building and stress reduction exercises with appropriate subject-area experts from the 74th Medical Group at Wright-Patterson Air Force Base.

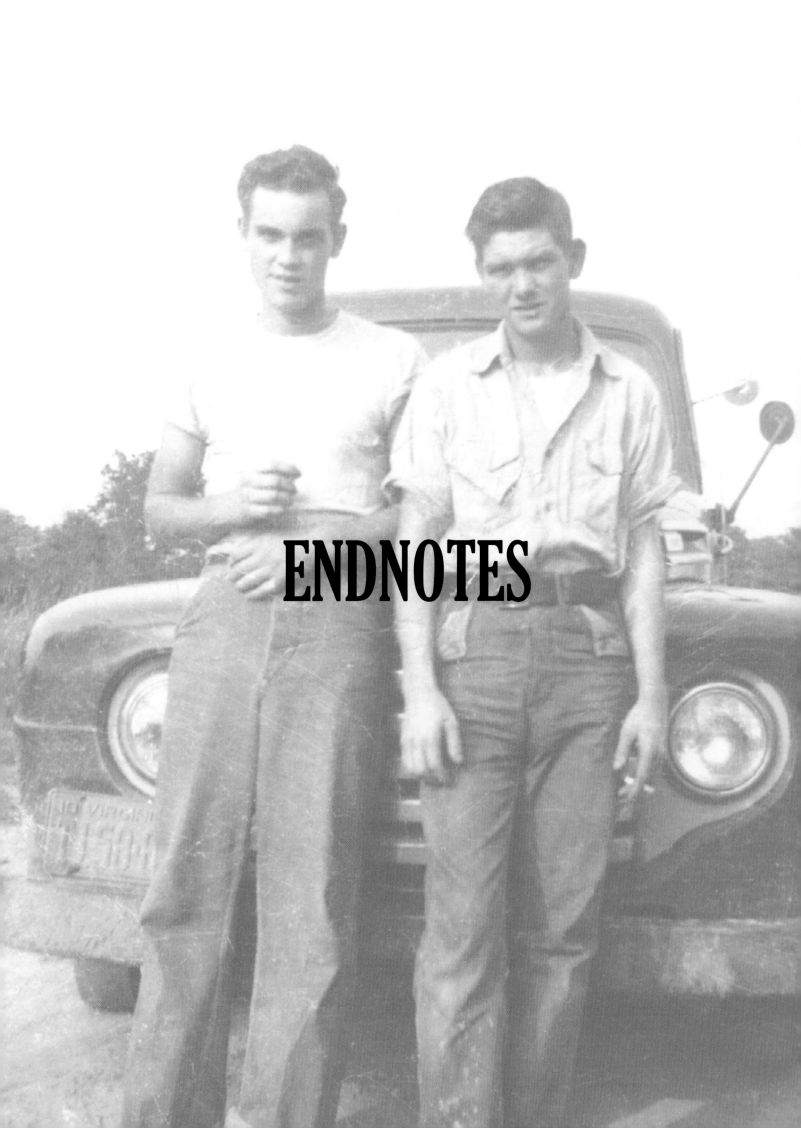

ENDNOTES

ENDNOTES

Chapter 1. BIRTHPLACE OF AVIATION

1. Fred C. Kelly, *The Wright Brothers: A Biography* (New York, 1943; reprint ed., New York, 1989), p 258.

2. Fred Howard, *Wilbur and Orville: A Biography of the Wright Brothers* (New York, 1987), p 296.

3. Kelly, *Wright Brothers*, p 258.

4. Orville's message read 59 seconds; this became 57 seconds during transmission; see Kelly, *Wright Brothers*, p 106.

5. *Ibid.*, p 107.

6. *Ibid.*

7. *Ibid.*, p 130.

8. *Ibid.*, p 137.

9. *Ibid.*, p 140.

10. *Ibid.*, pp 103-111, 123-125, 134-135, 139-141, 190-191, 216-225.

11. *Ibid.*, p 143.

12. *Ibid.* pp 142-143.

13. Cf. Matthew 13:57, Mark 6:4, and Luke 4:24. All passages from *The Bible* in this chapter are from the old Revised Standard Version.

14. Kelly, *Wright Brothers*, pp 234, 238.

15. *Ibid.*, pp 236-237, 241, 250.

16. *Ibid.*, p 250.

17. *Ibid.*, Chapter 14.

18. *Ibid.*, p 227.

19. The allusion is Amos Root's; see James F. Aldridge, *Wright from the Start: The Contributions of Dayton's Science and Engineering Community to American Air Power in the Twentieth Century* (Washington, D.C., 1997), p 37.

20. Howard, p 295.

21. Kelly, *Wright Brothers*, p 257.

22. *Ibid.*, p 258.

23. *Ibid.*

24. Tom Crouch, *The Bishop's Boys: A Life of Wilbur and Orville Wright* (New York and London, 1989), p 391.

25. Howard, p 295.

26. Ivonette Wright Miller, compiler, *Wright Reminiscences* (Wright-Patterson Air Force Base, 1978), p 10; Howard, p 295.

27. Fred C. Kelly, ed., *Miracle at Kitty Hawk: The Letters of Wilbur and Orville Wright* (New York, 1951; reprint ed., 1996), p 342.

28. Kelly, *Wright Brothers*, pp 258-259; Howard, p 296.

29. Crouch, pp 14, 119, 125, 476, 480-481.

30. Kelly, *Wright Brothers*, p 259.

31. *Ibid.*, p 259; Howard, pp 296-297.

32. Howard, p 297.

33. *Ibid.*

34. Kelly, *Wright Brothers*, p 259; Howard, p 297.

35. *Ibid.*; Miller, *Wright Reminiscences*, p 10.

36. Howard, p 297.

37. Kelly, *Wright Brothers*, p 259; Crouch, p 394.

38. Virginia and Bruce Ronald, *The Lands between the Miamis: A Bicentennial Celebration of the Dayton Area*, Volume 2 (Dayton, 1996), pp 207ff.

39. See Samuel Crowther, *John H. Patterson: Pioneer in Industrial Welfare* (Garden City, New York, 1924); and Mark Bernstein, *Grand Eccentrics. Turning the Century: Dayton and the Inventing of America* (Wilmington, Ohio, 1996), *passim*.

40. Bernstein, pp 97-98.

41. *Ibid.*, p 171.

42. Isaac F. Marcosson, *Colonel Deeds, Industrial Builder* (New York, 1947) does not mention Deeds' firing by Patterson.

43. Stuart W. Leslie, *Boss Kettering: Wizard of General Motors* (New York, 1983).

44. Bernstein, p 104.

45. Leslie, pp 36ff.; Marcosson, Chapter 6.

46. Leslie, pp 90ff.

47. Marcosson, pp 185ff.

48. *Ibid.*, pp 212ff.

49. *Ibid.*, pp 305ff.

50. *Ibid.*, pp 95ff.

51. James E. Cebula, *James M. Cox: Journalist and Politician* (New York and London, 1985) and James M. Cox, *Journey through My Years* (New York, 1946).

52. Cebula, Chapters 1-4.

53. *Ibid.*, Chapters 5, 6, and 7.

54. *Ibid.*, Chapters 9, 10, and 11.

55. *Ibid.*, p 33.

56. *Ibid.*, p 36.

57. *Ibid.*, pp 33, 36.

58. *Ibid.*, p 33.

59. *Ibid.*, Chapter 5.

60. Judith Sealander, *Grand Plans: Business Progressivism and Social Change in Ohio's Miami Valley, 1890-1929* (Kentucky, 1988).

61. Ronald, p 201.

62. Charlotte Reeve Conover, *Dayton, Ohio, An Intimate History*, ed. by Alexander Kaye; introduction by Roz Young (Dayton, 1995; based on earlier, complete edition, Dayton, 1932), pp 27-28.

63. *Ibid.*, p 26.

64. *Ibid.*, p 28.

65. *Ibid.*, p 33-34.

66. *Ibid.*, pp 34-35.

67. *Ibid.*, p 35.

68. John Bartlett, *Familiar Quotations*, 16th ed., (Boston, et al., 1992), p 278; the quotation, written by the son of Sir Christopher Wren, appears on a tablet in St. Paul's Cathedral, London, that Wren designed to replace the original gothic church destroyed in the great fire of 1688.

69. Conover, *Dayton*, p 48.

70. Crowther, p 26.

71. Conover, *Dayton*, pp 88-91.

72. *Ibid.*, p 48; Crowther, p 24

73. Crowther, p 28.

74. Conover, *Dayton*, pp 195-196.

75. Thomas Hobbes, *Leviathan*, 1651 ed., Part 1, Chapter 13, cited in John Bartlett, *Familiar Quotations*, 16th ed. (New York, 1992), p 239.

76. *For the Love of Dayton: Life in the Miami Valley, 1796-1996*, ed. by Ron Rollins (Dayton, 1996), p 20.

77. Conover, *Dayton*, p 71.

78. *Ibid.*, p 41.

79. *Ibid.*, p 42.

80. *Ibid.*, p 51.

81. *Ibid.*, p 67.

82. *Ibid.*, p 39.

83. *Ibid.*, p 52.

84. Biography, "William Henry Harrison," viewed online September 30, 2003, at www.whitehouse.gov/history/presidents/wh9.html.

85. Conover, *Dayton*, p 52.

86. *Ibid.*, p 79.

87. *Ibid.*, p 53.

88. Built ca. 1815-19; Conover, *Dayton*, p 54.

89. In 1819; *Ibid.*, p 54.

90. *Ibid.*, p 55.

91. *Ibid.*, p 58.

92. In 1849; *Ibid.*, p 60.

93. *Ibid.*, p 39.

94. *Ibid.*, p 48.

95. *Ibid.*, pp 54, 130-131, 218.

96. *Ibid.*, p 132.

97. *Ibid.*, p 135.

98. *Ibid.*, pp 132-134.

99. *Ibid.*, pp 107-109.

100. *Ibid.*, pp 81-82.

101. *Ibid.*, p 66.

102. *Ibid.*, p 97.

103. *Ibid.*, p 112.

104. Ronald, p 168.

105. Conover, *Dayton*, pp 117-118.

106. Ronald, p 168.

107. Conover, *Dayton*, p 116.

108. *Ibid.*, pp 117-118.

109. Crouch, pp 22-24.

110. In 1871; Crouch, p 44.

111. Crouch, p 476.

112. *Hippocrates on the Web: History of Medicine*, Faculty of Medicine, University of Manitoba, viewed online July 29, 2003, at www.umanitoba.ca/faculties/medicine/units/history/lister/pasteur.html.

113. Conover, *Dayton*, p 83.

114. *Ibid.*, p 149.

115. *Ibid.*, p 170.

116. *Ibid.*, pp 123, 125.

117. *Ibid.*, p 123.

118. *Ibid.*, p 181.

119. "Milestones in Telegraphic History," viewed online September 30, 2003, at www.members.tripod.com/morse_telegaph_club/images/newpage1.htm.

120. Conover, *Dayton*, pp 182-183.

121. *Ibid.*, p 183.

122. *Ibid.*, p 125.

123. Sealander, p 1.

124. Conover, *Dayton*, p 189.

125. *Ibid.*, p 141.

126. *Ibid.*, p 184.

127. *Ibid.*, p 168.

128. *Ibid.*, p 169.

129. *Ibid.*, p 64.

130. *For the Love of Dayton*, p 99.

131. *Ibid.*, p 121.

132. Conover, *Dayton*, p 169.

133. Charlotte Reeve Conover, *Some Dayton Saints and Prophets* ([Dayton], 1907), pp 221-228.

134. Conover, *Dayton*, p 167.

135. *Ibid.*, p 82.

136. Crowther, p 26.

137. Sealander, *passim.*

138. Crowther, pp 363-364.

139. Conover, *Dayton*, p 193.

140. Ronald, p 354.

141. See Mary Bellis, "The History of the Automobile: The Internal Combustion Engine and Early Gas-Powered Cars," viewed online September 30, 2003, at http://www.inventors.about.com/library/weekly/aacarsgasa.htm.

142. *For the Love of Dayton*, p 99.

143. *Ibid.*, p 100.

144. *Ibid.*, p 104.

145. *Ibid.*, p 107.

146. *Ibid.*, p 119.

147. *Ibid.*, p 118.

148. *Ibid.*, p 99.

149. *Ibid.*, p 104.

150. Crouch, p 394.

151. Kelly, *Wright Brothers*, p 230.

152. Crouch, p 398.

153. *Ibid.*, pp 397-399.

154. *Ibid.*, p 410.

155. Kelly, *Wright Brothers*, p 275.

156. *Ibid.*, pp 277ff.

157. *Ibid.*, p 271; Crouch, p 402.

158. *For the Love of Dayton*, p 104.

159. *Ibid.*, p 117.

160. *Ibid.*, p 107.

161. Bernstein, p 86; Marcosson, p 88.

162. Ronald, pp 219-220.

163. *For the Love of Dayton*, p 104.

164. Cebula, pp 39ff.

165. Bernstein, pp 135-137.

166. Crouch, pp 450.

167. Ronald, pp 249-250; Conover, *Dayton*, pp 263, 265.

168. *Ibid.*, p 250.

169. Conover, *Dayton*, p 98.

170. Crowther, pp 303-304; Bernstein, p 140.

171. Quoted in Curt Dalton, ed., *Through Flood, Through Fire: Personal Stories from Survivors of the Dayton Flood of 1913* (Dayton, 2001), p 48.

172. Cox, *Journey through My Years*, p 166.

173. Ronald, p 259.

174. "Titanic: A Special Exhibit from *Encyclopaedia Britannica*," viewed online September 30, 2003, at www.search.eb.com/titanic/01_01.html; "Titanic: A Special Exhibit from *Encyclopaedia Britannica*: Researcher's Note," viewed online September 30, 2003, at http://.search.eb.com/titanic/researchersnote.html.

175. Ronald, p 259.

176. *Ibid.*, p 260.

177. *Ibid.*, p 252.

178. Cox, pp 165, 167.

179. Ronald, p 252.

180. *Ibid.*, p 260.

181. Bernstein, p 151.

182. *Ibid.*, p 149.

183. Marcosson, pp 146-147.

184. Ronald, p 261.

185. Bernstein, pp 162-163.

186. *Ibid.*, pp 161.

187. Ronald, p 262; Bernstein, p 166.

188. Cox, pp 175, 177.

189. Conover, *Dayton*, p 268.

190. *Ibid.*, p 269.

191. Ronald, p 263.

192. *For the Love of Dayton*, p 125.

193. Phrase made famous by Barbara Tuchman, *The Guns of August* (New York, 1962).

194. H. W. Brands, *Woodrow Wilson* (New York, 2003), p 80.

195. Roger E. Bilstein, *Flight in America: From the Wrights to the Astronauts*, (Baltimore, 1984); rev. ed. (Baltimore and London, 1994), p 34.

196. Bilstein, *Flight in America*, p 33.

197. Alex Roland, *Model Research: The National Advisory Committee for Aeronautics, 1915-1958*, Volume 1 (Washington, D.C., 1985), pp 6ff.

198. Bilstein, *Flight in America*, p 31.

199. As told by Rosamund "Roz" Young in "Introduction to 1970 Edition," in Conover, *Dayton*, p 14.

200. Conover, *Dayton*, p 286.

201. Marcosson, pp 216-217.

202. *Ibid.*, p 218.

203. *Ibid.*, pp 213-214.

204. *Ibid.*, p 215.

205. *Ibid.*, p 215-216.

206. *Ibid.*, p 221.

207. Bilstein, *Flight in America*, p 36.

208. *Ibid.*, pp 28-29.

209. [Martin Classen], *Materiel Research and Development in the Army Air Arm, 1914-1945*, ([Washington, D.C.], November 1946), p 15.

210. Theodore von Karman, *The Wind and Beyond: Theodore von Karman, Pioneer in Aviation and Pathfinder in Space*, with Lee Edson (Boston and Toronto, 1967), p 7.

211. Marcosson, p 216.

212. Albert E. Misenko, *Aero Propulsion and Power Directorate: The McCook Field Years 1917-1927* (Wright-Patterson Air Force Base, 1995), pp 32-33.

213. Walter J. Boyne, "Treasures of McCook, Part I," in *The Best of Wings: Great Articles from* Wings *and* Airpower *Magazines* (Washington, D.C., 2001), pp 4-5.

214. Ronald, p 142.

215. Conover, *Dayton*, p 111.

216. Michael Farquhar, *A Treasury of Great American Scandals* (New York, 2003), p 204.

217. Cox, p 232.

218. *For the Love of Dayton*, p 128.

219. For a quick overview of the Wilson administration, see H. W. Brands, *Woodrow Wilson: Profiles in Power* (New York, 2003); for in-depth treatment, John A. Thompson, *Woodrow Wilson* (London, 2002).

220. Pundit and humorist, H.L. Mencken called Harding's utterances "Gamalielese," playing off his middle name.

221. Program, *International Air Races,* Dayton, Ohio, October 2-3-4, 1924, p 3.

222. *For the Love of Dayton*, p 132.

223. Walter Boyne, "The Treasures of McCook Field: America's Aero-Engineering and Testing Center, Part I," *Airpower* (July 1975),

republished in: Boyne, *The Best of Wings: Great Articles [by Walter J. Boyne] from* Wings *and* Airpower *Magazines* (Washington, D.C., 2001), p 4.

224. Bilstein, *Flight in America*, p 65.

225. Lois E. Walker and Shelby W. Wickam, *From Huffman Prairie to the Moon: The History of Wright-Patterson Air Force Base, Ohio* (Washington, 1986), p 105.

226. Program, *International Air Races,* 1924, p 27.

227. *A Little Journey to the Engineering Division, Army Air Service, McCook Field, Dayton, Ohio* ([Dayton, 1924]; reprint ed., Washington, D.C., 1988), p 4.

228. Walker and Wickam, *Huffman Prairie*, p 112.

229. *Ibid.*, pp 104f.

230. Roger E. Bilstein, *Orders of Magnitude: A History of the NACA and NASA, 1915-1990* (Washington, D.C., 1989), p 4.

231. *Ibid.*, p 5.

232. Walker and Wickam, *Huffman Prairie*, p 90.

233. *Ibid.*, p 105.

234. Bilstein, *Orders of Magnitude*, p 6.

235. Crowther, p 363.

236. *Ibid.*, p 350.

237. *Ibid.*, p 363.

238. *Ibid.*, p 364.

239. Walker and Wickam, *Huffman Prairie*, p 109.

240. *Ibid.*, p 109.

241. *Ibid.*, p 112.

242. Report, "Completion Report: Wright Field, Dayton, Ohio," Office of the Constructing Quartermaster, Dayton, Ohio, [July 1927], p 3.

243. Walker and Wickam, *Huffman Prairie*, pp 118ff.

244. *Ibid.*, p 257.

245. *Ibid.*, pp 121-122.

246. Karman, p 128.

247. Conover, *Dayton*, p 126.

248. *For the Love of Dayton*, p 102.

249. Conover, *Dayton*, p 243.

250. Ronald, pp 288-289.

251. *Ibid.*, pp 284-285.

252. *For the Love of Dayton*, p 128.

253. Roger E. Bilstein, *The Enterprise of Flight: The American Aviation and Aerospace Industry* (Washington and London, 2001), pp 26, 30.

254. Charles A. Lindbergh, *The Spirit of St. Louis* (New York, 1953), pp 456.

255. Lindbergh, *Spirit of St. Louis*, p 454.

256. *A General Remembers: Major General Fred J. Ascani, Wright Field's Father of Systems Engineering*, ed. by James F. Aldridge (Wright-Patterson Air Force Base, Ohio, 2001), pp I-1 to I-2.

257. *Ibid.*, p I-3.

258. *Ibid.*, p I-8.

259. *Ibid.*, pp I-12 to I-18; Albert E. Misenko and Philip H. Pollock, *Engineering History, 1917-1978: McCook Field to the Aeronautical Systems Division* (Wright-Patterson Air Force Base, Ohio, 1979), pp 27ff.

260. Walter J. Boyne, *The Leading Edge* (New York, 1986), p 77.

261. Boyne, *The Leading Edge*, "Martin B-10," p 86.

262. Peter L. Jakab, *Visions of a Flying Machine: The Wright Brothers and the Process of Invention* (Washington and London, 1990), pp 55-57.

263. Bill Gunston, *American Warplanes* (New York, 1986), p 52.

264. Gunston, p 54.

265. Harold Mansfield, *Vision: The Story of Boeing—A Saga of the Sky and the New Horizons of Space* (New York, 1966), pp 55-56.

266. Ronald, pp 199-200.

267. Crowther, Chapter 9.

268. Jim Nichols, *Dayton Album: Remembering Downtown*, (Dayton, 2000), p 22.

269. *For the Love of Dayton*, 144.

270. Ronald, p 294.

271. Marcosson, p 94; Ronald, p 298.

272. *Ibid.*, p 96.

273. Franklin D. Roosevelt, Inaugural Address, March 4, 1933.

274. James J. Niehaus, *Five Decades of Materials Progress, 1917-1967* (Wright-Patterson Air Force Base, Ohio, 1967), pp 56-57.

275. Karman, pp 226-227.

276. *Ibid.*, Chapter 25.

277. Bilstein, *The Enterprise of Flight*, p 70.

278. Gunston, *American Warplanes*, p 102.

279. Enzio Angelucci, ed., *The Rand McNally Encyclopedia of Military Aircraft, 1914-1980* (New York, 1980), p 464.

280. I.e., Building 65, Area B (Wright Field), Wright-Patterson Air Force Base.

281. Ronald, p 305.

282. Conover, *Dayton*, p 14.

283. Dalton, *Through Flood, Through Fire*, pp 48-51.

284. *Ibid.*, p 48.

285. Conover, *Dayton* (1932 ed.), pp 309-311.

286. Charlotte Reeve Conover, *History of Dayton and Montgomery County, Ohio*, 4 volumes (New York, 1932).

287. Rosamund "Roz" Young, in Conover, *Dayton*, pp 13-14.

288. Curt Dalton, *Home Sweet Home Front: Dayton During World War II* (Dayton, 2000), pp 7-8.

289. James H. Doolittle, *I Could Never Be So Lucky Again: An Autobiography*, with Carroll V. Glines (New York, 1991; reprint ed., Atglen, Pensylvania, n.d.), pp 260-275.

290. Conover, *Dayton*, p 95.

Chapter 2. MILITARY AVIATION COMES TO DAYTON

1. Charles deF. Chandler and Frank P. Lahm, *How Our Army Grew Wings: Airmen and Aircraft Before 1914* (New York, 1943), p 113. Signal Corps Specification Number 483, January 21, 1908, contained the data for the "Dirigible Balloon."

2. *Ibid.*, p 193.

3. Arthur Sweetser, *The American Air Service* (New York, 1919), p 11.

4. Chandler and Lahm, pp 277-278.

5. War Department General Order 75, December 4, 1913, as published in Chandler and Lahm, pp 314-315.

6. Juliette A. Hennessy, *The United States Army Air Arm, April 1861 to April 1917* (U.S. Air Force Historical Study 98, 1958), pp 236-237.

7. Chandler and Lahm, p 280.

8. Hennessy, p 233.

9. Robert Casari, "Number of U.S. Aircraft WWI," *AAHS Journal* (American Aviation Historical Society) 20 (Spring 1977), 36-38. The three squadrons were based, respectively, at Columbus, New Mexico; Manila, Philippine Islands; and Fort Sam Houston, Texas.

10. Alfred Goldberg, ed., *A History of the United States Air Force 1907-1957* (Princeton, 1957), p 14.

11. Sweetser, p 79.

12. *Ibid.*, p 81.

13. *Organization of Military Aeronautics 1907-1935* (Army Air Forces Historical Study 25, 1944), pp 26-27.

14. Sweetser, p 104.

15. *Ibid.*, pp 109-111.

16. *Ibid.*, pp 114-117.

17. *Ibid.*, p 98.

18. *Ibid.*, p 101.

19. *Story of the Miami Conservancy District* (Dayton, 1922), pp 8-10.

20. Isaac F. Marcosson, *Colonel Deeds, Industrial Builder* (New York, 1947), p 215.

21. Sweetser, p 106.

22. History of the Air Depot at Fairfield, Ohio 1917-1943 (Fairfield Air Service Command, Patterson Field, 1944), p 4. Hereafter cited as Hist, FAD, 1917-1943.

23. Copy of Lease, Purchase Request A-6951, Order No. 50214, between the Miami Conservancy District and Lieutenant Colonel C. G. Edgar, Signal Corps, July 1, 1917.

24. Memo, Office of the Chief Signal Officer, to all Divisions, no subj., June 6, 1917; *The Army's Order of Battle of United States Land Forces in the World War (1917-1919)*, Volume 3, Part 1 (Washington, 1949), p 897. There has been controversy through the years as to the original name of the Dayton installation. Although initial correspondence referred to "Dayton Aviation Field" or the "former Wright flying field," official memorandums were clearly marked "Wilbur Wright Field." No documentary evidence has been found that officially named the Huffman Prairie area as "Wright Field" or "Wright Flying Field." When asked in 1982, neither Mrs. Ivonette Wright Miller nor Mr. Horace Wright, surviving niece and nephew of the Wright brothers, could recall that any such title was applied to the Huffman Prairie area prior to World War I.

25. Annual Report, Signal Corps Aviation School, Wilbur Wright Field, Fairfield, Ohio, May 31, 1918, signed by Major Arthur E. Wilbourn, commanding officer. Hereafter cited as Wilbourn Report.

26. *Ibid.*

27. *Ibid.*

28. *Ibid.*

29. *Ibid.*

30. *Ibid.*

31. *Ibid.*

32. *Ibid.*

33. *Ibid.*

34. *Ibid.*

35. *Ibid.*

36. *Ibid.*

37. *Ibid.*, p 3.

38. *Ibid.*

39. *Ibid.*

40. *Ibid.*, p 22.

41. Sweetser, pp 347-348; Gilbert S. Guinn, "A Different Frontier: Aviation, The Army Air Forces and the Evolution of the Sunshine Belt," *Aerospace Historian* 29 (March 1982), pp 34-35.

42. R. K. McMaster, ed., "The Adventures of a Junior Military Aviator: Extracts from the Diary of Leo G. Heffernan," *Aerospace Historian* 25 (June 1978), pp 92-102.

43. Wilbourn Report, p 25.

44. *Ibid.*, pp 25-27.

45. *Ibid.*

46. *Ibid.*

47. *Ibid.*

48. *Ibid.*

49. Sweetser, pp 140-141.

50. *Ibid.*, p 144.

51. Wilbourn Report, pp 23-24.

52. Sweetser, p 126.

53. Maurer Maurer, ed., *The U.S. Air Service in World War I, Vol 1: The Final Report and a Tactical History* (Washington, 1978), pp 68-69, 70-72, 102-103, 105-106.

54. *Ibid.*, pp 234-235.

55. *The World Almanac and Book of Facts, 1980* (New York, 1979), p 333. Total U.S. armed forces were 4,743,826: Army (including Air Service) 4,057,101; Navy 599,051; Marines 78,839; and Coast Guard 8,835.

56. Letter, U.S. Army General Supply Depot Zone Seven, Chicago, Illinois, to Commanding Officer, Aviation School, Wilbur Wright Field, Fairfield, Ohio, subj: Stations, February 28, 1919, First Endorsement, Headquarters Wilbur Wright Air Service Depot, Fairfield, Ohio, to Zone Supply Officer, Zone 7, Chicago, March 4, 1919 (Hist, FAD, 1917-1943, Exhibit 11). An important adjunct to the former Fairfield Aviation General Supply Depot had been the Airplane Acceptance Park at Moraine City, four miles southwest of Dayton. This military organization was responsible for accepting or rejecting airplanes (mainly DeHavilland DH-4s) produced by the Dayton Wright Airplane Company.

57. Hist, FAD, 1917-1943, p 17.

58. Royal Frey, *Evolution of Maintenance Engineering 1907-1920*, Volume 1: Narrative (Air Materiel Command, Wright-Patterson Air Force Base, 1960), p 191.

59. Headquarters, Aviation General Supply Depot, Fairfield, Ohio (Wilbur Wright Field), SO 178, November 1, 1920 (Hist, FAD, 1917-1943, Exhibit 26).

Chapter 3. THE STORY OF AIR FORCE LOGISTICS

1. History of the Air Depot at Fairfield, Ohio, 1917-1943 (Fairfield Air Service Command, Patterson Field, 1944), pp 1-16. Hereafter cited as Hist, FAD, 1917-1943.

2. Telegram, Headquarters Air Service, to Major C. T. Waring, January 10, 1919 (Hist, FAD, 1917-1943, Exhibit 10).

3. Hist, FAD, 1917-1943, p 16. About 40 squadrons resided at Wilbur Wright Field and the Fairfield Aviation General Supply Depot for varying lengths of time between July 8, 1917, and February 20, 1919. Many of the squadrons came from Kelly Field, Texas, stayed about 90 days, then transferred to the Air Service Depot, Garden City, Long Island, New York, for probable reassignment to Europe.

4. Letter, Chief of Air Service to Commanding Officer, Aviation General Supply Depot, Fairfield, Ohio, citing "Orders 49, O.D.A.S., November 3, 1919," July 31, 1920 (Hist, FAD, 1917-1943, Exhibit 16).

5. Hist, FAD, 1917-1943, pp 16-17.

6. *Ibid.*, pp 35-36.

7. Edward O. Purtee, *History of the Army Air Service, 1907-1926* (Air Materiel Command, Wright-Patterson Air Force Base, 1948), p 111.

8. *Ibid.*; Alfred Goldberg, ed., *A History of the United States Air Force 1907-1957* (New York, 1957), p 29.

9. Letter, Office, Director of Air Service, to Commanding Officer, Aviation General Supply Depot, Fairfield, Ohio, subj: Removal of Aviation Repair Depot, Indianapolis, to Fairfield, July 16, 1920 (Hist, FAD, 1917-1943, Exhibit 22).

10. Report, Historical Data, Maintenance Division, Fairfield Air Service Command, Patterson Field, May 1, 1944.

11. Letter, Commanding Officer, Aviation Repair Depot, Speedway, Indianapolis, Indiana, to Commanding Officer, Aviation General Supply Depot, Fairfield, Ohio, subj: Move to Fairfield, Ohio, August 9, 1920; Letter, Commanding Officer, Aviation General Supply Depot, Fairfield, Ohio, to Commanding Officer, Aviation Repair Depot, Speedway, Indianapolis, Indiana, subj: Moving of Aviation Repair Depot to Fairfield, Ohio, August 12, 1920 (Hist, FAD, 1917-1943, Exhibits 23, 24).

12. Report, Historical Data, Maintenance Division, Fairfield Air Service Command, May 1, 1944.

13. *Ibid.*

14. Report, Chief of the Air Service to the Secretary of War for Fiscal Year 1920, October 4, 1921; Report, Aircraft Condition in the Air Service, June 1924 (Hist, FAD, 1917-1943, Exhibit 37).

15. Headquarters Aviation General Supply Depot, Fairfield, Ohio, SO 178, November 1, 1920 (Hist, FAD, 1917-1943, Exhibit 26).

16. War Department General Order 2, January 14, 1921 (Hist, FAD, 1917-1943, Exhibit 28).

17. War Department SO 179, August 4, 1921 (Hist, FAD, 1917-1943, Exhibit 30).

18. War Department SO 71, March 26, 1923 (Hist, FAD, 1917-1943, Exhibit 36).

19. Headquarters Materiel Division SO 101, May 19, 1927 (Hist, FAD, 1917-1943, Exhibit 32). The Materiel Division, commanded by Brigadier General William E. Gillmore, succeeded the Army Air Service Engineering Division when the Army Air Corps was established. At the time of its establishment, October 15, 1926, the Materiel Division was located at McCook Field.

20. War Department Circular 76, October 25, 1921 (Hist, FAD, 1917-1943, Exhibit 29).

21. Memo, Depot Supply Officer, to Commanding Officer, Fairfield Air Intermediate Depot, no subj, May 2, 1922 (Hist, FAD, 1917-1943, Exhibit 34).

22. *Ibid.*

23. Headquarters Fairfield Air Intermediate Depot General Order 3, February 12, 1924 (Hist, FAD, 1917-1943, Exhibit 31).

24. Lowell J. Thomas, *The First World Flight* (Boston, 1925), introduction by Major General Mason M. Patrick, chief of Air Service, p xxi.

25. Air Service *News Letter* VII (November 22, 1923), 6-8.

26. *Aircraft Year Book for 1924* (New York, 1925), p 237.

27. Lowell J. Thomas and Lowell Thomas, Jr., *Famous First Flights That Changed History* (Garden City, New York, 1968), pp 19-20; Air Force Pamphlet 190-2-2: *A Chronology of American Aerospace Events from 1903 through 1964* (September 1965), p 20.

28. Thomas and Thomas, *Famous First Flights*, pp 18-31.

29. Thomas, *First World Flight*, p 4.

30. Joe Christy, "That First Round-the-World Flight," *AIR FORCE Magazine* 57 (March 1974), pp 53-59. For a continuation of the story, see Alva Harvey's article, "Seattle Has Crashed in Alaska," *AIR FORCE Magazine* 57 (September 1974), pp 103-107.

31. Thomas, p 4.

32. Christy, p 53.

33. Lloyd Morris and Kendall Smith, *Ceiling Unlimited: The Story of American Aviation from Kitty Hawk to Supersonics* (New York, 1953), pp 236-237.

34. Morris and Smith, p 237.

35. Air Service *News Letter* VII (July 19, 1923), 4; Air Service *News Letter* VII (November 22, 1923), 6-8.

36. Thomas, p. 5.

37. Thomas, p 5; Air Service *News Letter* VIII (February 1, 1924), 12-13; booklet, *First Around the World* (Douglas Aircraft Corporation, Santa Monica, California, [1974]), p 2.

38. *Aircraft Yearbook for 1924*, p 239.

39. *Ibid.*

40. Thomas, pp 17-18.

41. *Aircraft Yearbook for 1924*, pp 72-73.

42. *Ibid.*, p 9.

43. Thomas and Thomas, p 63.

44. Harvey, pp 103-107. Sergeant Harvey graduated from pilot primary school in 1925, advanced through commissioned grades to the rank of colonel, and retired in that grade after World War II.

45. Thomas, pp 293-294.

46. *Ibid.*, pp 304-305.

47. *Ibid.*, p 315.

48. *Ibid.*, p 325.

49. National Aeronautic Association (NAA) *Review* 2 (Special Dayton Air Race Edition, September 18, 1924), 1.

50. Air Service *News Letter* VIII (October 31, 1924).

51. *Ibid.*

52. *NAA Review* 2 (September 18, 1924), 1.

53. Arthur G. Renstrom, *Wilbur & Orville Wright: A Chronology Commemorating the Hundredth Anniversary of the Birth of Orville Wright, August 19, 1871* (Washington, 1975), p 78; Hist, FAD, 1917-1943, pp 42-43.

54. Air Service *News Letter* VIII (July 31, 1924); NAA *Review* 2 (September 18, 1924), 4.

55. History of Wilbur Wright Field, 1917-1925, Volume 1 (2750th Air Base Wing, Wright-Patterson Air Force Base), p. 31.

56. Hist, FAD, 1917-1943, p 32.

57. *Ibid.*

58. *Ibid.*, p 48.

59. *AMC and Its Antecedents (1917-1960)* (Air Materiel Command Historical Study 329, 1960), pp 3-4.

60. Hist, FAD, 1917-1943, p 131.

61. *Ibid.*, p 57.

62. *Ibid.*, p 61.

63. Hist, 2750 Air Base Wing, July 1974-December 1975, Vol II: Wright-Patterson Air Force Base and 2750 Air Base Wing Heritage and Lineage 1917-1975, pp 45-46; Hist, FAD, 1917-1943, pp 66-67; "When Pilots Flew on Two Wings," Wright-Patterson Air Force Base *Skywrighter*, May 15, 1981, pp 16, 19.

64. Letter, Registrar Yale College [University], New Haven, Connecticut, to Frank S. Patterson, no subj., April 27, 1918; Certificate of Enlistment, Enlisted Reserve Corps (AGO Form 422-1).

65. War Department Special Order 110, paragraph 83, May 10, 1918.

66. Headquarters Signal Corps Aviation School, Wilbur Wright Field, SO 124, para 7, May 9, 1918.

67. Hist, FAD, 1917-1943, Exhibit 4.

68. Letter, Wilbur Wright Field Armorers School Commanding Officer (Major H. C. K. Muhlenberg) to Director of Military Aeronautics, Technical Section, Washington, D.C., subj: Report on Accident to DeHavilland Four Plane No. 32098, June 26, 1918.

69. LeRoy Amos Swan Collection, United States Air Force Museum Archives.

70. Air Force Pamphlet 190-2-2, *A Chronology of American Aerospace Events from 1903 through 1964*, September 1965, p 18.

71. Tom D. Crouch, *The Giant Leap: A Chronology of Ohio Aerospace Events and Personalities 1815-1969* (Columbus, 1971), p 32.

72. Hist, FAD, 1917-1943, p 69.

73. *Ibid.*, p 68.

74. *Ibid.*, p 22.

75. *Ibid*, pp 65-66.

76. "Historical Sketch, Supply Division, FASC, 1917-1938," no author, n.d.

77. *Ibid.*, pp 6-8.

78. Hist, FAD, 1917-1943, pp 73-74.

79. *Ibid.*, p 10.

80. *Ibid.*, p 78.

81. *Ibid.*, p 81.

82. *Ibid.*, p 83.

83. *Ibid.*, pp 83-84.

84. *A Pictorial Review, Wright-Patterson Air Force Base 1917-1967* (2750 Air Base Wing, Wright-Patterson Air Force Base, Ohio, 1967), p 26.

85. Hist, FAD, 1917-1943, p 89.

86. *Ibid.*, pp 78-79.

87. *Ibid.*, p 94.

88. *Ibid.*, p 92.

89. *Ibid.*

90. History of Fairfield Air Depot Control Area Command (FADCAC) and Fairfield Air Service Command (FASC), 1 February 1943-1 October 1944, p 35. (Hereafter cited as Hist, FADCAC and FASC, 1943-1944.)

91. "New Post Hospital Open," *Patterson Field Postings*, May 20, 1942, p 1.

92. Hist, FAD, 1917-1943, pp 95, 97.

93. Hist, FADCAC and FASC, 1943-1944, p 44.

94. "Photograph of Wright-Patterson Air Force Base and its Master Plan," 1948, p 5.

95. Hist, FAD, 1917-1943, pp 95-96.

96. "Service Awards to Civilian Employees Authorized by WD," *Patterson Field Postings*, November 5, 1943, p 2.

97. Hist, FAD, 1917-1943, p 141.

98. *Ibid.*, p 144.

99. Hist, FADCAC and FASC, 1943-1944, p 57.

100. Hist, FADCAC and FASC, 1943-1944, p 4.

101. *Ibid.*, p 112.

102. War Department General Order 68, August 18, 1944.

103. Maurer Maurer, ed., *Air Force Combat Units of World War II* (Maxwell Air Force Base, Alabama, 1960), pp 52-53.

104. Maurer Maurer, ed., *Combat Squadrons of the Air Force During World War II* (Maxwell Air Force Base, Alabama, 1969), pp 9-10.

105. Maurer, *Combat Units*, pp 129-139.

106. *Ibid.*, pp 193-194.

107. Maurer, *Combat Squadrons*, pp 63-64, 75, 261-262, 160.

108. Hist, FADCAC and FASC, 1943-1944, p 125.

109. *Ibid.*, pp 187-188.

110. *Ibid.*, p 383.

111. Hist, FAD, 1917-1943, pp 104, 109.

112. Notes on Joint Meeting of Air Service Command and Fairfield Air Depot Planning Board, p 5, February 17, 1943, as cited in "Comprehensive History of Patterson and Wright Field Planning, November 1942-February 1945."

113. Hist, FAD, 1917-1943, p 99.

114. *Ibid.*, p 101.

115. *Ibid.*, p 111.

116. Hist, FADCAC and FASC, 1943-1944, p 135.

117. Hist, FAD, 1917-1943, pp 113-114.

118. *Ibid.*, p 102.

119. Hist, FADCAC and FASC, 1943-1944, pp 137-138.

120. *Ibid.*, pp 141-145.

121. *Ibid.*, pp 145-146.

122. Hist, FAD, 1917-1943, pp 118-125.

123. *Ibid.*, p 121.

124. Doris A. Baker, *History of AMC Field Organization 1917-1955* (Air Materiel Command, Wright-Patterson Air Force Base, Ohio, 1956).

125. *The Post Script*, September 21, 1945; *The Post Script*, September 28, 1945; *The Post Script*, November 16, 1945; *The Wright-Patterson Post Script*, November 21, 1945; *The Wright-Patterson Post Script*, December 14, 1945; *The Wright-Patterson Post Script*, December 28, 1945.

126. "Integration of Wright and Patterson Fields Into a Single Unit."

127. Army Air Forces Technical Base Planning Board, Preliminary Master Plan Report, Approved March 18, 1947, pp 15-16.

Chapter 4. FULCRUM OF BASE SUPPORT

1.   Authority for the merger was General Orders No. 2, Headquarters U.S. Air Force, January 13, 1948.

2.   History of the 2750th Air Base Wing, July-December 1948, p 3.

3.   General Orders 32, Headquarters Wright-Patterson Air Force Base, August 27, 1948.

4.   Bernard J. Termena, Layne B. Peiffer, H. P. Carlin, *Logistics: An Illustrated History of AFLC and Its Antecedents 1921-1981* (Air Force Logistics Command, Wright-Patterson Air Force Base, 1983), pp 97-99; Alfred Goldberg, ed., *A History of the United States Air Force* (New York, 1957), p 241.

5.   *A Pictorial Review, Wright-Patterson Air Force Base, 1917-1967* (2750 Air Base Wing, Wright-Patterson Air Force Base, 1967), pp 41-42.

6.   History of the 2750th Air Base Wing, January-June 1951, pp 1-2, 4, 33.

7.   History of the 2750th Air Base Wing, January-June 1952, pp 63-64.

8.   Albert E. Misenko and Philip H. Pollock, *Engineering History, 1917-1978, McCook Field to the Aeronautical Systems Division*, 4th ed. (Aeronautical Systems Division, Wright-Patterson Air Force Base, 1979), p 16.

9.   History of the 2750th Air Base Wing, July-December 1953, pp 91-92.

10.   History of the 2750th Air Base Wing, January-June 1952, p 33, 51; History of the 2750th Air Base Wing, July-December 1952, p 77.

11.   *History of the AMC Contract Airlift System (LOGAIR) 1954-1955* (Air Materiel Command, Wright-Patterson Air Force Base, 1956), pp 17-18.

12.   *Ibid.*, p 35; *Logistics: An Illustrated History of AFLC*, pp 114-117.

13.   History of the 2750th Air Base Wing, January-December 1957, pp 22-23.

14.   Base Guide, 1956 (2750 Air Base Wing, Wright-Patterson Air Force Base, 1956), p 11.

15.   History of the 2750th Air Base Wing, July 1960-July 1961, pp 18, 41.

16.   History of the 2750th Air Base Wing, July 1967-June 1969, p 36.

17.   Wright-Patterson Air Force Base *Skywrighter*, June 17, 1960.

18.   Wright-Patterson Air Force Base *Skywrighter*, June 9, 1961.

19.   History of the 2750th Air Base Wing, July 1974-December 1975, p 156.

20.   *Ibid.*, p 34.

21.   Robert B. Clayton, chief, Civil Law, 88th Air Base Wing/JAC, personal communication with Lori S. Tagg, Aeronautical Systems Center History Office, August 2003.

22.   Henry M. Narducci, *Ghosts From the Past: The Recovery of M114 Biological Weapons at Wright-Patterson Air Force Base, Ohio, November 1995-September 1996* (Aeronautical Systems Center History Office, Wright-Patterson Air Force Base, 1999).

23.   "War Fighter Support Fact Sheet," News Release, ASC Public Affairs Office, Wright-Patterson Air Force Base, June 23, 2003; "Operation Iraqi Freedom—By The Numbers," Assessment and Analysis Division, U.S. Central Air Forces, Shaw Air Force Base, South Carolina, April 30, 2003.

24.   Henry M. Narducci, *Balkan Proximity Peace Talks and Wright-Patterson Air Force Base, 18 October – 22 November 1995*, Volume 1: Narrative (88 Air Base Wing, ASC, Wright-Patterson Air Force Base, 1997).

25.   Wright-Patterson Air Force Base *Skywrighter*, August 23, 2002.

26.   History of the 2750th Air Base Wing, July 1967-June 1969, p 12.

27.   Al Moyers, Air Force Weather History Office, Air Force Weather Agency, Offutt Air Force Base, Nebraska, personal communication with Lori S. Tagg, Aeronautical Systems Center History Office, July 25, 2003.

28.   General Orders 6, Headquarters Strategic Air Command, February 6, 1959.

29.   According to Master Sergeant David L. Wolf, Aeronautical Systems Center History Office.

30.   *Against the Wind: 90 Years of Flight Test in the Miami Valley* (Aeronautical Systems Center History Office, Wright-Patterson Air Force Base, 1994), p 125.

31.   History of the 2750th Air Base Wing, January-December 1976, p 48.

32.   Lois E. Walker and Shelby E. Wickam, *From Huffman Prairie to the Moon: the History of Wright-Patterson Air Force Base* (2750 Air Base Wing, Wright-Patterson Air Force Base, 1986), pp 440-442.

33.   *Ibid.*, pp 435-437.

34. *Ibid.*, pp 445-446.
35. Mike Wallace, "Band owes start to music man," Wright-Patterson Air Force Base *Skywrighter*, July 13, 2001.
36. Walker and Wickam, pp 433-434.
37. History of the 2750th Air Base Wing, January-June 1951, pp 2-3.
38. History of the 2750th Air Base Wing, January-June 1954, p 46.
39. James J. Niehaus, *Five Decades of Materials Progress 1917-1967* (Wright-Patterson Air Force Base, 1967), p 96.
40. Wright-Patterson Air Force Base *Skywrighter*, August 18, 1967.
41. Wright-Patterson Air Force Base *Skywrighter*, June 9, 1967.
42. Wright-Patterson Air Force Base *Skywrighter*, September 26, 1969.

Chapter 5. WRIGHT-PATTERSON AND THE MIAMI VALLEY

1. *For the Love of Dayton: Life in the Miami Valley, 1796*-1996, ed. by Ron Rollins (Dayton, 1996), p 160.
2. *Ibid.*, p 168.
3. *Ibid.*, p 165-166.
4. *Ibid.*, p 164.
5. *Ibid.*, p 168.
6. *Ibid.*
7. *Ibid.*
8. *Ibid.*, p 163.
9. *Ibid.*, p 168
10. *Ibid.*, p 164.
11. Reminiscence, [by Mrs. Corwin], "The Big Old Kitchen," Corwin papers, n.d.; provided courtesy of Mr. James Sprinkle.
12. See Chapter 1.
13. See Appendix 6
14. *For the Love of Dayton*, p 168.
15. Charlotte Reeve Conover, *Dayton, Ohio: An Intimate History* (Dayton, Ohio, 1995), p 233.
16. Tom Crouch, *The Bishop's Boys: A Life of Wilbur and Orville Wright* (New York and London, 1989), p 476.
17. Virginia and Bruce Ronald, *The Lands between the Miamis: A Bicentennial Celebration of the Dayton Area* (Dayton, Ohio, 1996), p 325.
18. *Ibid.*, p 326.
19. *Ibid.*, p 326; *For Love of Dayton*, p 182.
20. *Ibid.*, p 327.
21. *Ibid.*, p 330.
22. *Ibid.*, pp 329, 330, 330-331, and 327, respectively.
23. *Ibid.*, p 337.
24. Conover, p 34; Ronald, p 64
25. Ronald, p 337-8.
26. *Ibid.*, p 338; *For Love of Dayton*, p 246.
27. *American Historical Documents*, ed. by Harold C. Syrett (New York, 1960), p 413.
28. Ronald, p 332.
29. *For the Love of Dayton*, p 193.
30. *Ibid.*, p 188.
31. *Ibid.*, p 194.
32. Ronald, pp 402-3.
33. *Ibid.*, p 342.
34. *For the Love of Dayton*, p 222.
35. *Ibid.*, p 242.
36. *Ibid*, p 227; Ronald, 345.
37. Schuster Performing Arts Center, Dayton, Ohio, website, viewed online August 3, 2003, at http://www.schustercenter.org/about/general.html.
38. *For the Love of Dayton*, p 183.
39. *Ibid.*, 214.
40. Ronald, p 210.
41. *For the Love of Dayton*, p 226.
42. *Ibid.*, p 244.
43. *Ibid.*, p 243.
44. "Fifth Third Field: Fun, Facts & Trivia," viewed online October 2003, at http://www.daytondragons.com/stadium/facts.html.
45. "Riverscape," viewed online October 2003 at http://www.metroparks.org/Facilities/RiverScape/riverscape.html.
46. *For the Love of Dayton*, p 201.
47. Samuel Crowther, *John H. Patterson: Pioneer in Industrial Welfare* (Garden City, New York, 1924), pp 62-63.

48. "Benjamin & Marian Schuster Performing Arts Center," viewed online August 3, 2003, at http://www.schustercenter.org/about/general.html.

49. Conover, p 122.

50. Preservation Dayton website, viewed online October 2003 at http://www.preservationdayton.com.

51. *For the Love of Dayton*, p 233.

52. *Ibid.*, p 178.

53. *Ibid.*, p 215.

54. *Ibid.*, p 234.

55. *Ibid.*, p 168.

56. *Ibid.*, p 209.

57. *Ibid.*, p 222.

58. *Ibid.*, p 248.

59. *Ibid.*, p 243.

60. *Ibid.*, p 252.

61. "About NCR – History," viewed online October 2003 at http://www.ncr.com/history.htm.

62. *For the Love of Dayton*, p 224.

63. *Ibid.*, p 228.

64. *Ibid.*, p 234.

65. *Ibid.*, p 243.

66. *Ibid.*, p 167.

67. *Ibid.*, p 237.

68. *Ibid.*, p 251.

69. Press release, "Mead and Westvaco Complete Merger Creating Strong Global Organization," January 30, [2002].

70. *For the Love of Dayton*, p 178.

71. *Ibid.*, p 180.

72. *Ibid.*, p 210.

73. *Ibid.*, p 213.

74. *Ibid.*, p 226.

75. *Ibid.*, p 182.

76. *Ibid.*, p 196.

77. *Ibid.*, p 226.

78. *Ibid.*, p 242.

79. *Ibid.*, p 196.

80. Ronald, p 402.

81. *Ibid.*, p 402.

82. *For the Love of Dayton*, p 202.

83. Ronald, p 400.

84. *Ibid.*, p 400; University of Dayton Research Institute website, viewed online October 2003 at http://www.udri.udayton.edu.

85. *For the Love of Dayton*, p 234.

86. *Ibid.*, p 240.

87. "Economic Impact Analysis," September 30, 2002, Wright-Patterson Air Force Base, 88 Air Base Wing/FMA, p 6.

88. *Ibid.* p 11.

89. *Ibid.* p 9.

90. *Ibid.* p 14.

91. "July 2002 Population Estimates for Sub-County Areas in Ohio," viewed online August 14, 2003, at http://nodisnet1.csuohio.edu/nodis/2000reports/Ohio_sub_cnty_est_2002.pdf.

92. "EMTEC," viewed online October 18, 2003, at http://www.emtec.org/programs/programs.html; http://www.emtec.org/capabil/capabil.html; http://www.emtec.org/edtech/edtech.html; http://www.emtec.org/mem_shp/mem_shp.html; Letter dated August 12, 1987 by Lieutenant General William. E. Thurman to Dr. Ernest Moore, director, EMTEC; "Wright-Patterson Air Force Base Awarded Thomas Edison Award," *The WTN Innovator*, November/December 1997, p.1.

93. Fact Sheet, "Wright-Laboratory Technology Transfer from Wright-Patterson Air Force Base, Ohio," PAM #92-015.

94. Often, the non-federal entity paid royalties to the government for the licensing of a patented process developed by the government. By Air Force regulation, 20 percent of the royalties generated from the license of an Air Force patent went to the inventor and the remaining royalties went to the directorate from which the invention originated. Fact Sheet, "Cooperative Research and Development Agreements in Wright Laboratory," PAM #91-146.

95. Fact Sheet, "Continuing a Legacy of Innovation and Technology Development," AFRL/XPTT; "Deep Roots: Kettering Medical Center Network fosters new technology, community outreach," *Dayton Daily News*, Forecast 03 Health advertising supplement, April 13, 2003.

96. "Air Force SBIR" pamphlet, 2000.

97. "Air Force Independent Research and Development" pamphlet, AFRL/XPTT, n.d.

98. "TECH CONNECT" pamphlet, AFRL/XPTT, n.d.

99. Interview with Dr. Vince Russo, ASC executive director, June 2003.

100. National Composite Center website, viewed online June 11, 2003, at http://www.compositecenter.org/newsview2.php?ID=5.

101. "UD, National Composite receive nearly $4 million," *Dayton Daily News*, June 6, 2003, p D1. "Taft awards $1 million to NCC," *Dayton Daily News*, October 17, 2003, p D1.

102. Press Release, "Air Force Awards $7.2 Million Contract to Adapt Automotive Manufacturing Process for Aerospace Composites," PAM #97-117.

103. National Composite Center website, viewed online June 11, 2003, at http://www.compositecenter.org/history.php.

104. National Composite Center Fact Sheet, "The National Composite Center and Martin Marietta Composites Hold Special Ceremony to Unveil Largest Composite Bridge Installation to Date," viewed online May 2003, at http://www.compositecenter.org/newsview2.php?ID=17.

105. AFIT website, viewed online May 2003 at http://www.afit.edu.

106. SOCHE website, viewed online May 2003 at http://www.soche.org/index.htm.

107. DAGSI website, viewed online May 2003 at http://www/dagsi.org/index.shtml.

108. "AFIT Summer engineering program under way," Wright-Patterson Air Force Base *Skywrighter*, July 3, 2003, p.8A.

109. "AFIT's Systems, Logistics School to get temporary home," Wright-Patterson Air Force Base *Skywrighter*, July 3, 2003, p. 1A.

110. Sheila Wallace, program co-coordinator and Lincoln School teacher, stated "AFIT has been wonderful in their support of this program, and our students really enjoy the opportunity to share their scientific discoveries."

111. "Air Force Institute of Technology" section written by Kim Curry.

112. Air Force program manger for the University of Dayton curriculum is Rosalie Bonacci-Robert of the Aeronautical Systems Center's Directorate of Engineering.

113. Bill Kugel, chief of Human Resources for Aeronautical Systems Center FM, is the Air Force initiator and manager of the University of Cincinnati UC Mid-Day MBA program. University of Cincinnati website, viewed online May 2003 at http://www.business.uc.adv/mba; University of Dayton website, viewed online April 2003 at http://www.udayton.edu/ with links.

114. Fact Sheet, "Student Career Experience Program," 88 MSG/DPCXB, November 2002.

115. "Outreach office efforts reach Columbus," Wright-Patterson Air Force Base *Skywrighter*, August 1, 2003, p 3A.

116. *Ibid.*

117. "Wright Scholar intern program expands across base," Wright-Patterson Air Force Base *Skywrighter*, June 27, 2003, p 6A; Wright Scholar Research Assistant, viewed online July 2003 at http://www.pr.afrl.af.mil/jobs/scholar.htm.

118. "Rain Birds: Air Force display goes on despite heavy storm, " *Dayton Daily News*, May 11, 2003, p A1; "Open house draws weekend crowds," Wright-Patterson Air Force Base *Skywrighter*, May 16, 2003, p 7A.

119. According to Lieutenant Colonel Alan Sierichs, band commander.

120. "Bands orchestrate 100 years," Wright-Patterson Air Force Base *Skywrighter*, April 18, 2003, p 1A; "AF Band orchestrates aviation's first 100 years," *Leading Edge*, June 2003, p 26; 2nd Lieutenant Gailyn Whitman, personal communication with Helen Kavanaugh-Jones, summer 2003.

121. "'First Flight' stamp unveiled," Wright-Patterson Air Force Base *Skywrighter*, May 30, 2003, p 1A.

122. "World War I aviation to roar to life at Air Force Museum," Wright-Patterson Air Force Base *Skywrighter*, September 5, 2003, p 8A.

123. "Centennial activities take off at flying field," Wright-Patterson Air Force Base *Skywrighter*, July 3, 2003, p 3A

124. "It's Wheels Up," *Dayton Daily News*, July 18, 2003, p A2.

125. "AFRL presents century of flight awards," Wright-Patterson Air Force Base *Skywrighter*, July 25, 2003, p 7A.

126. "Exposition offers look at future of flight," *Dayton Daily News*, Inventing Flight special section, July 15, 2003, p 3.

127. "AIRBORNE: How Dayton's national park took flight," *Dayton Daily News*, October 19, 1992, p 9A.

128. "Park takes flight: Bush signs Dayton aviation park bill," *Dayton Daily News*, October 17, 1992, p 1A.

129. Dr. Jan Ferguson, 88 Air Base Wing, Office of Environmental Management, email to Helen Kavanaugh-Jones, Aeronautical Systems Center History Office, June 2, 2003.

130. Dave Egner, email dated May 23, 2003 and August 2003, to Helen Kavanaugh-Jones; "Open house draws weekend crowds," Wright-Patterson Air Force Base *Skywrighter*, May 16, 2003, p. 7A; Inventing Flight special section, *Dayton Daily News*, July 5, 2003; "AFRL presents Century of Flight Awards to top students," Wright-Patterson Air Force Base *Skywrighter*, July 11, 2003, p 8A; "Base support to Dayton air show doubles for larger event," Wright-Patterson Air Force Base *Skywrighter*, July 11, 2003, p 3A; "Community salutes base in centennial events today," Wright-Patterson Air Force Base *Skywrighter*, July 11, 2003, p 1A.

131. James H. Doolittle, *I Could Never Be So Lucky Again: An Autobiography*, with Carroll V. Glines (New York, 1991; reprint ed., Atglen, Pennsylvania, n.d.), p 87.

132. *For the Love of Dayton*, p 199.

133. *Ibid.*, pp 80-81.

134. Crowther, pp 118-19.

APPENDICES

APPENDIX 1

ORGANIZATIONAL EVOLUTION OF THE ACQUISITION AND LOGISTICS FUNCTIONS AT WRIGHT-PATTERSON AIR FORCE BASE AND ITS PREDECESSORS

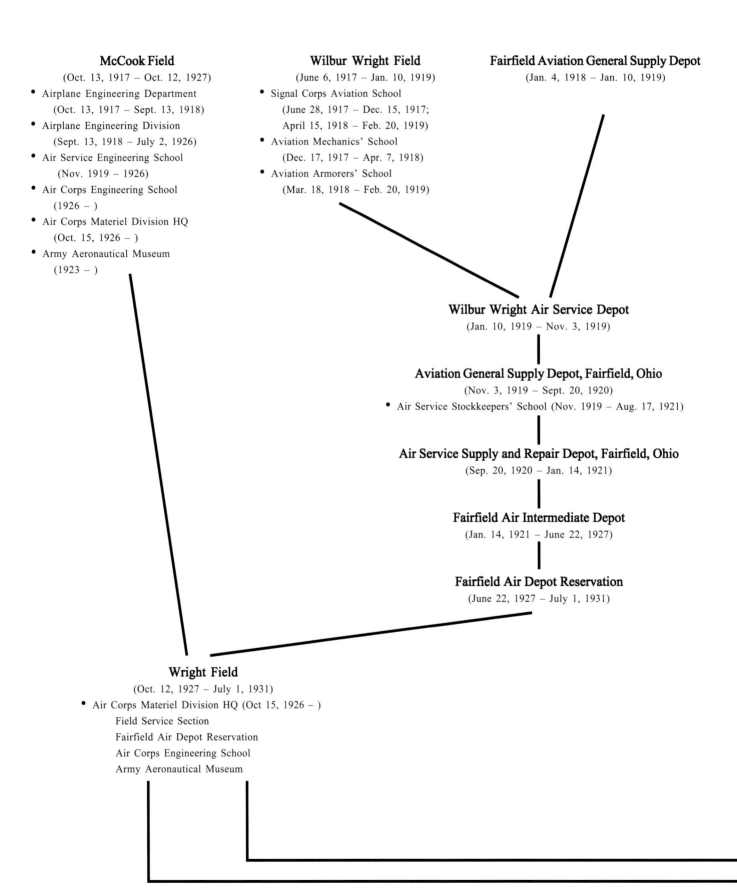

McCook Field
(Oct. 13, 1917 – Oct. 12, 1927)
- Airplane Engineering Department
  (Oct. 13, 1917 – Sept. 13, 1918)
- Airplane Engineering Division
  (Sept. 13, 1918 – July 2, 1926)
- Air Service Engineering School
  (Nov. 1919 – 1926)
- Air Corps Engineering School
  (1926 –)
- Air Corps Materiel Division HQ
  (Oct. 15, 1926 –)
- Army Aeronautical Museum
  (1923 –)

Wilbur Wright Field
(June 6, 1917 – Jan. 10, 1919)
- Signal Corps Aviation School
  (June 28, 1917 – Dec. 15, 1917;
  April 15, 1918 – Feb. 20, 1919)
- Aviation Mechanics' School
  (Dec. 17, 1917 – Apr. 7, 1918)
- Aviation Armorers' School
  (Mar. 18, 1918 – Feb. 20, 1919)

Fairfield Aviation General Supply Depot
(Jan. 4, 1918 – Jan. 10, 1919)

Wilbur Wright Air Service Depot
(Jan. 10, 1919 – Nov. 3, 1919)

Aviation General Supply Depot, Fairfield, Ohio
(Nov. 3, 1919 – Sept. 20, 1920)
- Air Service Stockkeepers' School (Nov. 1919 – Aug. 17, 1921)

Air Service Supply and Repair Depot, Fairfield, Ohio
(Sep. 20, 1920 – Jan. 14, 1921)

Fairfield Air Intermediate Depot
(Jan. 14, 1921 – June 22, 1927)

Fairfield Air Depot Reservation
(June 22, 1927 – July 1, 1931)

Wright Field
(Oct. 12, 1927 – July 1, 1931)
- Air Corps Materiel Division HQ (Oct 15, 1926 –)
  Field Service Section
  Fairfield Air Depot Reservation
  Air Corps Engineering School
  Army Aeronautical Museum

Wright Field
(July 1, 1931 – Dec. 15, 1945)
- Air Corps Materiel Division (– Oct. 2, 1939, transfer to D.C.)
    Army Aeronautical Museum (– June 1, 1940)
    Air Corps Engineering School (– Aug. 1941)
    Army Air Forces Engineering School (Aug. 1941 – Dec. 9, 1941)
- Materiel Center (Mar. 16, 1942 – Apr. 1, 1943)
- Materiel Command, transferred from D.C (Apr. 1, 1943 – Aug. 31, 1944)
    Army Air Forces Engineering School (Mar. 17, 1944 – Dec. 15, 1945)
- Air Technical Service Command (Aug. 31, 1944 – Mar. 9, 1946)*
- Army Air Forces Institute of Technology (Dec. 15, 1945 –)

Patterson Field
(July 1, 1931 – Dec. 15, 1945)
- Provisional Air Corps Maintenance Command (Mar. 15, 1941 – Apr. 29, 1941)
- Air Corps Maintenance Command (Apr. 29, 1941 – Dec. 15, 1942)
    Fairfield Air Depot (July 1, 1931 – Feb. 1, 1943)
        Air Corps Weather School (July 1, 1937 – June 1, 1940)
        Autogiro School (Apr. 20, 1938 – unknown)
- Air Service Command (Dec. 15, 1942 – Aug. 31, 1944)
    Fairfield Air Depot Control Area Command (Feb. 1, 1943 – May 17, 1943)
    Fairfield Air Service Command (May 17, 1943 – Dec. 6, 1944)
- Fairfield Air Technical Service Command (Dec. 6, 1944 –)

Army Air Forces Technical Base
(Dec. 15, 1945 – Dec. 9, 1947)
- Air Materiel Command (Mar. 9, 1946 –)
- Fairfield Air Technical Service Command (– Jan. 1, 1946)
- Air Force Institute of Technology (Dec. 5, 1947 –)

Air Force Technical Base
(Dec. 9, 1947 – Jan. 13, 1948)
- Air Materiel Command
- Air Force Institute of Technology

Wright-Patterson Air Force Base
(January 13, 1948 – Present)
- Air Materiel Command (– Apr. 1, 1961)
    Aeronautical Systems Center (Sept. 1958 – Apr. 1, 1961)
- Air Research and Development Command (Apr. 2, 1951 – June 1951, transferred to Baltimore)
    Wright Air Development Center (Apr. 2, 1951 – Dec. 15, 1959)
    Wright Air Development Division (Dec. 15, 1959 – Apr. 1, 1961)
- Air Force Logistics Command (Apr. 1, 1961 – July 1, 1992)
- Air Force Systems Command (Apr. 1, 1961 – July 1, 1992, headquarters at Andrews Air Force Base, Maryland)
    Aeronautical Systems Division (Apr. 1, 1961 – July 1, 1992)
- Air Force Materiel Command (July 1, 1992 – Present)
    Aeronautical Systems Center (July 1, 1992 – Present)
- Air Force Institute of Technology (– Present)
- National Air and Space Intelligence Center (May 21, 1951 – Present)
- United States Air Force Museum (Apr. 1954 – Present)
- United States Air Force Medical Center Wright-Patterson (July 1, 1969 – Present)
- 88th Air Base Wing (Apr. 1, 1944 – Present)

*The headquarters of the Air Technical Service Command originally was located on Patterson Field. When Air Service Command and Air Materiel Command merged in 1944, the acting commander of Air Service Command incorporated the part of Patterson Field occupied by headquarters into Wright Field. The headquarters portion of Wright Field then became known as Area A and the original Wright Field was known as Area B. The remainder of Patterson Field became Area C when it merged with Wright Field in 1948.

APPENDIX 2

DAYTON INDUSTRIES MOBILIZED TO HELP THE WAR EFFORT

With Wright and Patterson fields as neighbors, the industries of Dayton and surrounding cities adjusted their regular assembly lines into military supply lines during World War II. Some of the local companies that produced war materiel included:

⟨ **Acme Aluminum Alloys, Inc. (Dayton):** This machine-tool company turned to aluminum casting for the war effort, providing jigs, wood patterns, and engineering expertise.

⟨ **Aeronca Aircraft Corporation (Middletown):** Before the war, Aeronca supplied aircraft and aircraft parts. During the war, it produced the C-3, the PT-19, and the UC-64.

⟨ **American Rolling Mill Company (Middletown):** This company produced alloy and stainless steel ingots for propeller blades.

⟨ **Chandler-Evans Corporation (Dayton):** A new plant built in Dayton supplied the Army Air Forces with aircraft carburetors.

⟨ **Chrysler Corporation, Airtemps Plant (Dayton):** Originally a builder of refrigerators, it began to construct nose-cap assemblies for the A-20.

⟨ **City Engineering Company (Dayton):** This machine-tool company implemented new manufacturing processes, particularly automatic screw machines, to quicken assembly.

⟨ **Dayton Rubber Manufacturing Company (Dayton):** Once a supplier of tires and inner tubes for automobiles, Dayton Rubber began supplying the aircraft industry with tires, belts, pontoons, and rafts.

⟨ **Delco Products Division (Dayton):** Delco supplied the military long before the start of World War II. It constructed projectiles; 40-mm antiaircraft artillery; oleo struts; B-24, B-25, and B-26 aircraft engine parts; hydraulic actuating devices; and shock absorbers. When the war started, it stepped up production to meet the increased need.

⟨ **Duro Company (Dayton):** Duro stopped building pumps and began to supply the Army Air Forces with machine-gun mounts for airplanes.

⟨ **Frigidaire Division (Dayton):** Originally a manufacturer of air conditioners and refrigerators, Frigidaire built machine guns and .50 caliber ammunition.

⟨ **General Motors Corporation, Aeronautic Products Division (Moraine City and Vandalia):** General Motors produced the Hamilton propeller for commercial and military airplanes before, during, and after the war. It built a new plant in Vandalia to supply controllable pitch propellers.

⟨ **Harris-Seybold-Potter Company (Dayton):** This company switched from the production of paper mill machinery to projectiles.

⟨ **Hartzell Industries, Inc. (Piqua):** This supplier began production of Curtiss steel propeller blades.

⟨ **Hobart Manufacturing Company (Troy):** Hobart switched from the manufacturing of food-products machinery to producing mount telescopes and turret drivers.

⟨ **Inland Manufacturing Division (Dayton):** Once a supplier of automobile parts and plastics, Inland evolved into a manufacturer of carbines, .30 caliber M-1 submachine guns, .45 caliber M-2 guns, and Oerlikon gunsights.

⟨ **Joyce Cridland Company (Dayton):** This company halted the building of elevators and conveyers to build airplane wings and tail and wheel jacks.

⟨ **Lear, Inc. (Piqua):** Lear continued to provide the military with screw jacks, arresting hook retractors, compasses, and remote control direction finders, as well as aircraft actuators.

⟨ **Leland Electric Company (Dayton):** Once an electrical equipment supplier, Leland began supplying motor generator sets.

⟨ **National Cash Register Company (Dayton):** For years, NCR had supplied cash registers to the nation. When the war started, it began supplying mechanical time fuses, Sperry compensating sights for machine guns, bombsights, gun magazines, and 20-mm antiaircraft artillery shells.

⟨ **Northwest Airlines Incorporated, U.S. Government Modification Center (Vandalia):** This air transportation company helped the federal government run service tests.

⟨ **Plastic Metal, Inc. (Dayton):** A supplier to the Army Air Forces prior to the war, Plastic Metal continued to provide hydraulic packing rings and cups for aircraft engines.

⟨ **Sheffield Corporation (Dayton):** Sheffield ceased building machine tools during the war to produce gauging machines for the military.

⟨ **United Aircraft Products, Inc. (Dayton):** An aircraft-parts supplier, this company continued to supply aircraft oil-temperature regulators, fuel pumps and locks, solenoids, rod ends and dial assemblies, valves, and strainers.

⟨ **Wright Aeronautical Corporation (Lockland):** Wright built this new plant for R-2600 aircraft engine manufacturing.

Source: "World War II Industrial Facilities: Authorized Federal Funding," Heritage Research Center, Ltd., viewed online June 30, 2003, at http://www.heritageresearch.com/War%20Facilities5.html.

APPENDIX 3

LAND ACQUISITIONS

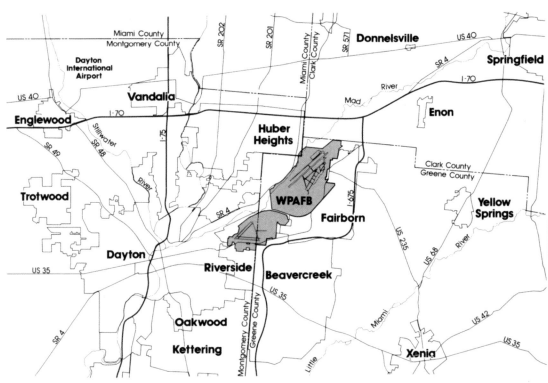

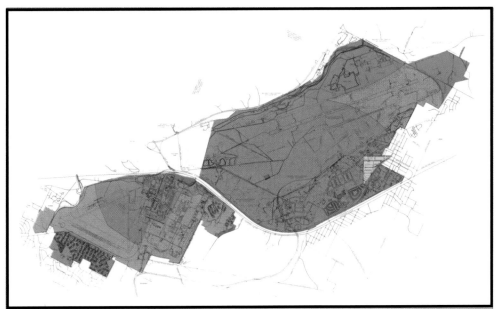

Land Acquisition Map

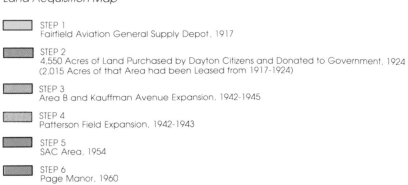

STEP 1
Fairfield Aviation General Supply Depot, 1917

STEP 2
4,550 Acres of Land Purchased by Dayton Citizens and Donated to Government, 1924
(2,015 Acres of that Area had been Leased from 1917-1924)

STEP 3
Area B and Kauffman Avenue Expansion, 1942-1945

STEP 4
Patterson Field Expansion, 1942-1943

STEP 5
SAC Area, 1954

STEP 6
Page Manor, 1960

STEP 7
Wright Memorial and Huffman Prairie Site Transferred to U.S. Air Force, 1978

APPENDIX 4

U.S. AIR FORCE UNITS ASSIGNED
AT WRIGHT-PATTERSON AIR FORCE BASE AND ITS PREDECESSORS

1917

12th Aero Squadron	July 1917-December 1917
13th Aero Squadron	July 1917-December 1917
19th Aero Squadron	August 1, 1917-October 31, 1917
20th Aero Squadron (I)	July 1917-November 1917
42nd Aero Squadron/Squadron I	August 25, 1917-February 21, 1919
43rd Aero Squadron	August 25, 1917-December 18, 1917
44th Aero Squadron/Squadron K/Squadron P	August 25, 1917-April 30, 1919
47th Aero Squadron	July 1917-August 1917
149th Aero Squadron	December 1917-February 1918
151st Aero Squadron	December 1917-February 1918
162nd Aero Squadron	December 1917-February 1918
163rd Aero Squadron	December 1917-February 1918
166th Aero Squadron	December 24, 1917-February 20, 1918
172nd Aero Squadron	December 1917-February 1918
211th Aero Squadron	December 1917-February 1918
246th Aero Squadron/807th Aero Squadron/Squadron A	November 1917-August 1918

1918

231st Aero Squadron/Squadron A	April 1918-December 1918
246th [II] Aero Squadron/Squadron L	May 1918-April 1919
255th Aero Squadron	March 1918-March 1918
256th Aero Squadron	March 1918-March 1918
257th Aero Squadron	March 1918-March 1918
258th Aero Squadron	March 1918-March 1918
259th Aero Squadron	March 1918-July 1918
260th Aero Squadron	March 1918-July 1918
265th Aero Squadron	February 1918-March 1918
287th Aero Squadron	May 1818-June 1918
288th Aero Squadron	May 1918-June 1918
342nd Aero Squadron/Squadron M/Squadron Q	August 1918-November 1918
507th Aero Squadron	July 1918-April 1919
669th Aero Squadron	May 1918-April 1919
678th Aero Squadron	February 1918-April 1919
827th Aero Squadron	February 1918-March 1918
851st Aero Squadron/Squadron B	March 1918-December 1918
874th Aero Squadron	April 1918-December 1918
881st Aero Squadron/Squadron B	July 1918-August 1918

1919

Air Service School of Application/Air Corps Engineering School/Army Air Forces Institute of Technology/Air Force Institute of Technology	November 1919-present

1922

88th Squadron/88th Observation Squadron	October 11, 1922-May 4, 1927

1926

Air Corps Materiel Division/Materiel Command/ Army Air Forces Materiel Command	October 15, 1926-August 31, 1944

1927

Fairfield Air Intermediate Depot/Fairfield Air Depot	c. June 1927-January 1, 1946

1935

1st Provisional Transport Squadron/1st Transport Squadron	July 15, 1935-May 20, 1942
Flight B, 1st Transport Squadron	June 25, 1935-May 19, 1942

1937

Aeromedical Laboratory	January 1937-unknown

1939

5th Transport Squadron	October 17, 1939-May 25, 1942

1940

9th Transport Squadron	December 1, 1940-September 17, 1941
11th Transport Squadron	December 1, 1940-July 3, 1941
13th Transport Squadron	December 1, 1940-July 9, 1941
19th Bomb Squadron	February 1, 1940-November 15, 1940
33rd Bomb Squadron	February 1, 1940-November 14, 1940
60th Transport Group	December 1, 1940-September 15, 1941

1941

2nd Air Corps Weather Region/2nd Weather Region	April 4, 1941-September 20, 1944
4th Air Depot Group	April 1, 1941-January 12, 1942
4th Repair Squadron	April 1, 1941-January 12, 1942
4th Supply Squadron/4th Depot Supply Squadron	January 1, 1941-January 12, 1942
50th Transport Wing	January 14, 1941-May 22, 1942
63rd Transport Group	February 17, 1941-September 15, 1941
Air Corps Maintenance Command	June 25, 1941-September 9, 1942
Second Weather Squadron/2nd Weather Squadron	April 4, 1941-September 7, 1944
Training Film Field Unit/Training Film Production Laboratory	April 19, 1941-February 7, 1944

1942

1st Radio Squadron	January 17, 1942-April 1, 1944
3rd Station Complement, Air Depot	June 1, 1942-January 13, 1943
6th Air Depot Group	January 5, 1942-June 6, 1942
6th Repair Squadron/6th Depot Repair Squadron	January 5, 1942-June 6, 1942
6th Supply Squadron/6th Depot Supply Squadron	January 5, 1942-June 6, 1942
IX AFSC	September 1, 1942-October 9, 1942
10th Air Depot Group	January 5, 1942-July 20, 1942
10th Repair Squadron	January 5, 1942-July 20, 1942
10th Supply Squadron	January 5, 1942-July 20, 1942
15th Statistical Control Unit	August 31, 1942-February 11, 1943
18th Air Depot Group	January 19, 1942-August 11, 1943
19th Air Depot Group	January 19, 1942-May 10, 1943

19th Depot Repair Squadron	January 19, 1942-May 10, 1943
19th Depot Supply Squadron	January 19, 1942-May 10, 1943
20th Air Depot Group	January 19, 1942-May 2, 1942
20th Depot Repair Squadron	January 19, 1942-May 2, 1942
20th Depot Supply Squadron	January 19, 1942-May 2, 1942
21st Air Depot Group	January 19, 1942-October 16, 1942
21st Depot Repair Squadron	January 19, 1942-October 16, 1942
21st Depot Supply Squadron	January 19, 1942-October 16, 1942
22nd Air Depot Group	January 19, 1942-August 6, 1942
22nd Depot Repair Squadron	January 19, 1942-August 6, 1942
22nd Depot Supply Squadron	January 19, 1942-August 6, 1942
23rd Air Depot Group	January 19, 1942-June 9, 1942
23rd Depot Repair Squadron	January 19, 1942-June 9, 1942
23rd Depot Supply Squadron	January 19, 1942-June 9, 1942
36th Transport Squadron	February 14, 1942-June 15, 1942
37th Transport Squadron	February 14, 1942-June 16, 1942
38th Transport Squadron	February 2, 1942-May 28, 1942
44th Transport Squadron	June 15, 1942-June 16, 1942
45th Transport Squadron	June 15, 1942-June 16, 1942
55th Air Depot Group	February 16 1942-August 14, 1943
55th Depot Repair Squadron	February 16 1942-June 4, 1943
55th Depot Supply Squadron	February 16 1942-April 23, 1943
56th Air Depot Group	February 16 1942-March 5, 1943
56th Depot Repair Squadron	February 16 1942-March 5, 1943
56th Depot Supply Squadron	February 16 1942-March 5, 1943
57th Air Depot Group	February 16 1942-July 20, 1943
57th Depot Repair Squadron	February 16 1942-July 20, 1943
57th Depot Supply Squadron	February 16 1942-July 20, 1943
98th Aviation Squadron	September 8, 1942-April 1, 1944
316th Transport Group	February 14,1942-June 17, 1942
361st Army Air Forces Band/661st Army Air Forces Band/661st Air Force Band/Air Force Band of Flight/ U.S. Air Force Band of Flight	Unknown 1942-present
477th Base Headquarters & Air Base Squadron	December 31, 1942-February 17, 1943
836th Military Police Company [Aviation]/836th Guard Squadron	May 19, 1942-April 1, 1944
923rd Guard Squadron	August 5, 1942-April 1, 1944
1072nd Guard Squadron	September 15, 1942-April 1, 1944
Air Service Command	September 10, 1942-April 1, 1944
Aircraft Radio Laboratory	November 15, 1942-October 16, 1944
HQ & HQ Squadron, Tenth Air Force	February 12, 1942-March 10, 1942
Signal Corps Aircraft Signal Service/Signal Corps Aircraft Signal Agency	June 12, 1942-October 16, 1944

1943

1st Plant Maintenance Squadron	April 15, 1943-April 1, 1944
15th Statistical Control Unit	February 12, 1943-April 1, 1944
18th Depot Supply Squadron	January 19, 1943-August 11, 1943
85th Depot Repair Squadron	March 8, 1943-August 11, 1943
85th Depot Supply Squadron	March 15, 1943-August 13, 1943
88th Depot Repair Squadron	March 15, 1943-December 6, 1943
97th Depot Supply Squadron	April 15, 1943-December 20, 1943
98th Service Group	April 23, 1943-August 6, 1943
308th Air Base Group	April 23, 1943-August 6, 1943
315th Depot Repair Squadron	April 22, 1943-October 28, 1943
345th Aviation Squadron	April 21, 1943-February 8, 1944
351st Aviation Squadron	January 29, 1943-April 1, 1944
371st Aviation Squadron	March 5, 1943-April 1, 1944
372nd Aviation Squadron	March 5, 1943-April 1, 1944
407th Service Squadron	August 25, 1943-March 25, 1944
435th Aviation Squadron	May 10, 1943-January 25, 1944
436th Aviation Squadron	May 10, 1943-January 25, 1944

437th Aviation Squadron | May 10, 1943-January 25, 1944
478th Base Headquarters & Air Base Squadron | January 2, 1943-April 1, 1944
Dayton Signal Corps Publications Agency | November 20, 1943-October 16, 1944
Fairfield Air Depot Control Area Command/Fairfield
 Air Service Command/Fairfield Air Technical Service Command | February 1, 1943-January 1, 1946
Finance Office, U.S. Air Force | July 19, 1943-February 29, 1948
HQ Army Air Forces Crystal Bank | December 15, 1943-July 1, 1945

1944

69th Army Air Forces Base Unit | September 7, 1944-October 1, 1945
345th Aviation Squadron | February 8, 1944-June 6, 1944
435th Aviation Squadron | January 25, 1944-February 11, 1944
436th Aviation Squadron | January 25, 1944-February 11, 1944
437th Aviation Squadron | January 25, 1944-February 11, 1944
463rd Aviation Squadron | November 21, 1944-May 25, 1945
4000th Army Air Forces Base Unit/4000th Air Force Base Unit | April 1, 1944-August 27, 1948
4001st Army Air Forces Base Unit | April 1, 1944-August 31, 1944
4020th Army Air Forces Base Unit/4020th Air Force Base Unit | August 31, 1944-August 26, 1948
4100th Army Air Forces Base Unit | April 1, 1944-February 21, 1946
4141st Army Air Forces Base Unit | August 31, 1944-February 21, 1945
5900th Army Air Forces Base Unit | April 1, 1944-April 15, 1945
5901st Army Air Forces Base Unit | April 1, 1944-April 15, 1945
Army Air Forces Materiel & Services/Army Air Forces
 Air Technical Service Command/Air Technical Service
 Command/Air Materiel Command/Air Force Logistics Command | July 14, 1944-July 1, 1992
Army Air Forces Regional Station Hospital | June 7, 1944-January 23, 1946
Disposal Board | June 29, 1944-October 16, 1944

1945

86th Depot Repair Squadron | September 13, 1945-November 12, 1945
91st Air Depot Operation | September 11, 1945-November 12, 1945
94th Depot Supply Squadron | September 13, 1945-November 12, 1945
4021st Army Air Forces Base Unit | November 8, 1945-January 31, 1946
4265th Army Air Forces Base Unit | September 16, 1945-July 1, 1946
4517th Army Air Forces Base Unit | July 15, 1945-August 1, 1945
Dayton Army Air Forces Procurement Field Office | October 29, 1945-unknown

1946

72nd Army Air Forces Base Unit | August 1, 1946-April 21, 1947
733rd Army Air Forces Base Unit/733rd Air Force Base Unit | October 21, 1946-June 3, 1948
Army Air Forces Regional Station Hospital | January 23, 1946-June 26, 1946
Army Field Printing Plant | c. May 1, 1946-unknown

1947

6th Photographic Technical Unit | September 22, 1947-November 15, 1948
7th Air Vehicle Repair Squadron | September 25, 1947-November 15, 1947
17th Motor Transport Squadron, Air Depot | September 25, 1947-November 15, 1947

1948

XVII Air Service Command, Special | March 21, 1948-April 24, 1951
103rd Airways and Air Communications Service
 Squadron/1914th Airways and Air Communications
 Service Squadron | June 1, 1948-July 18, 1956
901st Air Transport Squadron/1726th Air Transport Squadron | June 1, 1948-July 17, 1950
1702nd Air Transport Group | October 1, 1948-July 17, 1950

2790th Station Medical Squadron/2790th Base Medical
Complement/2790 Medical Group/2750th Medical
Group/2750th U.S. Air Force Hospital/2750th U.S.
Air Force Hospital, Wright-Patterson/U.S. Air Force
Medical Center, Wright-Patterson/Medical Center
Wright-Patterson Air Force Base/645th Medical Group/
74th Medical Group — August 28, 1948-present

2925th Base Supply/2925th Base Supply Group/2750th
Base Supply Group — August 28, 1948-May 1, 1954

2926th Base Maintenance/2926th Base Maintenance
Group/2750th Base Maintenance Group — August 28, 1948-May 1, 1954

3000th Women in the Air Force Squadron/2750th
Women in the Air Force Squadron — August 27, 1948-May 1, 1954

3051st Air Base Support Squadron — August 28, 1948-June 20, 1949

3060th Air Materiel Command Support Squadron/
3060th Support Squadron — August 28, 1948-January 1, 1953

4000th Air Force Base Unit/2750th Air Force Base
Unit/2750th Air Base Wing/645th Air Base Wing/
88th Air Base Wing — January 13, 1948-present

4140th Air Force Base Unit/3100th U.S. Air Force
Exhibit Unit/U.S. Air Force Exhibit Group/U.S.
Air Force Orientation Group/Orientation Group, U.S. Air Force — January 13, 1948-April 1, 1992

HQ Civilian Installations Group — November 1, 1948-January 11, 1949

1949

1st Weather Squadron	May 20, 1949-April 28, 1952
32nd Weather Squadron	October 3, 1949-June 23, 1951
342nd Air Base Group, Depot	July 27, 1949-April 24, 1951
342nd Communications Squadron, Depot	July 27, 1949-April 24, 1951
342nd Finance Disbursing Unit	December 2, 1949-April 24, 1951
3019th Base Services Squadron	October 5, 1949-May 20, 1950
3020th Air Police Squadron	October 4, 1949-May 20, 1950
3086th Air Base Group	October 5, 1949-May 20, 1950
3090th Installations Group	October 5, 1949-February 5, 1951
8512th Air Transport Group	September 1, 1949-May 1, 1950
8525th Air Transport Squadron	September 1, 1949-May 1, 1950
8548th Airways and Air Communications Service Squadron	September 1, 1949-July 3, 1951

1950

2nd Combat Camera Unit/2nd Photographic Squadron — September 1, 1950-September 5, 1951

97th Fighter-Interceptor Squadron — December 1, 1950-August 18, 1955

1062nd U.S. Air Force Communications Squadron/
2046th Airways and Air Communications Service
Squadron/2046th Communications Squadron/
2046th Communications Group/2046th Communications
& Installations Group/2046th Communications Group/
2046th Communications-Computer Systems Group/
645th Communications-Computer Systems Group/
88th Communications Group — December 18, 1950-present

Air Research and Development Command — November 16, 1950-June 24, 1951

HQ Air Engineering Development Division — January 1, 1950-November 14, 1950

1951

1125th U.S. Air Force Field Activities Group	June 1, 1951-July 1, 1961
1126th Air Intelligence Service Squadron	June 1, 1951-July 1, 1953
2750th Air Base Group	February 5, 1951-May 10, 1955
2750th Air Police Squadron	February 5, 1951-May 1, 1954
2750th Food Service Squadron	February 5, 1951-May 1, 1954

2750th Personnel Processing Squadron   February 5, 1951-June 1, 1953
2956th Air Intelligence Squadron   April 25, 1951-June 14, 1951
3061st Support Squadron/6500th Support Squadron   February 5, 1951-August 20, 1953
3062nd Support Squadron/6501st Support Squadron   February 5, 1951-August 20, 1953
3063rd Support Squadron   May 1, 1951-February 14, 1952
7300th Materiel Control Group   January 17, 1951-August 6, 1951
Air Development Center, Provisional   March 22, 1951-April 1, 1951
Air Development Force/Wright Air Development Center   April 1, 1951-December 15, 1959
Air Technical Intelligence Center   May 21, 1951-July 1, 1961

1952

1st Explosive Ordnance Disposal Squadron   June 16, 1952-May 7, 1954
6th Weather Group   April 20, 1952-June 18, 1958
3190th Materiel Control Group   January 5, 1952-May 28, 1952

1953

1350th Photographic Service Squadron/1350th
  Motion Picture Squadron   August 1, 1953-June 30, 1969
3090th Air Depot Wing   December 5, 1953-June 1, 1954
U.S. Air Force Technical Intelligence School   May 1, 1953-July 1, 1961

1954

2700th Explosive Ordnance Disposal Squadron   May 7, 1954-July 6, 1955
3500th U.S. Air Force Recruiting Wing   May 20, 1954-July 8, 1959

1955

56th Fighter-Interceptor Squadron   August 18, 1955-March 1, 1960
58th Air Division, Bomb/58th Air Division, Defense   September 8, 1955-February 1, 1959
1478th Flight Service/1289th Airways and Air
  Communications Service Squadron   July 1, 1955-April 18, 1957
3079th Aviation Depot Wing   February 6, 1955-July 1, 1962
4717th Ground Observer Squadron   September 8, 1955-September 1, 1958

1956

799th Aircraft Control and Warning Squadron   February 8, 1956-October 1, 1956
School of Business   July 23, 1956-June 27, 1966
School of Engineering   July 23, 1956-present

1958

2702nd Explosive Ordnance Disposal Squadron   July 1, 1958-February 1, 1962
Air Materiel Command, Aeronautical Systems Center   September 15, 1958-April 1, 1961

1959

66th Aviation Depot Squadron/66th Munitions
  Maintenance Squadron   September 15, 1959-September 20, 1972
922nd Air Refueling Squadron   December 1, 1959-September 30, 1975
4043rd Armament & Electronics Maintenance Squadron   October 1, 1959-February 1, 1963
4043rd Field Maintenance Squadron   October 1, 1959-February 1, 1963
4043rd Organizational Maintenance Squadron   October 1, 1959-February 1, 1963
4043rd Strategic Wing   April 1, 1959-February 1, 1963
4043rd Support Squadron/4043rd Combat Defense Squadron   October 1, 1959-February 1, 1963

U.S. Air Force Recruiting Service   July 8, 1959-June 15, 1965
Wright Air Development Division   December 7, 1959-April 1, 1961

1960

42nd Bomb Squadron   June 1, 1960-February 1, 1963
8401st Air Force Reserve Base Support Group/8318th
  Air Force Reserve Base Support Group   September 1, 1960-December 31, 1962
8401st Air Force Reserve Base Support Squadron   September 1, 1960-January 1, 1962
8401st Materiel Squadron   September 1, 1960-January 1, 1962
Air Force Museum/United States Air Force Museum   December 1, 1960-present
U.S. Air Force Central Motion Picture Film Depository/U.S.
  Air Force Central Audio-Visual Depository   September 1, 1960-October 15, 1960
U.S. Air Force Radiological Health Laboratory   April 1, 1960-February 8, 1969

1961

2726th Air Force Cataloging Group   September 6, 1961-March 1, 1964
6570th Aerospace Medical Research Laboratory   December 1, 1961-September 8, 1979
Foreign Technology Division/Air Force Foreign Technology
  Center/Foreign Aerospace Science and Technology
  Center/National Air Intelligence Center/National Air
  & Space Intelligence Center   July 1, 1961-present
HQ Aeronautical Systems Division/HQ Aeronautical Systems Center   April 1, 1961-present

1962

17th Armament & Electronics Maintenance Squadron/
  17th Avionics Maintenance Squadron   November 15, 1962-September 30, 1975
17th Bomb Wing   November 15, 1962-September 30, 1975
17th Field Maintenance Squadron   November 15, 1962-September 30, 1975
17th Organizational Maintenance Squadron   November 15, 1962-September 30, 1975
4043rd Airborne Missile Maintenance Squadron   November 1, 1962-February 1, 1963
Dayton Contract Management District   March 29, 1962-October 1, 1965

1963

17th Airborne Missile Maintenance Squadron   February 1, 1963-March 25, 1968
17th Armament & Elect Maintenance Squadron/
  17th Avionics Maintenance Squadron   February 1, 1963-September 30, 1975
17th Combat Defense Squadron/17th Security Police Squadron   February 1, 1963-September 30, 1975
17th Field Maintenance Squadron   February 1, 1963-September 30, 1975
17th Organizational Maintenance Squadron   February 1, 1963-September 30, 1975
34th Bomb Squadron   February 1, 1963-September 30, 1975
Aerospace Research Laboratories   August 13, 1963-July 1, 1975
Air Force Aero Propeller Laboratory   August 5, 1963-January 15, 1980
Air Force Avionics Laboratory   August 5, 1963-January 15, 1980
Air Force Flight Dynamics Laboratory   August 5, 1963-January 15, 1980
Air Force Materials Laboratory   August 5, 1963-January 15, 1980
Systems Engineering Group   August 21, 1963-November 15, 1967

1964

23rd Medical Service Squadron   March 8, 1964-January 31, 1973
Air Force Contract Law Center/Air Force Materiel
  Command Law Center/Air Force Materiel Command Law Office   October 15, 1984-present
Air Force Logistics Command ICBM Deactivation
  Task Force [Provisional]   December 10, 1964-August 1, 1965
HQ Air Force Packaging & Evaluation Activity/
  Air Force Packaging Evaluation Agency   May 22, 1964-September 23, 1994

1966

4th Maintenance Squadron [Mobile]   July 1, 1966-October 1, 1978
4th Supply Squadron, Mobile [Support]   July 1, 1966-October 1, 1978
2762nd Maintenance Squadron/2762nd Logistics
  Squadron/645th Materiel Squadron   April 1, 1966-present

1967

3030th Support Squadron   October 1, 1967-May 1, 1979
Air Force Logistics Command Advanced Logistics Systems Center   October 1, 1967-February 1, 1970

1968

17th Airborne Missile Maintenance Squadron   July 25, 1968-December 31, 1974
2863rd Ground Equipment Engineering and
  Installation Agency Squadron/1828th Electronic
  Installation Squadron   October 1, 1968-June 30, 1977
Nuclear Engineering Center   April 8, 1968-June 30, 1971

1969

Air Force Contract Maintenance Center   April 8, 1969-October 1, 1990
U.S. Air Force Radiological Health Laboratory   February 8, 1969-September 29, 1976

1970

1st Civil Engineering Group   March 16, 1970-November 1, 1971

1971

4950th Test Wing (Technical)/4950 Test Wing   March 1, 1971-June 30, 1994

1972

17th Munitions Maintenance Squadron   October 1, 1972-September 30, 1975
2763rd Supply Squadron   May 1, 1972-July 1, 1983
3070th Computer Services Squadron   January 1, 1972-July 1, 1983

1973

2750th Civil Engineering Squadron (combined
  with 2750th Services Squadron to form the 2750th
  Engineering & Service Group/645 Civil Engineering
  Group/645 Civil Engineering Group/88th Civil
  Engineering Group   January 1, 1973-December 31, 2001
2750th Materiel Squadron/2750th Logistics Squadron   January 1, 1973-September 10, 1990
2750th Security Police Squadron   January 1, 1973-March 1, 1975
3025th Management Engineering Squadron   October 1, 1973-September 30, 1993

1974

2732nd Acquisition Logistics Operations Squadron   July 1, 1974-July 1, 1976
2764th Transportation Operations Squadron   February 1, 1974-July 1, 1983

1975

4950th Armament & Electronics Maintenance Squadron/Avionics Maintenance Squadron	July 1, 1975-June 30, 1994
4950th Field Maintenance Squadron	July 1, 1975-June 30, 1994
4950th Organizational Maintenance Squadron	July 1, 1975-June 30, 1994
4951st Test Squadron	July 1, 1975-July 25, 1983
4952nd Test Squadron	July 1, 1975-June 30, 1994
4953rd Test Squadron	July 1, 1975-June 30, 1994
6592nd Management Engineering Squadron	February 1, 1975-January 1, 1978
Air Force Wright Aeronautical Laboratories/Wright Research and Development Center/Wright Laboratory	July 1, 1975-October 31, 1997

1976

15th Weather Squadron	January 1, 1976-April 1, 1980
35th Medical Service Squadron	January 1, 1976-unknown
87th Aerial Port Squadron	December 1, 1976-present
HQ Air Force Acquisition Logistics Division/ HQ Air Force Acquisition Logistics Center/HQ Acquisition Logistics Division/HQ Center Supportability & Technology Insertion	May 7, 1976-October 1, 1993

1977

3552nd U.S. Air Force Recruiting Squadron/338th U.S. Air Force Recruiting Squadron	December 16, 1977-present

1978

56th Tactical Training Squadron	October 1, 1979-unknown
401st Combat Logistics Support Squadron	October 1, 1978-present
Air Force Aerospace Medical Research Laboratory/ Armstrong Aerospace Medical Research Laboratory	September 8, 1979-December 13, 1990
Air Force Logistics Command International Logistics Center/Air Force Security Assistance Center	May 1, 1978-present

1980

2750th Security Police Squadron/645th Security Police Squadron/88th Security Police Squadron/88th Security Forces Squadron	March 16, 1980-present

1981

1815th Test & Evaluation Squadron/Air Force Communications Command Operational Test and Evaluation Center	June 1, 1981-March 29, 1991
Logistics Audit Region	September 15, 1981-unknown
Reserve Tactical Fighter Group Provisional 9906	October 1, 1981-July 1, 1982

1982

89th Tactical Fighter Squadron/89th Fighter Squadron	July 1, 1982-July 1, 1994
906th Combat Support Squadron	July 1, 1982-July 1, 1994
906th Consolidated Aircraft Maintenance Squadron	July 1, 1982-July 1, 1994
906th Tactical Fighter Group/906th Fighter Group	July 1, 1982-July 1, 1994

1983

906th Civil Engineering Squadron	October 31, 1983-July 1, 1994
Logistics Management Systems Center/Materiel Systems Center/Materiel Systems Group	July 1, 1983-October 26, 1990
Logistics Operations Center	July 1, 1983-October 26, 1990

1984

Wright-Patterson Contracting Center	November 15, 1984-June 26, 1992

1985

2750th Services Squadron (combined with 2750th Civil Engineering Squadron to form 2750th Engineering & Service Group/645th Civil Engineering Group/645th Civil Engineering Group/ 88th Civil Engineering Group	June 1, 1985-December 31, 2000
Air Force Logistics Center Inspection & Safety Center	July 1, 1985-July 1, 1992

1986

HQ Logistics Information Systems Division/ Logistics Communications Division	January 1, 1986-October 1, 1990

1987

HQ Air Force Distribution Agency	February 16, 1987-October 26, 1990

1989

3100th Specialized Mission Squadron/ 615th Specialized Mission Squadron	December 18, 1989-October 1, 1994

1990

2750th Comptroller Squadron	September 10, 1990-October 1, 1992
2750th Logistics & Operations Group/645th Logistics & Operations Group/88th Logistics Group/88th Logistics & Operations Group	September 10, 1990-October 1, 2002
2750th Mission Support Squadron/645th Mission Support Squadron/88th Mission Support Squadron	September 10, 1990-present
2750th Supply Squadron/645th Supply Squadron/ 88th Supply Squadron	September 10, 1990-August 31, 1998
2750th Transportation Squadron/645th Transportation Squadron/88th Transportation Squadron	September 10, 1990-August 31, 1998

1991

Air Force Reserve Officer Training Corps Field Training Unit	April 1, 1991-July 31, 1991

1992

645th Morale, Welfare, Recreation & Services Squadron	October 1, 1992-October 1, 1993
645th Weather Squadron/645th Weather Flight/ 88th Weather Flight/88th Weather Squadron	October 1, 1992-present
Air Force Materiel Command	July 1, 1992-present
HQ Joint Logistics Systems Center	March 2, 1992-April 30, 1999

1993

47th Airlift Flight	May 31, 1993-present
645th Civil Engineering Maintenance Squadron/788th Civil Engineering Squadron	October 1, 1993-December 31, 2000
645th Civil Engineering Operations Squadron/ 88th Civil Engineering Squadron	October 1, 1993-December 31, 2000
645th Operations Support Squadron/88th Operations Support Squadron	October 1, 1993-present
645th Support Group/88th Support Group/ 88th Mission Support Group	October 1, 1993-present
907th Airlift Group	April 1, 1993-September 30, 1994

1994

74th Aero Medical Squadron	October 1, 1994-present
74th Dental Squadron	October 1, 1994-present
74th Medical Operations Squadron	October 1, 1994-present
74th Medical Support Squadron	October 1, 1994-present
88th Communications Squadron	October 1, 1994-present
88th Equipment Maintenance Squadron/88th Maintenance Squadron	October 1, 1994-August 31, 1998
89th Airlift Squadron	October 1, 1994-present
356th Airlift Squadron	October 1, 1994-present
445th Aircraft Maintenance Squadron	October 1, 1994-present
445th Aircraft Squadron	October 1, 1994-present
445th Aeromedical Evacuation Squadron	October 1, 1994-present
445th Aeromedical Staging Squadron	October 1, 1994-present
445th Aerospace Medical Squadron	October 1, 1994-present
445th Airlift Wing	October 1, 1994-present
445th Civil Engineering Squadron	October 1, 1994-present
445th Combat Logistics Support Squadron	October 1, 1994-present
445th Logistics Readiness Squadron	October 1, 1994-present
445th Maintenance Group	October 1, 1994-present
445th Maintenance Operations Squadron	October 1, 1994-present
445th Maintenance Squadron	October 1, 1994-present
445th Mission Support Group	October 1, 1994-present
445th Operations Group	October 1, 1994-present
445th Operations Support Squadron	October 1, 1994-present
445th Security Forces Squadron	October 1, 1994-present
AFMC Civil Engineering Squadron	October 1, 1994-present

1997

HQ Air Force Research Laboratory	April 1, 1997-present

2003

74th Surgical Operations Squadron	September 10, 2003

APPENDIX 5

WRIGHT-PATTERSON AIR FORCE BASE
ON-BASE TENANT ORGANIZATIONS, 2003

Department of Defense (DOD)
Civilian Personnel Management Service, Priority Placement Branch, Systems Support Branch and Office of Complaints Investigation
  Defense Commissary Agency (DECA)
  Defense Contract Audit Agency (DCAA)
  Defense Contract Management Agency (DCMA) - Dayton
Defense Courier Service - Louisville
  Defense Information Systems Agency (DISA):
  Defense Enterprise Computing Center (DECC) - Detachment Dayton
  Defense Technical Information Center (DTIC) - Midwestern Regional Office
  Defense Institute of Security Assistance Management (DISAM)
  Defense Intelligence Agency (DIA):
  National Capital Region Operating Base, Detachment 430 - Dayton Element
  Defense Logistics Agency (DLA):
  Defense Automatic Addressing System Center (DAASC)
Document Automation and Production Service (DAPS)
Defense Energy Support Center
  Defense Reutilization and Marketing Office (DRMO)
  Defense Security Service (DSS)
  Joint Group on Depot Maintenance:
  Joint Depot Maintenance Analysis Group (JDMAG)
TRICARE Heartland – Department of Defense Health Service Region 5

Headquarters United States Air Force
Air Force Audit Agency:
  Acquisition and Logistics Directorate
  Area Audit Office
Air Force C2ISR Center - Operation Location-C
Air Force Financial Systems Management Office
Air Force Legal Service Agency (AFLSA) - United States Air Force Judiciary:
Area Defense Counsel (ADC)
United States Air Force Band of Flight
338th U.S. Air Force Recruiting Squadron

Air Combat Command (ACC)
Air Combat Command Systems Office (ACCSO-A)
Air Intelligence Agency (AIA):
  National Air and Space Intelligence Center (NAIC)
  18th Intelligence Squadron, Detachment 1
Consolidated Mobility Bag Control Center
55th Wing – Operating Location-WP:
National Airborne Operations Center (NAOC)

Air Education and Training Command (AETC)
Air University (AU):
  Air Force Institute for Advanced Distributed Learning (AFIADL):
    Air Technology Network Program Management Office – Operating Location-A
  Air Force Institute of Technology (AFIT)
  Air Force Reserve Officer Training Corps (AFROTC) - Northeast Region
  Civil Air Patrol (CAP) - USAF, Detachment 3 Great Lakes Liaison Region (GLLR)

Air Mobility Command (AMC)
Headquarters Air Mobility Command - Detachment 2
47th Airlift Flight

Air Force Materiel Command (AFMC)
Aeronautical Systems Center (ASC):
  311th Human Systems Wing (311 HSW/YA WPAFB)

Air Armament Center - 46th Test Wing:
>46th Operations Group, Munitions Test Division, Aerospace Survivability and Safety Flight Operating Location-AC (46 OG/OGM/OL-AC)

Air Force Flight Test Center - 412th Test Wing:
>412th Aircraft Maintenance Squadron Test Squadron Operating Location (LGHSK-OL)
>413th Flight Test Squadron Operating Location-HN

Air Force Research Laboratory (AFRL)
Air Force Security Assistance Center (AFSAC)
Electronic Systems Center (ESC):
>Materiel Systems Group (MSG)

Headquarters Air Force Materiel Command
United States Air Force Museum (USAFM)
Warner Robins Air Logistics Center - Air Force Petroleum Office:
>Detachment 3 (WR-ALC/AFT)

Air Force Office of Special Investigations (AFOSI)

Office of Special Investigations 1st Field Investigations Region Operating Location-A
Office of Special Investigations Region 1
Office of Special Investigations Detachment 101
Office of Special Investigations Detachment 702

Air Force Reserve Command (AFRC)

445th Airlift Wing

United States Army (USA)

Army Medical Department:
>Veterinary Services

Corps of Engineers (MCLARE)
656th Transportation Company
731st Ordnance Company (EOD)

United States Navy (USN)

United States Naval Reserve:
>Naval Contingency Engineering Management Cell - Europe East
>Naval Health Research Center - Toxicology Detachment
>Naval Test Squadron 0195

United States Naval Reserve and United States Marine Corps Reserve:
>Naval and Marine Corps Reserve Center

Non-Department of Defense Federal Organizations

Department of the Treasury - Internal Revenue Service (IRS):
>Cincinnati Submission Processing Center

Federal Emergency Management Agency (FEMA):
>Ohio Task Force 1 (Miami Valley Urban Search and Rescue)

General Accounting Office (GAO)
Library of Congress:
>Motion Picture Conservation Center

National Park Service (NPS):
>Dayton Aviation Heritage National Historical Park

United States Postal Service (USPS):
>Wright-Patterson Air Force Base branches

Civilian Organizations

American Red Cross:
>Dayton Area Chapter

Bank One NA (National Association)
Civil Air Patrol (CAP) Wright-Patterson Composite Squadron
National Aviation Hall of Fame
Wright-Patt Credit Union, Inc.

Other

Army Air Force Exchange Service (AAFES)

Sources: On-Base Organizations List, 88 MSG/LGRRP, October 20, 2003; Quick Reference Guide [tenant-host agreements], 88 MSG/LGRRP, October 20, 2003; Mr. James F. Miller, Support Agreement Manager, 88 MSG/LGRRP, personal communication with James R. Ciborski, Aeronautical Systems Center History Office; Directory, Defense Metropolitan Area Telephone System Dayton, 2003, 88th Communications Group.

APPENDIX 6

WRIGHT-PATTERSON AIR FORCE BASE
PERSONNEL STRENGTH, 1918-2002

Year	Military	Civilian	Total	Year	Military	Civilian	Total
1918	4,683	1,985	6,668	1971	7,875	17,744	25,619
1920	452	1,565	2,017	1972	8,167	17,520	25,687
1938	212	2,222	2,434	1973	8,323	16,920	25,243
1939	652	3,059	3,711	1974	8,694	17,037	25,731
1940	708	7,455	8,163	1975	7,548	15,975	23,523
1941	2,125	15,398	17,523	1976	7,182	15,812	22,994
1942	9,592	36,908	46,500	1977	7,607	15,523	23,130
1943	14,821	30,926	45,747	1978	7,686	15,879	23,565
1944	16,119	29,356	45,475	1979	7,636	15,832	23,468
1945	11,100	20,180	31,280	1980	7,992	17,031	25,023
1946	8,261	19,358	27,619	1981	7,608	15,662	23,270
1947	3,192	17,588	20,780	1982	8,919	17,549	26,468
1948	5,082	20,108	25,190	1983	9,374	16,754	26,128
1949	4,434	20,443	24,877	1984	9,347	17,973	27,320
1950	4,745	23,781	28,526	1985	9,336	18,766	28,102
1951	8,946	25,738	34,684	1986	9,169	18,561	27,730
1952	8,284	22,144	30,428	1987	10,692	18,493	29,185
1953	6,805	20,478	27,283	1988	10,206	19,144	29,350
1954	7,098	20,264	27,362	1989	10,190	20,353	30,543
1955	7,248	20,627	27,875	1990	10,143	18,994	29,137
1956	6,762	21,701	28,463	1991	10,222	18,264	28,486
1957	6,313	19,557	25,870	1992	9,578	17,316	26,894
1958	6,284	18,893	25,177	1993	9,159	15,782	24,941
1959	6,471	18,331	24,802	1994	7,479	14,023	21,502
1960	6,948	20,966	27,914	1995	9,185	13,929	23,114
1961	7,364	20,714	28,078	1996	8,980	13,757	22,737
1962	7,301	20,301	27,602	1997	10,386	12,654	23,040
1963	12,185	19,273	31,458	1998	8,337	13,121	21,458
1964	11,431	19,112	30,543	1999	7,817	11,194	19,011
1965	7,082	19,170	26,252	2000	6,600	10,353	16,953
1966	7,531	19,234	26,765	2001	7,264	11,109	18,373
1967	7,290	19,299	26,589	2002	7,506	10,358	17,864
1968	7,207	19,163	26,370				
1969	7,557	18,568	26,125				
1970	7,596	17,761	25,357				

APPENDIX 7

WRIGHT-PATTERSON AIR FORCE BASE MEMORIALIZATION PROGRAM

The nation's bicentennial celebration in 1976 awakened interest in the history and heritage of Wright-Patterson Air Force Base, giving rise to a Base Memorialization Committee. It established policies and procedures for dedicating facilities on base and also identified and recommended appropriate streets, buildings, recreational areas, and medical facilities to honor, as well as individuals to be honored. While several buildings and streets had been named in recognition of notable individuals, over the remainder of the decade a wave of memorial dedications swept the base. The first one came on October 27, 1976, when Building 262, Area A, was named Gillmore Hall for Brigadier General William E. Gillmore, first chief of the Air Corps Materiel Division at McCook Field.

As its second project, the Memorialization Committee recommended renaming the lettered and numbered streets in the Brick Quarters in Area A. It selected 10 officers, all Ohio natives, for this honor. Five of the officers had spent a portion of their careers at Wright-Patterson. The collective span of service represented by these men spread from August 1, 1907, when the U.S. Army Signal Corp's Aeronautical Division was established, to July 31, 1957, when the last individual among them retired. The honorees represented the following Ohio hometowns:

Major General Robert G. Breene	Dayton
Lieutenant General George H. Brett	Cleveland
Colonel Charles deF. Chandler	Cleveland
General Benjamin W. Chidlaw	Cleves
Colonel Gerald R. Johnson	Akron
Brigadier General Frank P. Lahm	Mansfield
2nd Lieutenant William E. Metzger	Lima
Lieutenant General David M. Schlatter	Fostoria
Brigadier General Nelson S. Talbott	Dayton
Lieutenant General Barton K. Yount	Troy

Dedication ceremonies were held on July 22, 1977.

On September 23, 1977, a beautiful living memorial was dedicated to the memory of Major General Frank G. Barnes. General Barnes served as deputy chief of staff for Engineering and Services at Headquarters Air Force Logistics Command (AFLC) from February 1973 until his death in 1976. The Frank G. Barnes Memorial Park was located adjacent to Building 266, Area A, on a site that General Barnes could see from his office window. The area had been occupied by World War II warehouses, housing units, and visiting officers' quarters, which were removed in the 1970s to make way for a parking lot. When General Barnes found out about the plan, he replied, "We need this land for trees and grass—not asphalt!" A park was built instead and, appropriately, named after its sponsor. Barnes Park featured a Japanese *tori* gate, reflective pond with fountain, and 21 varieties of deciduous, flowering, and conifer trees, flowering bushes and shrubs, and 11 species of perennial flowers.

The dining hall and five dormitories in Kittyhawk Center were dedicated June 22, 1979, in honor of six Ohio airmen who died from enemy action in South Vietnam. This event was the first formal base dedication honoring deceased enlisted men. General Bryce Poe II, the AFLC commander, and Colonel James Rigney, the 2750th Air Base Wing commander, presided at the ceremonies where they unveiled six bronze plaques bearing the names of the honorees:

Sergeant James D. Locker	Building 1217
Staff Sergeant James R. Lute	Building 1216
Airman First Class William H. Pitsenbarger	Building 1214 (dining hall)
Airman First Class James E. Pleiman	Building 1212
Technical Sergeant Roy D. Prater	Building 1213
Staff Sergeant Frederick Wilhelm	Building 1215

Walnut plaques with sketches and biographies of the airmen were placed on permanent display in the dayrooms of the respective dormitories and in the dining hall.

More memorializations followed in the 1980s. On June 19, 1981, streets in the vicinity of the United States Air Force Museum were named in honor of General Carl A. Spaatz, Major Richard I. Bong, and 1st Lieutenant Edward Ward. On August 28, 1981, the 2750th Air Base Wing named its headquarters, Building 10, Area C, in honor of Brigadier General Joseph T. Morris, the first commander of the air base wing. On November 18, 1982, the Air Force Institute of Technology (AFIT) dedicated its School of Engineering in Building 640, Area B, to Colonel Thurman H. Bane, the first, post-World War I commander of McCook Field and founder of the Air School of Application, forerunner of AFIT.

The main auditorium in Building 262, AFLC headquarters, was named on June 13, 1984, for Mr. Aristides Sarris, a former assistant deputy chief of staff for Plans and Programs and senior civilian at the command. Building 248 in Area B was dedicated as the Lieutenant Colonel Paul M. Fitts Human Engineering Laboratory on June 6, 1985. Lieutenant Colonel Fitts organized and directed the first program in the field of human aeronautical engineering and founded the design discipline. General Nathan F. Twining was honored on September 25, 1985, when Building 641 in Area B, AFIT's School of Systems and Logistics, was named for him. General Twining had been the wartime commander of 15th and 20th Air Forces, a commanding general of the Air Materiel Command from December 1945 to October 1947, chief of staff of the Air Force, and chairman of the Joint Chiefs of Staff. Wright-Patterson honored another of its own on August 28, 1986, when the new gymnasium in Kittyhawk Center, Building 1245, was dedicated for Brigadier General Irby B. Jarvis, the 2750th Air Base Wing and Wright-Patterson Air Force Base commander from June 1972 to January 1975.

Air Force Logistics Command celebrated its 45th anniversary in 1989 by erecting a flag presentation in front of its headquarters, Building 262. A static display of two retired F-4D aircraft donated by the 906th Tactical Fighter Group flanked the flags. The flags projected the command's professional image, served as a statement of pride and confidence, and were a reflection of the command's impact on the public. The aircraft symbolized the command's emphasis on quality support to operational units. One of the aircraft, tail number 66-7554, was named the *City of Fairborn*. The two red stars on its fuselage, representing the two MiG kills it recorded during the Vietnam War, were displayed in memory of the achievements and fallen comrades. The command also renamed the streets in Area A surrounding its headquarters in honor of its 10 field units.

Throughout the 1990s, Wright-Patterson's memorialization program continued to honor individuals who had contributed substantially to the nation's defense. The most significant of these occurred on September 16, 1994, when the new Acquisition Management Complex in Area B became the General James H. Doolittle Acquisition Management Complex. The complex's first structure, Building 557, was dedicated the same day as Lieutenant General James T. Stewart Hall. Three more structures were named on April 30, 1997: Lieutenant General Kenneth B. Wolfe Hall (Building 558), Major General William L. Mitchell Hall (Building 556), and Frederick T. Rall, Jr., Hall (Building 560). Major General Franklin Otis Carroll Hall (Building 553) was memorialized in June 2001.

In October 1995, the National Air Intelligence Center named its new Foreign Materiel Exploitation Facility (Building 4023, Area C) in honor of Major General Harold E. Watson. The street in front of NAIC headquarters was also named for General Watson. Watson led a group of pilots who flew captured German aircraft out of Europe following World War II, and later commanded Air Technical Intelligence Center, predecessor of NAIC. During 1997, NAIC dedicated its special operations facility, Building 858 in Area A, in honor of Dr. Anthony J. Cacioppo, a senior analyst who passed away while on official travel.

Other memorial actions took place as well. When Building 271, Area A, opened in 1990 as AFLC's logistical system operations center, it was named Warrior Hall to honor the command's civilian workforce. The same year, AFIT opened its new Science, Engineering, and Support Facility, Building 642 in Area B, and memorialized it in memory of General George C. Kenney. The achievements of Lieutenant General George L. Monahan, Jr., were recognized at a March 30, 1993, ceremony when Building 12 and Third Street in Area B were named in his honor. The American Society of Mechanical Engineers designated the Five-Foot Wind Tunnel in Building 19, Area B, a National Historic Mechanical Engineering Landmark in 1995. Finally, a November 21, 1997 ceremony commemorated the Dayton Peace Agreement with the dedication of the Peace Walk in Area A.

The following is a list of memorials on Wright-Patterson Air Force Base:

Name	Memorialized For	Building	Date
Estabrook Drive	Brigadier General Merrick G. Estabrook, Jr.	Street	Unk
Pearson Road	1st Lieutenant Alexander Pearson, Jr.	Street	Unk
Skeel Avenue	Captain Burt E. Skeel	Street	Unk
Wright Avenue	Wilbur and Orville Wright	Street	Unk
Firehouse No. 3	Frank A. Smith	76	1932
Patterson Swimming Pool	Lieutenant Frank Stuart Patterson	Pool	1936
Wright Memorial	Wilbur and Orville Wright	Monument	1940
Dodge Gymnasium	Delphine Dodge Godde	849	1947
Aerial Photoreconnaissance Personnel	Captain Eugene Leger, Master Sergeant Paul L. Hayes, Private First Class Dorothy E. Kimmel, and Mr. Joseph H. Britain	17	1952
Hadden Park (original)	Mr. William Hadden	Park	1953
Page Manor	Brigadier General Edwin R. Page	Housing	1953
Hadden Park (relocated)	Mr. William Hadden	Park	1960
Gillmore Hall	Brigadier General William E. Gillmore	262	1976
Barnes Park	Major General Frank G. Barnes	Park	1977
Breene Drive	Major General Robert G. Breene	Street	1977
Brett Drive	Lieutenant General George H. Brett	Street	1977
Chandler Drive	Colonel Charles deForest Chandler	Street	1977
Chidlaw Road	General Benjamin W. Chidlaw	Street	1977
Johnson Drive	Colonel Gerald R. Johnson	Street	1977
Lahm Circle	Brigadier General Frank P. Lahm	Street	1977
Metzger Drive	2nd Lieutenant William E. Metzger, Jr.	Street	1977
Schlatter Drive	Lieutenant General David M. Schlatter	Street	1977
Talbott Road	Brigadier General Nelson S. Talbott	Street	1977
Yount Drive	Lieutenant General Barton K. Yount	Street	1977
Locker Hall	Sergeant James D. Locker	1217	1979
Lute Hall	Staff Sergeant James R. Lute	1216	1979
Pitsenbarger Hall	Airman First Class William H. Pitsenbarger	1214	1979

Pleiman Hall	Airman First Class James E. Pleiman	1212	1979
Prater Hall	Technical Sergeant Roy D. Prater	1213	1979
Wilhelm Hall	Staff Sergeant Frederick Wilhelm	1215	1979
Bong Drive	Major Richard I. Bong	Street	1981
Spaatz Circle	General Carl A. Spaatz	Street	1981
Ward Road	1st Lieutenant Edward Ward	Street	1981
Morris Hall	Brigadier General Joseph T. Morris	10	1981
Bane Hall	Colonel Thurman H. Bane	640	1982
Sarris Auditorium	Mr. Aristides Sarris	262	1984
Fitts Hall	Mr. Paul M. Fitts	248	1985
Twining Hall	General Nathan F. Twining	641	1985
Arnold House	General Henry H. "Hap" Arnold	8	1986
Jarvis Gymnasium	Brigadier General Irby B. Jarvis	1245	1986
POW/MIA Memorial	Prisoners of War and Missing in Action	Memorial	1986
Hope Hotel	Mr. Bob Hope	824	1989
Robins House	Brigadier General Augustine Warner Robins	700	1989
Fire Station No. 1	Mr. Dale V. Kelchner and		
	Mr. William J. Collins	163	1989
Foulois House	Major General Benjamin D. Foulois	88	1989
Warrior Hall	Civilian Work Force (Warriors)	271	1990
Kenney Hall	General George C. Kenney	642	1990
Monahan Hall	Lieutenant General George L. Monahan	12	1993
Monahan Way	Lieutenant General George L. Monahan	Street	1993
Doolittle Acquisition Management Complex	General James H. Doolittle	Complex	1994
Stewart Hall	Lieutenant General James T. Stewart	557	1994
Watson Way	Major General Harold E. Watson	Street	1995
Watson Hall	Major General Harold E. Watson	4023	1995
Patterson Parkway	1st Lieutenant Frank S. Patterson	Street	1996
Peace Walk	Balkan Proximity Peace Talks	Sidewalk	1997
Wolfe Hall	Lieutenant General Kenneth B. Wolfe	558	1997
Mitchell Hall	Major General William Mitchell	556	1997
Rall Hall	Mr. Frederick T. Rall, Jr.	560	1997
Cacioppo Annex	Dr. Anthony J. Cacioppo	858	1997
Carroll Hall	Major General Franklin O. Carroll	553	2001
Hobson Way	Congressman Dave Hobson	Street	2003

Aerial view of Wright Memorial. The laboratory and acquisition complex of old Wright Field, now Area B, is visible in the background.

The original Hadden Park in Area C, was dedicated on September 19, 1953, as a tribute to the services of William Hadden, who served as a noncommissioned officer in charge of utilities at Wilbur Wright Field in 1917 and, later, as chief of Maintenance in the Directorate of Base Air Installations. Prior to retiring in 1952, Hadden was instrumental in developing the park that bore his name. The park was relocated twice, each time growing in physical size and facilities, until it was closed in April 1985 due to environmental concerns associated with adjacent landfills. Hadden Park's second location, shown here, was between National and Zink roads.

Memorialization ceremony for the Brick Quarters streets, July 22, 1977. Ten lettered and numbered streets in the housing area were renamed in honor of outstanding officers who were native sons of Ohio.

The Peace Walk led from the Hope Hotel to the Five-Plex where the delegates were housed for the Balkan Proximity Peace Talks.

Frank G. Barnes Memorial Park was dedicated in September 1977 in honor of Major General Frank Barnes, former deputy chief of staff for Engineering and Services, Headquarters Air Force Logistics Command.

Secretary of the Air Force F. Whitten Peters presented the Medal of Honor posthumously to Airman First Class William H. Pitsenbarger during a December 8, 2000, ceremony at the United States Air Force Museum. William F. Pitsenbarger, the late airman's father, accepted on his son's behalf. Next to Mr. Pitsenbarger are his wife Irene and Air Force Chief of Staff General Michael E. Ryan. The base dining hall is named in honor of the Ohio hero.

APPENDIX 8

HUFFMAN PRAIRIE REDEDICATION CEREMONY— EXCERPTS FROM OPENING REMARKS BY LIEUTENANT GENERAL RICHARD V. REYNOLDS (COMMANDER OF AERONAUTICAL SYSTEMS CENTER) JULY 16, 2003

Ladies and gentlemen, distinguished guests and visitors, good afternoon and welcome to today's rededication of this historic sight.

Today's ceremony is the simple result of the hard work and cooperation of a lot of people from a lot of different organizations, and I'd like to personally thank everyone who has had a hand in restoring this field to the conditions and appearance it would have had while the Wright brothers were learning to fly here nearly a hundred years ago.

Special thanks to the National Park Service for everything they've done and for the tremendous cooperation they've shown with Team Wright-Patt in making this possible. And to the Ohio Historical Society, AIAA [American Institute of Aeronautics and Astronautics], and the Bicentennial Commission as well for their role in this rededication.

In the time that it's been my privilege to serve as the commander of the Aeronautical Systems Center and installation commander for Wright-Patt, I've taken our vision—the Birthplace, Home, and Future of aerospace—to heart.

And I've done what I can to instill in those I serve with an appreciation of the rich heritage of aviation history that was nurtured here at the Huffman Prairie flying field—instill it so vigorously, sometimes to the point of annoying them!

Though most of the world is more familiar with the famous events that took place at Kitty Hawk on December 17th, 1903, history tells us that the Wrights immediately began looking for a more suitable location closer to home to conduct their flying experiments.

They found it when Torrence Huffman loaned out this field "rent free" to the two brothers. Though this 100-acre plot gave them some troubles due to the fact that the trees on the west and the north cut off the wind, and cattle and horses frequently wandered onto their takeoff and landing paths, the prairie did provide them the room, and the privacy they needed, to build, and test, and perfect the first *practical* airplane, and to really *learn how to fly.*

In 1904 and 1905, with the Wright Flyer II and the Wright Flyer III, the two brothers *perfected* the capacity for controlled flight—launching hundreds of *ever* longer and more ambitious flights, and eventually learning to turn, bank, and fly circles, as well as figure eights....

I'd...like to remind everyone that though we're in the middle of an enormous celebration of the centennial of manned, controlled, powered flight, the true history of aviation goes back much further than that. You could trace the history of the successful invention of powered, controlled flight back as far as 1878, when Bishop Milton Wright brought home a small toy helicopter powered by a rubber band which would lift itself into the air.

Eleven-year-old Wilbur and seven-year-old Orville built several copies of the toy, including larger versions, with the idea of maybe *one day* building a toy big enough to lift them up into the air.

But I think the history of aviation goes back even further. The dream of flight has captivated all mankind from the dawn of history. There is something in the sky overhead that has always called out to the human spirit. To mimic the birds in their graceful freedom and darting eloquence. And only the infiniteness of the sky can match the limitless desires of the human heart.

To men and women of every age and every era, the sky has been an unmistakable symbol of freedom, freedom from the tyranny of gravity, as well as from oppression and cruelty. In the Psalms, David cries out to the Lord, *"Oh that I had wings like a dove! for then would I fly away, and be at rest."* The Greeks tell the dramatic story of Icarus and Daedalus, who used wings made of wax and feathers to escape their imprisonment on Crete.

How fitting then that it should be in America, the land of the free, that man first successfully conquered the skies. And how much more fitting that it should be here in Dayton, the heart of America, that the dream was finally realized. This centennial of flight celebration will end, but Dayton will *always* be the birthplace, home, and future of aerospace.

In 1942, December 17th, 39 years after the Wrights' grand achievement and one year and a few days after the attack at Pearl Harbor, a creed was written and affirmed for Wright Field, to guide the endeavors of all who followed afterwards. I like it for what it says, for its poetry, and for how it invokes the heavens' role in measuring our success here on the place where the Wrights gave mankind such a marvelous gift, but most of all, how it captures the character and spirit of those two brothers. It goes:

> Here on this ground where Wilbur and Orville Wright brought to full life man's age-old dream of rising in flight above the earth, we of Wright Field consecrate ourselves to the splendid vision and unswerving purpose which motivated those great and honored pioneers of the sky. Their patience, their firm determination, their untiring devotion to their aim—these we take as a light to guide and inspire us.
>
> We hold in all humility to the faith that man *can* if man *will.* We believe that there is no true failure save the failure of the human spirit. We have met uncounted times, and shall meet uncounted times again, the little failures that try and break the souls of little men. We have tried, and we have failed; we shall try again tomorrow *not* to fail. But these defeats we shall meet undaunted, as we have met the defeats of the past. We know that at last we shall have a measure of success; and in that moment all the failures shall become paltry things indeed.

Yet even in success shall our humility be maintained. For it is the essence of our creed that perfection is an elusive myth that never shall man do *well*, but that *he may do better*—that success is but the stepping-stone to new trial and new success—that there shall be no final triumph, but only the long and glorious record of those who have given themselves, in life and death, to the pursuit of *knowledge* and *advancement*, to the crusade of learning which shall never end.

Our hands and eyes lifted, in war and in peace, toward the heavens which alone shall measure our hopes and aspirations—we make this solemn confession of our faith.

For a hundred years, and longer—long before this creed was even written—the spirit of these words, the example set down by those two brave brothers who taught mankind to fly here so long ago, has guided the efforts of every man or woman who has looked towards the sky and dreamed of the freedom of flight.

I firmly believe that that spirit is alive and well today, and that it will help to bear the light of freedom on for a hundred years more....

APPENDIX 9

OFF-BASE SITES SUPPORTING AIR FORCE RESEARCH

From time to time, researchers at Wright-Patterson Air Force Base have required support facilities off-site, but relatively nearby, to accomplish their research goals. These facilities were on Air Force property, staffed by Air Force personnel, and any structures were considered Air Force property. A few major sites in the Dayton area included the telescope facility at John Bryan State Park, the sensors targeting facility on Trebein Road, and the 4950th Test Wing's facility in Jamestown.

Perhaps the most well-known research site was the Air Force Celestial Guidance Research Facility in John Bryan State Park, approximately 14 miles from Wright-Patterson. The Air Force leased the 1.8 acres of land from the state of Ohio, on which the Avionics Laboratory constructed a 1,200-square-foot structure (Building 835), completed in 1967 at a cost of more than $130,000, for exploratory development of celestial guidance techniques. Personnel conducted tests to verify the ability of the then current, and future, star sensors and tracking systems to meet established performance requirements. The facility included seismic mounts; an optical dark room; laboratory areas; and a 20-foot by 20-foot room with a sliding roof, observatory-type dome of 15 feet, and a 12-inch Cassegrain astronomical telescope.

In the 1970s, the Air Force returned the land to the state with a stipulation that the public be allowed periodic access. In 1977, the Miami Valley Astronomical Society secured a lease with the Ohio Department of Natural Resources for use of the observatory. The society holds regular, scheduled stargazes for the campers at the facility, now called the John Bryan State Park Observatory.

The Targeting Systems Characterization Facility (Building 5238), on Trebein Road east of the base, was established in the late 1950s as a communications research site. Equipment from Wright Field, where the radio-frequency environment was unsuitable, was moved to the new facility. In the 1970s, the Avionics Laboratory used the 6,954-square-foot facility as a "target" area for testing various kinds of sensors from the top of the twin-towers building (Building 620) approximately five miles southwest on Wright-Patterson. On a balcony atop the towers, researchers directed electronic sensors, like those on reconnaissance and fighter aircraft, toward the targeting facility to measure weather and atmospheric effects on sensor performance. Much of this research related to Air Force operations in Vietnam.

Equipment at the Trebein facility included the sensors targets (aluminum radiance panels and simulated "thermal targets" representing tanks, trucks, and machinery that radiate heat) and an automatic, weather-sampling pole to record rainfall, fog, wind velocity and humidity. Identical weather poles were located on Building 620 and midway between Building 620 and the Trebein facility. The data from the three weather poles were automatically transmitted to a computer center in an on-base building near Building 620. Sensors tested at the site included television, low-light-level television, forward-looking infrared, an electro-optical identification system, and CATIES—a Common Aperture Technique for Imaging Electro-optical Sensors.

In 2001, the Air Force declared the Trebein facility excess to its needs, and the secretary of the Air Force signed a quit-claim deed transferring ownership to Greene County.

The 4950th Test Wing and the base's Air Traffic Control personnel began using a facility near Jamestown, Ohio, in 1956, when the Air Force acquired 8.2 acres of land on Plymouth Road, about 25 miles from Wright-Patterson, for aircraft identification and control. The Jamestown site, with its AN/CPS-6 early-warning radar, was part of the base's air traffic control network. It fed information to the RAPCON (Radar Approach Control) Center at the base, served as a backup to Air Defense Command (ADC), and performed research functions. The facility was part of an electronic network that included a centralized Air Traffic Control Center for long- and short-range radar control and final-approach facilities at Sand Hill (23 miles from the Jamestown site) near Area C at Wright-Patterson.

An AN/FPS-3 system later replaced the AN/CPS-6 radar, and a new Operations and Maintenance building and an AN/FPS-6 Height Finder Radar Tower were constructed. Fencing around the facility also was extended to include more land, and in 1958, a new power building, sub-station and parking facilities were added.

The 4950th Test Wing also used the Jamestown site for range control. The test wing later set up a facility at the Trebein site to demonstrate techniques for test-range control that did not require range radar. After that, in 1989, the Jamestown site became excess to Air Force needs. The site, still owned by the government, is now home to the Southwest Ohio Weather Radar.

Occasionally, research programs required space—but not a dedicated structure or facility—as part of a test regimen. For example, in 1992, the Armstrong Aerospace Medical Research Laboratory worked on a highly computerized manikin for testing high-speed ejection seats. ADAM (Advanced Dynamic Anthropomorphic Manikin), developed under contract with Systems Research Laboratories (later Veridian Engineering) of Beavercreek, Ohio, made parachute jumps over Skydive Greene County, a privately owned, skydiving area near Xenia, Ohio. ADAM's 30-plus jumps were from a Beechcraft E-18 aircraft flying at an altitude of 10,000 feet. The dummy's instrumentation, including 128 electrical channels, each taking 1,000 samples per second, measured pitch, roll, and yaw accelerations; head and neck movements; and the position of its limbs, head, and neck as parachute forces impacted its body.

During the skydiving, one technician pulled ADAM from the aircraft and descended with it, while a second filmed the event with a helmet-mounted video recorder. Occasionally, if it drifted outside the drop zone, ADAM had to be retrieved from corn and soybean fields surrounding the 150-acre complex. Besides helping researchers study parachute opening forces and acceleration effects on pilots and crewmembers, the data collected by ADAM helped improve the design of aircraft ejection seats and eliminated or minimized injury to aircrews when they had to escape from their high-speed aircraft.

In September 1987, another site near Wright-Patterson was used temporarily for test purposes. On the property of Systems Research Laboratories, the Air Force erected Survivable Collective Protection System-Medical (SCPS-M) emergency treatment shelters for use during chemical and/or conventional attacks. Researchers from the Air Force's Human Systems Division conducted the testing, while members of the Minnesota Air National Guard medical squadrons simulated being casualties receiving medical treatment. Participants garbed with protective chemical ensembles used stretchers and other rescue equipment. The testing was supported by personnel from Aeronautical Systems Division (ASD), especially the Office of Public Affairs, which alerted area residents about the simulated nature of the events. When testing ended, the shelter system was taken to the Air National Guard Combat Readiness Training Center at Alpena, Michigan, where it is still used for medical training.

Sources: Ron Kaehr, Roger Cranos, Ray Rang, e-mails to Helen Kavanaugh-Jones, June 2003; Andrew Kididis, telephone communication with Helen Kavanaugh-Jones, May 2003; ASD Information Office, News Release 78-177, "New AFAL Facility Studies Sensor Performance;" ASD Office of Public Affairs, News Release 88-147, "Meet ADAM: He'll Go Anywhere;" Aeronautical Systems Center Office of Public Affairs, Caption 92-196, "Dummy 'ADAM' Parachutes for Air Force Researchers;" Office of Public Affairs, Captions CPAM #87-184/5/6; Bruce Hess, ASC History Office, personal communication with Helen Kavanaugh-Jones, June 2003.

Researchers garbed in chemical defense ensembles simulate carrying casualties into a SCPS-M (Survivable Collective Protection System-Medical) shelter for "chemical decontamination and medical treatment." The evaluation occurred in the fall of 1987 at Systems Research Laboratories' facility in Beavercreek, Ohio, and was sponsored by the Human Systems Division, Brooks Air Force Base, Texas.

Early-warning radar at the Jamestown site was used by the 4950th Test Wing and the base's Air Traffic Control. Still government-owned, the site is now home to the Southwest Ohio Weather Radar.

The Air Force Celestial Guidance Research Facility in John Bryan State Park was used in exploratory development of celestial guidance techniques in the late 1960s and 1970s. The 15-foot dome houses a 12-inch Cassegrain telescope. Kenneth Kissel, right, managed the facility when the Avionics Laboratory at Wright-Patterson used it for satellite tracking experiments. Currently the Miami Valley Astronomical Society holds regular, scheduled stargazes for campers in the park.

ADAM and Mike Spurgeon skydive at privately owned Skydive Greene County. The pair jumped from a Beechcraft E-18 aircraft flying at 10,000 feet so that ADAM's highly computerized body could study opening parachute forces and determine improved designs of aircraft ejection seats.

ADAM and skydivers at Skydive Greene County where the manikin made more than 35 jumps to determine forces on its body. Left to right: (standing) Jim West and John Brinkman, (kneeling) Mike Spurgeon and John Murphy

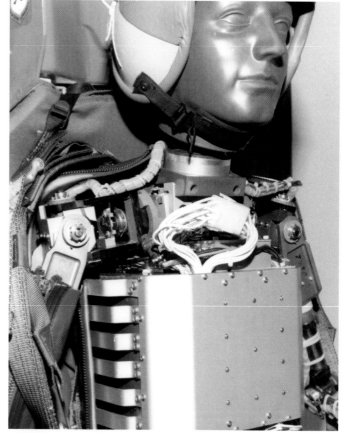

ADAM's exposed chest cavity shows the computer that gathers and stores 56 channels of test data to help researchers study the forces of parachute openings and the effects of g forces on pilots ejecting from high-speed aircraft.

GLOSSARY

GLOSSARY

A

AAF	Army Air Forces
AAFES	Army Air Force Exchange Service
AAFTB	Army Air Forces Technical Base
ABW	Air Base Wing
ACCSO	Air Combat Command Systems Office
ADAM	Advanced Dynamic Anthropomorphic Manikin
ADC	Air Defense Command
AEF	American Expeditionary Forces
AETC	Air Education and Training Command
AFA	Air Force Association
AFALD	Air Force Acquisition Logistics Division
AFIADL	Air Force Institute for Advanced Distributed Learning
AFIT	Air Force Institute of Technology
AFLC	Air Force Logistics Command
AFLSA	Air Force Legal Service Agency
AFMC	Air Force Materiel Command
AFOG	Air Force Orientation Group
AFOSI	Air Force Office of Special Investigations
AFOUA	Air Force Outstanding Unit Award
AFRC	Air Force Reserve Command
AFRL	Air Force Research Laboratory
AFROTC	Air Force Reserve Officer Training Corps
AFS	Air Force Station
AFSAC	Air Force Security Assistance Center
AFSC	Air Force Systems Command
AFTB	Air Force Technical Base
AFWAL	Air Force Wright Aeronautical Laboratories
AGSD	Aviation General Supply Depot
AIA	Air Intelligence Agency
AIAA	American Institute of Aeronautics and Astronautics
AICUZ	Air Installation Compatible Use Zone
AID	Air Intermediate Depot
AMC	Air Materiel Command
ARDC	Air Research and Development Command
ARIA	Advanced Range Instrumentation Aircraft Apollo Range Instrumentation Aircraft
ASC	Air Service Command Aeronautical Systems Center
ASC/HO	Aeronautical Systems Center History Office
ASD	Aeronautical Systems Division
ASSRD	Air Service Supply and Repair Depot
ATEGG	Advanced Turbine Engine Gas Generator
ATSC	Air Technical Service Command
AU	Air University
AV	Air Vehicle
AVHC	American Veterans Heritage Center, Inc.
AW	Airlift Wing

B

BEEF	Base Engineering Emergency Force
BOQ	Bachelor Officers' Quarters

C

C4I	Composites For Infrastructure
CAA	Civil Aeronautics Administration
CAP	Civil Air Patrol
CATIES	Common Aperture Technique for Imaging Electro-optical Sensors
CCAFB	Clinton County Air Force Base
CDF	Cargo Deploying Function
CEO	Chief Executive Officer
CFC	Combined Federal Campaign
CIAPS	Customer Integrated Automated Procurement System
CONUS	Continental United States
CORE	Congress of Racial Equality
CRADA	Cooperative Research and Development Agreement
CWA	Civil Work Administration

D

D-WC	Douglas World Cruiser
DAASC	Defense Automatic Addressing System Center
DAGSI	Dayton Area Graduate Studies Institute
DANFS	Dictionary of American Naval Fighting Ships
DAPS	Document Automation and Production Service
DCAA	Defense Contract Audit Agency
DCG	Disaster Control Group
DCJD	Dayton Christian Jewish Dialogue
DCMA	Defense Contract Management Agency
DECA	Defense Commissary Agency
DECC	Defense Enterprise Computing Center
Delco	Dayton Engineering Laboratories Company
DESC	Defense Electronics Supply Center
DET	Detachment
DIA	Defense Intelligence Agency
DISA	Defense Information Systems Agency
DISAM	Defense Institute of Security Assistance Management
DLA	Defense Logistics Agency
DMATS	Defense Metropolitan Area Telephone System
DMS	Digital Multiplex System
DOD	Department of Defense
DP	Disaster Preparedness
DRMO	Defense Reutilization and Marketing Organization
DSS	Defense Security Service
DTIC	Defense Technical Information Center
DWF	Dayton Wright Field

E

EAA	Experimental Aircraft Association
EM	Office of Environmental Management

EMTEC	Edison Materials Technology Center
EOD	Explosive Ordnance
EPA	Education Partnership Agreements
ESC	Electronic Systems Center

F

FAA	Federal Aviation Administration
FAD	Fairfield Air Depot
FADCAC	Fairfield Air Depot Control Area Command
FADO	Fairfield Air Depot Operations
FADR	Fairfield Air Depot Reservation
FAGSD	Fairfield Aviation General Supply Depot
FAID	Fairfield Air Intermediate Depot
FASC	Fairfield Air Service Command
FATSC	Fairfield Air Technical Service Command
FCC	Federal Coordinating Centers
FEAF	Far East Air Forces
FEMA	Federal Emergency Management Agency
FFO	Designator for Wright-Patterson Air Force Base Airfield
FIRST	For Inspiration and Recognition of Science and Technology
FIS	Fighter-Interceptor Squadron
FMA	Financial Management and Analysis
FSTR	Full Spectrum Threat Response
FTD	Foreign Technology Division

G

GAO	General Accounting Office
GCA	Ground Controlled Approach
GDAFD	Greater Dayton Area Fire Departments
GHQ	General Headquarters
GLLR	Great Lakes Liaison Region
GM	General Motors
GMC	General Motors Corporation

H

HSW	Human Systems Wing

I

I	Interstate
IG	Inspector General
ILC	International Logistics Center
IR&D	Independent Research and Development
IRE	Initial Response Element
IRR	Individual Ready Reservists
IRS	Internal Revenue Service
IWSM	Integrated Weapon System Management

J

JDMAG	Joint Depot Maintenance Analysis Group
JPATS	Joint Primary Aircraft Training System

L

LANTIRN	Low Altitude Navigation and Targeting Infrared system for Night
LLC	Limited Liability Company
LOGAIR	Logistics Airlift

M

MAC	Military Airlift Command
MARS	Military Amateur Radio System
	Military Affiliate Radio System
MAS	Military Airlift Squadron
MASDC	Military Aircraft Storage and Disposition Center
MDC	Mead Data Central
MISSE	Materials on the International Space Station Experiments
MMC	Martin Marietta Composites
MOU	Memorandum of Understanding
MRC	Medical Reserve Corps
MRI	Magnetic Resonance Imaging
MSG	Materiel Systems Group
MSRC	Major Shared Resource Center
MvPTA	Miami Valley Pet Therapy Association

N

NA	National Association
NAA	National Aeronautic Association
NACA	National Advisory Committee for Aeronautics
NAECON	National Aerospace and Electronics Conference
NAHF	National Aviation Hall of Fame
NAIC	National Air Intelligence Center
	National Air and Space Intelligence Center
NAOC	National Airborne Operations Center
NARA	National Archives and Records Administration
NASA	National Aeronautics and Space Administration
NASP	National Aerospace Plane
NASTC	National Aeronautical Systems and Technology Conference
NATO	North Atlantic Treaty Organization
NCC	National Composite Center
NCCJ	National Conference for Community and Justice
NCCST	National Center for Composite Systems Technology
NCO	Noncommissioned Officer
NCR	National Cash Register (Company)
NDMS	National Disaster Medical System
NHDVS	National Home for Disabled Volunteer Soldiers
NORTHCOM	U.S. Northern Command
NPS	National Park Service
NRL	Naval Research Laboratory

O

OCAS	Office of the Chief of the Air Service
OG	Operations Group
OH-TF-1	Ohio Task Force 1
OHS	Ohio Historical Society

P

P4	Programmable Powdered Preform Process
P4A	P4 Process to Aerospace Components
PMCCS	Property, Maintenance, and Cost Compilation Section
PMEL	Precision Measurement Equipment Laboratory
PMGA	Provisional Maintenance Group Areas
POW	Prisoner of War

R

R&D	Research and Development
RAM	Rapid Area Maintenance
RAPCON	Radar Approach Control
RMA	Reserve Military Aviator
ROTC	Reserve Officer Training Corps
RTD	Research and Technology Division

S

SAC	Strategic Air Command
SBIR	Small Business Innovative Research
SCC	Sinclair Community College
SCPS-M	Survivable Collective Protection System-Medical
SEMEDS	Scanning Electron Microscope Educators
SERC	Signal Enlisted Reserve Corps
SFS	Security Forces Squadron
s.g.	Senior Grade
SGLPFE	Strange Green Little Person From Elsewhere
SIA	Societa Italiana Aviazione
SOCHE	Southwestern Ohio Council for Higher Education
SP	Security Police
SPO	System Program Office
SR	State Route
SRIM	Structural Reaction Injection Molding
STARBASE	Science and Technology Academies Reinforcing Basic Aviation and Space Exploration
STEM	Science, Technology, Engineering, and Mathematics
STEPP	Science, Technology, and Engineering Preparatory Program
SVA	Societa Verduzio Ansaldo

T

T^2	Technology Transfer
TVA	Tennessee Valley Authority

U

UAV	Unmanned Aerial Vehicle
UC	University of Cincinnati
UD	University of Dayton
UDRI	University of Dayton Research Institute
UFO	Unidentified Flying Object
UHF	Ultra High Frequency
UNIVAC	Universal Automatic Computer
US&R	Urban Search and Rescue
USA	United States Army
USAC	U.S. Army, Combat
USAF	United States Air Force
USAFM	United States Air Force Museum
USCAR	United States Consortium for Automotive Research
USEPA	United States Environmental Protection Agency
USN	United States Navy
USPS	United States Postal Service

V

VA	Veterans Affairs
	Veterans Administration
VAMC	Veterans Affairs Medical Center
VHB	Very Heavy Bomber
VHF	Very High Frequency
VOQ	Visiting Officers' Quarters

W

WAC	Women Army Corps
WACO	Weaver Aircraft Company
WADC	Wright Air Development Center
WADD	Wright Air Development Division
WASP	Women's Airforce Service Pilots
WAVES	Women Appointed for Voluntary Emergency Service
WBAC	Wright Brothers Aeroplane Company
WOW	Wizards Of Wright
WPA	Works Progress Administration
WPAFB	Wright-Patterson Air Force Base
WR-ALC/AFT	Warner Robins Air Logistics Center - Air Force Petroleum Office
WRA	War Relocation Authority
WRDC	Wright Research and Development Center
WS	Weather Squadron
WSU	Wright State University
WTC	World Trade Center
WWASD	Wilbur Wright Air Service Depot

Y

YMCA	Young Men's Christian Association
YWCA	Young Women's Christian Association

BIBLIOGRAPHY

1ST SERGEANT—

BIBLIOGRAPHY

BOOKS

Aircraft Yearbook for 1924. New York: Manufacturers Aircraft Association, 1925.

Aldridge, James F. *Wright from the Start: Contributions of Dayton's Science and Engineering Community to American Air Power in the Twentieth Century.* WPAFB: ASC History Office, 1999.

Angelucci, Enzio, ed. *The Rand McNally Encyclopedia of Military Aircraft, 1914-1980.* New York: Military Press, 1980.

The Army's Order of Battle of United States Land Forces in the World War (1917-1919). Vol. 3, Part 1. Washington: Government Printing Office, 1949.

Balmer, Joe, and Davis, Ken. *There Goes a Waco: An American Classic Aircraft.* Troy, Ohio: Little Otter Productions, 1992.

Bartlett, John. *Familiar Quotations,* 16th ed. Boston: Little, Brown, 1992.

Bernstein, Mark. *Grand Eccentrics. Turning the Century: Dayton and the Inventing of America.* Wilmington, Ohio: Orange Frazer Press, 1996.

The Bible. Revised Standard Version. New York: Oxford University Press, 2002.

Bilstein, Roger E. *The Enterprise of Flight: The American Aviation and Aerospace Industry.* Washington: Smithsonian, 2001.

Bilstein, Roger E. *Flight in America: From the Wrights to the Astronauts.* Baltimore: Johns Hopkins UP, 1984; rev. ed., Baltimore and London: Johns Hopkins UP, 1994.

Bilstein, Roger E. *Orders of Magnitude: A History of the NACA and NASA, 1915-1990.* Washington, D.C.: NASA, 1989.

Bluth, John A. *Stinson Aircraft Company.* Chicago: Arcadia Publishing, 2002.

Boyne, Walter J. *The Best of "Wings": Great Articles from "Wings" and "Airpower" Magazines.* Washington: Brasseys, 2001.

Boyne, Walter J. *The Leading Edge.* New York: Stewart, Tabori & Chang, 1986.

Brands, H. W. *Woodrow Wilson.* New York: Times Books, 2003.

Cebula, James E. *James M. Cox: Journalist and Politician.* New York: Garland, 1985.

Chandler, Charles deF. and Lahm, Frank P. *How Our Army Grew Wings: Airmen and Aircraft Before 1914.* New York: Ronald Press, 1943.

Conover, Charlotte Reeve. *Dayton, Ohio: An Intimate History.* Dayton: Landfall Press, 1995.

_____. *Dayton, Ohio: An Intimate History.* New York: Lewis Historical Publishing Co., 1932.

_____. *History of Dayton and Montgomery County, Ohio.* 4 volumes. New York: Lewis Historical Publishing, 1932.

_____. *Some Dayton Saints and Prophets.* Dayton, Ohio: United Brethren Publishing House, 1907.

Cox, James M. *Journey through My Years.* New York: Simon and Schuster, 1946.

Crouch, Tom D. *The Bishop's Boys: A Life of Wilbur and Orville Wright.* New York: Norton, 1989.

_____. *The Giant Leap: A Chronology of Ohio Aerospace Events and Personalities 1815-1969.* Columbus: Ohio Historical Society, 1971.

Crowther, Samuel. *John H. Patterson, Pioneer in Industrial Welfare.* Garden City, New York: Doubleday, 1924.

Dalton, Curt. *Home Sweet Home Front: Dayton During World War II.* Dayton: Dalton, 2000.

Dalton, Curt, ed. *Through Flood, Through Fire: Personal Stories from Survivors of the Dayton Flood of 1913.* Dayton: Dalton, 2001.

Doolittle, James H. *I Could Never Be So Lucky Again: An Autobiography.* With Carroll V. Glines. New York: Bantam, 1991.

Dyson, Emma J. H., et al. *The Engineering of Flight: Aeronautical Engineering Facilities of Area B, Wright-Patterson Air Force Base, Ohio.* National Park Service: Government Printing Office, 1993.

Farquhar, Michael. *A Treasury of Great American Scandals.* New York: Penguin, 2003.

Gheen, Dottie Grey. "The Villages of Fairfield and Osborn, Ohio," in Lois E. Walker and Shelby E. Wickam, *From Huffman Prairie to the Moon: The History of Wright-Patterson Air Force Base,* WPAFB: Office of History, 2750th Air Base Wing, 1986.

Goldberg, Alfred, ed. *A History of the United States Air Force, 1907-1957.* New York and Princeton: D. Van Nostrand, 1957.

Gunston, Bill. *American Warplanes.* New York: Salamander, 1986.

Howard, Fred. *Wilbur and Orville: A Biography of the Wright Brothers.* New York: Knopf, 1987.

Ingells, Douglas J. *They Tamed The Sky.* New York: D. Appleton-Century, 1947.

Ingler, Charles W. *Founding and Fulfillment: 1964-1984.* Dayton: Wright State University, 1987.

Jakab, Peter L. *Visions of a Flying Machine: The Wright Brothers and the Process of Invention.* Washington: Smithsonian, 1990.

Johnson, Mary Ann. *A Field Guide to Flight: On the Aviation Trail in Dayton Ohio.* Rev. ed. Dayton: Landfall Press, 1996.

Karman, Theodore von, and Edson, Lee. *The Wind and Beyond: Theodore von Karman, Pioneer in Aviation and Pathfinder in Space.* Boston: Little Brown, 1967.

Kelly, Fred C., ed. *Miracle at Kitty Hawk: The Letters of Wilbur and Orville Wright.* New York: Farrar, Straus and Young, 1951; reprint ed., New York: Da Capo, 1996.

Kelly, Fred C. *The Wright Brothers: A Biography.* New York: Harcourt, Brace, 1943; reprint ed., New York: Dover, 1989.

Leslie, Stuart W. *Boss Kettering: Wizard of General Motors.* New York: Columbia University Press, 1983.

Lindbergh, Charles A. *The Spirit of St. Louis.* New York: Scribner, 1953.

Mansfield, Harold. *Vision: The Story of Boeing—A Saga of the Sky and the New Horizons of Space.* New York: Popular Press, 1966.

Marcosson, Isaac F. *Colonel Deeds, Industrial Builder.* New York: Dodd, Mead, 1947.

Martin, Herbert Woodward, and Primeau, Ronald. *In His Own Voice: the Dramatic and Other Uncollected Works of Paul Laurence Dunbar.* Athens, Ohio: Ohio University Press, 2002.

Maurer, Richard. *The Wright Sister.* Brookfield, Connecticut: Roaring Brook Press, 2003.

Miller, Ivonette Wright, compiler. *Wright Reminiscences.* WPAFB: Air Force Museum Foundation, 1978.

Morris, Lloyd and Smith, Kendall. *Ceiling Unlimited: The Story Of American Aviation from Kitty Hawk to Supersonics.* New York: Macmillan Co., 1953.

Nichols, Jim. *Dayton Album: Remembering Downtown.* Dayton: Viewpoint Publications, 2000.

Peters, Margaret E. *Dayton's African American Heritage.* Virginia Beach, Virginia: The Donning Company, 1995.

A Pictorial Review, Wright-Patterson Air Force Base, 1917-1967. WPAFB: 2750th Air Base Wing, 1967.

Renstrom, Arthur G. *Wilbur & Orville Wright: A Chronology Commemorating the Hundredth Anniversary of the Birth of Orville*

Wright, August 19, 1871. Washington: Library of Congress, 1975.

Roland, Alex. *Model Research: The National Advisory Committee for Aeronautics, 1915-1958.* 2 vols. Washington: Government Printing Office, 1985.

Rollins, Ron, ed. *For the Love of Dayton: Life in the Miami Valley, 1796-1996.* Dayton: Dayton Daily News, 1996.

Ronald, Virginia and Bruce. *The Lands between the Miamis: A Bicentennial Celebration of the Dayton Area.* Volume 2. Dayton: Landfall Press, 1996.

Sealander, Judith. *Grand Plans: Business Progressivism and Social Change in Ohio's Miami Valley, 1890-1929.* Lexington: University Press of Kentucky, 1988.

Splendid Vision, Unswerving Purpose: Developing Air Power for the United States Air Force during the First Century of Powered Flight. WPAFB: ASC History Office, 2002.

The Story of the Miami Conservancy District. Dayton: Miami Conservancy District, 1922.

Sweetser, Arthur. *The American Air Service.* New York: Appleton, 1919.

Syrett, Harold C., ed. *American Historical Documents.* New York: Barnes and Noble, 1960.

Thomas, Lowell J. *The First World Flight.* Boston: Houghton, Mifflin Co., 1925.

Thomas, Lowell J. and Thomas, Lowell, Jr. *Famous First Flights That Changed History.* Garden City, New York: Doubleday, 1968.

Thompson, John A. *Woodrow Wilson: Profiles in Power.* London: Pearson Longman, 2002.

Tuchman, Barbara. *The Guns of August.* New York: Macmillan, 1962.

The University of Dayton Research Institute Sponsored Research: 40 Years of Progress, 1956-1996. Dayton: UDRI, 1998.

Walker, Lois E. and Wickam, Shelby E. *From Huffman Prairie to the Moon: The History of Wright-Patterson Air Force Base.* WPAFB: Office of History, 2750th Air Base Wing, 1986.

Winegarten, Debra L. *Katherine Stinson: The Flying Schoolgirl.* Austin, Texas: Eakin Press, 2002.

The World Almanac and Book of Facts, 1980. New York: Newspaper Enterprise Association, 1979.

OFFICIAL USAF UNIT HISTORIES

ARMY AIR FORCES HISTORICAL STUDIES

Army Air Forces Historical Study No. 25, *Organization of Military Aeronautics 1907-1935,* 1944.

Army Air Forces Historical Study No. 50, *Materiel Research and Development in the Army Air Arm, 1914-1945,* 1946.

USAF HISTORICAL STUDIES

Hennessey, Juliette A. USAF Historical Study No. 98, *The United States Army Air Arm, April 1861 to April 1917,* 1958.

USAF HISTORIES

Gropman, Alan L. *The Air Force Integrates, 1945-1964.* Washington, D.C.: Office of Air Force History, 1978.

Maurer, Maurer, ed. *Air Force Combat Units of World War II.* Maxwell AFB, Alabama: Albert F. Simpson Historical Research Center, 1960.

_____. *Aviation in the U.S. Army, 1919-1939*. Washington: Office of Air Force History, 1987.

_____. *Combat Squadrons of the Air Force During World War II*. Maxwell AFB, Alabama: Albert F. Simpson Historical Research Center, 1969.

_____. *The U.S. Air Service in World War I*. Vol. 1: *The Final Report and a Tactical History*. Washington: AF Historical Office, 1978.

COMMAND HISTORICAL STUDIES

AMC Historical Study No. 329, *AMC and Its Antecedents, 1917-1960*, 1960.

COMMAND HISTORIES

Baker, Doris A. *History of AMC Field Organization, 1917-1955*. Historical Office, Air Materiel Command, 1956.

Frey, Royal. *Evolution of Maintenance Engineering, 1907-1920*. Historical Office, Air Materiel Command, 1960.

History of the Air Depot at Fairfield, Ohio, 1917-1943. Patterson Field: Fairfield Air Service Command, 1944.

History of the AMC Contract Airlift System (LOGAIR), 1954-1955. Historical Office, Air Materiel Command, 1956.

History of Fairfield Air Depot Control Area Command (FADCAC) and Fairfield Air Service Command (FASC), 1 Feb 1943-1 Oct 1944. Historical Office, Air Materiel Command, n.d.

Purtee, Dr. Edward O. *History of the Army Air Service, 1907-1926*. Historical Office, Air Materiel Command, 1948.

Termena, Bernard J.; Peiffer, Layne B.; and Carlin, H.P. *Logistics: An Illustrated History of AFLC and Its Antecedents, 1921-1981*. WPAFB: Air Force Logistics Command, 1983.

AERONAUTICAL SYSTEMS CENTER HISTORIES AND STUDIES

Cornelisse, Diana G., et al., *Against the Wind: 90 Years of Flight Test in the Miami Valley*. WPAFB: ASC History Office, 1994.

Cornelisse, Diana G. *The Foulois House: Its Place in the History of the Miami Valley and American Aviation*. WPAFB: ASC History Office, 1991.

Misenko, Albert E. *Aero Propulsion and Power Directorate: The McCook Field Years 1917-1927*. WPAFB: ASC History Office, 1995.

Moutoux, Bonnie. *An Historic Tour of Wright-Patterson Air Force Base*. WPAFB: ASC History Office, 1995.

Narducci, Henry M. *Ghosts From the Past: The Recovery of M114 Biological Weapons at Wright-Patterson AFB, Ohio, November 1995-September 1996*. WPAFB: ASC History Office, 1999.

Splendid Vision, Unswerving Purpose: Developing Air Power for the United States Air Force during the First Century of Powered Flight. WPAFB: ASC History Office, 2002.

AERONAUTICAL SYSTEMS DIVISION HISTORIES AND STUDIES

Misenko, Albert E. and Pollock, Philip H. *Engineering History, 1917-1978, McCook Field to the Aeronautical Systems Division*. 4th ed. WPAFB: Aeronautical Systems Division History Office, 1979.

WING HISTORIES AND STUDIES

Economic Impact Analysis, WPAFB: 88th Air Base Wing/FMA.
  Sep 30, 2000; Sep 30, 2001; Sep 30, 2002.

Histories of the 2750th Air Base Wing
  Jul-Dec 1948; Jan-Jun 1951; Jan-Jun 1952; Jul-Dec 1952; Jul-Dec 1953; Jan-Jun 1954; Jan-Dec 1957; Jul 1960-Jul 1961; Jul 1967-
  Jun 1969; Jul 1974-Dec 1975; Jan-Dec 1976.

History of Wilbur Wright Field, 1917-1925. Volume 1. WPAFB: 2750th Air Base Wing.

Narducci, Henry M. *Balkan Proximity Peace Talks and Wright-Patterson AFB, 18 October – 22 November 1995.* Volume 1. WPAFB:
88th Air Base Wing, 1997.

Walker, Lois E. and Wickam, Shelby E. *From Huffman Prairie to the Moon: The History of Wright-Patterson Air Force Base.* WPAFB:
Office of History, 2750th Air Base Wing, 1986.

OTHER HISTORIES

"Comprehensive History of Patterson and Wright Field Planning, Nov 1942-Feb 1945." ASC History Office Archives.

"History of 302D Special Operations Wing 1 Jan - 31 Mar 1971, including closing of CCAFB" Folder, ASC History Office Archives.

A Little Journey to the Home of the Engineering Division, Army Air Service, McCook Field, Dayton, Ohio. Dayton: McCook Field,
Headquarters Engineering Division, 1926; reprint ed., Washington: n.p., 1988.

Niehaus, James J. *Five Decades of Materials Progress: Air Force Materials Lab, 1917-1967.* WPAFB: 1967.

A Pictorial Review, Wright-Patterson Air Force Base, 1917-1967. WPAFB: 2750th Air Base Wing, 1967.

REPORTS

Annual Report, Signed by Major Arthur E. Wilbourn, Signal Corps Aviation School, Wilbur Wright Field, Fairfield, Ohio, May 31, 1918.

Army Air Forces Technical Base Planning Board, Preliminary Master Plan Report, approved Mar 18, 1947.

Completion Report, "Wright Field, Dayton, Ohio." Vol. 1. Office of Constructing Quartermaster, Dayton, Ohio, ca. Jul 1927.

Report, "Armed Service Participation in the National Aircraft Show, Dayton, Ohio, 4-6 September 1954." General E. W. Rawlings to
Director, Office of Public Information.

Report, "Exercise Evaluation Team Report: SABER SHIELD 00-2." ASC/IG, Jun 26-30, 2000.

Report, "Exercise Evaluation Team Report: STRONGWIND 99-1." ASC/IG, Mar 8-11, 1999.

Report, Historical Data, Maintenance Division, Fairfield Air Service Command, Patterson Field, May 1, 1944.

Shiman, Philip. *Forging the Sword: Defense Production During the Cold War.* CERL Special Report 97/77. Washington: Construction
Engineering Research Laboratory, 1997.

DOCUMENTS

Engineering Division, Air Service. Specification for Barling Bombardment Airplane, May 15, 1920.

Enlisted Reserve Corps. Certificate of Enlistment, AGO Form 422-1.

Headquarters U.S. Air Force. General Orders No. 2, Jan 13, 1948.

Headquarters Signal Corps Aviation School, Wilbur Wright Field. Special Orders 124, May 9, 1918.

Headquarters Wright-Patterson Air Force Base. General Orders 32, Aug 27, 1948.

"Historical Sketch, Supply Division, FASC, 1917-1938," n.d.

"Integration of Wright and Patterson Fields Into a Single Unit." ASC History Office Archives.

Miami Conservancy District and Lieutenant Colonel C. G. Edgar. Lease, Purchase Request A-6951, Order No. 50214, Jul 1, 1917.

Morgan, Ernest. Letter to *Dayton Daily News*, Jul 10, 1989, on file at ASC History Office.

Muhlenberg, Major H. C. K., Wilbur Wright Field Armorers School Commanding Officer. Letter to Director of Military Aeronautics, Technical Section, Washington, D.C., subj: Report on Accident to DeHavilland Four Plane No. 32098, Jun 26, 1918.

"Mutual Aid Agreement for Fire Protection." Agreement, ASC History Office Archives, Jan 24, 2001.

National Aircraft Show Official Program and Log, Dayton, Ohio, 1953. ASC History Office Archives.

Orenstein, Julie. "David A. Sinclair: A Biography." Sinclair Community College Archives, 2003.

Otero-Stinson Family Papers. University of New Mexico, Center for Southwest Research. Letters between Katherine Stinson Otero and Emma Stinson, re: Dayton, OH property, 1923-25.

Registrar Yale College [University], New Haven, Conn. Letter to Frank S. Patterson, no subj., Apr 27, 1918.

Signal Corps. Specification Number 483, Jan 21, 1908.

Thurman, Lieutenant General William E. Letter to Dr. Ernest Moore, Director, Edison Materials Technology Center (EMTEC), Aug 12, 1987.

War Department. General Orders 68, Aug 18, 1944.

War Department. Special Orders 110, May 10, 1918.

NEWSPAPERS AND PERIODICALS

9-11 Response Articles and Photos. WPAFB *Skywrighter*, Sep 14, 2001, p 1.

"Acme, Setting Meet Record With 80-51 Win, Faces Zollners In Finals." *Dayton Daily News*, Mar 24, 1945, Sports Section, p 8.

"AF Band orchestrates aviation's first 100 years." *Leading Edge*, Jun 2003, p 26.

"AFIT Summer engineering program under way." WPAFB *Skywrighter*, Jul 3, 2003.

"AFIT's Systems, Logistics School to get temporary home." WPAFB *Skywrighter*, Jul 3, 2003.

"AFRL presents Century of Flight awards." WPAFB *Skywrighter*, Jul 25, 2003, p 7A.

"AFRL presents Century of Flight Awards to top students." WPAFB *Skywrighter*, Jul 11, 2003, p. 8A.

Air Service News Letter
  Jul 19, 1923, Vol. VII
  Nov 22, 1923, Vol. VII
  Feb 1, 1924, Vol. VIII
  Jul 31, 1924, Vol. VIII
  Oct 31, 1924, Vol. VIII

"Air-Tec Travels End but May Play Here." *Dayton Daily News*, Mar 3, 1945, p 3.

"The Air Techs." *Wright Field Take-Off*, Dec 30, 1944, p 10.

"Air-Tecs Get Service 'Invite.'" *Dayton Daily News*, Feb 9, 1945, p 25.

"Air-Tecs Put on Display of Cage Prowess Here." *Dayton Daily News*, Feb 12, 1945, p 14.

"Air Tecs Put on Show." *Dayton Daily News*, Feb 12, 1945, p 14.

"AIRBORNE: How Dayton's National Park took Flight." *Dayton Daily News*, Oct 19, 1992, p 9A.

"Aircraft Production in Dayton." *NCR World* (Sep-Oct 1970), pp 21-24.

"Bands orchestrate 100 years." WPAFB *Skywrighter*, Apr 18, 2003, p1A.

"Base officials gave briefing at town hall meeting." WPAFB *Skywrighter*, May 9, 2003, p.8A.

"Base support to Dayton air show doubles for larger event." WPAFB *Skywrighter*, Jul 11, 2003, p 3A.

Batz, Bob. "Beauty Squadron: Women served their country with pride at 1941 air show." *Dayton Daily News*, Aug 3, 2003, p E3.

"Bodily remains buried, but spirit lives on." *Dayton Daily News*, Jul 21, 2003, p. A1.

"Capt. Chuck Taylor Guest At Muni Party." *Dayton Daily News*, Apr 10, 1945, p 21.

"Capt. Chuck Taylor Is Shifted To Army." *Dayton Daily News*, Mar 19, 1945, p 15.

Casari, Robert. "Number of U.S. Aircraft WWI." *AAHS Journal* 20 (Spring 1977), pp 36-38.

"Centennial activities take off at flying field." WPAFB *Skywrighter*, Jul 3, 2003, p 3A.

Christy, Joe. "That First Round-the-World Flight." *Air Force Magazine* 57 (Mar 1974), pp 53-59.

"Community salutes base in centennial events today." WPAFB *Skywrighter*, Jul 11, 2003, p 1A.

"Cyclists weather rain to coast into Kitty Hawk." *Dayton Daily News*, Aug 15, 2003.

"Dayton International Airport History: Not Always Smooth Flight." *Bowling-Moorman Newspapers, Inc.*, 1990.

"Deep Roots: Kettering Medical Center Network fosters new technology, community outreach." *Dayton Daily News*, Forecast 03 Health advertising supplement, Apr 13, 2003.

"Employee '5' Books Fast Air Tecs." *Postings*, Jan 26, 1945, p 3.

"Exposition offers look at future of flight." *Dayton Daily News*, Inventing Flight Special Section, Jul 15, 2003, p 3.

"'First Flight' stamp unveiled." WPAFB *Skywrighter*, May 30, 2003, p 1A.

"Flying Field To Be Divided For National Air Olympics." *Dayton Herald*, Aug. 25, 1941, n.p.

"From the Editors Desk." *Aerospace Historian* 27(1), Spring/March 1980, pp 55-57.

Guinn, Gilbert S. "A Different Frontier: Aviation, The Army Air Forces and the Evolution of the Sunshine Belt." *Aerospace Historian* 29 (Mar 1982), pp 34-35.

Harvey, Alva. "Seattle Has Crashed in Alaska." *Air Force Magazine* 57 (Sep 1974), pp 103-107.

"Huffman Prairie First Airfield." *Dayton Daily News*, Dec 13, 1953, n.p.

Inventing Flight Special Section. *Dayton Daily News*, Jul 5, 2003.

"It's Wheels Up." *Dayton Daily News*, Jul 18, 2003, p A2.

"Jumper takes to skies in Wright 'B' Flyer" WPAFB *Skywrighter*, Jun 27, 2003, p B1.

Leslie, Stuart W. "The Bug: 'Boss' Kettering's Cruise Missile." *Timeline* Aug/Sep 1991: 42-51.

Lincoln, Marshall. "The Barling Bomber." *Air Classics* 2(5) Feb 1965.

McMaster, R. K., ed. "The Adventures of a Junior Military Aviator: Extracts from the Diary of Leo G. Heffernan." *Aerospace Historian* 25, Jun 1978, pp 92-102.

"Memo: Columbus sculptor has Wright design." *Dayton Daily News*, Oct 10, 1994, p B1.

"Monuments to Times Past." *Dayton Daily News*, Aug 14, 2003, p A15.

Morgan, Arthur E. *The Christian Register*, May 1945.

"Mrs. Buda Simms Dies; Country Club Founder." *Dayton Daily News*, Jan 17, 1978, p 22.

National Aeronautic Association (NAA) Review 2 (Special Dayton Air Race Edition), Sep 18, 1924.

"New Acme Lineup To Be Seen Sunday In National Preview." *Dayton Daily News*, Mar 18, 1945, Sports Section, p 2.

"New Post Hospital Open." *Patterson Field Postings*, May 20, 1942, p 1.

"Open house draws weekend crowds." WPAFB *Skywrighter*, May 16, 2003, p 7A.

"Outreach office efforts reach Columbus." WPAFB *Skywrighter*, Aug 1, 2003, p 3A.

"Park takes flight: Bush signs Dayton Aviation Park Bill." *Dayton Daily News*, Oct 17, 1992, p 1A.

"Rain Birds: Air Force display goes on despite heavy storm." *Dayton Daily News*, May 11, 2003, p A1.

"Research Institute Breaks Record." *University of Dayton Quarterly*, Autumn 2003, p 3.

Roberts, Carl V. "Birth of a City." Camerica, *Dayton Daily News*, Dec 18, 1949, pp 4-11.

"Saving History." *Government Executive*, Jul 2003, pp 45-52.

"Service Awards to Civilian Employees Authorized by WD." *Patterson Field Postings*, Nov 5, 1943, p 2.

"Simms Station is Topic at Aero Chapter 536 Meeting." *Fairborn Daily Herald*, Mar 15, 1984, p 12.

"Soaring: The Sculpture of John Safer." *The Dayton Art Institute Member Quarterly*, Jul 2003, p 9.

STRONGWIND 99-1 Photo Captions. WPAFB *Skywrighter*, Mar 19, 1999, p 15.

"Taft awards $1 million to NCC." *Dayton Daily News*, Oct 17, 2003, p. D1.

"Taylor [   ] Schedule [   ]." Dayton *Journal-Herald*, Dec 3, 1944, n.p.

"Tecs Breeze To 67-41 Win." *Dayton Daily News*, Feb 10, 1945, p 5.

"Tecs, Subbing For Buckeyes, Beat Aviators." *Dayton Daily News*, Mar 5, 1945, p 14.

"Test pilots evaluate Wright brothers' gliders." WPAFB *Skywrighter*, May 2, 2003, p 3A.

"Three Regular Players In Acme Lineup Against Chicago Here." *Dayton Daily News*, Mar 11, 1945, Sport Section, p 2.

Tilford, Captain Earl H., Jr. "The Short, Unhappy Life of the Barling Bomber." *Air Force Magazine*, Feb 1978, pp 68-70.

Turner, Brett. "Air Force Marathon a Success." WPAFB *Skywrighter*, Sep 26, 2003, p 1A.

"UD, National Composite receive nearly $4 million." *Dayton Daily News*, Jun 6, 2003, p D1.

"UD researchers get $31.5M contract." *Dayton Daily News*, May 12, 2003, p D1.

Wallace, Mike. "Band owes start to music man." WPAFB *Skywrighter*, Jul 13, 2001.

"When Pilots Flew on Two Wings." WPAFB *Skywrighter*, May 15, 1981, pp 16, 19.

"World War I aviation to roar to life at Air Force Museum." WPAFB *Skywrighter*, Sep 5, 2003, p 8A.

"Wow! Off Went Lid On Victory Day!" *The Wright Idea*, Wilbur Wright Field, Nov. 16, 1918.

"Wright 'B' Flyer Circles again." *Dayton Daily News*, May 26, 2003, p A1.

"Wright Scholar intern program expands across base." WPAFB *Skywrighter*, Jun 27, 2003, p 6A.

"Wright-Patterson Air Force Base Awarded Thomas Edison Award." *The WTN Innovator*, Nov/Dec 1997, p.1.

Wright-Patterson Air Force Base *Skywrighter*
   Jun 17, 1960
   Jun 9, 1961
   Jun 9, 1967
   Aug 18, 1967
   Sep 26, 1969
   Aug 23, 2002

Wright-Patterson *Post Script*
   Sep 21, 1945
   Sep 28, 1945
   Nov 16, 1945
   Nov 21, 1945
   Dec 14, 1945
   Dec 28, 1945

"Zollners Swamp Acme 78-[52]." *Dayton Daily News*, Mar 25, 1945, Sports Section, p 1.

INTERVIEWS/PERSONAL COMMUNICATIONS

Mr. Abe Aamidor, personal communication with James Ciborski, ASC History Office, 2003.

Mrs. Deborah Anderson, email to Robin Smith, ASC History Office, Jul 28, 2003.

Major General Fred J. Ascani. *A General Remembers: Major General Fred J. Ascani, Wright Field's Father of Systems Engineering*. Edited with narrative by James F. Aldridge. WPAFB: ASC History Office, 2001.

Mr. Jim Beisner, Waco Learning Center, personal communication with Helen Kavanaugh-Jones, ASC History Office, Apr 2003.

Mr. Oscar Boonshoft, personal communications with Robin Smith, ASC History Office, Jun-Oct 2003.

Mr. Robert B. Clayton, Chief, Civil Law, 88th ABW/JAC, personal communication with Lori S. Tagg, ASC History Office, Aug 2003.

Mr. Gary R. Chandler, Wright-Patterson Air Force Base Airfield Supervisor, personal communication with Master Sergeant David L. Wolf, ASC History Office, May 13, 2003.

Mr. Roger Cranos, email to Helen Kavanaugh-Jones, ASC History Office, Jun 2003.

Mr. Chris Doyle, Converse Public Relations, personal communication with James Ciborski, ASC History Office, Oct 14, 2003.

Mr. Robert Durrum, interview with Robin Smith, ASC History Office, Sep 24, 2003.

Mr. Dave Egner, Lead Air Traffic Control Specialist, 88 OSS Base Operations, email to Helen Kavanaugh-Jones, ASC History Office, May 23, 2003 and Aug 2003.

Dr. Jan Ferguson, Cultural Resources Program Manager, 88 ABW/EMO, email to Helen Kavanaugh-Jones, ASC History Office, Jun 2, 2003.

Mrs. Shirley Flacks, personal communication with Robin Smith, ASC History Office, Jun 26, 2003.

Mr. Jeff Hull, VA Medical Center, personal communication with Diana Cornelisse, ASC History Office, Jul 14, 2003, and Sep 15, 2003.

Ms. Toni Jeske, Archivist, Wright State University Libraries Special Collections and Archives, email to Robin Smith, ASC History Office, dated Jun 27, 2003.

Mr. Ron Kaehr, email to Helen Kavanaugh-Jones, ASC History Office, Jun 2003.

Mr. Andrew Kididis, telephone conversation with Helen Kavanaugh-Jones, ASC History Office, May 2003.

Ms. Nancy Brown Martinez, Reference Coordinator, Center for Southwest Research, email to Robin Smith, ASC History Office, Jun 9, 2003.

Mr. Robert J. McKee, former OH-TF-1 task force leader, personal communication with James R. Ciborski, ASC History Office, Jun 2003.

Mr. Bill Meixner, Society of Air Racing Historians, personal communication with Lori S. Tagg, ASC History Office, Jul 28, 2003.

Mr. Al Moyers, Air Force Weather History Office, Air Force Weather Agency, Offutt Air Force Base, Nebraska, personal communication with Lori S. Tagg, ASC History Office, Jul 25, 2003.

Ms. Karen Pittman, personal communication with Helen Kavanaugh-Jones, ASC History Office, summer 2003.

Mr. Ray Rang, email to Helen Kavanaugh-Jones, ASC History Office, Jun 2003.

Mr. R. Michael Rives, chief, Medical Readiness Flight, personal communication with Mr. James R. Ciborski, ASC History Office Archives, Oct 7, 2003.

Dr. Vince Russo, ASC Executive Director, interview with Helen Kavanaugh-Jones, ASC History Office, Jun 2003.

Mr. Jim Sandegren, Woodland Cemetery historian, personal tour for ASC History Office Staff, Fall 2002.

Ms. Marion Schniegenberg, WBAC Public Relations, telephone conversation with Helen Kavanaugh-Jones, ASC History Office, Oct 15, 2003.

Technical Sergeant Greg Sell, Band of Flight Director of Operations, email to Helen Kavanaugh-Jones, ASC History Office, Jul 25, 2003.
Ms. Bessie Shina, Wright State University Records Archivist, email to Robin Smith, ASC History Office, dated Jul 1, 2003.

Mr. Joseph M. (Jerry) Slade, telephone conversation with Robin Smith, ASC History Office, Jul 10, 2003.

Mr. Stephen M. VanDegrift, Disaster Preparedness Specialist, interview with Mr. James R. Ciborski, ASC History Office, Aug 26, 2003.

2nd Lieutenant Gailyn Whitman, email to Helen Kavanaugh-Jones, ASC History Office, summer 2003.

ONLINE SOURCES

"About Dayton International Airport." Viewed online Oct 15, 2003, at http://www.flydayton.com/airport_info/airport_info_frmst.htm.

"About NCR – History." Viewed online Aug 9, 2003, at http://www.ncr.com/history.htm.

Air Force Institute of Technology website. Viewed online May 2003 at http://www.afit.edu.

"Appendix 17: The Navy in Space." Viewed online Oct 15, 2003, at http://www.history.navy.mil/avh-1910/APP17.PDF.

"Arthur Morgan." Viewed online Jul 8, 2003, at http://www.phd.antioch.edu/Pages/APhDWeb_Prospects/arthurmorgan.

"Bath Township—Greene County, Ohio," Green County Public Library. Viewed online Sep 26, 2003, at http://www.gcpl.lib.oh.us/bicentennial.asp.

Bellis, Mary. "The History of the Automobile: The Internal Combustion Engine and Early Gas-Powered Cars." Viewed online Sep 30, 2003, at http://inventors.about.com/library/weekly/aacarsgasa.htm.

"Centenarian Booksellers II." Viewed online at PublishersWeekly.com, Aug 14, 2000.

"Charles Kettering." Viewed online May 5, 2003, at http://www.nationalaviation.org/museum_enshrinee.

Civil Air Patrol website. Viewed online Oct 15, 2003, at http://www.cap.gov/, with links.

Clayton J. Brukner Biographical Sketch. Viewed online Jun 19, 2003, at http://www.libraries.wright.edu/special/manuscripts/ms109biosketch.html.

"Clinton County Air Force Base, Ohio." Viewed online Oct 15, 2003, at http://www.globalsecurity.org/wmd/facility/clinton_county.htm.

Community Profile. "Fairborn, Ohio." Viewed online Jul 18, 2003, at http://www.nationjob.com/showcomp.cgi/faoh.html.

Csutoras, Deborah S. "Family Day at the Air Force Museum." Viewed online at Skywrighter.com, Mar 14, 2003.

Dayton Area Graduate Studies Institute (DAGSI) website. Viewed online May 2003 at http://www.dagsi.org/index.shtml.

"Dayton Cycling Club pedals more than 900 miles to the Outer Banks for the flight Centennial." Viewed online Aug 25, 2003, at http://kittyhawkfreepress.com/Daytonbicycle.htm.

"A Dedication to Aviator and Test Pilot Lieutenant Giovanni (John) Pirelli," John Pirelli Lodge #1633. Viewed online Jul 21, 2003, at http://www.johnpirelliosia.org/johnpirelli_dedication.htm.

Dictionary of American Naval Fighting Ships (DANFS). Viewed online Aug 22, 2003, at http://www.hazegray.org/danfs/cruisers/cl105.txt.

The Early Birds website. Viewed online Sep 23, 2003, at http://www.earlybirds.org/.

Edison Materials Technology Center (EMTEC) website. Viewed online Oct 18, 2003, at http://www.emtec.org/programs/programs.html and http://www.emtec.org/capabil/capabil.html and http://www.emtec.org/edtech/edtech.html and http://www.emtec.org/mem_shp/mem_shp.html.

Fact Sheet. "The National Composite Center and Martin Marietta Composites Hold Special Ceremony to Unveil Largest Composite Bridge Installation to Date," National Composite Center. Viewed online May 2003, at http://www.compositecenter.org/newsview2.php?ID=17.

Ferns, Susan. "Huffman Prairie Flying Field to be Honored Next Week." Viewed online at Skywrighter.com, Jul 11, 2003.

"Fifth Third Field: Fun, Facts & Trivia." Viewed online Jun 11, 2003, at http://www.daytondragons.com/stadium/facts.html.

Fisher, Dr. Eugene. "National Workshop on Christian-Jewish Relations." Viewed online Oct 15, 2003, at http://www.ibiblio.org/bgreek/archives/96-08/1004.html.

"Genealogy of the Flying Stinsons." Viewed online Jul 20, 2003, at http://www.natchezbelle.org/oldtime/stinson.htm#jack.

Hickcock, Ralph. "History of the International Air Races." Viewed online Jul 16, 2003, at http://www.hickoksports.com/history/airrace.shtml.

"Hippocrates on the Web," Faculty of Medicine, University of Manitoba. Viewed online Jul 29, 2003, at http://www.umanitoba.ca/faculties/medicine/units/history/lister/pasteur.html.

"The History of Air Racing," European Sport Pilot Association. Viewed online Jul 17, 2003, at http://www.esparacing.com/Air%20Racing%20History/MAIN%20MENU.htm.

"July 2002 Population Estimates for Sub-County Areas in Ohio." Viewed online Aug 14, 2003, at http://nodisnet1.csuohio.edu/nodis/2000reports/Ohio_sub_cnty_est_2002.pdf.

"The Major Trophy Races of the Golden Age of Air Racing," U.S. Centennial of Flight Commission. Viewed online Jul 17, 2003, at http://www.centennialofflight.gov/essay/Explorers_Record_Setters_and_Daredevils/trophies/EX10.htm.

McGee, Chris. "Band of Flight in Concert at Museum." Viewed online at Skywrighter.com, Mar 21, 2003.

"Milestones in Telegraphic History." Viewed online Sep 30, 2003, at http://www.members.tripod.com/morse_telegaph_club/images/newpage1.htm.

Naismith Memorial Basketball Hall of Fame website. Viewed online Sep 22, 2003, at http://www.hoophall.com/halloffamers/TaylorC.htm.

National Aviation Hall of Fame website. Viewed online May 2003 at www.nationalaviation.org.

National Composite Center website. Viewed online Jun 11, 2003, at http://www.compositecenter.org/newsview2.php?ID=5 and http://www.compositecenter.org/history.php.

"National Military Home, Dayton, Ohio: Virtual Tour and Exhibits." Viewed online Sep 25, 2003, at http://www.dayton.med.va.gov/museum.

"Our History: Volunteers Serving in Times of Need." Viewed online Oct 15, 2003 at http://www.cap.gov/about/history.html.

Preservation Dayton website. Viewed online Aug 10, 2003, at http://www.preservationdayton.com.

"Reibold Building." Viewed online Sep 25, 2003, at http://www.skyscrapers.com.

Riverscape website. Viewed online Jul 2, 2003, at http://www.metroparks.org/Facilities/RiverScape/riverscape.html.

Schuster Performing Arts Center website. Viewed online Aug 3, 2003, at http://www.schustercenter.org/about/general.html.

Southwestern Ohio Council for Higher Education (SOCHE) website. Viewed online May 2003 at http://www.soche.org/index.htm.

"A Starting Point: Dayton's GM Story." Viewed online May 5, 2003, at http://www.thinktv.org/program/archive/gmstory/gmstory.html.

"Stinson." Viewed online Jul 3, 2003, at http://www.aerofiles.com/_stin.html.

"Stinson Aircraft Corporation." Viewed online Jun 25, 2003, at http://www.centennialofflight.gov/essay/GENERAL_AVIATION/stinson/GA2.htm.

"The Story of the Dayton Air Show." Viewed online Jul 21, 2003, at http://www.usats.org/html/history.htm.

"Titanic: A Special Exhibit from *Encyclopaedia Britannica*." Viewed online Sep 30, 2003, at http://www.search.eb.com/titanic/01_01.html.

"Titanic: A Special Exhibit from *Encyclopaedia Britannica*: Researcher's Note." Viewed online Sep 30, 2003, at http://www.search.eb.com/titanic/researchersnote.html.

"The U.S. Aircraft Industry During World War I." Viewed online Jul 1, 2003, at http://www.centennialofflight.gov/essay/Aerospace/Wwi/Aero5.htm.

U.S. Northern Command website. Viewed online Jul 29, 2003, at http://www.northcom.mil/index.cfm?fuseaction=s.whoweare.

"United Brethren Building." Viewed online Sep 25, 2003, at http://www.skyscrapers.com.

University of Cincinnati MBA website. Viewed online May 2003 at http://www.business.uc.adv/mba.

University of Dayton website. Viewed online Apr 2003 at http://www.udayton.edu/ with links.

University of Dayton Research Institute website. Viewed online Sep 29, 2003, at http://www.udri.udayton.edu.

"William Henry Harrison." Viewed online Sep 30, 2003, at http://www.whitehouse.gov/history/presidents/wh9.html.

"Wilmington Square." Viewed online Oct 15, 2003, at http://www.rgprop.com/Wilmington_square_overview.htm.

"World War II Industrial Facilities: Authorized Federal Funding," Heritage Research Center, Ltd. Viewed online Jun 30, 2003, at http://www.heritageresearch.com/War%20Facilities5.html.

Wright Brothers Aeroplane Company website. Viewed online Oct 15, 2003, at http://www.wright-brothers.org.

"Wright Scholar Research Assistant." Viewed online Jul 2003 at http://www.pr.afrl.af.mil/jobs/scholar.htm.

Wright State University website. Viewed online Apr 2003, at http://ww.libraries.wright.edu/special/1902/.

OTHER

Base Guide. WPAFB: 2750th Air Base Wing, 1956.

Booklet. "First Around the World." Douglas Aircraft Corporation, Santa Monica, California, 1974.

Brochure. "Aviation Trail." Aviation Trail, Inc.

Brochure. Stinson School of Aviation. 1943.

Caption 87-184/5/6. ASC/PAM, 1987.

Caption 92-196. "Dummy 'ADAM' Parachutes for Air Force Researchers," ASC Office of Public Affairs.

Caption 97-125. ASC/PAM, Jul 25, 1997.

Corwin Papers. Courtesy of Mr. James Sprinkle.

James Custer Collection, ASC History Office Archives.

"Dayton Army Air Field, Vandalia, Ohio" Folder, ASC History Office Archives.

The Dayton Public Art Catalogue. City of Dayton Public Arts Commission, 2001.

Fact Sheet. "1903 Wright Flyer Project at WSU Special Collections & Archives," Brown & Bills Architects, n.d.

Fact Sheet. "Addendum to Award Submittal," Brown & Bills Architects, n.d.

Fact Sheet. "Continuing a Legacy of Innovation and Technology Development," AFRL/XPTT, n.d.

Fact Sheet. "Cooperative Research and Development Agreements in Wright Laboratory," U.S. Air Force, PAM #91-146, 1991.

Fact Sheet. "Engineers Club Tour," Dayton Engineers Club, ver. 10-15-2002.

Fact Sheet. "Student Career Experience Program," 88 MSG/DPCXB, Nov 2002.

Fact Sheet. "Wright-Laboratory Technology Transfer from Wright-Patterson Air Force Base, Ohio," U.S. Air Force, PAM #92-015, 1992.

Folder. "Barling Bomber." ASC History Office Archives.

Folder. "Clinton County AFB Guides." ASC History Office Archives.

Folder. "Clinton County Army Air Field, Wilmington, Ohio." ASC History Office Archives.

Folder. "Event: Dayton Air Show 1953." ASC History Office Archives.

Folder. "Giovanni Pirelli." ASC History Office Archives.

Folder. "Wright Brothers Huffman Prairie Location." ASC History Office Archives.

"Full Spectrum Threat Response (FSTR) Planning and Operations." Air Force Instruction 10-205-1, Dec 24, 2002.

LeRoy, Amos Swan Collection, U.S. Air Force Museum Archives.

News Release. "Air Force Awards $7.2 Million Contract to Adapt Automotive Manufacturing Process for Aerospace Composites," U.S. Air Force, #97-117, 1997.

News Release. "Mead and Westvaco Complete Merger Creating Strong Global Organization," Jan 30, 2002.

News Release. "Meet ADAM: He'll Go Anywhere," Aeronautical Systems Division Office of Public Affairs, #88-147, 1988.

News Release. "New AFAL Facility Studies Sensor Performance," Aeronautical Systems Division Information Office, #78-177, 1978.

News Release. "Tech 21 Team Promotes First All-Composite Bridge in Ohio," ASC/PAM, #97-124, Aug 6, 1997.

News Release. "War Fighter Support," Aeronautical Systems Center Public Affairs Office, Wright-Patterson Air Force Base, Jun 23, 2003.

"Operation Iraqi Freedom—By The Numbers," Assessment and Analysis Division, U.S. Central Air Forces, Shaw Air Force Base, South Carolina, Apr 30, 2003.

Pamphlet. "1911 Wright 'B' Flyer." Wright "B" Inc.

Pamphlet. "Air Force Independent Research and Development," AFRL/XPTT, n.d.

Pamphlet. "Air Force SBIR," U.S. Air Force, 2000.

Pamphlet. "Centennial Flyer at the Dayton International Airport." City of Dayton, Department of Aviation, and Wright Brothers Aeroplane Company of Dayton, Ohio.

Pamphlet. "A Chronology of American Aerospace Events from 1903 through 1964," U.S. Air Force, #190-2-2, Sep 1965.

Pamphlet. "Tech Connect." AFRL/XPTT, undated.

"Photograph of Wright-Patterson Air Force Base and its Master Plan," ASC History Office Archives, 1948.

Program. "International Air Races." Dayton, Ohio, Oct 2-3-4, 1924.

Franklin D. Roosevelt, Inaugural Address, Mar 4, 1933.

INDEX

INDEX

1,000-ton molding press, **256**
12th Air Force Service Command, Italy, 141
1478th Flight Service, 317
18th Constitutional Amendment, 40
18th Medical Supply Platoon, Patterson Field, 132
1913 Flood, see Dayton, 1913 Flood
1924 International Air Races, see International Air Races, 1924
22 South Williams Street, 285
2750th U.S. Air Force Hospital, Building 830, 204, 316; also see Wright-Patterson Air Force Base, Medical Center
2nd Air Corps Weather Region, 313
2nd Weather Region, 313
4-H, and World War II home front, 42
  visit to Fairfield Air Depot, 101
6570th Aerospace Medical Research Laboratory, 318
7 Hawthorne Street, parade in honor of Wrights, 4
74th Medical Center, Mental Health Flight responds to 2000 Xenia tornado, 176
75th Anniversary of Powered Flight, 167, 170
862nd Army Air Forces Specialized Depot, 257
8th Air Force Service Command, England, 141
9-11 Terrorist Attacks, see September 11, 2001
91st Air Depot Operation, 315
A-76 Cost Comparison Studies, 157, 165, 174
Abolition Society, Dayton, 234
Acme Aluminum Alloys, Inc., 310
Acme team, 229
Acquisition and Logistics Directorate of Air Force Audit Agency. 323
Acquisition Logistics Division, Air Force Logistics Command, 159, 204, 320
Adams, Wade, Dr., **266**
Adamski, David, 189
Adena Culture, 168, 243
Adler, Elmer E., 88, **89**
  assist with air races, 95
  assistance with Round-the-World flight, 92
AdTech Systems Research, 254
Advanced Composites, Wright-Patterson development of, 203; also see Composites
Advanced Dynamic Anthropomorphic Manikin (ADAM), 332, **334**
Advanced Range Instrumentation Aircraft (ARIA), 198, 201; also see Apollo Range Instrumentation Aircraft
Advanced Strategic Tactical Expendable Flares, 183
Advanced Turbine Engine Gas Generator (ATEGG), 203
Aero Club of America, 4, 30
Aero Club of the Sarthe, 4
Aero Club of the United Kingdom, 4
Aero Propeller Laboratory, 318
Aero Propulsion Laboratory, Wright-Patterson Air Force Base, 157, 203, 260
Aeromedical Laboratory, 313
Aeronautical Board, **49**, 153
Aeronautical Division, Signal Corps, becomes Aviation Section, 48
  establishment of McCook Field, 47
  establishment of Wilbur Wright Field, 47
  establishment of, 47
  personnel strength, 48
Aeronautical Museum, U.S. Army, see United States Air Force Museum
Aeronautical Research Laboratory, 157
Aeronautical Society of Great Britain, 4
Aeronautical Systems Center, **245**, **251**, 258, 318, 323
  2750th Air Base Wing assigned to, 174
  311th Human Systems Wing assigned to, 173
  A-76 impact, 174
  and 4950th Test Wing, 201

and Munitions Directorate at Eglin, 183
and Ohio Task Force One, 178
annual budget, 246
as tenant organization on Wright-Patterson Air Force Base, 190
assumes duties of installation commander, 158, 172, 174
B-1 System Program Office, 221
B-2 System Program Office, 221
building renovations, 221
C-17 System Program Office, 183, 221
Combat Electronics Division, 182
Contracting Directorate, 172
Developmental Manufacturing and Modification Facility, 175
Directorate of Engineering, 221, 264
F/A-22 System Program Office, 221, 244, **245**
F-117 System Program Office, 221
F-16 System Program Office, 221
formerly Aeronautical Systems Division, 172
Global Hawk System Program Office, **281**
headquarters in Building 14, **245**
History Office, **259**, 269, 280
mission, 172
places McCook Field marker, 187
receives award for role in Balkan Peace Talks, 185
role in Balkan Proximity Peace Talks, 184
role in Operation Allied Force, 182
role in Operation Iraqi Freedom, 183
transfer munitions systems to Eglin, 173
workforce, 246
Aeronautical Systems Division, **235**, 318
  Aero Medical Division, Aerospace Medical Laboratory, 203
  as tenant organization on Wright-Patterson Air Force Base, 190
  development of aircraft in post-Vietnam period, 159
  Directorate of Flight Test, 198
  flight testing, 201
  formerly Wright Air Development Division, 151, 203
  improved warfighter systems for Vietnam, 153
  Limited War Office, 153
  Office of Aerospace Research, Aerospace Research Laboratory, 203
  redesignated Aeronautical Systems Center, 172
  Research and Technology Division, Materials Laboratory, 203
    Aero Propulsion Laboratory, 203
    Avionics Laboratory, 203
    Flight Dynamics Laboratory, 203
  responsibility for AFWAL, 157
  role in Operation Desert Storm, 180
  Stewart commander of, 167
  supported SCPS-M research, 333
Aeronca Aircraft Corporation, 310
Aeroshell Aerobatic Team, **282**
Aerospace composites, development of, 255; also see advanced composites, composites
Aerospace Guidance and Meteorology Center, 173
Aerospace Medical Research Laboratory, 320
  Toxic Hazards Research Facility, 221
Aerospace Physiological Training Unit, Wright-Patterson Air Force Base, 175
Aerospace preform composite parts company, 256
Aerospace Research Laboratories, 157, 318
Aerospace Technical Intelligence Center, in Building 12, 220
Aerothermodynamic machines, 260
Africa, 234
African-American community, Dayton, 234
Agnew, John, 188
Air Armament Center, 46th Test Wing, 324
Air Combat Command Systems Office, 323

Air Combat Command, 172
  activation of, 172
Air Corps Act, July 2, 1926, 102, 110
Air Corps Anti-Aircraft Exercises, 1933, Patterson Field, 112
Air Corps Carnival, 1930, 104
Air Corps Engineering School, 77, 199, 312; also see Air Force Institute of Technology
Air Corps Expansion Program, 1940, 116
Air Corps Five-Year Expansion Program, 1926, 110
Air Corps Maintenance Command, 125, 134, 313
Air Corps Maneuvers, Fairfield Air Depot Reservation, 1931, 103, 105
Air Corps Materiel Division, see Wright Field, Materiel Division
Air Corps Technical School, Chanute Field, 113
Air Corps Training Center, Kelly Field, 153
Air Corps Weather School, Patterson Field, 112, 113
Air Corps, assumed responsibility for mail, 102, 112-113
  creation of, 76
  planned force of 1939, 128
Air Defense Command, 148, 193
Air Depot Control Areas, established, 125
Air Development Center, Provisional, 317
Air Development Force, 317
Air Engineering Development Division, Headquarters, 316
Air Force Academy, 276
Air Force Acquisition Logistics Center, Headquarters, 320
Air Force Acquisition Logistics Division, Headquarters, 320
Air Force Art Collection, 137, 208
Air Force Association, 267
Air Force Audit Agency, 323
Air Force Band of Flight, see Band of Flight
Air Force C2ISR Center, 323
Air Force Celestial Guidance Research Facility, 332, **333**
Air Force Centennial of Flight Display, 279
Air Force Centennial of Flight Office, Washington, D.C., 190, 270
Air Force Combat Command, formerly General Headquarters Air Force, 125
Air Force Communications Command Operational Test and Evaluation Center, 320
Air Force Contract Law Center, 318
Air Force Contract Maintenance Center, 172, 319
Air Force Data Systems Design Center, 166
Air Force Day Celebration, 1948, **140**
Air Force Distribution Agency, Headquarters, 321
Air Force Financial Systems Management Office, 323
Air Force Flight Safety Certificate, awarded to 2750th Air Base Wing, 163
Air Force Flight Test Center, Edwards Air Force Base, 200
  412th Test Wing, 324
Air Force Flying Safety Award, awarded to 2750th Air Base Wing, 155
Air Force Foreign Technical Center, 318
Air Force Institute for Advanced Distributed Learning, 323
Air Force Institute of Technology Foundation, 274
Air Force Institute of Technology, **202**, **247**, 248; also see Air Corps Engineering School; Air Service Engineering School; Air Service School of Application; Army Air Force Institute of Technology; and Army Air Forces Engineering School
  Air Corps Engineering School, 141, **149**
  Air Force Logistics Command Education Center, 202
  Air Force Institute of Technology, and Icarus statue, 274
  and Nuclear Engineering Test Facility, 218
  as tenant organization, 190, 323
  Civil Engineer and Services School, 202
  Civil Engineering Center, 202
  Civil Engineering School, 202
  Civilian Institution Programs, 202
  College of Engineering Sciences, 199
  College of Industrial Administration, 199
  Defense Institute of Security Assistance Management (DISAM), 202
  educational links with Dayton community, 262, 264
  effect of educational programs, 202
  extension courses in Southeast Asia, 153
  facilities, 202
  Graduate School of Engineering and Management, 202, 248, 262, 264, **268**
  history of, 199, 202
  logistics education program with Ohio State University, 202
  member of Dayton-Miami Valley Consortium, 202
  mural, **264**
  origins as Air Service School of Application, 77, 312
  Project Corona Harvest, 153
  Resident College, 199, 202
  role in Wright Scholar program, 268
  School of Business, 202
  School of Civil Engineering Special Staff Officer's Course, 199
  School of Engineering, **202**
  School of Logistics, 202
  School of Systems and Logistics, 202, 248, 264
Air Force Intelligence Command, 172
Air Force Legal Service Agency, United States Air Force Judiciary, 323
Air Force Logistics Center, Inspection & Safety Center, 321
  Logistics Operation Center, 174
Air Force Logistics Command, 269, 315
  2732nd Acquisition Logistics Operations Squadron, 204
  Acquisition Logistics Division, 159, 204
  Advanced Logistics Systems Center, 319
  and Gentile Air Force Station, 257
  assignment of 2750th Air Base Wing, 151
  celebration for 45th anniversary, 327
  changes in management techniques, 159
  Command Center, 174
  Deputy Chief of Procurement and Production, 166
  Deputy Chief of Staff for Acquisition Logistics, 204
  disestablished, 158, 172
  Education Center, 202
  Flying Safety Award to Wright-Patterson Air Force Base, 155
  formerly Air Materiel Command, 151
  headquarters at Wright-Patterson Air Force Base, 151, 190
  ICBM Deactivation Task Force [Provisional], 318
  International Logistics Center, 204, 320
    becomes Air Force Security Assistance Center, 204
    Operations Office, 204
    Plans and Procedures Office, 204
    Programs and Resources Office, 204
  Office of the Assistant for International Logistics, 204
  Pacer Energy program, 159
  responsibility for museum, 209
  role in Persian Gulf War, 180
  role in Vietnam War, 152
  static airplane display, 327
  support for EMTEC, 255
Air Force Marathon, **188**
Air Force Materials Laboratory, 318
Air Force Materiel Command, 88, 247, 321, 324
  activated, 158, 172
  adopts Integrated Weapon System Management, 173
  Air Force Security Assistance Center, 204, 247, 320, 324
  Civil Engineering Squadron, 322
  closed depots, 173
  headquarters in Building 262, 190
  Law Center, 318
  Law Office, 318
  Lester Lyles, commander of, **235**, 281
  long-term strategic plans, 173
  mission, 172
  responsibility for Air Force Research Laboratory, 204
  role in Operation Allied Force, 181-182
  role in Operation Enduring Freedom, 182
  role in Operation Iraqi Freedom, 182
  technology transfer, 173

Air Force Museum Foundation, 167, 210
Air Force Museum, see United States Air Force Museum
Air Force Office of Scientific Research, 173, 247, 264
Air Force Office of Special Investigations, 324
Air Force Organizational Excellence Award
    for role in Balkan Proximity Peace Talks, 185
    awarded to AFOG, 208
Air Force Orientation Group, see Orientation Group, U.S. Air Force
Air Force Outstanding Unit Awards
    for role in Balkan Proximity Peace Talks, 185
    for role in Operation Homecoming, 154
    2750th Air Base Wing, 154, 161, 163, 174
    445th Airlift Wing, 198
Air Force Packaging & Evaluation Activity, Headquarters, 318
Air Force Packaging Evaluation Agency, 318
Air Force Personnel Center, Randolph Air Force Base, Texas, 174
Air Force Research Laboratory, 247, 322
    activation, 172, 204
    Air Vehicles Directorate, 247
    and Centennial of Flight Awards, **282**
    and educational links with Dayton, 262, 267,
    and technology transfer, 251, **254**
    and University of Dayton Research Institute, 260
    and Wright Scholar, **268**
    annual budget, 247
    as tenant organization, 190, 324
    Eagle Sector Program, 182
    history of, 202-204
    opened Major Shared Resource Center, 177
    role in Operation Enduring Freedom, 182
    role in Operation Iraqi Freedom, 182
    Sensors Directorate facilities, 221
    transfer wind tunnels, 175
Air Force Reserve Command, 194, 324
    445th Airlift Wing, 324; also see Wings, 445th Airlift Wing
Air Force Reserve Officer Training Corps, 174
    encampments at Wright-Patterson Air Force Base, 149, 159
    Field Training Unit, 321
    Northeast Region, 323
Air Force Reserve, fighter groups, 194
Air Force Scientific Advisory Board, see Scientific Advisory Board
Air Force Security Assistance Center, 247, 320, 324
    formerly International Logistics Center, 204
Air Force Security Forces, evolution of, 186
Air Force Systems Command, 255
    Aeronautical Systems Division, see Aeronautical Systems Division
    assignment of AFWAL, 157
    developed UNIVAC 1050-II, 152
    Director of Laboratories, 203
    disestablished, 158, 172
    Foreign Technology Division, 207
    formation of, 145
    formerly Air Research and Development Command, 151, 203
    headquarters assignment of Area B labs, 151
    reorganization of laboratories, 157
    reorganized 4950th Test Wing, 198
Air Force Technical Base, Dayton, Ohio, 108, 136
    date established, 140
    Morris as commander, 140, 141
Air Force Technical Museum, Wright-Patterson Air Force Base, 209;
also see United States Air Force Museum
Air Force Thunderbirds, 167, 280, **281**
Air Force Wright Aeronautical Laboratories (AFWAL), 255, 320
    activated, 157, 203
    Aero Propulsion Laboratory, 157, 203, 260
    Avionics Laboratory, 157
    Flight Dynamics Laboratory, 157, 203, 318
    Materials Laboratory, 157
    reassigned to ASD, 157
    reorganized as Wright Research and Development Center, 157, 204

Air Force, United States
    1961 restructuring, 151
    40th anniversary celebration, 170-171
    50th anniversary celebration, 187, 251
    A-76 cost comparison studies, 157, 165, 174
    acquisition reform, 173
    downsizing in post-Vietnam period, 157
    first birthday celebration, 139
    ill-prepared for World War II, 143
    independence from Army, 136
    post-Cold War changes, 171-172
    recruiting done by Army, 198
    restructuring laboratories in 1990s, 172
    status on eve of Korean War, 143
Air Foundation of Cleveland, 196
Air Installation Compatible Use Zone, 164
Air Intelligence Agency, 249, 323
Air Materiel Command, 232, 315
    Aeronautical Systems Center, 151, 317
    changes following Korean War, 144
    Experimental Test Pilot School, Patterson Field, 200
    Flight Test Division, 139
    formation of, 136
    headquarters in Building 262, 190
    host of Army Air Forces Institute of Technology, 199
    joint tenancy agreement with SAC, 192
    Korean War expenditures, 144
    Logistics Airlift (LOGAIR) System, 148
    personnel strength, 1950s, 144
    procurement functions transfer to Air Research and Development
Command, 151
    receives first computer, 145
    redesignated Air Force Logistics Command, 151
    role in Berlin Airlift, 142
    role in Korean War, 144
    separation of research and development, 145
    Wright-Patterson assigned to, 139
Air Mobility Command, 172, 175
    Detachment 2, Headquarters, 323
Air National Guard
    Combat Readiness Training Center, Alpena, Michigan, 333
    participation in Armed Forces Day, 140
Air Power 2003, 190, 270, **272**, 276
Air Racing, 95-100
    history of, 96-97
Air Research and Development Command, 316
    and Nuclear Engineering Test Facility, 218
    becomes Air Force Systems Command, 145, 151, 203
    Directorate of Research, 218
    formation of, 145, 203
    moved headquarters to Andrews Air Force Base, 145
    moved headquarters to Baltimore, 145
    responsibility for new systems, 151
    use of LOGAIR system, 148
    Wright Air Development Center, 145; also see Wright Air Development
Center
Air Reserve Training Branch, Wright-Patterson Air Force Base, 144
Air Service Airplane Engineering Department, see Airplane Engineering
Department
Air Service Airplane Engineering Division, see Airplane Engineering
Division
Air Service Armorers' School, 74, 80; also see Armorers' School
Air Service Command Areas, 125
Air Service Command, 314
    assumed logistics function from Materiel Division, 107
    conference of control area commanding officers, 126
    establishes Air Depot Control Areas, 125
    headquarters complex, Patterson Field, 118, **120**, 125, **127**
    internal reorganization, 126
    newspaper, *Take-off,* 169

newspaper, *The Mascot*, 169
newspaper, *The Wright Flyer*, 169
pre-service training programs established, 134
recruitment of women employees, 124
redesignated Army Air Forces Technical Service Command, 126
reorganization of Air Corps Maintenance Command, 125
separated from Materiel Division, 125
Air Service Engineering Division; see Engineering Division
Air Service Engineering School, 199; also see Air Force Institute of Technology
Air Service Information Division, 98
Air Service Liquidation Board, 82
Air Service School of Application, 77, 199, 262, 312; also see Air Force Institute of Technology
Air Service Stock-Keepers' School, 82
Air Service Supply and Repair Depot, see Fairfield Air Depot
Air Service Supply Division, see Supply Division
Air Service
    name change to Air Corps, 76
    plans for postwar aviation force, 80, 82
    postwar materiel surplus, 82
    research and development in Dayton, 7
    V Corps, 74
    Wilbur Wright Field contributions to, 67
Air Show, Dayton, 276, 280; also see Dayton Air Show; Vectren Dayton Air Show
Air Technical Intelligence Center, 317
Air Technical Service Command, 229, 315
    extended 1945 Air Forces Fair, 207
    newspaper, *Wright Field Air-Tec Skylines*, 169
    ordered to create touring exhibit, 207
    received activity report of Supply Division, 124
    redesignated Air Materiel Command, 136
Air Technology Network Program Management Office, 323
Air Training Command, 198
Air University, Maxwell Air Force Base, Alabama, 199, 323
    responsibility for museum, 209
Airborne Express, 232
Airborne Laser Laboratory, 199, **200**, 246
    and 4950th Test Wing, 201
    retired to United States Air Force Museum, 201
Aircraft Branch, Engineering Section, Materiel Division, 77
    test of pressurized cabin aircraft, 41
Aircraft Engines
    Allison V-1710, testing at Wright Field, **123**
    Curtiss D-12 engine, 100
    Curtiss OX-5, 59, **62**
        postwar inventory, 84
        problems with, 63
    Fiat, 69
    Hall-Scott A-7, 59
    Hall-Scott A-7A, 59
    J33, 123
    J33/I-40, 123
    Liberty Engine, 40, **52**, 69, 71
        developed at McCook Field, 72
        for World Cruisers, 92
        manufactured for DH-4 airplanes, 32
        postwar inventory, 84
        powered Barling Bomber, 70
    Liberty V-12, **86**
    research on jet-engine technology, 145
    turbojet engines tested at Wright Field, 123
    V-1719-89/91, 123
    Wright R-2600, 310
Aircraft Laboratory, 260
    Glider Branch, tests at Clinton County Army Air Field, 200
Aircraft Production Board, approval of site of McCook Field, 72
    Edward Deeds a member of, 31, 33, 52
Aircraft Radio Laboratory, 314

Aircraft Tire and Storage Distribution Point, 175, 180
Aircraft
    707, Boeing, converted to EC-18B's, 198
    A-10 Thunderbolt, Fairchild-Republic, 153
        in Persian Gulf War, 180
        titanium armor cockpits, 181
    A-20 Havoc, Douglas, 128, 310
    Advanced Range Instrumentation Aircraft, 201
    Apollo Range Instrumentation Aircraft, 201; also see Advanced Range Instrumentation Aircraft
    B-1A, Rockwell, 153, **235**
    B-1B Lancer, Rockwell, **157,** 159, 246
    B-2 Condor, Curtiss,
        at 1933 Air Corps Anti-Aircraft Exercises, 112
        Spirit of Texas, **172**
    B-2 Spirit, Northrop
        AV-5, christened *Spirit of Ohio*, 187
        deployment, 173
        development of, 246
        flyover at Wright-Patterson Air Force Base, 187
        role in Operation Allied Force, 182
        portable shelters, 183
    B-4A, Keystone, 112
    B-6A, Keystone, **112**
    B-7, Douglas, at 1933 Air Corps Anti-Aircraft Exercises, 112
    B-9, Boeing, 40, 112
    B-10, Martin, 40
        YB-10, and 1934 Alaskan flight, 113
    B-17 Flying Fortress, Boeing, 40, **44, 117**, 128
        on Patterson Field, 147
    B-18 Bolo, Douglas, 131
    B-24 Liberator, Consolidated Vultee, 128, 310
    B-25 Mitchell, North American, 45, 310
    B-26 Marauder, Martin, 310
    B-29 Superfortress, Boeing, 41, 128
        modified to carry atomic bomb, 44
        range of, 44
        XB-29, on Wright Field, **129**
    B-36 Peacemaker, Consolidated-Vultee, 44, 139, 260
        runway, see Very Heavy Bomber Runway
    B-52 Stratofortress, Boeing
        Litening Target Pods, 183
        Red Scramble, 148
        of 17th Bombardment Wing, **192**
        wing at Wright-Patterson Air Force Base, 192, 193
        B-52E, with 4043rd Strategic Wing, Wright-Patterson Air Force Base, 192
        B-52H, dropping M-117 bombs, **177**
        B-52H, with 17th Bombardment Wing, Wright-Patterson Air Force Base, 193
    Baby Blimp, **73**
    C-3, Aeronca 310
    C-5 Galaxy, Lockheed, **152**
        C-5A Galaxy, 153
    C-12 Super King Air, Beech, 175
    C-14B, Fokker, and first automatic landing, 115
    C-17 Globemaster III, Lockheed, 159, **172**, 246
        deployment, 173
        System Program Office, 183
    C-21, Learjet, **173**, 175
        of 47th Airlift Flight, 184
    C-27 Airbus, Bellanca, 130
    C-33, Douglas, **109**, 130
    C-34, Douglas, 130
    C-39, Douglas, 130
    C-40B Communications Command aircraft, 183
    C-45 Expediter, Beechcraft, 129, 149
    C-46 Commando, Curtiss, 129, 148
    C-47 Skytrain, Douglas, 129, 131, 142
        AC-47 gunship, 153

cadet training, 149
on Patterson Field, 147
replaced by T-39 and T-29 on Wright-Patterson Air Force Base, 151
supported by AFSAC, 204
C-50, Douglas, 130
C-54 Skymaster, Douglas, 129, **142**
in LOGAIR system, 148
on Patterson Field, 147
VC-54C, *Sacred Cow* Presidential Aircraft, 187
C-119 Flying Boxcar, Fairchild aircrew training, 232
C-130 Hercules, Lockheed
Battle-Damage Repair Team, 180
AC-130 gunship, 153
AC-130U, gunship, 246
MC-130H, 246
C-131, Convair, 198, 199
C-135 Stratolifter, Boeing
Advanced Range Instrumentation Aircraft (ARIA), 198
EC-135N ARIA, 199, **199**
KC-135, 148
KC-135, with 4043rd Strategic Wing, Wright-Patterson Air Force Base, 192
KC-135, zero-g testing, 199
modified for Open Skies, 173
NKC-135, Airborne Laser Laboratory, **200**, 201
OC-135B, 175
of 4950th Test Wing, 198
WC-135B, 175
C-141, Starlifter, Lockheed, 153, **182**
Hanoi Taxi, 195, 198
of 445th Airlift Wing, **195**
tail number 65-0258, 198
tail number 66-0132, 198
tail number 67-0031, 198
with 445th Airlift Wing, 194
CG-3A, Waco, 85
CG-4A, Waco, glider, **85**
CG-13A, Waco, 85
CG-15A, Waco, 85
CH-3, Sikorsky, 159
Cloudster, Davis-Douglas, 93
Curtiss Model N, 59
DC-3, Douglas, 130
DH-4, DeHavilland, **52, 71**
accidents, 107
at McCook Field, **99**
crash of Frank Stuart Patterson, 106
engines for, 86
in 1924 Air Races, 99
manufactured at Dayton Wright Airplane Company, 32, 52, 71
mated with Liberty Engine at McCook Field, 72
participation in model Airway System, 100
postwar inventory, 84
testing at Wilbur Wright Field, 71
DH-9, DeHavilland, 73
Douglas observation airplane, **103**
DT-2, Douglas, 93
D-WC World Cruiser, Douglas, 90, 91, 92, 93
with pontoons, **92**
Boston II, 94
Boston, 94
Chicago, 94, 95
four on display, **93**
in flight, **94**
New Orleans, 94, 95
Seattle, 94
E-4, 175
E-4B, **174**
E-18, Beechcraft, 332

EC-18B, converted Boeing 707s for ARIA, 198
F/A-22 Raptor, Lockheed Martin, 246
test flights, 173
as Advanced Tactical Fighter, 159
F-4 Phantom, McDonnell, Air Force Reserve fighter units, 194
City of Dayton, display Building 262, 194
display at Gate 12A, **218**
F-4C, 153
F-4D Phantom II, *City of Fairborn*, 194
modified by 4950th Test Wing, **198**
F-5, Fokker, 93
F-15 Eagle, McDonnell Douglas, 153, 246
aerial demonstration for Festival of Flight, 171
F-15E, APG-70 radar, 181
in Persian Gulf War, 180
Litening Target Pods, 183
upgrade, 159
F-16 Fighting Falcon, General Dynamics, **182**, 246
APG-68 radar, 181
flyover for Festival of Flight, 171
in Persian Gulf War, 180
supported by AFSAC, 204
System Program Office, in Building 12, 220
upgrade, 159
valve protection, 183
with 906th Tactical Fighter Group, **194**
F-35 Joint Strike Fighter, Lockheed Martin, 246
F-80 Shooting Star, Lockheed, 143, 147, 260; also see P-80
F-84 Thunderjet, Republic, 144
F-86 Sabre, North American, 144, 260
at All-Weather Flying Center, **201**
crash at National Aircraft Show, 196
F-86D, with 97th Fighter-Interceptor Squadron, 194
XF-86, **143**
F-89 Scorpion, Northrop, flutter model, **145**
F-104 Starfighter, Lockheed, 194, **193**
F-111 Aardvark, General Dynamics, **153**
F-117 Nighthawk, Lockheed,
F-117A, **157**, 159, 246
in Persian Gulf War, 180
static display at Wright-Patterson Air Force Base, 187
stealth technology, 181
Floyd Bennett, 101
Greyhound, Stinson Aeroplane Company, 23
H-13, Bell, 149
JN-4 Jenny, Curtiss
JN-4A, **66**
JN-4D, 59, **60, 66, 68**
JN-4D, problems with, 63
postwar inventory, 84
JN-6, Curtiss, postwar inventory, 84
Keystone biplane, modified for airmail, 113
LB-1 Pegasus, Huff-Daland, **72**
Liberty Airplane, 73
Lighter-than-Air
Dirigible balloons, 47
Goodyear *Puritan* airship, 101
Hydrogen balloons, 47, **48**
Signal Corps Balloon No. 10, **48**
Signal Corps Dirigible Number One, 47, 48
TC-5 non-rigid dirigible, 99
MB-2, Martin, 82
in 1924 Air Races, 99
overhauled at Fairfield, 84
postwar inventory, 84
Messenger, Sperry, at 1924 Air Races, 99
Model 299, Boeing, **77**
NBS-1, Martin, at 1924 Air Races, **97**, 99
NC-4 Flying Boats, Curtiss, (U.S. Navy), 90
P-35, Seversky, 129

P-36, Hawk, Curtiss, 129
P-38 Lightning, Lockheed, 123, **129**
P-39 Airacobra, Bell, 129
P-40 Warhawk, Curtiss, 123, 129
P-47 Thunderbolt, Republic, 129
P-51 Mustang, North American, 123, 129
    aerial demonstrations, 207
    crash in air show, 97
P-61 Black Widow, Northrop, 129
P-80 Shooting Star, Lockheed, 123; also see F-80
Presidential Aircraft, 159
PT-19, Aeronca, 310
PW-8, Curtiss, in 1924 Air Races, 100
R-4, Vought-Sikorsky, helicopter, 128
R-6, Curtiss, crash of, 98
R-8, Curtiss, in 1924 Air Races, 100
RQ-1 Predator, General Atomics, **173**, 182, 246, 270
RQ-4 Global Hawk, Teledyne Ryan, 173, **177**, 183, 246, 270
    Global Hawk Program, Weather Squadron assistance, 175
Sacred Cow, VC-54C, Presidential Aircraft, 187
SE-5, Standard, postwar inventory, 84
Signal Corps Airplane Number One, 47, 48
SJ-1, Standard, 59, **60**
    manufactured at Dayton Wright Airplane Company, 32, 52
SM-1 Detroiter, 23
Space Shuttle *Columbia*, 179
Space Shuttle *Enterprise*, **168**
Spirit of St Louis, 101
SR-71 Blackbird, Lockheed, **153**
T-1A Jayhawk, Beech, 159, 173, 246
T-6 Texan, Beech, 246, **282**
T-29, Convair, 151
T-33, Lockheed, cadet training, 149
    jet-transition training, 147
    transferred from Wright-Patterson to Davis-Monthan Air Force Base, 163
T-39 Sabreliner, North American, 151
    at Wright-Patterson Air Force Base, **163**
    flyover at First Flight Ceremony, 170
    of 4950th Test Wing, 198
    transfer from Wright-Patterson to Davis-Monthan Air Force Base, 163
T-40 Jetstar, Lockheed, 163
Thomas-Morse pursuit, 69
UC-64, Aeronca, 310
USAC-1, 73
USD-9, 73
    USD-9A, 73
Verville-Sperry racer, in 1924 Air Races, 100
Vickers-Vimy biplane, 90
Vought pursuit biplane, **86**
Waco 10, 85
Waco biplane, **85**
Wright B Flyer, **244**, 280, 287
    flyover at Wright-Patterson Air Force Base, 187
    flyover for Festival of Flight, 171
    flyover Woodland Cemetery, wreath-laying ceremony, 17
Wright Flyer III, 1905, 285
    sculpture of, **275**,
    in Dayton Aviation Heritage National Historical Park, 189
Wright Flyer, 1903 replica, **245**, **275**
Wright Flyer, 274
    crash of, 24
    deadline extension from War Department, 1
    Fort Myer demonstration, 1, 2, 6, 24
    in France, **3**
    Kill Devil Hill, **2**
    Kitty Hawk, at 1924 Air Races, 99
Wright glider, 1902, 275
Wright Military Flyer, 1909, **22**

    sculpture, **274**
Wright Model B, 17, **49**, **50**
    at 1924 Air Races, 99
Wright Model C, **50**
    accidents, 48
X-20 Dyna Soar, Boeing, 219
X-30A National Aerospace Plane, 159
XB-45 Tornado, North American, **143**
XB-70 Valkyrie, North American, 153, 219
XC-35, Lockheed, first successful pressurized cabin aircraft, 41
XNBL-1 Barling bomber, at Wilbur Wright Field, 35, 76
    escorts World Cruisers, 94
    special hangar, **70, 105**
XPS-1, 52
YG-1, Kellett, autogiro, **112**
    YG-1B, Kellett, 113
Foreign
    France, LePere pursuit, 71
    Germany, Messerschmitt 262, 206
    Great Britain, Bristol fighter, 71, 73
    Italy, Ansaldo Balilla, crash at Wilbur Wright Field, **150**
    Italy, Pomilio fighter, 69
    Italy, Scout, SVA (Societa Verduzio Ansaldo), 69, **71**
    Italy, SIA (Societa Italiana Aviazione), 69
    Russia, Tu-154, Open Skies aircraft, **174**
    Soviet Union, MiG-21UM, on display at NAIC, **206**, 207
    Soviet Union, MiG-29 Fulcrum B, on display at NAIC, 207
    Soviet Union, MiGs in Korean War, 144
Airplane Engineering Department, Air Service
    merger with Engineering Production Department, 72
    operations begin at McCook Field, 72
    Signal Corps Equipment Division, 69
Airplane Engineering Division, Air Service, 74, 86
    created at McCook Field, 72, also see Engineering Division
Air-Tecs, Wright Field, 229
Akron, University of, 258
Al Qaeda, 182, 183, 249
Alabama, University of, at Birmingham, 258
Alamo Flight, see 2750th Air Base Wing, Alamo Flight
Alaskan Flight, 1934, 113
Alcock, John, 90
Alfred P. Murrah Federal Building, Oklahoma City, 178
Allen, Eddie, test pilot, 200
Allen, James, 6
Alliance for Education, agreement with Wright-Patterson Air Force Base, 177
Alliance Sub-depot, Nebraska, 135
All-metal airplane wing, development of, 40
All-Weather Flying Center, Wright-Patterson Air Force Base, 201
All-Weather Flying Division, 232
Allyn Hall, Wright State University, **269**
Allyn, Stanley, 269
Altus Air Force Base, Oklahoma, 192
Amann, J. R., **201**
America Firsters, World War II, 44
American Airlines, 148
American Association of Blacks in Energy, 268
American Bicentennial celebrations, 166, 167
American Expeditionary Forces, France, 91
American Export and Import Company of Miami, Florida, 148
American Institute of Aeronautics and Astronautics, 263, 267, 287
American Red Cross, 107, 144
    and winter storms, 161
    Dayton Area Chapter, 324
American Revolution Bicentennial Commission, 167
American Revolution, 13
American Rolling Mill Company, 310
American Society of Mechanical Engineers, 267, 327
Ames, James S., 51
Andrews Air Force Base, Washington, D.C., 145

Andrews, Bob (C.R.), **261**
Anger, Frank G., 210
Annual American Radio Relay League, 149
Anthrax, *Brucella suis*, 176
Antioch College Antiochiana Collection, Yellow Springs, Ohio, 28
Apollo Range Instrumentation Aircraft, 201, also see Advanced Range Instrumentation Aircraft
Apollo Space Missions, 198, 200
Appalachia, 122
Apple Automobile Company, Dayton, 22
Apple, Nick, **153**
Appropriation Act for Fiscal Year 1912, 48
Arc Light Missions, Southeast Asia, 193
ARD-21 Air Rescue Hovering Set, 199
Area A, see Wright-Patterson Air Force Base, Area A
Area Audit Office, 323
Area B, see Wright-Patterson Air Force Base, Area B
Area C, see Wright-Patterson Air Force Base, Area C
Area D, see Wright-Patterson Air Force Base, Area D
Area Defense Counsel, 323
ARIA Program, 201; also see Advanced Range Instrumentation Aircraft
Armament Branch, Engineering Section, Materiel Division, 77
Armament Laboratory, 253
Armed Forces Communications and Electronics Association, 267
Armed Forces Day, 140, **149**
Armorers' School, Wilbur Wright Field, 80
Armstrong Aerospace Medical Research Laboratory, 172, 320
    and ADAM, 332
Armstrong Laboratory, 172
Armstrong Museum, 263
Armstrong, Harry G., 41
Armstrong, John L, 196
Armstrong, Neil, 17, 278, 287
    attends First Flight Banquet, 170
    in National Aviation Hall of Fame, 211
Army Aeronautical Museum, 77, also see United States Air Force Museum
Army Air Corps, 153, 253
Army Air Force Exchange Service, 324
Army Air Forces Air Technical Service Command, 315
Army Air Forces Crystal Bank, Headquarters, 315
Army Air Forces Engineering School, 199; also see Air Force Institute of Technology
Army Air Forces Fair, 1945, at Wright Field, **129**, 207
Army Air Forces Institute of Technology, 199, 312; also see Air Force Institute of Technology
Army Air Forces Materiel & Services, 315
Army Air Forces Materiel Command, 313
Army Air Forces Nurses Training Detachment No. 6, training at Patterson Field, 133
Army Air Forces Regional Station Hospital, 315
Army Air Forces Technical Base, 108
    established, 136, 140
    Morris as commander, 141
Army Air Forces Technical Service Command, established, 126
Army Air Forces
    basketball team, 229
    establishment of, 125
Army Air Service, 243
Army Corps of Engineers, office on Wright-Patterson Air Force Base, 142
Army Field Printing Plant, 315
Army Relief Association, 104
Army Reorganization Act of 1920, 82
Army Signal Corps; see Signal Corps
Army, U.S., 324
    competition with U.S. Navy, 91
    contract with Wright brothers, 24
    first airplane, 47
    first provisional aero squadron, 48

Medical Department, 324
    participation in Armed Forces Day, 140
    procurement of aircraft, problems, 33
    recognition of military aviation as part of field force, 48
    Veterinary Services, 324
Arnold House Heritage Center; also see Buildings, Area C, Building 8
    history of, **222**
    dedication of, 328
    proximity to golf course, 217
Arnold, Henry H., "Hap", 81, **102**
    and 1934 Alaskan flight, 113
    and Barling bomber, 70
    chairman of Air Corps Carnival, 104
    chief of Air Corps, 128
    commander of Field Service Section, 102
    consultation with Theodore von Karman, 41
    dedication of house in honor, 328
    lived in Building 8, 222
    opinion of Robins, 87
    opinion of Waco gliders, 85
    orders to create traveling fair, 207
    visit to FAD, 116
Arnold, Leslie P., **93**, 94
Ascani, Fred, 40
Ashburn Field, Chicago, Illinois, 53
Ashworth, Earl, **201**
AT&T, 242
Augusta, Georgia, Signal Corps Aviation School, 48
Aural Display and Bioacoustics Branch, **254**
Auster, Arthur and Judith, 265
Austria-Hungary, pre-World War I aviation, 49
Autogiro School, first opened at Patterson Field, 113
Automated Notification Network, 166
Automobile manufacturing, Dayton, 7, 22
    Apple Automobile Company, 22
    Barney & Smith Car Company, 22
    Courier Car Company, 22
    Darling Motor Car Company, 22
    Dayton Electric Car Company, 22
    Jitney Transportation Company, 22
    Speedwell Motor Car Company, 22
    Stoddard, 22
Automobile, invention of, 22
Automotive Composites Consortium, 259
Automotive industry, 255
Aviation archives, 269
Aviation Armorers' School, Wilbur Wright Field, **67, 68**
    Squadron B, **68**
    demobilization, 74
    JN-4D, **68**
Aviation armorers' training, 67-68
Aviation Art Worldwide, 276
Aviation General Supply Depot, Fairfield, Ohio, see Fairfield Air Depot
Aviation Learning Center, 85; also see Weaver Aircraft Company
Aviation mechanics, private industry training of, 66
    training, 66
Aviation Mechanics' School, Wilbur Wright Field, 65-67, **66**
Aviation Repair Depot (Speedway), Indianapolis, Indiana, 63
    move to Fairfield, 82
    staff officers, **83**
Aviation Schools
    Augusta, Georgia, 48
    Barron Field, Texas, 60
    Carruthers Field, Texas, 60
    Chanute Field, Illinois, 79, 80
    College Park, Maryland, 48, **49, 50**
    Kelly Field, Texas, 60
    logistical support, 79
    Love Field, Texas, 60
    North Island, San Diego, 48, **50, 51**

Scott Field, 79, 80
Selfridge Field, Michigan, 79, 80
Taliaferro Field, Texas, 60
Wilbur Wright Field, 55, 59-60, **69**, 79
    demobilization, 74, 80
Aviation Section, Army Air Corps, 262
Aviation Section, Signal Corps, creation of, 48
    establishment of Wilbur Wright Field, 54
    expanded flight training, 53
    George O. Squier, officer in charge, 51
    need for mechanics, 66
    on eve of World War I, 51
    requirements for pilot training, 53
    selection of sites for aviation schools, 54
Aviation Trail Visitor Center, 275
Aviation Trail, 285
Aviation Trail, Inc., 263, **284**
    designated aviation landmarks in Miami Valley, 190
Aviation Weekly, newspaper, 169
Aviation/Space Writers Association, 263
Aviators, 229
    friendship with dogs, 61
Avionics Laboratory, 318
    and Air Force Celestial Guidance Research Facility, 332
    Electromagnetic Warfare Applications Branch, 221
    Electromagnetic Warfare Branch, 221
    facility expansion, 221
Avionics Maintenance Squadron, 320
Baby Blimp, at McCook Field, **73**
Bachmann, Joyce, 180
Baer Field, Indiana, 131
Baer Sub-depot, Indiana, 135
Baker, Newton D., secretary of war, 52, 54
Baldwin, Thomas Scott, 47
Balkan Proximity Peace Talks, 174, 183-185
    held at Wright-Patterson Air Force Base, 181
    presidents sign agreement, **185**
Balloon Celebration, **277**
Band of Flight, U.S. Air Force, 314, 323
    Concert Band, receives Howard Citation of Musical Excellence, 209
    performs at Dayton Air Show, **281**
    performs for Music Power 2003, 272, 276
    performs for Ohio bicentennial, **249**, 280
    history of, 208-**209**
Bands
    361st Army Air Forces Band, 208, **209**, 314
    661st Air Force Band, 314
        formerly 661st Army Air Forces Band, 209
    661st Army Air Forces Band, 314
        formerly 361st Army Air Forces Band, 208, 209
Bane, Thurman H., 199, 287
    dedication of Building 640, 326
    commander of McCook Field, 100
Bank One NA (National Association), 324
Bank, Leah, **289**
Barksdale, Eugene "Hoy", test pilot, 200
Barling Bomber, see Aircraft, XNBL-1 Barling bomber
Barling, Walter J., 70
Barnes, Frank G., Memorial Park, 326, **329**
Barney & Smith Car Company, Dayton, 22
Barns Project, 280
Barron Field, Texas, 60
Barrow, James, 180
Barton, Paul L., dates of Wright-Patterson command, 158
Bartron, Harold A., 81, **88, 108**
    and 1924 Air Races, 98
Base Memorialization Committee, Wright-Patterson Air Force Base, 167, 326-329
Basketball clinic, first, 229
Bath Township, Ohio, 56, 79

Battle of Missisinewa, War of 1812, 13
Battle of the Thames, War of 1812, 13
Battlefield Air Operations Kit, 182
Bauer, Barbara, **277**
Baumgartner, Ann, test pilot, 200
Beacom, Edwin, 180
Beale Air Force Base, California, 193
Beard, Luther, managing editor of Dayton *Journal*, 2
Beavercreek, Ohio, 231, 250, 271
Becher, Eugene M., 114
Beery, Todd, 180
Bell Telephone Building, Dayton, 38
Bell Telephone Laboratories, 144
Bella Vista, Arkansas, 275
Bellamy, Edward, 28
Bellbrook, Ohio, 250
Bells Project, 280
Belmont Park, New York, 96
Benedic, Michael, 180
Benjamin and Marian Schuster Performing Arts Center, 237; also see Schuster Center
Benjamin, Scott, 180
Bennett Trophy, 96
Bennett, Floyd, 101
Bergamo, Mount St. John's, 265
Berlin Airlift, 142, 232, **272**, 244
Berlin Crisis, 1961, 180
Berlin Wall, tearing down, 157
Berry Sub-depot, Tennessee, 135
Bethesda, Maryland, 205
Bettis, Cyrus, 100
Bicentennial of Ohio, 280
Bicycle Shop, The, 276
Bicycles, Dayton, 20, 22
Big Brothers/Big Sisters organizations, 253
Big Four Railroad Company, 79
Big press, 256
Big three automakers, 259
Bigger Road Veterinary Clinic, **258**
Bildt, Carl, 185
Biltmore Hotel, Dayton, 40
Birth of Aviation, 245
Black Ben, 16
Black cultural renaissance, 235
Black Knights, 192
Black Leadership Program, Dayton, 235
Black mayors, Dayton, 235
Black slaves, 234
Black, David, **274**, 275
Black, Kim, 272
Blacks in Dayton, first mention of, 234
Blair, Shiras A., 83, 84
Blake, Lawrence, 285, 286
Blanton, Dr. John C., 287
Bleriot, 245
Bleriot, Louis, 96
Blot, Jacques, 185
Blue Ash facility, University of Cincinnati, 264
Blum, Joshua, 180
Boeing, 258
    Phantom Works, 256
Boger, Erwin F., **84**
Bolger, William F., 170
Bolling Field, Washington, D.C.
    launch of 1934 Alaskan flight, 113
    stop on Model Airway, 100
    stopover of Round-the-World flight, 94
Bombe program, 43
Bomford, George N., 64
Bonebrake Theological Seminary, 228

Bong, Richard I., dedication of street in honor, 326
Boone County, Missouri, 178
Boone, Timothy, **241**
Boonshoft Museum of Discovery, 244, **253**, 267
Boonshoft, Oscar, **253**
Bordosi, Fred, test pilot, 200
Borglum, Gutzon, 37
Bosnian-Croat Federation, 185
Bourquin, Thomas, 180
Bowling, Rob, **162**
Bowman Field, Kentucky, 316th Transport Group assigned to, 131
Bowman Sub-depot, Kentucky, 135
Boy Scouts, and World War II home front, 42
Boyd, Albert, 196, 200, **201**
Boyette, Terri, 180
Bradford Elementary School, **289**
Bradley Tech, 229
Bradley, Mark E., Jr., 269
Breckner, William, **154**
Breene, Robert G., 326
Brett, George H., **88**, 95, 326
Brick Officers' Quarters, see Wright-Patterson Air Force Base, housing
Bridge deck installation, Greene County, 259
Bridge of Wings, mural in Building 262, **137**
Brindley, Oscar, 107
Brinkman, John, **334**
Britain, Joseph H., 327
Brookley Air Force Base, Alabama, weekly flights to Wright-Patterson Air Force Base, 148
Brookley Field, Alabama, 63rd Transport Group assigned to, 130
Brookley, Wendell H., 100
Brooks Air Force Base, Texas, 244
    headquarters of Armstrong Laboratory, 172
    Human Systems Center, 173
Broomfield, Tyree, 235
Brown & Bills Architects of Dayton, 274
Brown, C. Pratt, 158, 212
Brown, Clarence J., **162**, 170
Brown, George S., **235**
Brown, Robert J., Jr., 91
Brucella suis, anthrax, 176
Brucker, Lynn, **271**
Brucker, Roger, 271
Brukner Nature Center, Troy, Ohio, 85
Brukner, Clayton, 85
Brundidge, Alabama, 275
Buby, Father Bert, 265
Buckeye Iron and Brass, 242
Buckeye Phantoms, 194
Buffalo, New York, 194
Buildings
    Alfred P. Murrah Federal Building, Oklahoma City, 178
    Area A
        Building 262, 119, **120**, 121, **151**
            AFMC headquarters, 172, 190
            Air Materiel Command headquarters, 136
            Air Service Command headquarters, 125, **127**
            aircraft display, **171**, 194
            Bridge of Wings mural, **137**
            dedicated Gillmore Hall, 327, 328
            Sarris Auditorium, 183
        Building 262A, 119, **120**, 326
            destroyed by fire, 216, **218**
        Building 266, 218
        Building 271, dedicated Warrior Hall, 327, 328
            Warrior Hall, **218**
        Building 274, Civilian Club, **133**, 215
        Building 280, Prisoners of War Murals, **126**, 176
        Building 287, 198
        Building 288, 198

Building 700, Brick Quarters, **111**
    dedicated Robins House, 87, 328
Building 800, Officers' Club, **113**
Building 824, Hope Hotel, **215**, **328**
    and Balkan Proximity Peace Talks, 183
    dedicated in honor of Bob Hope, 328
    Hope Hotel and Conference Center, construction of, 215
Building 825, housing, 212
Building 826, **212**
Building 828, **206**, 207, 220
Building 829, NAIC, 207
Building 830, 2750th U.S. Air Force Hospital, 204
    hospital, 214
    Medical Center, **204**
Building 831, Fisher House, 215
Building 832, in Operation Homecoming, 154
    and Balkan Proximity Peace Talks, 183
Building 833, in Operation Homecoming, 154
    and Balkan Proximity Peace Talks, 183
Building 834, and Balkan Proximity Peace Talks, 183
Building 835, and Balkan Proximity Peace Talks, 183
Building 836, and Balkan Proximity Peace Talks, 183
Building 849, dedication of, 327
    Dodge Gymnasium, in Operation Homecoming, 154
Building 856, NAIC, 207
Building 858, dedicated in honor of Cacioppo, 327, 328
    NAIC, 207
Building 1419, 193
Building 1420, 193
Building 1421, 193
Building 1445, 194, 212
Building 1446, 194
Building 1447, 194
Building 1448, 194, 212
Building 1449, 194
Building 1450, 194
Building 1451, 194
Building 4023, dedicated in honor of Watson, 327, 328
    NAIC, 207
Area B
    Building 8, flight operations, 147
    Building 11, 221
    Building 11A, 176
    Building 12, 209, 220
        Aeronautical Museum, 209
        dedicated Monahan Hall, 220, 327, 328
        history of, **220**
    Building 14, **152**
        Aeronautical Systems Center headquarters, **246**
        WADC headquarters, **145**
    Building 15, AFRL headquarters, 204
    Building 17, dedication of, 327
    Building 18, 123
    Building 19, Five-Foot Wind Tunnel, 327
        Five-Foot Wind Tunnel, National Historic Mechanical Engineering Landmark, 327
    Building 22, 221
    Building 32, 176, **219**, 221
    Building 33, 221
    Building 40, dispensary, 204
    Building 65, 219, 221
    Building 76, dedication of, 327
    Building 125, 202
    Building 126, *Wright Field* mural, **137**
    Building 145, 221
    Building 146, 221
    Building 188, 163
    Building 248, 221
        dedicated to Fitts, 326, 328
    Building 288, 202

Building 470, Nuclear Engineering Test Facility, 218, **203**
Building 489, United States Air Force Museum, 210
Building 553, 221, 327, 328
Building 556, 221, 327, 328
Building 557, 221, 327, 328
Building 558, 221, 327, 328
Building 560, 221, 327, 328
Building 620, **157**, 221, 332
Building 630, child development center, 216
Building 640, and AFIT, 202, 326, 328
Building 641, and AFIT, 202, 326-327, 328
Building 642, AFIT Science, Engineering, and Support Facility, 202
      dedicated in honor of Kenney, 327, 328
Building 643, AFIT School of Civil Engineer and Services, 202
Building 655, Materials Processing Laboratory, 221
Building 675, Occupational Health Clinic, 204
Building 821, **203, 212**
Building 824, 221
Hangar 2, 221
Hangar 3, 221
Hangar 10, 221
Hangar 22, fitness center, 216
Area C
  Air Transport Terminal, **117**
  Building 1, **79**, 100, **109**
    at 1930 Air Corps Carnival, **104**
    Aviation General Supply Depot, Fairfield, Ohio, storage, 82
    FAD administrative personnel, 109
    FAD, construction, 79
    FAID, 1923, **47**
    railroad switch, 79
    Receiving Department, 116
    renovations, 224
    storage system, 82
    supply operations prior to World War II, 115
    trainway, **80**
  Building 2, 111, **116**
  Building 3, 111, **116**
  Building 4, 111, **116**
  Building 6, 117
  Building 8, Arnold House Heritage Center, history of, **222**
    residence of Robins, 87
    dedication of, 328
    proximity to golf course, 217
  Building 10, 90, **152**
    dedicated Morris Hall, 141, 326, 328
    Fairfield Air Service Command headquarters, 118, **126**
    Fairfield Air Service Command staff, **127**
    headquarters of FADCAC, 126
  Building 11, FAD new headquarters, 109
    Patterson Field headquarters, 118, **126**
    temporary FADCAC offices, 125
  Building 13, at 1930 Air Corps Carnival, **104**
    expansion of, 118
    FAD engine overhaul facility, **102**, 103
    maintenance on B-17, **117**
    Patterson Field headquarters, **109**
    transfer to 4950th, 163
  Building 25, Pirelli crash near, 150
  Building 52, 109
  Building 54, 103, 109
  Building 70, 118
  Building 71, 118
  Building 72, 118
  Building 80, 117
  Building 88, 121
    dedication of, 328
    Foulois House, history of, **223**

Building 95, 118
Building 105, 163
Building 110, 224
Building 114, 118
Building 142, 224
Building 143, 174, 224
Building 145, 104
Building 146, 117, 224
  Air Terminal, 117, **147**
  facilities for 97th FIS, 194
Building 148, 163
Building 152, 163
Building 153, 193-194, **194**
Building 163, dedication of, 328
  Fire Station No. 1, **218**
Building 169, 163
Building 174, 118
Building 206 North, 163
Building 206, **221**, 224
  Base Operations, **116**, 270
  construction, 117
  facilities for 97th FIS, 194
  flight operations, 147
  link trainer, **131**
  new aircraft control tower, 224
  Passenger Terminal, 221
  receives new radar, 147
  space used by Federal Aviation Administration, 164
Building 207, 84
Building 219, 121, **118**
  Post Hospital, 204
  WASP barracks, 204
Building 247, 224
Building 252, 62, 118
Building 253, 118
Building 254, 118
Building 255, 118
Building 256, 163
Building 257, 118
Building 258, **117**, 118
Building 841, space used by Federal Aviation Administration, 164
Building 884, 163
Building 1084, Aero Repair Training School, **131**
Building 1404, child development center, 216
Building 4041, 224
Building/Hangar 89, **118, 210**
  museum opened, 209, 210, 220
  temporary office for Federal Disaster Assistance Team, 161
Buildings/Hangar 145, oldest at Wright-Patterson Air Force Base, 244, 270
Control Tower, **118**
Hangar 5, 115
Hangar 152, plaque honoring Giovanni Pirelli, 150
Warehouse 209, **213**
  Base exchange, 214
Dayton
  Citizens Federal Centre Tower, 233, **240**
  Dayton arcade, **236**, 237
  Arcade Centre Tower, 233, **240**
  Bell Telephone Building, 38
  Caldwell Street Center, 260
  Callahan Building, 234
  Canfield house, 228
  Centre City Building, formerly United Brethren Building, **21**
  City of Dayton Building, 266
  Convention and Exhibition Center, 170, 171, 274, 282
  Dave Hall Plaza, 266
  Dayton Daily News building, **8, 236**

Dunbar home, **285**
Gem Savings Building and Plaza, 233
Hulman Building, 39
Kettering Tower, 233, **240**
Kuhns building, **236**
Masonic Temple, 36, **38**
Mead Tower, 233, **240**
New NCR headquarters building, **239**
Post Office building, **236**
Reibold Building, **21**
Shroyer Park Center, 260
United Brethren Building, later Centre City Building, **21**
Winters Tower, 233
YMCA, **237**
John Bryan State Park, Building 835, Air Force Celestial
Guidance Research Facility, 332
Kittyhawk Center
Building 199, 147
Building 1212, dedication of, 326, 328
housing, 212
Building 1213, dedication of, 326, 328
housing, 212
Building 1214, dedication of, 326, 327
housing, 212
Building 1215, dedication of, 326, 328
dining hall, 212
Building 1216, dedication of, 326, 327
Building 1217, dedication of, 326, 327
Building 1235, childcare center, 214
Building 1245, dedicated in honor of Jarvis, 327, 328
Building 1250, 215
Trebein Road, Building 5238, Targeting Systems Characterization
Facility, 332
Wood City
Building 50, 134
Building 118, destroyed by fire, 214
Building 1044, 134, 215
Building 1045, 134, 214
Building 1046, 134, 214
Building 1084, damaged by fire, 214
Building 1087, damaged by fire, 214
Building 1089, damaged by fire, 214
Building 1113, hospital, 204
Building 1220, Chapel 2, 214
Bureau of Aircraft Production, 72
Bureau of Fisheries, U.S., assistance with Round-the-World flight, 92
Burger, Joshua, 283
Burial mounds, 168, 176, **243**
Burns, Michael, **154**
Burroughs B-3500 Computer, 166
Burte, Dr. Harris M., 255
Busch, Jerry, **261**
Bush, George H. W.
ordered forces to Persian Gulf, 177
signs Strategic Arms Reduction Treaty, 171
Bush, George W., **276**, 283
creation of U.S. Northern Command, 191
identified Iraq as threat to world peace, 182
visit to Wright-Patterson Air Force Base, **162**
Butler County Engineer's Office, 252
Butler County, Ohio, bridge, 252
Byrd, Richard E., **101**
Byrne, Ronald, **153**
Cacioppo, Anthony J., dedication of Building 858, 327
Cadillac XLR sportscar, 259
Cahoon, Renee, 271
California, 259, 270
Cambridge Research Laboratory, 200
Cambridge, Ohio, 134
Camp Kelly, Texas, see Kelly Field

Camp X-Ray, Guantanamo Bay, Cuba, 183
Campbell, Burton, **153**
Campbell, Douglas, 155
Campbell, Glenn, 171
Candy bomber, **272**
Canfields, 228
Caniff, Milton, 170, 278
Cannell, James, 180
Capital Airlines of Nashville, Tennessee, 148
Carbon-Carbon Development, 203
Carillon Historical Park, 31, 275
and Dayton Aviation Heritage National Historical Park, 189, **285**
on Aviation Trail, 284
Time Flies program, 280
Carnell, Julia Shaw Patterson, **38**, 113
and Dayton Air Institute, 38
mother of Frank Stuart Patterson, 105, 107
Carroll, Franklin Otis, 219
dedication of Building 553, 327
Carroll, Kay, 42
Carroll, Patrick, **219**
Carruthers Field, Texas, 60
Carter, William, **234**
Car-Vet, **258**
Cash register, **233**, 239
electrification of, 7
invention of, 237
James Ritty's, **7**
Catholic Calvary Cemetery, Dayton, 17
Catton, Jack J., 154, **162**
Cayse, Mike, 180
CBS Radio, "Cheers from the Camps" broadcast, **130**
Celebration Central, **279**
Celeste, Richard F., 255
Cemeteries, Dayton
Catholic Calvary Cemetery, 17
Fifth Street Cemetery, 18
Woodland Cemetery, 4, 5, 15, 17, 18, 25
Cemeteries, Wright-Patterson Air Force Base
Cox Cemetery, 155
Hebble Creek Cemetery, 125
Landis-Shank Cemetery, **125**
Centennial Celebration of Aviation Art, 276
Centennial Flyer, 244
Centennial of Flight awards, 281, **282**
Centennial of Flight celebration, preparations for, 189
Centennial of Flight Office, U.S. Air Force, 190
Center Supportability & Technology Insertion, Headquarters, 320
Centerville, Ohio, 231, 250
Central Branch of the National Home for Disabled Volunteer Soldiers, 10
Central Civilian Personnel Office, 218
Central Department, Chicago, Illinois, Headquarters, 80
Central High School, 234
Central State University, 250
Centre City Building, formerly United Brethren Building, **21**
Century of Flight, A, 276
Chabrian, Dr. Peggy, 272
Chandler, Charles deForest, **49**, 326
Chandler-Evans Corporation, 310
Chanute Field, Rantoul, Illinois, 55, 66, 84
aviation school, 79
Stock-Keepers' School, 82
stop on Model Airway System, 100
Chanute Sub-depot, Illinois, 135
Chanute, Octave, 4
Charleston Air Force Base, South Carolina, 172
Cheers from the Camps broadcast, Patterson Field, **130**
Chemineer Company, 242
Chess, Edwin R., 214

Chevy Blazers, 242
Chicago World Fair, 234
Chidlaw, Benjamin W., dedication of street in honor, 326
Children's Museum Board, 253
Chilstrom, K. O, **201**
Chinese-American service squadrons, trained at Patterson Field, 133
Chisum, J. Y., 81
Christians, 265
Christie, Arthur R.
    as commander of Wilbur Wright Field, 59, **64**
    in France, World War I, 74
    made Wilbur Wright Field's first test flight, 59
Christine, F. F., **88**, 98
Christopher Award, 263
Christopher, Warren, 183, **184**, 185
Chrysler Corporation, Airtemps Plant, 310
Chrysler, 256
Chuck Taylor All-Star "Chucks", **229**
Church of the United Brethren in Christ, 4, 263
Cincinnati Bell Foundation, 264
Cincinnati, Ohio, 237, 267, 274
Cincinnati, University of, 264
City Engineering Company, 310
City of Dayton, F-4 Phantom, 194
City of Fairborn, F-4D Phantom, 194
Civil Aeronautics Administration, Air Traffic Control Division, 147
Civil Air Patrol, 272
    and World War II home front, 42
    cadets, **273**
    encampments on Wright-Patterson Air Force Base, 149
    Great Lakes Region, 273
    participation in Armed Forces Day, 140
    poster, **273**
    USAF, Detachment 3 Great Lakes Liaison Region, 323
    work with Weaver Aircraft Company, 85
    Wright-Patterson Composite Squadron, 324
Civil Defense Organization, participation in Armed Forces Day, 140
Civil Engineer and Services School, see Air Force Institute of Technology
Civil Engineering, Wright-Patterson Air Force Base, 286
Civil War, 16, 18, 19
Civil Work Administration, work on Fairfield Air Depot, 111
Civilian Conservation Corps, 111, 168
Civilian Installations Group, Headquarters, 316
Civilian Institution Program, 262
Civilian Personnel Management Service, 323
Civilian Welfare Association, 104
Clark Air Base, Philippines, 154, 195
Clark County, Ohio, 246, 250
Clark, Kelly, 180
Clement, Robert W., dates of Wright-Patterson command, 158
Clementz, Chad, 279
Cleveland Foundation, 264
Cleveland, Ohio, 237, 242, 267
    host of Air Races, 97
Clifton Campus, University of Cincinnati, 264
Clinical studies, 1970s, 288
Clinton County Air Force Base, **232**
    closing ceremony, **232**
    part of Dayton RAPCON, 147
Clinton County Army Air Field, 232
    glider tests, 200
Clinton County, Ohio, 259
Clinton, Bill, 263
Coast Guard, U.S., assistance with Round-the-World flight, 92
Cochran, James C., dates of Wright-Patterson command, 158
CocKayne, Robert, 180
Coffee, Kevin, 180
Coffeyville Sub-depot, Kansas, 135
Coffin, Howard E., chairman of Council of National Defense, 33
Coffyn, Frank, **24**

Cold War, 142, 157, 171
Cole, Richard E., dog Jinx, **61**
Coleman, Fred H., 81, 108, 116
College of Business Administration, University of Cincinnati, 264
College Park, Maryland, Signal Corps Aviation School, 48, **49, 50**
College Professors' Tour, **259**, 262
Collier Trophy, 41
Collins, William J., 218, 328
Colonel George S. Howard Citation of Musical Excellence for Military Concert Bands, 209
Colonel Glenn Highway, 121, 213, **269**
Columbus, Ohio, 237, 267, 275
Combat Lightning, Southeast Asia operations, 193
Combined Federal Campaign, 220, 250
Command Publications Distribution Center, 175
Commission on Base Realignment and Closures, 257
Common Aperture Technique for Imaging Electro-Optical Sensors, 332
Community Service, Inc., founded by Arthur Morgan, 28
Community welcome signs, **248**
Companies
    656th Transportation Company, United States Army, 324
    731st Ordnance Company (EOD), United States Army, 324
    836th Military Police Company [Aviation], 314
    912th Engineer Headquarters Company, Patterson Field, 132
    1157th Signal Depot Company, Springfield, Illinois, 132
    1916th Quartermaster Truck Company, Patterson Field, 132
    2007th Quartermaster Truck Company, Patterson Field, 132
    Company K, 3rd Ohio National Guard, 59
    Headquarters 2054th Ordnance Company, 132
Compaq ES-45 computer, 177
Complements
    2790th Base Medical Complement, 316
    3rd Station Complement, Air Depot, 313
Composites, 251, 255; also see advanced composites
    airframe and automotive parts, 258
    bridge decks, 259
    bridge, first in state, 252
    for commercial products, 251
    reducing cost of manufacture, 256
    scaled, 283
Comprehensive Environmental Response, Compensation and Liability Act, 165
Condo craze, 237
Congressional appropriations for military aviation
    1911, 48
    1915, 49
    1916, 51, 54
    1919, 82
    1939, 115, 128
    lack of, 47
    for Korean War, 144
Connolly, Maurice, 64
Conover, Charlotte Reeve, **41**, 237
    Civil War remembrances, 18
    death of, 44-45
    education, 30, 33
    impression of Phillips Hotel library, 15
    nineteenth-century Dayton remembrances, 19
    observer of aviation progress, 30, 33
    opinion of lack of Cooper statue, 12
    reaction to Phillips House demolition, 33
Conservancy Act of Ohio, 29
Consolidated Mobility Bag Control Center, 323
Construction Department, Signal Corps
    lease for air depot, 79
    road construction at Wilbur Wright Field, 54, 58
Construction, services, materials, equipment, and supplies, Wright-Patterson Air Force Base, 249
Continuing Education program, 262
Converse shoe company, 229

Coolidge, Calvin, 34, 94
Cooper Lofts, **237**
Cooper Park, 12, 20, **231**, 237
Cooper townhouses, **237**
Cooper, Daniel C., **12**, 13, **231**, 233
Cooper, Maggie, **289**
Cooperative Research and Development Agreement (CRADA), 254
Cope, Douglas, 180
Cord Corporation, 23
Cord, Erret L., 23
Corning Glass, 268
Corvette, 259
Council of National Defense, 33
Courier Car Company, Dayton, 22
Courthouse Square, 237
Cover, C. A., 99
Cox, James M., **27**, 239
    and creation of Dayton Wright Airplane Company, 52
    and Dayton airport, 241
    as congressman, 7
    as governor of Ohio, 7, 8
    Dayton Daily News, publisher, 7, 25
    Democratic candidate for presidency, 7, 25, 34
    gravesite of, **242**,
    reaction to flood of 1913, 27
Craig Air Force Base, Alabama, 147
Crane, Carl J., and first automatic landing, 115
Crane, Joseph, 12, 15
Crawford, Don L., 235
Croatia, 185
Crossroads of America, 241
Crouch, Harold, 263
Crouch, Tom D., **263**
Cruciger, Scott, 283
Crumrine, Clarence E., 91, 92
Cuba, 232
Culbertson, Raymond E., 108
Culver, Douglas, 70
Curran, Chuck, 275
Curran, Jack, 275
Currituck County Airport, 245
Curtiss Aeroplane & Motor Company, 52, 59, 63
    patent war with Wright brothers, 24, 33
    schools of aviation, 53
Curtiss propellers, 310
Curtiss, Glenn, winner of Bennett Trophy, 96
Custer Specialty Company, 22
Custer, Luzern, **21**
Customer Integrated Automated Procurement System, 166
Cyclorama, Gettysburg Avenue, 11
Damm, Henry J., 107
Daniels, E. A., 266
Darling Motor Car Company, 22
Davis, James, **24**
Davis-Monthan Air Force Base, Arizona, 163
Dawn Patrol Rendevous World War I Fly-In, **277**, 279
Dayton Abolition Society, 16, 234
Dayton Air and Trade Show, 197
Dayton Air Fair, 197
Dayton Air Force Depot, Gentile Air Force Station, **257**
Dayton Air Service Committee, formation of, 36, 200, 222
Dayton Air Show, 197, 244
    history of, 196-197
    Vectren, 241, 245; also see Vectren Dayton Air Show
Dayton Airport, **241**
Dayton Airport, Inc., 241
Dayton and Montgomery County MetroParks, **233**
Dayton Area Graduate Studies Institute (DAGSI), 262
Dayton Army Air Field, Vandalia, Ohio, 200, **241**
Dayton Army Air Forces Procurement Field Office, 315

Dayton Art Institute, 36, **38**, **238**
    Pathway sculpture, **275**
    Summer Camp, **289**
Dayton Aviation Heritage National Historical Park, 284, **285**, 324
    creation of, 187, 283
    Huffman Prairie Flying Field, 187, 285, 286
    Paul Laurence Dunbar State Memorial, 189, 285
    Wright Cycle Shop, 187, 285
    Wright Flyer III at Carillon, 189, 285
Dayton Bicycle Club, 22, **242**
Dayton Chamber of Commerce Trophy Race, 1924 Air Races, 99
Dayton Chamber of Commerce, hosts National Aircraft Show, 196
Dayton Christian-Jewish Dialogue, **265**
Dayton Contract Management District, 318
Dayton Country Club, 3
Dayton Cycling Club, 270, 271
Dayton Daily News, 7, **8**, 241
    advice during World War II, 42
    formerly *The Evening News*, 3
    James M. Cox, publisher, 25
Dayton Dragons baseball team, 237, **238**
Dayton Electric Car Company, 22
Dayton Engineering Laboratories Company (Delco), 22, **42**, 242
    automotive inventions, 32
    Delco Products Division, 310
    formation of, 7, 31
    shut down during air races, 99
    World War II efforts on home front, 43, 227, 310
Dayton Engineers Club, **37**
Dayton Flood Prevention Committee, 27
Dayton Foundation, McCook Field marker, 187
Dayton General Airport South, 196, 244
Dayton *Herald*, support for Wrights, 2, 4
Dayton International Air Show and Trade Exhibition, 197
Dayton International Airport, 164, **241**
Dayton Inventing Flight, 190
Dayton *Journal*
    building arson over slavery issue, 18
    first flight not printed, 1
    John Van Cleve publisher, 15
    questions Orville Wright, 2
Dayton Lyceum Association, 15
Dayton Metal Products Company, 31, 35
Dayton Municipal Airport, 197, 241
Dayton Peace Agreement, Balkan Proximity Peace Talks, 185, 327
Dayton Philharmonic Orchestra, 235, 272
Dayton Plan for city government, 8
Dayton Power and Light Company, 65
Dayton Public Arts Commission, 275
Dayton Public Library, 20, 27, 237
Dayton RAPCON, 147
Dayton Rubber Manufacturing Company, 310
Dayton Signal Corps Publications Agency, 315
Dayton Society of Natural History, **253**
Dayton Soldiers' Home, 7, **10-11**, 223
Dayton Stamp Club, 112, 170
Dayton State Hospital, **20**
Dayton Steel Foundry, 22
Dayton Tire and Rubber Corporation, 242
Dayton Triangle, 228
Dayton View, Dayton, settlement, 33
Dayton Wright Airplane Company, 32, **52**
    manufactured DH-4, 71
    manufactured Liberty Airplane, 73
Dayton Wright Brothers Airport, **244**
Dayton, 250
    1811 earthquake, 12-13
    1913 Flood, 20, 25, **26-27**, 29, 54, 56
        comparison to Titanic, 27
    activities of Ku Klux Klan, 40

African-American community, 234
as "Air City of America," 34
as "City of Factories," 34
as "Gem City," 8, 33
as birthplace of flight testing, 200
automobile manufacturing, 7, 22
    Apple Automobile Company, 22
    Barney & Smith Car Company, 22
    Courier Car Company, 22
    Darling Motor Car Company, 22
    Dayton Electric Car Company, 22
    Jitney Transportation Company, 22
    Speedwell Motor Car Company, 22
    Stoddard, 22
bicycles as means of transportation, 20
birthplace of Christian-Jewish dialogue movement, 265
Buildings
    Arcade Centre Tower, 233, **240**
    Bell Telephone Building, 38
    Centre City Building, **21**
    Corwin home, 228
    Courthouses, **16**
    Dave Hall Plaza, 266
    Dayton Daily News building, **8**
    downtown, 38, 39
    Hulman Building, 39
    Masonic Temple, 36, **38**
    Reibold Building, **21**
    United Brethren Building, later Centre City Building, **21**
business and industry in transition, 239
celebration of end of World War I, 33, 75
Cemeteries
    Catholic Calvary Cemetery, 17
    Fifth Street Cemetery, 18
    Woodland Cemetery, 4, 5, 15, 17, 18, 25
center of invention and industry, 7
Central Branch of the National Home for Disabled Volunteer Soldiers, 10
communications systems, 19
Dayton Air Service Committee, 36
diffusion of wealth, 25
disease and illness, 18
Dixie Highway, establishment of, 14
donation of land for Wright Field, 36, 200, 222
downtown revitalization, 233
during World War II, 227
early settlement, 12
early transportation routes, 13
electric lighting, 19
establishment of Navigation Board, 14
exodus to suburbia, 25, 228
first hospital, 20
first natural history museum, 15
first professional fire department, 20
first professional police force, 20
first superhighway, 233
flight testing, 200-201
flood control, early attempts, 20, 26
flooding, 25
growth of slums, 233
historic districts, **230**, 237
home building boom, 25, 40
Homecoming Celebration, see Wright Brothers Homecoming Celebration
Hotels
    Biltmore Hotel, 40
    Miami Hotel, **39**
    Phillips House, 15, **16**, 33
housing shortage during World War II, 228
in Great Depression, 40-41

industrial workforce, 228
industries mobilized to help World War II war effort, 310
inventions, **233**
newspapers
    Dayton Daily News, 7
    Dayton *Herald*, 2, 4
    Dayton *Journal*, 1, 2, 15, 18
    Empire, 18
    The Evening News, 3
    The *Journal Herald*, 42
No. 1 critical labor area, 228
photographic views, **13**, **14**, **225**, **231**, **243**
    nineteenth century, **15**
    1923, **35**
    1930, **19**
    1950, **45**
    late 1930s, **44**
population boom in World War II, 44
population, 228
publishing in, 24
Radio Dog, 34-35
reunion of Civil War veterans, 19
revitalization of downtown neighborhoods, 237
Rike-Kumler Department Store, 6, 24, **39**, 42
rise of Greater Dayton, 228
role in Balkan Proximity Peace Talks, 184
schools desegregated, 235
settlement of Dayton View, 33
settlement of North Dayton, 33
settlement of West Dayton, 33
skyscrapers, 233
Soldiers' Monument, 19
Sons of Italy, 150
Southern Ohio Lunatic Asylum, **20**
street improvements, 19-20
streets, **14**
suburbanization, 231
telegraph service, 19
telephone service, 19
Temperance Movement, 20, 40
winter storms, 161, 163
Women's Bicycle Club, 22
World War II, home front efforts, 42-43
Dayton, Jonathon, **12**
Dayton-Miami Valley Consortium, 202, 262
Dayton-Springfield Metropolitan Statistical Area, 250
Dayton-Springfield Turnpike, 222, 223
Dayton, University of, 243, 264, 283
  and Research Institute, 260,
  educational partnerships with Wright-Patterson, 262, 268
Deeds Legacy Plaza, 279
Deeds Point, 237, **243**, **279**
Deeds, Edith Walton, **31**
Deeds, Edward A, **7**, **30**, **31**, 239
  and creation of Dayton Wright Airplane Company, 52
  and Carillon Historical Park, 31
  and creation of McCook Field, 72, 243
  and Delco, 22, 31
  and Engineers Club, 37
  and establishment of Wilbur Wright Field, 54
  and NCR, 7, 40
  as chief of Signal Corps Equipment Division, 31, 33, 52
  as member of Aircraft Production Board, 31, 33, 52
  at 1924 International Air Races, **34**
  commissioned Colonel, 33
  first airfield in country, 33
  founding of Miami Conservancy District, 7
  grave of, 17
  influence on aviation in Dayton, 54
  mausoleum, **242**

Moraine Farm, **31**, 33
    on Munitions Standards Board, 33
    sent to Memphis to hire engineer for flood control, 29
Defense Acquisition University Midwest Region, 264
Defense Act of 1956, 233
Defense Automatic Addressing System Center, 323
Defense Commissary Agency, 323
Defense Contract Audit Agency, 323
Defense Contract Management Agency, Dayton, 323
Defense Contract Management Center International, 172
Defense Courier Service, Louisville, 323
Defense Electronics Supply Center (DESC), 208, 256, **257**
Defense Energy Support Center, 323
Defense Enterprise Computing Center, Detachment Dayton, 323
Defense Finance and Accounting Service, 174
Defense Information Systems Agency, 323
Defense Institute of Security Assistance Management (DISAM), 202, 323
Defense Intelligence Agency, 323
Defense Logistics Agency, 323
Defense Metropolitan Area Telephone System, 166
Defense Reutilization and Marketing Office, 176, 323
Defense Security Assistance Management Education Program, 202
Defense Security Service, 323
Defense Technical Information Center, 254
    Midwestern Regional Office, 323
Defense, Italian Air Command, 149
Defiance County, Ohio, 259
Delcher, Harry, **154**
Delco, see Dayton Engineering Laboratories Company
Delphi Automotive Systems, 259
Delta Air Lines, 176
Demaret, Jimmy, 217
Denman, Dr. Gary L., 255
Department of Commerce, U.S., 113, 197, 256
Department of Defense Health Service Region 5, 204, 247, 323
Department of Defense Supply Agency, **257**
Department of Defense
    educational programs effect on, 202
    policy on air races, 97
Department of Energy, U.S., 256, 268
Department of Health, Education, and Welfare, U.S., 212
Department of Homeland Security, U.S., 191
Department of the Treasury, Internal Revenue Service, 324
Depression, see Great Depression
Desegregation of Dayton schools, 235
DeSpain, Lisa, 276
Detachments
    1st Mobile Rubber Repair Detachment, Patterson Field, 132
    64th Ordnance Detachment, Fort Benjamin Harrison, 180
    71st Ordnance Detachment (Army), Wright-Patterson Air Force Base, 180
    838th Army Air Forces Specialized Depot Detachment, Marion, Ohio, 132
    Detachment 2, 1401st Military Airlift Squadron, 163
    Detachment 15, 15th Weather Squadron, 164, 190-191
    Detachment 859th Signal Service Company, 132
    Detachment 905th Quartermaster, 132
Detroit Pistons, 229
Dhahran, Saudi Arabia, 180
Diconix, 243
Diddel, William, 214, 217
Dietz, James, 279
Digital Multiplex System (DMS-100), 166
Directional traffic lights, 233
Directorate of Engineering, Aeronautical Systems Center, 264, 268
Dirigible balloons in Signal Corps fleet, 47
Disaster Preparedness Office, Wright-Patterson Air Force Base, 191
Disposal Board, 315
Distinguished Flying Cross, for first automatic landing, 115

Distinguished Service Medal, given to Round-the-World flight crews, 95
Dittrick, Dale, 180
Divisions
    42nd Rainbow Division, 82
    58th Air Division, Bomb, 317
    58th Air Division, Defense, 317
    58th Air Division, Wright-Patterson Air Force Base, 193
    83rd Division, U.S. Army, 107
    First Provisional Air Division, 103
Division of Health Systems Management, 253
Dixie Highway, establishment of, 14
Dixieland Flight, see 2750th Air Base Wing, Dixieland Flight
Dixon, Richard Clay, 235
Document Automation and Production Service, 323
Dogs
    "Wright Rudder," 42
    friendship with aviators, 61
    Orville Wright's Scipio, **61**
    Richard Cole's Jinx, **61**
    Wilbur Wright's Flyer, **61**
Doherty, James F., 81
Dole Pineapple Company, 101
Dole Prize, 101
Domestic Engineering Company (Delco-Light), 31
Donegia, Scott, 180
Donne, Thomas J., 161
Donnelly, W. E., **88**
Doolittle Acquisition Management Complex, , **219**, 221
    dedication of buildings, 327
Doolitte, James H., "Jimmy", 287
    attends First Flight Banquet, 170
    and Tokyo Raid, 45
    in National Aviation Hall of Fame, 211, 278
    test pilot, 200
Doremus, Robert, 195
Dorothy Jean, steamboat, **29**
Dorst, Debbie, **277**,
Douglas Airplane Company, 93
Douglas, Donald, 93
Dual use technologies, 227
Dufour, Howard E., **275**
Duke, D. G., 99
Dunbar Library, Wright State University, **275**
Dunbar, Joshua and Matilda, 234
Dunbar, Paul Laurence, **234**, 237, 275, 283
    burial place of, 17
Duncan, Thomas E., 64
Dunlap, L. H., **88**
Dunlop, Dawn, 245
Dunn, R. E., 66
DuraSpan system, 259
Duro Company, 310
EAA AirVenture, Oshkosh, Wisconsin, **283**
Earhart, Amelia, 23, **101**
    in National Aviation Hall of Fame, 210-211, 278
East High School, 244
Eastern Air Defense Force, 193
Eastern Air Defense Region, 193
Eastern Kodak Company, 243
Eaton, Samuel G., 98
Edgar, C. G., signs lease for Wilbur Wright Field, 54, 55
Edinger, Harold, **244**,
Edison Materials Technology Center (EMTEC), 255
Education Partnership Agreements (EPAs), 251
Educational Committee of the Dayton YMCA, 266
Educational Outreach office, 262, **267**
    and centennial of flight, **272**, 279
Educational Outreach Program, at Wright-Patterson Air Force Base, 265, **266**

Edwards Air Force Base, California
  412th Test Wing, 199
  Air Force Flight Test Center, 200
  Airborne Laser Laboratory, 201
  flight test mission transferred, 172, 198
Edwards, Michael, **289,**
Edwards, Thomas A., **188**
Eglin Air Force Base, Florida, 173, 198
  Armament Laboratory, 157
  Munitions Directorate, 183, 247
Eglin Field, Florida, sub-depot in, 125
Egress Recap, 154; also see Operation Homecoming
Eickmann, Kenneth E., 158, 178, **251**
Elder-Beerman, 233
Electronic Systems Center, 247, 324
  Materiel Systems Group, 324
Eleftherios Karkadoulias Bronze Company, 274
Ellington Field, Houston, Texas, 67, 68, 70
  and Barling bomber, 70
Ellis, Ed, 6
Emery Forwarding, 241
Empire newspaper, Dayton, 18
Employees' Monument, Wright-Patterson Air Force Base, 167
Energy Crisis, 1973, 159
Engine overhaul process, 84, 86
Engineering Data Automated Logistics Program, 152
Engineering Department, Air Service Supply and Repair Depot, functions of, 84
Engineering Division
  80th Anniversary celebration, 187
  and Barling bomber, 70
  and Oscar Boonshoft, 253
  at McCook Field, 35, 72, 77
  recommendation for World Cruiser, 93
  test of World Cruiser, 93
Engineering Maintenance Officers' Training School, Fairfield Air Depot, 133
Engineering Procurement Branch, Engineering Section, Materiel Division, 77
Engineering Production Department, merger with Airplane Engineering Department, 72
Engineering Repair Section, Air Service Supply and Repair Depot, 77, 84, 86 Engineering Section, Materiel Division, 77
Engineers Club of Dayton, 36, **37**
Engle, Joe, 270
Engler, Nick, 244
Englewood, Ohio, 250
Enlisted Men's Fund, 104
Enlisted Reserve Corps, U.S. Army, 106
Enon, Ohio, 250
Environmental Management, Office of, Wright-Patterson Air Force Base, 280, 286
Eppright, Lieutenant, **201**
Equipment Branch, Engineering Section, Materiel Division, 77
Equipment Division, Signal Corps, 33, 69
  aircraft production in 1920s, 40
  Deeds as chief, 31, 52
  purchase land for air depot, 79
  responsibility for depots, 80
Estabrook Drive, dedication in honor of Estabrook, 327
Estabrook, Merrick G., Jr., 81, 108
  career of, 114
  as Patterson Field commander, **114**
Estes, Amanda, **289,**
Eugene Kettering Gallery, 190
Evangel University, Springfield, Missouri, 265
Executive director, Aeronautical Systems Center, **256,**
Exhaust, newspaper, 169
Exon, Arthur E., 158
Experimental Aircraft Association (EAA) AirVenture, 283

Explorers, encampment on Wright-Patterson Air Force Base, 149
Fain, James A., Jr., 158, 199
Fairborn, Ohio, 56, 161, 231, 250
  early history of, 165
  Morris resided, 141
  moved sewage and waste treatment plant, 192
Fairchild, Muir S., 70, 76
Fairfield Air Depot Control Area Command, 125, 131, 315
  renamed Fairfield Air Service Command, 126
Fairfield Air Depot Operations Hotel, Patterson Field, 117
Fairfield Air Depot, 313; also see Fairfield Air Depot Control Area Command; Fairfield Air Service Command; Fairfield Air Technical Service Command
  1922, **89**
  1927 air races, timer's stand, **97**
  1931 work, 109
  1939 personnel strength, 115
  1942 personnel strength, 121
  1st Provisional Air Transport Squadron, 107
  1st Transport Squadron, 107
  4-H visits, 101
  Air Service Supply and Repair Depot, 1921, **84**
    aero-repair function, 83, **84**
    Engineering Repair Section, 84, 86
    formation of, 82, 140
    parachute testing, 86
  aircraft engine storage, **111**
  and Patterson Field, **109**
  assigned to Field Service Section, Materiel Division, 125
  assigned to Provisional Air Corps Maintenance Command, 125
  Aviation General Supply Depot, Fairfield, Ohio, 82
    merger with Aviation Repair Depot, 82-83
    living quarters, 83
    chronology, 81, 140
  becomes Area C, 139
  chronology, 81, 86
  civilian training programs, 133-134
  command assignments, 125-126, 130
  continued function on Patterson Field, 105
  creation of new depots, 134-135
  distinguished visitors in 1927-1928, 101
  Engineering Department, 109
    1939 personnel strength, 115
    challenges of flying mail, 112
    civilian training, 134
    met supply challenges in World War II, 122
    training moved to Wood City, 134
  Engineering Maintenance Officers' Training School, 133
  Estabrook commander of, 114
  FADO Hotel, **116**
  Fairfield Air Depot Control Area Command, chronology, 81
  Fairfield Air Depot Reservation, Air Corps Carnival, **104**
    chronology, 81
    established, 102, 103, 140
    in 1931 maneuvers, 103, **105**
    soldiers on flightline, **103**
  Fairfield Air Intermediate Depot, 313
    1923, **47**
    1925, airplane repair, **84**
    assistance with Round-the-World flight, 90, 92
    aviation records, 91
    chronology, 74, 81, 140
    Engineering Department
      Airways Branch, 100
      Radio Branch, 100
    extent of support, 86
    host air races, 95, 99
    insignia of, **86**
    mission, 86, 88, 89
    Model Airway System, 100, 102

permanent unit designation, 86
Property, Maintenance, and Cost Compilation Section, 88-90
staff photo, **88**
stock keeping records, 89
Supply Department, 89
Fairfield Air Service Command
Aero Repair Training School, **131**
chronology, 81
civilian employees, **133**
staff, **127**
Fairfield Air Technical Service Command, chronology, 81
Estabrook commander of, 108
inactivation, 135
Fairfield Aviation General Supply Depot, 1918, **79**
chronology, 81, 135, 140
civilian employees, 80, 82
commander of, 64
construction of, 79
diagram of, **55**
storage of war surplus, 82
incorporated into FADR, 105
map, 311
merger with Wilbur Wright Field, 74
mission, 80
post-World War I changes, 80
proximity to Fairfield, 56
squadrons at, 80
telephone service, 65
Fairfield Provisional Maintenance Group Area, 125
Fairfield Provisional Maintenance Group, 134
FFO airfield designator, 165
first golf course, 217
flying mail, 112-113
heart of logistical support, 76
lack of space for training, 134
leased storage space in Springfield, Ohio, 118
Liberty V-12 engines in storage, **86**
Maintenance Division, personnel replacements, 134
major portion of Patterson Field, 107
needed construction, 1939, 116
newspaper, *Exhaust,* 169
officer strength
operations during Great Depression, 110
packing and crating shop, **124**
parachute testing, **103**
part of Materiel Division, 76
Personnel and Training Division, 134
problems in operations in World War II, 122
problems with transportation of supplies, 124
proving ground for new supply ideas, 122
radio-mechanics training program with University of Wisconsin, 134
Receiving Department, expansion of, 116
recruiting for training, 134
redesignated Fairfield Air Depot Control Area Command, 125
Robins commander of, 87
shipment of aircraft engines, **124**
shops training school, 133
stockpile of war materiel during World War II, 115
storage of Barling Bomber, 70
sub-depots under control, 1942, 134-135
Supply Department, 115, 116, **124**
Supply Division, 124, 133
support of 1933 Air Corps Anti-Aircraft Exercises, 112
support of 1934 Alaskan flight, 113
Traffic Section, 124
training programs during World War II, 131, 133
Transportation Corps, 124
U.S. Army Materiel Division Supply School, 133
units and organizations trained at, 132
Wilbur Wright Air Service Depot, chronology, 81

Fairfield Air Intermediate Depot, see Fairfield Air Depot
Fairfield Air Service Command, 315
established, 126
headquarters in Building 10, 118
renamed Fairfield Air Technical Service Command, 126
Fairfield Air Technical Service Command, 124, 126, 315
Fairfield Aviation General Supply Depot, see Fairfield Air Depot
Fairfield High School, 287
Fairfield Provisional Maintenance Group Area, see Fairfield Air Depot
Fairfield, Ohio, 54, 56-**57**, 79
1923, **47, 76**
early history, 165
Fairmont West High School, 244
Family Day, 280
Far East Air Forces, Korea, 143
Far Hills, 228
Farmer, Alfred G., 65, 204
Farrell, Lawrence P., 183
Fasure, Julia, **285,**
Fawcett, Dr. Novice G., 269
Federal Aviation Administration, 164, **244**, 281
Federal Disaster Assistance Team, assistance after Xenia tornado, 161
Federal Emergency Management Agency, 178, 324
Federal Employees Retirement System, 265
Federal Facility Agreement, 176
Federal Highway Administration, 252
Federal Housing Administration, 211
Federal Housing Authority, 42
Federal Technology Transfer Act of 1986, 254
Feeney, John, **261,**
Fels Institute for the Study of Human Development, 28
Ferguson, Thomas R., Jr., 158
Festival of Flight, 171
Fiber Form, Newark, Ohio, 258
Fiberglass bridge, **252**
Field of Dreams, 274
Field Service Section
Air Service, 89
Army Air Corps, 88
assistance with Round-the-World flight, 92
becomes component of Materiel Division, 89, 102
discontinued, 125
headquarters at Wright Field, 125
headquarters moved to Patterson Field, 125
Materiel Division, 77, 125
mission, 89
offices, **90**
Robins commander of, 87
under Provisional Air Corps Maintenance Command, 125
Fifth Street Cemetery, Dayton, 18
Fifth Third Field, 237, **238**
Fifth U.S. Army, Fort Knox, Kentucky, 161
Fight Flight Stamp, **277**
Fighting McCooks, 33
Finance Office, U.S. Air Force, 315
Fires
Airmen's Service Club, 214
Building 262A, 1961, 216
Wood City, 1961, 214
Firestone Tire and Rubber Company, 242
First Division, Ohio Militia, War of 1812, 13
First Flight Centennial Federal Advisory Board, 263
First Flight Ceremony, 170, 190
First Flight, 276
First LEGO League, 267
Fisher House Foundation, 215
Fisher House, Wright-Patterson Air Force Base, 215
Fisher, Zachery and Elizabeth, 215
Fisk University, 262
Fitts Human Engineering Laboratory, 326

Fitts, Paul M., 326
Flack, Dr. Harvey, 235
Flacks, Paul and Shirley, 265
Flights
    47th Airlift Flight, **173**, 175, 322, 323
        role in Balkan Proximity Peace Talks, 184
        role in Operation Uphold Democracy, 175-176
    645th Weather Flight, 191, 321
    88th Weather Flight, 191, 321
    Flight B, 1st Transport Squadron, 313
Flight Dynamics Laboratory, 318
Flight Test Division, Wright Air Development Center, 145
Flight Test Section, McCook Field, 73, 76, 200
Flight Test Training Unit, Patterson Field, 200
Flight training, World War I, 53
Flood, 1913, see Dayton, 1913 Flood
Florence, Italy, 274
Florida Adirondack Preparatory School, 106
Florida, 265
Floyd Bennett, 101
Fly Me to the Future, 276
Flying Branch, Field Service Section, Materiel Division, 77
Flying Tigers, 123
Flynn, Gary, 180
Flyover sculpture, downtown Dayton, **274**, 275
Focal Project II, 259
Fogleman, Ronald R., 187
Ford, Gerald, **208**, 256
Ford, Harrison, 278
Foreign Aerospace Science and Technology Center, 318
Foreign Data Section, 205; also see National Air and Space Intelligence Center
Foreign military sales, 204
Foreign Technology Division, 318
    became National Air Intelligence Center, 175
    office space in Building 219, 204
    Readix computer, 207
    reassignment to Air Force Intelligence Command, 172; also see National Air and Space Intelligence Center
    Wright Field, 207
Fort Benjamin Harrison, Indiana, 180
Fort Bragg, North Carolina, 115
Fort Knox, Kentucky, 161
Fort Monmouth, New Jersey, 113
Fort Monroe, Virginia, 115
Fort Myer, Virginia, demonstration flights, 2, **22**, 24, 153
Fort Sam Houston, Texas, 48, **50**
Fort Sill, Oklahoma, 115
Fort Wayne Zollners, 229
Forward Looking Advanced Multi-mode Radar, 181
Foster, Master Sergeant Ronald, 272
Foulois House, **34**, **223**; also see Buildings, Area C, Building 88
    dedicated in honor of Foulois, 328
    history of, 223
Foulois, Benjamin D., **50**, **105**, **223**
    as Army's only active pilot, 47
    dedication of house in honor of, 328
    on Aeronautical Board, **49**
    tour of Dayton, 54
    attends World War I pilot reunion, 155
    ceremony honoring Giovanni Pirelli, 150
    commendation of 1933 Air Corps Anti-Aircraft Exercises, 112
    demonstration of Wright Flyer, 24
    lived in Building 88, 223
    organized First Provisional Air Division for 1931 maneuvers, 103
Fox, Charles E., Jr., 158
France, pre-World War I aviation, 48, 49
Frank, Walter H., 126, 137
Franklin, Ohio, 134
Fraze, Ermal C., 237

FRC-19 Console, installed in Area C control tower, 147
Free states, 234
Freedom Flight America, 187
Freeman Field, Indiana, role in Project Lusty, 206
French Legion of Honor, and Round-the-World flight crews, 95
Friedland, Dr. Eric, 265
Friendship Flyer, 245
Frigidaire Division, 227, 242, 310
Fujifilm, 279
Fusari, Lodovico, 150
Gabriel, Ron, **261**,
Gaddis, Thomas P., 52
Gamble, Greg, 276
Geauga County, Ohio, 259
Gem City, immigration to during World War I, 33
General Accounting Office, 324
General Billy Mitchell Field, Wisconsin, 130
General Framework Agreement for Peace in Bosnia and Herzegovina, 185
General Headquarters Air Force, later Air Force Combat Command, 125
General Management Plan, 285
General Motors Corporation
    Aeronautics Products Division, 310
    and USCAR, 256
    changes following Depression, 239
    employment in Dayton, 242
    funding for Wright STEPP program, 268
    purchase of Dayton Wright Airplane Company, 52
    purchase of Kettering business interests, 32
    shut down for air races, 99
    war materiel production, 44
General Motors Research Corporation, Dayton, 7, 32
General Motors, Inland Manufacturing Division, 227, 242, 310
General Services Administration, transfers Skyway Park to state of Ohio, 121
General Thomas D. White Fish and Wildlife Conservation Award, 155
General Thomas D. White Historic Building Preservation Award, 176
Geneseo, Illinois, **277**
Gentile Air Force Depot, **257**
Gentile Air Force Station, **257**
    facilities for AFOG, 208
Gentile, Don S., **257**
George Sub-depot, Illinois, 135
Georgetown University, 229
Germany
    bombing of Great Britain, World War II, 44
    pre-World War I aviation, 49
Gerstner Field, Essington, Pennsylvania, 53
GI Gertie, 137
Gibson, Jeffrey, 188
Gillmore Hall, see Buildings, Area A, Building 262
Gillmore, William E., 77, **101**, 326
Gimbel, Richard, 135
Girl Scouts, and World War II home front, 42
Giroux, Julie, 276
Gittleman, Neal, 272
Glenn, John, 17, 287
    enshrined in National Aviation Hall of Fame, 211, 278
    proponent of Dayton Aviation Heritage Historical Park, 285
Glider Branch, Wright Field, 232
Glider tests, 232
Glyphidocera wrightorum, on Huffman Prairie, 189-190
Gnorimoschema huffmanellum, **190**
Godde, Delphine Dodge, 327
Godwin, Larry, 274, 275
Goebel, Arthur, 101
Gold, Dave, 284
Golden Age of Air Racing, 97
Goldwater, Barry M., 167

Golf Courses, Wright-Patterson Air Force Base
    East Course, **217**
    Prairie Trace Golf Club, **217**
    Twin Base Golf Club, 214, **217**
    Wilbur Wright Field Officers' Golf Club, 214
Goodier, L. E., 81
Goodyear, 279
Gorbachev, Mikhail, signs Strategic Arms Reduction Treaty, 171

Grafton Hill Historic District, 237
Grafton Hill, **230**
Great Blimp Meet, 279
Great Britain, 44, 49, 90
Great Depression, 7, 228, 239
    Hoovervilles, 40
    in Dayton, 40-41
    operations at Fairfield Air Depot, 110
Great Flood of 1913, see Dayton, 1913 Flood,
Great Miami River, 228
    aircraft crashes, 35
    and black settlement, 234
    and RiverScape, **233**, 237, 275
    as transportation route, 13-14
    early Dayton, 12
    first bridge over, 14
    flood control system, **28**
    flooding, 13, 25, also see Dayton, 1913 Flood
    Great Flood of 1913, see Dayton, 1913 Flood
    Hoovervilles, 40
    parade in honor of Wrights, 4
Great Migration, 234
Great War Aeroplanes Association, 279
Greater Dayton Area Chamber of Commerce, 167, 244
Greater Dayton Area Fire Departments, 191
Greek Orthodox Church, **230**
Green Acres Housing Complex, see Wright-Patterson Air Force Base, housing
Greene County, Ohio, 246, 250, 259
Greene Memorial Hospital, Xenia, Ohio, 159
Greene, Carl, 40
Greene, Duff W., 223
Gregory, H. F., 115
Gridley, Craig W., 146
Grief counseling, 288
Griffiss Air Force Base, New York, 198
Griffith, J. S., **201**
Grimmy Award, 267
Ground Controlled Approach, Area C, 144
Groups
    1st Civil Engineering Group, 319
    1st Transport Group, Patterson Field, 130
    2nd Bombardment Group (medium), Mitchel Field, 131
    2nd Transport Group, Patterson Field, 130
    3rd Transport Group, Patterson Field, 130
    4th Air Depot Group, 313
    4th Transport Group, Patterson Field, 130
    5th Transport Group, Patterson Field, 130
    6th Air Depot Group, 313
    6th Weather Group, 317
    10th Air Depot Group, 313
    10th Observation Group, Patterson Field, 130
    10th Transport Group, Patterson Field, 130
    17th Bomb Group, connection with 34th Bombardment Squadron, 193
    18th Air Depot Group, Patterson Field, 132, 313
    19th Air Depot Group, 313
    20th Air Depot Group, 314
    21st Air Depot Group, 314
    22nd Air Depot Group, 314
    22nd Bombardment Group (Medium), Patterson Field, 131

23rd Air Depot Group, 314
46th Operations Group, Munitions Test Division, Aerospace Survivability and Safety Flight Operating Location-AC, 324
55th Air Depot Group, Patterson Field, 132, 314
56th Air Depot Group, 314
57th Air Depot Group, 314
60th Transport Group, Patterson Field, 131, 313
61st Transport Group, Patterson Field, 131
63rd Transport Group, Patterson Field, 130, 313
74th Medical Group, 316
    A-76 impact, 174
    and disaster preparedness, 191
    assigned to Aeronautical Systems Center, 172
    deployments for Operation Enduring Freedom, 183
    deployments for Operation Iraqi Freedom, 183
    formerly 645th Medical Group, 205
88th Civil Engineering Group, 319, 321
88th Logistics & Operations Group, 321
88th Logistics Group, 321
88th Mission Support Group, 322
88th Support Group, 322
96th Service Group, Oscoda, Michigan, 132, 133
98th Service Group, 314
155th Tactical Reconnaissance Group, Lincoln, Nebraska, 160-161
308th Air Base Group, 314
316th Transport Group, Patterson Field, 130, 131, 314
332nd Fighter Group, Tuskegee Airmen, 17
342nd Air Base Group, Depot, 316
445th Maintenance Group, 322
445th Mission Support Group, 322
445th Operations Group, 322
645th Civil Engineering Group, 319, 321
645th Communications-Computer Systems Group, 316
645th Logistics & Operations Group, 321
645th Medical Group, 205, 316
645th Support Group, 322
906th Fighter Group, 320
906th Tactical Fighter Group, 194, 320
    401st Combat Logistics Support Squadron, 180, 320
    aircraft, **194**
    and static airplane display for AFLC, 327
    transition to F-16, 194
907th Airlift Group, Wright-Patterson Air Force Base, 194, 322
1125th U.S. Air Force Field Activities Group, 316
1702nd Air Transport Group, 315
2046th Communications & Installations Group, 316
2046th Communications Group, **165**, 316
    end agreement with Federal Aviation Administration, 164
    realigned with 88th Air Base Wing, 174
    assistance after Xenia tornado, 160
2046th Communications-Computer Systems Group, 316
2726th Air Force Cataloging Group, 318
2750th Air Base Group, 316
2750th Base Maintenance Group, 316
2750th Base Supply Group, 316
2750th Engineering & Service Group, 319, 321
2750th Logistics & Operations Group, 321
2750th Medical Group, 316
2851st Air Base Group, Kelly Air Force Base, 152
2925th Base Supply Group, 316
2926th Base Maintenance Group, 316
3086th Air Base Group, 316
3090th Installations Group, 316
3190th Materiel Control Group, 317
7300th Materiel Control Group, 317
8318th Air Force Reserve Base Support Group, 318
8401st Air Force Reserve Base Support Group, 318
8512th Air Transport Group, 316
First Pursuit Group, Selfridge Field, 101
Joint Depot Maintenance Analysis Group, 323

Grove City, Ohio, **277**
Gruenberg, James, 180
Guantanamo Bay, Cuba, U.S. Naval Base, 183, 249; also see Camp X-Ray
Gunnery Section, Signal Corps Air Division, 68
Gunnery testing, **67**-68
Hadden Park, 165
    dedication of, 327, **329**
    moved for West Ramp construction, 192
Hadden, William, 327
Hall, Titus C., 158, **235**
Hall, Tony, 285
Halsey, Admiral, and USS *Dayton*, 42
Halvorsen, Gail, **272**
Hamilton propellers, 310
Hamilton, Gary, 180
Hamilton, Ohio, 134
Hanoi Taxi, **195**, 198, **282**
Hanscom Air Force Base, Massachusetts, 247
Hansen, Lauren, 281
Hanson, Chris, 229
Hardcore Composite LLC, 259
Harding, John, 94
Harding, Warren Gamaliel, 34
Hardy, Donald L., 158
Harer, R. J., **201**
Harris, Bill, 217
Harris, Harold R., 76, **201**
    and Barling bomber, 70
    escort World Cruisers, 94
    test pilot, 200
Harrison, William Henry, 13
Harris-Seybold Company, 242
Harris-Seybold-Potter Company, 310
Harry B. Combs Research Center, 278
Harry B. Combs Resource Laboratory, 278
Harshman, Jonathan, early settler, 12
Hartman, David, 270
Hartsville College, Indiana, 4
Hartzell Industries, Inc., 275, 310; also Hartzell Propellers, Inc.
Harvey, Alva, **93**, 94
Haskell, Henry J., 5
Haug, P. P., **201**
Hawes, 228
Hawkins, Randall, 179, 180
Hawthorn Hill, Wright home, **18**, 228
Hayes, Paul L., 327
Hazelhurst Field, Minneola, Long Island, New York, 53, 66
Hazen, Michael W., 189
He, Ray, 281, **282**
Headquarters & Headquarters Squadron, Tenth Air Force, 314
Headquarters Squadron 2750th Air Force Base, see 2750th Air Force Base
Health Resources Management and Emergency Operations Act, 11
Hebble Creek Bridge, 259
Hebble, Henry, 222
Heffernan, Leo G., 59, 62, **64**
Heiliger, Donald, **153**
Helton, Elbert, 158
Hempfield High School, Landisville, Pennsylvania, 281
Herr, Andrew R., **188**
Herrington-Curtiss Company, Hammondsport, New York, 96
Hessinger, Robert, 180
High-tech products and services, 227
Hilberg, Christopher, 180
Hill Air Force Base, Utah
    Detachment 6, 15th Weather Squadron, 190
    in LOGAIR system, 148
Hills and Dales Park, Dayton, **75**
Hippert, R. D., **201**

Hi-Reach Water Tower Vehicle, 214
Historical Aerospace Site, 287
Historical Markers Project, 280
Historically black colleges, 250
Hoban, Richard M., 154
Hobart Manufacturing Company, 310
Hobley, A. H., 68
Hoboken, New Jersey, 253
Hobson, David L., 177, 285
    and Ohio Task Force One, 178
    dedication of street in honor, 328
Hoelle, Father Phillip, 265
Hoke, Dr. John, 283
Holbrooke, Richard C., 185
Holloman, George V., and first automatic landing, 115
Holly Water Company, Dayton, 18
Holycross, Casey, **268**,
Home Field Advantage, 270
Homecoming celebration, 1909, see Wright Brothers, Homecoming Celebration
Honaker, Hiram, 155
Hoover Block, 275, **284**, 285
Hoover, Herbert Clark, 34, 40, 103
Hoovervilles, 40
Hope, Bob, **130**, 215, 328
Hot air balloon celebration, 279
Hotels, Dayton
    Biltmore Hotel, 40
    Miami Hotel, **39**
    Phillips House, 15, **16**, 33
Houghton, Junius H., 81, **108**
Hovatter, Den, **174**
Hovey, Bill, **261**,
Howe, John D., 158
Howe, R. M., **201**
Howells, William Dean, 234
Huber Heights, Ohio, 228, 250
Huber, Charles H., 228
Hudson-Fulton Celebration, 244
Huffman Dam, **53**, 149
    1926, **77**
    and location of Wright Field, 36, 105
    construction of, 3
Huffman Prairie Flying Field Interpretive Center, **189**, 286
    groundbreaking, 189
    official opening, 190
Huffman Prairie Flying Field, 259, 280, **285**, **286**
    and moth species, 189
    dedication of, 187
    Festival of Flight, 171
    listed on National Register of Historic Places, 155
    on National Millennium Trail, 190
    on regional bike path, 189
    preservation of, 176
    site in Dayton Aviation Heritage National Historical Park, 187
    Wright brothers' 1905 hangar, **189**
Huffman Prairie League, and the Arnold House Heritage Center, 222
Huffman Prairie Rededication Ceremony, opening remarks, 330-331
Huffman Prairie, 54, 270, 280; also see Huffman Prairie Flying Field
    and establishment of Army airfield, 33
    became National Historic Landmark, 187
    dedicated as Huffman Prairie Flying Field, 187
    flyers trained at, 168
    flying school, **3**, 24, also see Wright School of Aviation
    map of, **54**, 311
    part of FADR, 105
    Stinsons' flight training, 23
    transportation to, 3
    Wright brothers' flight experiments, 1, 17
Huffman, Torrence, 17, 19

Huffman, William, 19
Huffy Corporation, 256
Hull, William, 13, 26
Human Effectiveness Directorate, Air Force Research Laboratory, 247, **254, 282**
Human Systems Center, Brooks Air Force Base, 173
   reorganized as 311th Human Systems Wing, 173
Human Systems Division, Brooks Air Force Base, 332
Humphreys, Frederic E., on Aeronautical Board, **49**
Hunt Building Corporation, 213
Hunter, Richard, 235
Hureau, Ed, 276
Hurricane Florence, Florida, 149
Hussein, Saddam, 177, 182
Hutchins, Chuck, **261,**
Hydrogen balloons, 47, **48**
I Could Never Be So Lucky Again, 287
IAMS pet food company, 242
Iams, Paul, 242
Icarus, statue of in United States Air Force Museum, **274**
Ice-cube tray, **233**
Idaho, 259
Ignico, R. V., **89**
Illinois, 259
Impact aid, 250
Independent Research and Development (IR&D) program, 254
Indianapolis, Indiana
   Aviation Repair Depot (Speedway), 63
   considered for research and development center, 33
Industrial War Plans Section, Materiel Division, 77
Information Directorate, Rome Laboratory, 247
Innovative Science Solutions, 283
Integrated Weapon System Management, 173
Internal Revenue Service, Cincinnati Submission Processing Center, 324
International Air and Space Symposium and Exposition, 282, **283**
International Air Races, 1924, **95**-100
   advertising poster, **96**
   hangar of Wright Flyer, **95**
   Wilbur Wright Field, **34**, 35, 97
   Dayton Chamber of Commerce Trophy Race, 99
   John L. Mitchell Trophy Race, 100
   Liberty Engine Builders Trophy, 99
   On-To-Dayton race, 99
   Pulitzer Trophy Race, 100
International Civil Aviation Organization, 165
International Logistics Center, 204
   becomes Air Force Security Assistance Center, 203
Internet, 280
Internet-based instruction, 264
Interpretive Center, Huffman Prairie Flying Field, 286; also see Huffman Prairie Flying Field
Interstate (I)-75, 233
Interurban, Dayton-Springfield line, 1, 3, 33
Inventing Flight, 267, 271, 276, **277**, 279
Ion Propulsion, 281
Iowa, 259
Ironton, Ohio, 134
Irvin, Frank G., **201**
Irvin, Stanley, 180
Ischinger, Wolfgang, 185
Italian Military Mission, Pirelli, member of, 150
Italy, pre-World War I aviation, 49
Ivanov, Igor S., 185
IX AFSC, 313
Izetbegovic, Alija, 185
J.C. Penney, 233
Jackson, Isaiah, 235
Jacques Schneider Trophy, 96
James Gordon Bennett Trophy, see Bennett Trophy

James M. Cox Dayton International Airport, 241
James M. Cox Municipal Airport, 241
James River Corporation, 242
Jamestown Facility, 4950th Test Wing, 332, **333**
Jamestown, Ohio, 250
Japanese-Americans relocation, World War II, 42
Jarvis, Irby B. Jr.
   dates of Wright-Patterson command, 158
   during Xenia tornado response, 159
   greets prisoners of war, 154
   briefs Nixon on Xenia tornado damage, **162**
   dedication of Building 1245, 327
Jeffrey, John, **269**
Jews, 265
Jim Crow, 234
Jitney Transportation Company, Dayton, 22
Jobs created, indirect, Wright-Patterson Air Force Base, 249
John Bryan State Park Observatory, 332
John Bryan State Park, 332
John L. Mitchell Trophy Race, 97
   at 1924 Air Races, 100
John Pirelli Lodge #1633, Sons of Italy, 151
Johnson Airplane and Supply Company, 241
Johnson, D. A., **201**
Johnson, E. A. "Al", **241**
Johnson, Gerald R., 326
Johnson, Lyndon, 235
Johnstone, Ralph, **24**
Joint Group on Depot Maintenance, 323
Joint Logistics Systems Center, 172, 321
Jones, Butterscotch, **258**
Jones, Byron Q., 64
Jones, Pauline Neville, 185
Jones, Thomas Z., 217
Jose, Elmer H., 81, **108**
Joyce Cridland Company, 242, 310
JP-8A fuel, 181
Judge, Warren, 271
Junior Achievement, 267
Junkin, James Elwood, 85
Kane, Clarence P., 81
Karl Taylor Orchestra, 196
Kauffman Avenue, 121, **269**
Keenan, Lieutenant, **68**
Kelchner Environmental of Centerville, Ohio, 189
Kelchner, Dale V., 218, 328
Kelley, Father John (Jack), 265
Kellogg Field, Michigan, 131
Kellogg Sub-depot, Michigan, 135
Kelly Air Force Base, Texas
   closed San Antonio Air Logistics Center, 173
   Detachment 7, 15th Weather Squadron, 190
   in LOGAIR system, 148
   2851st Air Base Group, 152
Kelly Field, Texas, 59, 80
   Camp Kelly, Texas, and 42nd Aero Squadron, 192
   Air Corps Training Center, 153
   and Barling bomber, 70
   aviation school, 60
   Stinson as flight instructor, 23
Kelly, M. J., **144**
Kelly, Oakley G., 21
Kelsey, Benjamin, test pilot, 200
Kennedy, Stephen, 281
Kenney, F. P., **89**
Kenney, George C., 99, 327
Kenney, Michael, 180
Kent Elementary School, Columbus, Ohio, **267**
Kettering "Bug," see Kettering Aerial torpedo
Kettering Aerial torpedo ("Bug"), **32**

Kettering Business Park, 254, **255**, 256, 275
Kettering Laboratories, 260
Kettering Medical Center, cooperative agreement with Wright-Patterson medical center, 175
Kettering Memorial Hospital, 253, **254**
Kettering, Charles F., **7**, **32**
   and 1924 Air Races, 99
   and Dayton Wright Airplane Company, 32, 52
   and Delco, 22, 31, 32
   and Engineers Club, 37
   and General Motors Research Corporation, 7, 32
   and NCR, 7, 25
   automobile industry, 7
   birth of, 32
   Charles, saved Winters Bank from closure, 40
   developed anti-knock gas with Midgely, 32
   development of the "Bug," **32**
   enshrined in National Aviation Hall of Fame, 32
   Foundation, 210
   inspection of land for airfield, 33
   member of "Barn Gang," 32
   owner of land that became McCook Field, 72
   sells business interests to General Motors, 32
Kettering, Eugene W., chairman of Museum Foundation, 210
Kettering, Ohio, 228, 239, 250, **255**, 257
   50th anniversary, **258**
Kettering, Virginia, **210**
Kids' Art Takes Flight, **289**
Kill Devil Hill, North Carolina, **2**, 271
Kimmel, Dorothy E., 327
King, Kevin, 180
King, Paul, Dr., **268**,
Kirk, Maxwell, 59
Kirtland Air Force Base, New Mexico, 247, 260
Kitty Hawk Police, 271
Kitty Hawk, North Carolina, 1, 245, 270, 271, 274, 285
Kittyhawk Center, Community Shopping Center Complex, 214
   memorializations, 326
   previously Wood City, 214; also see Buildings, Kittyhawk Center; Wood City
Kittyhawk Flight, see 2750th Air Base Wing, Kittyhawk Flight
Kneisly, John, 3
Knerr, Hugh J., **88**, 95
Kochensparger, Danny, 180
Koenigsberg, Harry and Eleanor, 265
Kollar, Stephen F., Dates of Wright-Patterson command, 158
Korean War Commemorative Community, 187
Korean War Veterans Memorial Association, 187
Korean War, 142, 143, 144
   50th Anniversary Commemoration, 187
   and 17th Bombardment Wing, 192
Kosovo, 181, 249
Krug, J., **201**
Ku Klux Klan, 40, 235
Kuhns, Ezra M., 54
Kuwait, invasion of, 177
La Chapelle, Duval, **24**
Lackland Air Force Base, Texas, 186
LaGuardia, Fiorello, 273
Lahm, Frank P., **50**
   attends World War I pilot reunion, 155
   dedication of street in honor, 326
   first Army pilot, 48
   on Aeronautical Board, **49**
   recognition ceremony at Wright-Patterson Air Force Base, 153
Laird, Melvin R., 154
Lamar, William E., 280
Lambert, David, 180
Landisville, Pennsylvania, 281, **282**
Lane, Amanda Wright, 17, 276, **286**, 287

Lane, G. V., **201**
Langford, Frances, at Patterson Field, **130**
Langin Field, Moundsville, West Virginia, 100
Langley Field, Virginia, 84
   and Barling bomber, 70
   and World Cruiser, 93-94
   move experimental activities to McCook Field, 74
   stop on Model Airway, 100
Langley Memorial Laboratory, location of first wind tunnel, 35
Laughlin, Edward, **88**, 95, 98
Law Enforcement Training Course, Lackland Air Force Base, 186
Lawless, Trace, 180
Le Mans, France, 2
Lear, Inc., 310
Leger, Eugene, 327
LEGO League Championship, 267
LEGO Mindstorms Robotic Kits, 267
Leland Electric Company, 310
LeMay, Curtis E., 202
Lexington Blue Grass Army Depot, Kentucky, 153
LexisNexis, 242
Liberty Engine Builders Trophy, 1924 Air Races, 99
Library of Congress, 324
   Motion Picture Conservation Center, 324
Light, Grant, 180
Lincoln Elementary School, Dayton, 264
Lincoln, Abraham, 10, 18
Lindbergh, Charles A., 263
   America Firster, World War II, 44
   flight across Atlantic, 96
   in National Aviation Hall of Fame, 210, 278
   visit McCook Field, 40
   visit to Fairfield Air Depot, **101**
Litecast™, 259
Liteflex Limited Liability Company (LLC), 259
Liteflex™, 259
Litening Target Pods, 183
Liu, Sam, **261**
Living room dialogues, 265
Lockbourne Army Air Base, Ohio, 131, 232
Lockbourne Sub-depot, Ohio, 135
Locker, James D., 326
Lockheed Martin, 258
Lockwood, Jones and Beals, Inc., 252
LOGAIR, see Logistics Airlift
Logan County, Ohio, 101
Logistics Airlift (LOGAIR) System, 148, **175**
Logistics Audit Region, 320
Logistics Communications Division, 172, 321
Logistics Information Systems Division, Headquarters, 321
Logistics Management Systems Center, 321
Logistics Operations Center, 321
Lombardi, Timothy, 180
London Daily Mail Prize, 96
Long-EZ, **283**
Lotz, Michael, II, 180
Love Field, Texas, 60
Low Altitude Navigation and Targeting Infrared system for Night (LANTIRN), 180
Lower Dayton View, 228, 237
   architecture of, 36
   settlement of, 14
   Wattendorf purchases property, 41
Lowes, 228
Luedtke, Louis A., **256**
Lueker, Rano E., 158
Luftwaffe, bombing of Great Britain, World War II, 44
Luneke, Douglas, 180
Lute, James R., 326
Luttrell, Zach, **272**,

Lykins, Joseph, 180
Lyles, Lester L., **235**, 272, 281
Lynn, James T., 161
M-1 tank, 252
M-114 Cluster bombs found on Wright-Patterson Air Force Base, 176
MacArthur, Douglas, 145
Machine Guns
    Lewis, 68
    Marlin, 68
    Nelson, test of, 106
    synchronization, 67
Mackay Trophy
    awarded to Benjamin Foulois, 223
    for first automatic landing, 115
Macready, John A., 21, 99, 200
Mad River and Lake Erie Railroad, 5, 56
Mad River Township Board of Education, 212
Mad River Township School District, 212
Mad River, site of Wilbur Wright Field, 54
Madison Sub-depot, Wisconsin, 135
MAGNA, 260
Magnetic laboratory, UDRI, **261**
Magnetic resonance imaging (MRI) machine, **254**
Magnetoplasmadynamics, 281
Magnificent Flyers, Those, 276
Mahle, Bernie, **261**
Mahnken, Corporal John, 229
Main Street Bridge, in flood of 1913, 27
Mainella, Fran, 190
Maintenance Division, Air Service Command
    Joseph Morris assistant chief, 141
    Technical Data Section, newspaper, *Skyliner*, 169
Major Shared Resource Center, **175**, 177, 246
    Simulation and Analysis Facility, 246
Malfunction junction, 233
Malone, Anthony, 180
Manlove, Clifford T., 146
Manuel, Timothy, 180
Manufacturing magazines for 20-mm anti-aircraft guns, **228**
Mara, Bill, 23
March Field, California, 113
Marionist Order, 243
Mark, Hans M., 202
Markow, Tanya, 245
Marl Road, 286
Marlin-Rockwell Company, New Haven, Connecticut, 68
Marras, E., 148-149
Marriott, William, 180
Mars, James A., 79-80, **81**
Martin Marietta Composites (MMC), 259
Martin Marietta Materials, Inc., **252**, 259
Martin, Dave, 180
Martin, Frederick L, **93**, 94
Martin, Marion, **196**
Martin, Scott, 244
Maryland, 259
Masonic Temple, Dayton, 36, **38**, **230**
Massachusetts Institute of Technology, 106, 107
Massachusetts, 270
Massive Ordnance Blast Munitions, 183
Master of McCook, 287
Materials and Manufacturing Technology Directorate, 247, 254, 262, **266**
Materials Branch, Engineering Section, Materiel Division, 77
Materials Laboratory, 218, 221, 255
Materiel Command, 253, 313
    directs Provisional Air Corps Maintenance Command, 125
    established Flight Test Training Unit at Patterson Field, 200
    establishment of, 125
    headquarters moves from Washington to Wright Field, 125

redesignated Army Air Forces Technical Service Command, 126
Materiel Division, Air Corps
    aircraft production for World War II, 44
    assignment of Patterson Field, 107
    at McCook Field, 76, 77
    at Wright Field, 40, 77
    becomes Materiel Command, 125
    creation of, 102
    first automatic landing, 115
    Foulois chief of, 50
    guidance of Maintenance Command, 125
    jurisdiction for Fairfield Air Depot, 125
    loses maintenance and logistics functions, 125
    move to Wright Field from McCook, 76, 89, 103
    oversight of Field Service Section, 89
    six sections of, 77
    Technical Data Branch, in Building 12, 220
    three pillars of, 145
    wind tunnels, 41
Materiel Systems Center, 247, 321
Math Resource program, 267
Mathis, Lieutenant, **68**
Matter of Balance, A, 280
Maxwell Air Force Base, Alabama, 244, 273
    and Air University, 199
    responsibility for museum, 209
Maxwell Field, Alabama, sub-depot in, 125
McCabe, Dr. Mickey, 260
McCall (Printing) Corporation, 42, 243
McChord Air Force Base, Washington, in LOGAIR system, 148
McClellan Air Force Base
    closed Sacramento Air Logistics Center, 173
    Detachment 8, 15th Weather Squadron, 190
McClellan Field, California, 86
McClellan, Major, **201**
McClernon, Glen J., 158, 164
McConnell, John P., 221
McCook Field, **30**, **69**, **72**-73
    70th Anniversary celebration, 170-171
    80th Anniversary celebration, 187
    Air Service Engineering School, 199
    Air Service School of Application, 199, 262
    air traffic, 40
    aircraft crashes, 35
    Aircraft Radio Technical Section, 202
    Airplane Engineering Department, 69
        Foreign Data Section, 205
    Airplane Engineering Division, 86
    Airplane Technical Section, 202
    and Barling bomber, 70
    and flight testing, 200
    and Round-the-World flight, 92, 94
    Armament Technical Section, 202
    aviation records, 91
    beginnings of museum, 209
    bread supply, 65
    closure, 76
    deficiencies, 35, 69
    development of aircraft, 72-73
    dynamometer, 123
    Engineering Division, 35, 74
    Equipment Technical Section, 202
    establishment of, 31, 35, 47, 140, 202
    experimental work, 145
    Flight Test Section, 73, 76, 200
    historical marker, 187
    inspected by General Mitchell, 100
    Jimmy Doolittle at, 287
    Lighter-Than-Air Technical Section, 202
    Materials Technical Section, 202

Materiel Division, 76, 77
military pay, 63
Model Airway System, 99, 100, 102
motto, 35, 69
move to Wright Field, 36, 69, **72**, 123
naming of, 33
night-landing light developed, **100**
parachute development, 86
postwar changes
Power Plant Technical Section, 202
Propeller Unit, **73**
proximity to Weaver Aircraft Company, 85
Radio Dog, 34-35
request hangar space at Wilbur Wright Field, 69
research facilities, **30**
tests World Cruiser, 93
Thurman Bane, 199
value of, 243, 246
McCuddy, Bill, **283**,
McDonald Charles C., 187
McDonnell Douglas, 258
McGee, James H., Mayor, **235**,
McGuire Air Force Base, New Jersey, 179
  15th Weather Squadron, 190
McIntosh, L. W., 64
McIntyre, J. J., 75
McKee, J. Robert, 179, 180
McKinley, William, 12
McLin, C. J., Jr., 235
McLin, Rhine, 235, 245, **286**, 287
McMullen, Kevin, 180
McNeal, Don, **112**
McNeil, Joseph, 180
McPeak, Merrill A., 146, 158, 173
McPhersontown Historic District, 237
McPhersontown, Dayton, 41
McPike, George, V., 89
Mead Technology Laboratories, 274
Mead, 242
Mead, Dudley, 245
Mead, George, 52
MeadWestvaco, 242
Mechanics Institute, Dayton, 15
Mechenbier, Edward, **153**, **195**
Media empires, 239
Medical Center Wright-Patterson Air Force Base, see Wright-Patterson Air Force Base, Medical Center
Medical Reserve Corps, 59
Medway, Ohio, 263
Memorial Road, 286
Menard, Dave, **208**
Menoher, Charles T., 82
Merrell, Jack G., **210**
Merrill, William G., 59
Meske, Sergeant, **154**
Met Life, 279
Metcalf, Charles D., 276
Meteorological School, Fort Monmouth, New Jersey, move to Patterson Field, 113; also see Weather School, Air Corps
Meteorology, developments in, 113
MetroParks, of Dayton and Montgomery County, **233**
Metzenbaum, Howard, 285
Metzger, William E., 326
Metzler, Eric, 189
Meyers, Tony, 188
Miami Conservancy District, 269
  Board of Directors, 170
  construction of Huffman Dam, 53
  creation of, 7, 28, 29, 31
  decision to move Osborn, 56

Dorothy Jean steamboat, **29**
  headquarters of, 36
  jurisdiction over land chosen for Wilbur Wright Field, 54
  land purchased by Patterson Field, 118
  map showing flood control work, **29**
  negotiations with Signal Corps, 79
  photographic archive, 28
  plan for Wright Memorial, 168
  purchased land of Foulois House, 223
  transfer of land parcels to Wright-Patterson Air Force Base, 168
Miami County, Ohio, 246, 250
Miami Hotel, Dayton, **39**
Miami University (Ohio), 263
Miami University (Oxford), 269
Miami University, 243, 269
Miami Valley Astronomical Society, 332
Miami Valley Interactive Distance Learning Consortium, 267
Miami Valley Military Affairs Association, 222
Miami Valley Pet Therapy Association (mvPTA), 288
Miami Valley READS, 267
Miami Valley Research Park, 243, 255, 264
Miami Valley Urban Search and Rescue, 178; also see Ohio Task Force One
Miami Valley
  Flood of 1913, see Dayton, 1913 Flood
  flooding, 28, also see Dayton, 1913 Flood
  mention, 263, 265, 269, 274, 278, 284
Miami-Erie Canal, 9, **13**, 14-15, 234
Miami-Jacobs College, Dayton, 265
Miamisburg, Ohio, 161, 231, 250
Michelin Cup, 96
Michigan, University of, 265
Mid-Day MBA Program, 264
Middletown Air Depot, 3rd Transport Squadron, 107
Middletown Air Intermediate Depot, Pennsylvania, 86
Middletown, Pennsylvania, 273
Midgely, Tom, developed anti-knock gas with Kettering, 32
Mignogno, Craig, 180
Mildrum, Herb, **261**,
Military Affiliate Radio System, 149
  assistance after Xenia tornado, 160
Military Aircraft Storage and Disposition Center, Davis-Monthan Air Force Base, 163
Military Airlift Command, 163, 172
  and 15th Weather Squadron, 190
Military Amateur Radio System, 149
Military Flight Service Station, 280
Military techology in commercial aviation, 251
Miller, Ivonette Wright, **269**, 274
  attends transfer of Wright Memorial to Wright-Patterson Air Force Base, 170
  loaned Orville's scissors for museum ribbon cutting, 167
  at First Flight Ceremony, **170**
Miller, Joe, **283**
Miller, Jon, 258
Miller, Ken, 276
Miller, Lester T., **197**
Millett, Dr. John W., 269
Mills, Gordon, **261**,
Mills, H. H., 100
Milosevic, Slobodan, 185
Minardi, Mark, **271**
Minnesota Air National Guard, 333
Minnesota, 265
Mirabal, Eva, 137
Mississippi River, 243
Mississippi, 270
Mistretta, John, 252
Mitchel Field, Long Island, New York, 89

2nd Bombardment Group (medium), assigned to Patterson Field, 131
  and Barling bomber, 70
  endurance flight from to Alaska, 91
  First Provisional Air Division headquarters, 103
  squadrons assigned, 131
  stop on Model Airway, 100
Mitchell, William "Billy", 70, **91**
  awarded Special Congressional Medal of Honor, 91
  concept of Round-the-World flight, 91
  dedication of Building 556, 327
  on Model Airway, 100
  with World Cruisers crew, **94**
Mobile Air Materiel Area, Alabama, 151
Model Airway System, **99**, 100, 102
Modlich, Matthew, 180
Moen, Rusty, 279
Mojave Airport, California, 283
Mollen, Rich, 256
Monahan, George L., 220, 327
Monsanto Chemical Company, 228
Montgomery County Airport, 196
Montgomery County Fairgrounds, 6, 235
Montgomery County Historical Society, 187
Montgomery County, Ohio, 246, 250
Montgomery, Alabama, sub-depot in, 125
Monument Avenue, 12, 19, 233, **275**
Moore, Dr. Frank, 255
Moorman, Beth, **289**,
Moorman, Eileen, 265
Moraine City, Ohio, 52
Moraine Farm, **31**, 33
Moraine, Ohio, 242, 250
Moreau, John, **261**,
Morgan Engineering Company, 28
Morgan, Arthur E., **28**, 29, 57
Morgan, Ernest, on moving Osborn, 57
Morris, Greg, 180
Morris, Joseph T., 81, **136, 140, 141**
  as "father of Wright-Patterson Air Force Base", 141
  career at Wright-Patterson Air Force Base, 108, 136, 140, 158
  career of, 141
  dedication of Building 10, 326
  directed racial integration of Wright-Patterson Air Force Base, 141
  inspection of Police Department, **141**
Moseley, Corliss, 97
Mound builders, 243
Mount St. John's Bergamo, 265
Mueller, Eric, 281
Muhl, Michael, 180
Mullins, Matt, **268, 283**
Munitions Directorate, 247
Munitions Standards Board, 33, 52
Munroe, Sarah, **272**,
Muroc, California, 200
Murphy, John, **334**
Muslim-Croat Federation, 185
Mutual benefits of research and development, 251
MV Communities, 213
Myers, D. M., 99
Myers, Mike, **281**,
Naismith Memorial Basketball Hall of Fame, 229
Nanocomposite magnets, **261**,
Nanomaterials, Wright Brothers Institute Endowed Chair in, **261**,
National Advisory Committee for Aeronautics, 30, 35
National Aerobatic Championship, 196
National Aeronautic Association, 95, 99, 223
National Aeronautical Systems and Technology Conference (NASTC), 251

National Aeronautics and Space Administration (NASA), 263, 270, 280
  and ARIA program, 201
  Glenn Research Center, 280
National Aerospace and Electronics Conference (NAECON), 251
National Air and Space Intelligence Center, **206**, 249, **250**, 318, 323; also see National Air Intelligence Center
  aircraft display, 207
  as tenant organization, 190
  facilities, 207
  Foreign Materiel Exploitation Facility, 207, 327
  history of, 205-207
  Photo Laboratory, 207
  Special Operations Intelligence Facility, 207
National Air and Space Museum (NASM), 95, 263
National Air Intelligence Center, 318; also see National Air and Space Intelligence Center
  Foreign Material Exploitation Facility, 327
  formerly Foreign Technology Division, 175
  role in Open Skies Treaty, 175
  role in Operations Enduring and Iraqi Freedom, 183
National Air Meet, 97
National Air Museum, Smithsonian Institution, 209
National Air Olympics, 196, 241
National Air Races, 97
National Air Racing Group, 97
National Airborne Operations Center, at Wright-Patterson Air Force Base, **174**, 175, 323
National Aircraft Show, 196, 241
  1953, program, **197**
National Arbor Day Foundation, 155
National Aviation Day, 284
National Aviation Hall of Fame, 274, 278, 324
  Brukner enshrined, 85
  Earhart enshrined, 101
  Kettering enshrined, 32
  Learning Center, 278
  Lindbergh enshrined, 101
  opened in United States Air Force Museum, 210
National Capital Region Operating Base, Detachment 430, Dayton Element, 323
National Cash Register (NCR) Company, **8**, 20, 227, **228, 239**, 310; also see NCR Corporation
  and Charles Kettering, 32
  and Edward Deeds, 31, 40
  and the Great Depression, 7
  blacks admitted to apprentice program, 235
  commercial success, 24
  Dayton's first microeletronics research laboratory, 239
  founded by John Patterson, 7
  in World War II, **43**, 44
  purchases California-based computer company, 239
  role in Bombe project, 43
  Schoolhouse, 9, 25
  shut down during air races, 99
  Sugar Camp, 43
  welfare of employees, 8
National Center for Composite Systems Technology (NCCST), 256
National Championship Air Races, Reno, Nevada, 97
National Command Authorities, 248
National Composite Center, 254, **255-256, 275**, 280
  and composite airframe parts, 258-259
  and composite automotive parts, 258-259
  and Wright glider display, **275**
National Conference for Community and Justice (NCCJ), 253
National Defense Act of June 3, 1916, 49, 51
National Disaster Medical System, 191
National Guard, participation in 1931 maneuvers, 103
National Historical Landmark Plaque, 287

National Home for Disabled Volunteer Soldiers, 10, also see Dayton Soldiers' Home
National Homes Construction Corporation, Inc., 213
National Housing Agency, 121
National Institute of Standards and Technology, 256
National Millennium Trail, Huffman Prairie Flying Field on, 190
National Museum of American History, 263
National Parachute Jumping Contest, 196
National Park Service
    and Dayton Aviation Heritage National Historical Park, **285**, 286
    as tenant on Wright-Patterson, 324
    at dedication of Huffman Prairie Flying Field, 187
    plan for Wright Memorial, 168
National Professional Basketball Tournament, 229
National Register of Historic Places, 155
National Road (U.S. Highway 40), 241
National Science Foundation, 268
National Security Act of 1947, 187
National Service Invitational Basketball Tournament, 229
National Teachers Training Institute, 267
National Workshop on Chrisitan-Jewish Relations, 265
Natural History Museum, Los Angeles, California, 95
Nature Conservancy, Ohio Chapter, and Dayton Aviation Heritage National Historical Park, 189
Naval Aircraft Factory, 35
Naval Research Laboratory, established, 35
Naval Reserve, U.S., 229, 324
    Naval Contingency Engineering Management Cell, Europe East, 324
    Naval Health Research Center, Toxicology Detachment, 324
    Naval Test Squadron 0195, 324
Navigation Board, Dayton, establishment of, 14
Navstar Global Positioning System, 199
Navy, U.S., 324
    Bombe program, 43
    competition with U.S. Army, 91
    participation in Armed Forces Day, 140
NCR Corporation, **239**, 242; also see National Cash Register Company
    support during Vietnam War, 153
Neil Armstrong Museum, 263
Nelson, Byron, 217
Nelson, Erik H., **91**, 93, 94
Nesselbush, L. K., **201**
New archeology, 245
New Carlisle, Ohio, 161, 250
New York Celtics, 229
New York, 259, 265
Newark Air Force Base, closure, 173
Newcom, George, 12, **13**, 17
Newspapers, Dayton
    Dayton Daily News, 7
    Dayton *Herald*, 2, 4
    Dayton *Journal*, 1, 2, 15, 18
    Empire, 18
    The Evening News, 3
    The *Journal Herald*,, 42
Newsweek, 243
Nielsen, Paul, 281
Niergarth, Omar, **89**
Nightingale House, Wright-Patterson Air Force Base, 215
Ninth Virginia Regiment, 167
Nixon, Richard M.
    visits Xenia, 161, **162**
    dedication of United States Air Force Museum, 210
    hosted dinner for prisoners of war, 154
Noonan, Fred, 101
Norfolk, Virginia, intended as center of aeronautical research and development, 35
Normandy Elementary, **289**
Norrod, David, 180
North Atlantic Treaty Organization, 175, 181, 247

North Carolina Highway Patrol, 271
North Carolina State University, 229
North Carolina, 259
North Dayton, settlement of, 33
North Dixie Highway, 241
North Field, Triangle Park, potential for research and development center, 33
North Island, San Diego, California
    aviation school, 48, **50, 51**
    early Army depot, 33
Northrop Grumman, 258
Northrop, John, 93
Northwest Airlines Incorporated, U.S. Government Modification Center, 310
Norton Air Force Base, California, in LOGAIR system, 148
Norton Field, Ohio, 100
Norton, Gail, 190
Norwood, Ohio, 134
Nuclear Engineering Center, 319
Nuclear Engineering Test Facility, 218, **203**
Nunn-Lugar Exchange Act, 176
Nutt, Clifford C., 81, 92
Oakwood, Ohio, 25, 228, 250
Ober, Stephen, 180
Oberlin College, 5
Odom, Bill, 97
Oelman, Robert S., 170, 269
Office of Chief Signal Officer, announcement of new aviation schools, 55
Office of Civilian Defense, 42
Office of Naval Research, 232
Office of Research and Technology Applications, 255
Office of Special Investigations
    1st Field Investigations Region Operating Location-A, 324
    Detachment 101, 324
    Detachment 702, 324
    Region 1, 324
Office of the Chief of the Air Service, 88
Office of the Chief Signal Officer, 47
Office of the Department of Quartermaster, Central Department, Chicago, Illinois, 63
Officers Reserve Corps, U.S. Army, 106
Officers' Club, Wright-Patterson Air Force Base, 269
Ogden Air Logistics Center, Utah, 247
Ogden, Henry H., **93**, 94
Ogg, Jon, 274
Ohain, Dr. Hans von, 260
Ohio Advanced Technology Center, 255
Ohio Aerospace Institute, 255
Ohio American Revolution Bicentennial Advisory Commission, 263
Ohio Bicentennial Commission, 280, 287
Ohio Bicentennial, **249**, 280, 287
Ohio Board of Regents, 268
Ohio Department of Development, 255
Ohio Department of Education, 212
Ohio Department of Natural Resources, 332
Ohio Division of Forestry and Reclamation, 155
Ohio Environmental Protection Agency, 166, 176
Ohio Historic Preservation Office, work with 2750th Air Base Wing, 176
Ohio Historical Center, Columbus, Ohio, 263
Ohio Historical Marker, 280
Ohio Historical Society (OHS), 263, 287
Ohio Lepidopterist Society, 189
Ohio Memory Project, 280
Ohio National Guard, 162nd Fighter Squadron, 139
Ohio National Guard, 59, 241
    assistance after Xenia tornado, 160
    called in for flood of 1913, 27
    participation in Armed Forces Day, 140

Ohio National Landmark, 286
Ohio Region Bicentennial Boy Scout Jamboree, 167
Ohio Space Grant Consortium, 267
Ohio State Archeological and Historical Society, 243
Ohio State Big Ten Championship team, 229
Ohio State Board of Vocational Education, 134
Ohio State University (Columbus), 243, 262, 263, 265, 269
   location of pre-flight school, 53
   logistics education program with AFIT, 202
   responsibility for wind tunnels, 175
Ohio Task Force 1 (Miami Valley Urban Search and Rescue), **178, 179,** 180, 324
   list of members, 180
   responds to World Trade Center, 179
   Saber Shield 00-2 exercise, 178
   Strongwind 99-1 exercise, 178
Ohio Technology Transfer Organization, 255
Ohio University, 263
Ohio, celebration of bicentennial, 280
Ohio, first all-composite bridge, **252,**
Ohman, Klas, 245
Oklahoma City Air Logistics Center, Oklahoma, 247
Old North Dayton Historic District, 237
Old River Park, **239**
Oliver, Prince A., 64, 81
Olmsted Air Force Base, Pennsylvania, 148
Olmsted Brothers, 168
Online scrapbook, 280
On-to-Dayton Race, 1924 Air Races, 99
Open Skies Treaty, 173, 175
   On Site Inspection Agency, 175
Operation Allied Force, 181-182, 249
Operation Castle, **261**
Operation Desert Fox, 181
Operation Desert Shield, 177, 194
Operation Desert Storm, 172
   beginning, 177
   equipment handling at Wright-Patterson Air Force Base, **177**
   new technologies
      APG-68 Attack Radar, 181
      APG-70 Attack Radar, 181
      Forward Looking Advanced Multi-mode Radar, 181
      JP-8A fuel, 181
      Low Altitude Navigation and Targeting Infrared system for Night (LANTIRN), 180
      Protective Integrated Hood/Masks, 181
   role of 445th Airlift Wing, 194
   soldiers await deployment, **176**
   veterans' use of Veterans Affairs Medical Centers, 11
Operation Enduring Freedom, 175, 182
   new technologies
      Battlefield Air Operations Kit, 182
      laser eye protection, 182
      Massive Ordnance Blast Munitions, 183
      panoramic night vision goggles, 182
   role of 445th Airlift Wing, 194
Operation Green Rush, 155
Operation Homecoming, and Wright-Patterson Air Force Base, **154**
   prisoners of war, **153**
   role of 445th Airlift Wing, 195, 198
   role of Wright-Patterson Air Force Base, 153
Operation Iraqi Freedom, 175
   beginning of, 182
   new technologies
      "bunker buster" munitions, 183
      Advanced Strategic Tactical Expendable Flares, 183
      C-40B Communications Command aircraft, 183
      dust eater valves, 183
      Litening Target Pods, 183
      portable aircraft shelters, 183

RQ-4 Global Hawk, 183
   role of 445th Airlift Wing, 194
   role of Aeronautical Systems Center, 183
Operation Just Cause, 194
Operation New Life, 194
Operation Noble Eagle, 175, 182, 183
Operation Paperclip, 206; also see Project Lusty
Operation Provide Hope, 194
Operation Provide Promise, 249
Operation Sun Run, 148
Operation Uphold Democracy, role of 47th Airlift Flight, 175-176
Operation Vittles, see Berlin Airlift
Oregon Historic District, 237
Oregon, 259
Orellana, William B., **158**
Orientation Group, U.S. Air Force, **207, 208,** 316
   Armed Forces Bicentennial Caravan Project, 208
   facilities in Wood City, 208
   move because of fire, 214
   move to Defense Electronics Supply Center, Gentile Air Force Station, 208
   as tenant organization, 190
Osborn High School, 134
Osborn Removal Company, 56
Osborn, E. F., 56
Osborn, Ohio, **56, 57,** 58
   1923, **76**
   and Big Four Railroad Company, 79
   early history, 165
   nearest railroad to Wilbur Wright Field, 58
Oscars of Aviation, 278
Oshkosh, Wisconsin, 283
Overland Park, Kansas, 260
Owens Corning Company, 256
P4, see Programmable Powdered Preform Process, 256
P4A machine, **256,**
Pacer Energy Fuel-Conservation Program, 159
Pacific Proving Grounds, **261,**
Page Manor Elementary School, 212
Page Manor, see Wright-Patterson Air Force Base, housing
Page, Dwayne, **261**
Page, Edwin R., **89,** 211, 327
Paleoindians, 243
Palmer, Arnold, 214, 217
Palumbo, Michael, 180
Panama Canal Zone, 89, 90
Panic of 1893, 40
Parachute Museum, 284
Parachutes, testing at Fairfield Air Depot, 103
Park Field, Memphis, Tennessee, 53
Parks Masonry, 189
Parsons, S. P., **201**
Pasadena, California, 244
Pathway sculpture, Dayton Art Institute, **275,**
Patrick Air Force Base, Florida, 198
Patrick, Mason M., **91,** 94, 100
   approved reconfiguration of World Cruiser, 93
   at International Air Races, 1924, **98**
   promotes Round-the-World flight, 91
Patterson Field Army Air Base, commander of, 126
Patterson Field Postings, newspaper, 169
Patterson Field, 228
   "Cheers from the Camps" broadcast, **130**
   1st Provisional Transport Squadron, 130
   1st Transport Group, 130
   2nd Transport Group, 130
   3rd Transport Group, 130
   4th Transport Group, 130
   5th Transport Group, 130
   10th Observation Group, 130

10th Transport Group, 130
11th Transport Squadron, 131
13th Transport Squadron, 131
19th Bombardment Squadron (medium), 131
22nd Bombardment Group (Medium), 131
33rd Bombardment Squadron (medium), 131
60th Transport Group, 131
61st Transport Group, 131
63rd Transport Group, 130
316th Transport Group, 130, 131
4100th Army Air Force Base Unit, 126, 140
4265th AAF Base Unit Separation Center, 135
1933 Air Corps Anti-Aircraft Exercises, 112
1940s construction, 116-117
1942 personnel strength, 121aerial photos, **106**, **115**, **136**
Aero Repair Training School, **131**
Air Corps Weather School, 112
Air Service Command, headquarters complex, 118, **120**, **127**
air traffic controllers, **119**
Air Transport Terminal, **117**
and 1934 Alaskan flight, 113
and 96th Service Group, 133
and Flight Test Training Unit, 200
assigned squadrons, 130
assigned to Air Service Command, 107
assigned to Materiel Division, 107
autogiro school, 112, 113
Building 206, construction, 117
CCC camp, 111
Civilian Club, Building 274, Area A, **133**
civilian employees, 122, 133
concrete runways, 44
construction of housing in World War II, 119, 121
construction of temporary barracks, **120**
Control Tower, **118**
date established, 140
depot supply operation prior to World War II, 115
developments in meteorology, 113
effects of Depression on, 41
Estabrook commander of, 114
expansion, map, 311
Experimental Test Pilot School, 200
Fairfield Air Depot Operations Hotel, 117
first automatic landing, 115
flight testing, 200
formed air depot groups for training, 131
function of, 107
German prisoners of war, murals in Building 280, **126**
headquarters of Maintenance Command
headquarters of Provisional Air Corps Maintenance Command, 134
Housing, see Wright-Patterson Air Force Base, housing
list of commanders, 108
major logistics center, 135
merger with Wright Field, 108, 136, 139
morale building during war, 130
naming of, 36, 105
newspaper, *Patterson Field Postings*, 169
origins of Band of Flight, 208
payroll in World War II, 122
personnel strength, 1931, 107
personnel strength, 1940, 121
Post Hospital, opened, 204
postwar changes, 135
postwar separation center, 135
pre-service training programs, 134
runways, 118, **119**
Skeel Avenue, extension of, 119
south end becomes Area A, 139
Taxiway No. 12, see runways
Taxiway No. 8, see runways

Test Pilot School, Class of 1949, **201**
training of Army Air Forces Nurses Training Detachment No. 6, 133
training of Chinese-American service squadrons, 133
training of enlisted mechanics, **134**
Transient Camp, **110**
Very Heavy Bomber Runway constructed, 118, **136**
visit of Bob Hope and Frances Langford, 130
WAC Supply School, 131
Weather School, 113
women employees, 121, **122**, **124**
Women's Army Corps officers at WAC Supply School, **131**
workforce demographics, 121
World War II land acquisition, 118
Patterson Knoll, Woodland Cemetery, 17
Patterson, Frank Jefferson, 105
Patterson, Frank Stuart, **107**
  assigned to Wilbur Wright Field, 106
  birth of, 105
  death of, 35, 36, 106
  dedication of swimming pool in honor, 327
  funeral procession, **107**
  grave of, 17
  joining Signal Corps, 106
  naming of Patterson Field, 105
  schooling, 106
  with dog, **61**
  with fellow pilots, **107**
Patterson, Frederick B., **35**, **36**, 241
  and Dayton Air Service Committee, 36
  attended banquet for 1924 Air Races, 99
  cousin of Frank Stuart Patterson, 107
  president of National Aeronautic Association, 95
Patterson, Jefferson, 107
Patterson, John H., **7**, **35**, 228, 239, 287
  and 1913 flood, 9
  and flood relief efforts, 20, 26, 27
  and NCR, 7, 25, 31
  and the Dayton Plan, 8
  as a manager, 9
  bid for Air Service's research and development work in Dayton, 36, 243
  brother of Frank Jefferson Patterson, 105
  civic improvements, 20
  death of, 36
  education, 9
  indictment for restraint of trade, 25
  job with Miami-Erie Canal, 9
  prediction of Panic of 1893, 40
  Progressive politics, 8
  purchase of cash register business, 9
  saving family property, 12
  welfare of employees, 8, 9
Patterson, Mary, 107
Patterson, Robert, 12, 106
Patton, Eliza, 223
Patton, F. S., 223
Pau, France, 2
Paul Laurence Dunbar State Memorial, site in Dayton Aviation Heritage National Historical Park, 189
Paul, Charles H., 95
Paul, Richard R., 177
Peace Walk, Area A, dedicated, 184, 327, **329**
Peace Wall, 184
Pearl Harbor, Hawaii, 45
Pearson Road, 327
Pearson, Alexander, **98**, 100, 327
Peggy Sue, **288**
Pei, I. M. and Partners, 233
Peirce, Joseph, 12, 18
Pelli, Cesar, 233

Pena, Tony, 217
Pennsylvania, 259
Pentagon, Washington, D.C., terrorist attacks, 178, 182
People for People Festival, Wright-Patterson Air Force Base, 167
Perkins, Scott, 180
Persian Gulf War, 172, also see Operation Desert Storm
　　beginning, 177
　　role of 2750th Air Base Wing, 174
　　veterans' use of Veterans Affairs Medical Centers, 11
Pestian, Fred, **261**
Pet Therapy Program, **288**
Peters, F. Whitten, **329**
Peterson, Leonard R., 158
Phillips Field, 99
Phillips House Hotel, Dayton, 15, **16**, 33
Phillips Laboratory, 172
Phillips, George, 19
Phillips, Horatio G., 15
Physician Leadership Development Program, 253
Pickering, David, 180
Pine Estates Housing Complex, see Wright-Patterson Air Force Base, housing
Piqua, Ohio, 250, 257, 275
Pirelli, Giovanni, 149, **150**
Pirelli, John, see Pirelli, Giovanni
Pitsenbarger, Irene, **329**
Pitsenbarger, William F., **329**
Pitsenbarger, William H., 214, 326
Plastic Metal, Inc., 310
Pleiman, James E., 326
Poe, Bryce, II, at First Flight Ceremony on 75th Anniversary of powered flight, 170
Pop-top can, **233**, 237
Portsmouth, Ohio, 134
Posse Comitatus Act, 191
Post Field, Fort Sill, Oklahoma, 106, 107
Post Script, newspaper, 169
Postal Service, U.S., **277**, 279, 324
　　Air Corps assumed responsibility for mail, 112-113
POW/MIA Memorial, dedication of, 328
Power Plant Branch, Engineering Section, Materiel Division, 77
Powers, J. B., **88**, 98
Prairie Trace Golf Club, Wright-Patterson Air Force Base, 217
Prairies at Wright Field, see Wright-Patterson Air Force Base, housing
Prater, Roy D., 326
Precision Measurement Equipment Laboratory (PMEL), 163, 175
Presidential campaign of 1920, 34
Pretzinger, Albert and Freeman, 220
Prevost, Maurice, 96
Prime Base Engineering Emergency Force (BEEF), 152
Prisoners of War and Missing in Action Memorial, Wright-Patterson Air Force Base, 328
Private Fair, 233
Prix de la Vitesse, 96
Procurement Section, Materiel Division, 77
Professional Race Pilots Association, 97
Professors' Tour, **259**,
Programmable Powdered Preform Process, 256
Project 100, Phase I, 259
Project Castle, 260
Project Corona Harvest, 153
Project Greenhouse, 260
Project Hardtack, 260
Project Lusty, 206; also see Operation Paperclip
Project Materials on the International Space Station Experiments (MISSE), 267
Project Plumbob, 260
Project Realign, 198
Project Redwing, 260
Project Rough Rider, 198, 200

Project Teapot, 260
Project Tumbler-Snapper, 260
Project Upshot-Knothole, 260
Propeller Unit, McCook Field, **73**
Property Requirement Division, Office of the Chief of the Air Service, 88
Property, Maintenance, and Cost Compilation Section, Fairfield Air Intermediate Depot, 88-90
　　name changed to Field Service Section, 89
　　Supply Division, staff, **89**
Propulsion Directorate, Air Force Research Laboratory, 247, 268, 282, **283**
Protective Integrated Hood/Masks, 181
Provisional Air Corps Maintenance Command, see Air Corps Maintenance Command
Provisional Maintenance Group Areas, under Provisional Air Corps Maintenance Command, 125
Public Law 102-419, 187; also see Dayton Aviation Heritage National Historical Park
Public Law 81-874, 250
Pueblo Incident, 198
Puerto Rico, 232, 262
Pulitzer Trophy Race, 1924 Air Races, 100
Pulitzer Trophy, 95, **97**
Pulitzer, Ralph, 97
Pulsed detonation engine program, **268**, 282, **283**
Punitive Expedition, 49
Purcell, Robert, **153**
Purdue University, 265
Puritan, 101
Purkey, Tom, **271**
Putt, Donald, 200, **201**
Quarles, Donald A., **144**
Quisenberry, G. B., **201**
Rabbit Patch, history of, **123**
Radar
　　AN/CPS-5 radar, 232
　　AN/CPS-6 Radar, 332
　　AN/FPS-3 radar, 332
　　AN/FPS-6 Height Finder Radar Tower, 332
　　APG-68 Attack Radar, 181
　　APG-70 Attack Radar, 181
Radio Dog, McCook Field, 34-35
Radio Section, Signal Corps, 115
Rafalko, Edmund A., 158
Raggio, Robert F., 158, 189, **219**
Rail Strike, 1951, 147
Railroad, switch in Building 1, 79
Raleigh, North Carolina, **252**,
Rall, Frederick T., Jr., dedication of Building 560, 327
Ranch-style homes, 228
Randolph Air Force Base, Texas, and 3500th U.S. Air Force Recruiting Wing, 198
Randolph Field, Texas, Robins commander of, 87
Rapid Area Maintenance Teams, 152
Rapid City Sub-depot, South Dakota, 135
Rapid Deployment Force, 159
Ravenna Ordnance Plant, Ohio, 153
Rawlings, Edwin E., 144, 196
RE/MAX, **277**, 279
Read, Albert C., 90
Readers Digest, 243
Readix computer, Foreign Technology Division, 207
Reagan, Ronald, 220
Reall, Jack, 180
Recruitment from Hispanic and other minority populations, 262
Red Scramble, 148
Red-Bar Battery Corporation, 42
Reece, Jane, photographer, 38
Reed Elsevier, 242

Reed, Thomas C., 167
Reeve, John C., physician, 15, 20, 26
Reeve, Scott, 256
Regiment, 5th Regiment, U.S. Colored Cavalry, 155
Reibold Building, Dayton, **21**
Reif, Louis, **277,**
Reinburg, George E. A., **81**
 and merger with Aviation Repair Depot, 83
 as commander of Wilbur Wright Field, 64
 commander of Fairfield Air Intermediate Depot, 88
 commander of Fairfield depot, 83
Remington Rand UNIVAC, 145
Repair and Maintenance Section, Materiel Division, 77
Republic of Moldova, 207
Republic of Yugoslavia, 181
Research and Technology Division, see Aeronautical Systems Division
Reserve Officer Training Corps (ROTC), 262
Reserve Tactical Fighter Group Provisional 9906, 320
Resort Airlines, 148
Retirees in area, Wright-Patterson Air Force Base, 249
Reynolds, Richard V., 270, 274, 282
 at Huffman Prairie Rededication Ceremony, **286**, 330-331
 attendance at wreath-laying ceremony, Woodland Cemetery, 17
 dates of Wright-Patterson command, 158
Rhein-Main Air Base, Germany, 142
Rhombic array antennas, 232
Rib-workshop program, 245
Rice, Judge Walter, 285
Rickenbacker Air Force Base, Ohio, 193, 194
Rickenbacker, Eddie, 155, 197
Ridenour, James, 187
Ridenour, Louis N., 145
Riedel, Doug, 180
Riemar, Thomas, 180
Rigney, James H. Jr., Dates of Wright-Patterson command, 158
Rike Family Foundation of Dayton, 274
Rike, Frederick H., 99
Rike's department store, 233, 235
Rike-Kumler Company, 99
Rike-Kumler Department Store, Dayton, 6, 24, **39**
 recruiting posters, World War II, 42
Ripley, Robert, 241
Ripley's "Believe It or Not", 241
Ritty, James, 7, 9, 237
 cash register, **7**
RiverScape, **233**, 237
Riverside, Ohio, 250
Robins Air Force Base, Detachment 13, 15th Weather Squadron, 190
Robins Air Force Base, in LOGAIR system, 148
Robins Board on Supply Accountability, 87
Robins Field, Georgia, named in honor of Robins, 87
Robins House, dedicated in honor of Robins, 328
Robins, Augustine W., 81
 and 1924 Air Races, 95, **98**
 and golf, 217
 and Model Airway System, 100, 102
 as commander of Fairfield Air Depot, **87**
 as commander of Wilbur Wright Field, 64
 career of, 87
 commander of Fairfield Air Intermediate Depot, **88**
 commander of Field Service Section, 102
 dedication of house in honor, 328
 lived in Building 88, 223
 proposed transport system, 107
 with family, **87**
Robotic fiber placement machine, 256
Rock and Roll Hall of Fame, 280
Rockwell Air Depot, California, 141
 4th Transport Squadron, 107
Rockwell Air Intermediate Depot, California, 86

Rodgers, Cal, 96
Rogers Dry Lake, Muroc, California, 200
Rogers, F. Michael, 161
Rollason, Sherry, 271
Rome Laboratory, New York, 172
Rome, New York, 247
Roosevelt High School, 234
Roosevelt, Eleanor, **162**
Roosevelt, Franklin D., 241
 budget authorization, 115
 call for aircraft production, World War II, 44
 elected as president, 40
 running mate of Cox during 1920 presidential campaign, 34
 Tennessee Valley Authority, 28
 war message to Congress, 45
Root, Amos, 2
Rose Bowl stadium, 244
Rosengarten, Irma, 262
Rosengarten, Nathaniel R., Colonel, 262
Ross, Jerry, 287
Ross, Mitchell, 180
Ross, Phil, 287
Roth, R. M., **201**
Round-the-World Flight, 1924, 90-95; also see Aircraft, D-WC, Douglas World Crusiers
 committee, 91
 crew photo, **93**
 logistical support, 92
 pilots, 93-94
 route map, **90**
Route 25, 233
Rowe, Joseph, 256
Royal Canadian Air Force, 91
Royal Flying Corps, 90
Royce, Ralph, 113
Rubenstein, Harold and Sophie, 265
Rubin, Lieutenant, **68**
Rubinstein, Harold, **265,**
Rudolph, Joseph H., **88**, 95
 and Aviation Repair Depot, 83
 commander of Speedway Aviation Repair Depot, **83**
Rumsfeld, Donald, **208**, **235**
Rusin, Michael, 180
Russia, pre-World War I aviation, 49
Russo, Dr. Vincent J., **256**, 258, **266**
Rwanda, 249
Ryan, Michael E., **329**
Ryterband, Dr. Louis, 265
Saber Shield 00-2, Ohio Task Force One, exercise, 178
Sachs, Edward "Red", 137
Sacramento Air Depot, California, 85
Sacramento Air Logistics Center, McClellan Air Force Base, 173, also see Sacramento Air Depot
Sacramento Air Materiel Area, California, role in Vietnam War, 152
Sacred Heart church, **236**
Sadler, Thomas, 186
Safer, John, 275
Saint Charles School, **289**
Salina Sub-depot, Kansas, 135
Saltzman, Charles McK., on Aeronautical Board, **49**
San Antonio Air Depot, 2nd Transport Squadron, 107
San Antonio Air Intermediate Depot, Texas, 86
San Antonio Air Logistics Center, Kelly Air Force Base, 173, also see San Antonio Air Depot
San Antonio Provisional Maintenance Group, 125
Sand Hill Road, 147
Sand Hill, 332
Sandia Laboratories, 200
Sarazen, Gene, 217
Sargent, Dr. Gordon, 256

Sarris, Aristides, 326
Saturn Ion, 279
Savage Arms Corporation, Utica, New York, 68
Scaled Composites LLC, 283
Scanning Electron Microscope Educators (SEMEDs), 265
Schantz, Adam, 17, 40
Schauer, Dr. Fred, 283
Schemmel, Rita, **277,**
Schenck, Robert C., **14,** 15
Schick, Johnny, 229
Schlatter, David M., **144,** 326
Schlei, Ed, **261,**
School of Business, 317
School of Engineering, 317
School of Engineering, University of Dayton, 264
Schoolhouse, NCR, 9, 25
Schraut, Dr. Kenneth, 260
Schroeder, Rudolph "Shorty", test pilot, 200
Schultze, LeClaire, 93, 94
Schuster Performing Arts Center, 233, **237, 238,** 272
Science and Technology Academies Reinforcing Basic Aviation and Space Exploration, **266**
Science Technology Engineering and Mathematics (STEM), 268
Scientific Advisory Board, 145
Scofield, Richard B., **158**
Scopes for Students program, 267
Scotland, 266
Scott Air Force Base, Illinois, 15th Weather Squadron, 190
Scott Field, Illinois, 66, 79
Scott Sub-depot, Illinois, 135
Seabold, Brian, 180
Sears Roebuck Corporation, 126
Sedalia Sub-depot, Missouri, 135
Segregation of military, 121
Sehorn, James E., and Hanoi Taxi, **195**
Seiler, Heidi, 180
Selfridge Air Force Base, Michigan, 56th Fighter Wing, 194
Selfridge Field, Mt. Clemens, Michigan, 55, 66, 67, 84
  aviation school, 79
  First Pursuit Group, 101
  Spaatz commander of, 100
  stop on Model Airway System, 100
Selfridge Sub-depot, Michigan, 135
Selfridge, Thomas, 24
Selma, Alabama, sub-depot in, 125
Selma, Indiana, **277**
Semkin, Fedor, **174**
Sensors Directorate, Air Force Research Laboratory, 221, 247
Sentinel Echo, 154; also see Operation Homecoming
September 11, 2001, terrorist attacks, 178-179, 182, 191, 288
Sharon, L. E., **88,** 98
Shartle, Alexander, 180
Shartle, Paul, 208
Shaw Sub-depot, South Carolina, 135
Sheffield Corporation, 310
Sherer, Carl, 52
Sherlock, Donald, 180
Shops Branch, Engineering Section, Materiel Division, 77
Shuey, Edwin, 266
Shupert, Steven, 180
Sidney, Ohio, 250
Siebenthaler, Bob, 275
Sierichs, Alan, 272
Signal Corps Aeronautical Division, see Aeronautical Division
Signal Corps Air Division Gunnery Section, see Gunnery Section
Signal Corps Aircraft Signal Agency, 314
Signal Corps Aircraft Signal Service, 314
Signal Corps Aviation Schools
  Augusta, Georgia, 48
  College Park, Maryland, 48, **49, 50**

curriculum, 53
  logistical support, 79
  North Island, San Diego, California, 48, **50, 51**
  Wilbur Wright Field, 55, 59-60, **69**
Signal Corps Aviation Section, see Aviation Section
Signal Corps Construction Department, see Construction Department
Signal Corps Equipment Division, see Equipment Division
Signal Corps Reserve, 31, 51
Signal Corps Supply Division, formerly Construction Division, see Supply Division
Signal Corps, Army, 257
  acquisition of airplanes, 48
  negotiations with Miami Conservancy District, 79
Signal Enlisted Reserve Corp, training of civilian pilots, 53
Simms Station, **3,** 54, 189
Simms, Buda, founding of Dayton Country Club, 3
Simms, Charles H., 3
Simms, W. A., 3
Simon, Chase, **259**
Sinclair Community College, 233, 243, **266,** 267
Sinclair, David A., **266**
Sines, Ed, 6
Sioux City Sub-depot, Iowa, 135
Sioux Falls Sub-depot, South Dakota, 135
Skeel, Burt E., **98,** 100, 327
Skinner, Lieutenant, **68**
Skydive Greene County, 332
Skyhook project, 232
Skyliner, newspaper, 169
Skyway Park, see Wright-Patterson Air Force Base, Housing
Skywrighter, newspaper, 169
Skywriter, newspaper, 169
Slavery
  "Black Ben," 16
  issue in Dayton, 16, 17
  meeting of Dayton Abolition Society, 16
Sloan, William A., Jr., 244, 287
Small Business Innovative Research (SBIR) program, 254
Smart dipstick, 260
Smith Road Bridge #03.730. 252
Smith, Al, II, 180
Smith, Casey, 271
Smith, Charles M. "Smitty", **114**
Smith, Chuck, **271**
Smith, Frank A., 327
Smith, Frank Hill, 241
Smith, Lowell H., **93,** 94
Smith, Wanda, **196**
Smithsonian Distinguished Lecturer Award, 263
Smithsonian Institution, 48, 189, 209, 263
  display of *Chicago* aircraft, 95
Smokestack industries, 227
Smyrna Sub-depot, Tennessee, 135
Sneed, Albert L., 81, 108
Snyder, E. Eugene, Dr. and Mrs, **258,**
SOARING, 275
Society of Mary, 265
Soldiers' Monument, Dayton, 19
Somalia, 249
Sons of Italy, Dayton Chapter, 150
South Carolina, 259
South Charleston, Ohio, 263
South Park Historic District, 237
South Vietnamese Air Force officers, **232**
South Williams Street, 285
Southard, Fred, 107
Southeast Asia
  Arc Light operations, 193
  Combat Lightning operations, 193
  Young Tiger operations, 193

Southern Ohio Lunatic Asylum, Dayton, **20**
Southwest Ohio Instructional Technology Association/Greater Miami Valley Educational Technology Council, 267
Southwest Ohio Weather Radar, 332
Southwestern Ohio Council for Higher Education (SOCHE), 262
    formerly Dayton-Miami Valley Consortium, 202
Spaatz, Carl A., 100, 155, 326
Space Command, activation of, 172
Space Shuttle *Columbia*, 179
Space Shuttle *Enterprise*, **168**
Spaulding, Reid, 180
Special Order 178, War Department, 76
Speedway Aviation Repair Depot, see Aviation Repair Depot (Speedway)
Speedwell Motor Car Company, Dayton, 22
Sperry, Lawrence, Sr., **21**
Spirit of Flight Award, 278
Spirit of Ohio, B-2 bomber, 187
Spirit of St Louis, 101
Splendid Vision, Unswerving Purpose, 246, **270**
Spokane Air Depot, Washington, Morris commander of, 141
Sporting Goods Industry Hall of Fame, 229
Spring Valley, Ohio, 250
Springboro, Ohio, 250
Springfield High School, Ohio, 134
Springfield Municipal Airport, 160
Springfield, Ohio, 250, 273, 285
    celebration of end of World War I, 75
    Fairfield Air Depot leased storage space in, 118
Spurgeon, Mike, **334**
Squadrons
    1st Aero Squadron, Chanute Field, 84
        Punitive Expedition, 49
        Signal Corps, 51
    1st Explosive Ordnance Disposal Squadron, 317
    1st Plant Maintenance Squadron, 314
    1st Provisional Air Transport Squadron, Fairfield Air Depot, 107
    1st Provisional Transport Squadron, Patterson Field, 130, 313
    1st Radio Squadron, 313
    1st Transport Squadron, Fairfield Air Depot, 107, 313
    1st Weather Squadron, 316
    2nd Aero Squadron, Signal Corps, 51
    2nd Photographic Squadron, 316
    2nd Provisional Squadron, Armorers' School, Wilbur Wright Field, 80
    2nd Transport Squadron, San Antonio Air Depot, 107
    2nd Weather Squadron, 313
    3rd Aero Squadron, Signal Corps, 51
    3rd Transport Squadron, Middletown Air Depot, 107
    4th Depot Supply Squadron, 313
    4th Maintenance Squadron [Mobile], 319
    4th Provisional Squadron, Armorers' School, Wilbur Wright Field, 80
    4th Repair Squadron, 313
    4th Supply Squadron, 313
    4th Supply Squadron, Mobile [Support], 319
    4th Transport Squadron, Rockwell Air Depot, 107
    5th Provisional Squadron, Armorers' School, Wilbur Wright Field, 80
    5th Squadron, Langley Field, 84
    5th Transport Squadron, 313
    6th Depot Repair Squadron, 313
    6th Depot Supply Squadron, 313
    6th Repair Squadron, 313
    6th Supply Squadron, 313
    7th Air Vehicle Repair Squadron, 315
    9th Transport Squadron, 313
    10th Repair Squadron, 313
    10th Supply Squadron, 313
    11th Squadron, Langley Field, 84
    11th Transport Squadron, Patterson Field, 131, 313
    12th Aero Squadron, combat credits, World War I, 71, 74
        Wilbur Wright Field, 59, 312

    13th Aero Squadron, combat credits, World War I, 74
        Wilbur Wright Field, 59, 313
    13th Transport Squadron, Patterson Field, 131, 313
    15th Squadron, Chanute Field, 84
    15th Weather Squadron, 190-191, 320
        Detachment 15, **192**
    17th Airborne Missile Maintenance Squadron, 318, 319
    17th Armament & Electronics Maintenance Squadron, 318
    17th Avionics Maintenance Squadron, 318
    17th Combat Defense Squadron, 318
    17th Field Maintenance Squadron, 318
    17th Motor Transport Squadron, Air Depot, 315
    17th Munitions Maintenance Squadron, 319
    17th Organizational Maintenance Squadron, 318
    17th Security Police Squadron, 318
    17th Squadron, Selfridge Field, 84
    18th Depot Supply Squadron, 314
    18th Intelligence Squadron, Detachment 1, 323
    19th Aero Squadron, missions in France, World War I, 74
        Wilbur Wright Field, 59, 312
    19th Bombardment Squadron (medium), Patterson Field, 131, 313
    19th Depot Repair Squadron, 314
    19th Depot Supply Squadron, 314
    20th Aero Squadron, combat credits, World War I, 74
        Langley Field, 84
        Wilbur Wright Field, 59, 312
    20th Depot Repair Squadron, 314
    20th Depot Supply Squadron, 314
    21st Depot Repair Squadron, 314
    21st Depot Supply Squadron, 314
    22nd Depot Repair Squadron, 314
    22nd Depot Supply Squadron, 314
    23rd Depot Repair Squadron, 314
    23rd Depot Supply Squadron, 314
    23rd Medical Service Squadron, 318
    27th Pursuit Squadron, Selfridge Field, 100
    32nd Weather Squadron, 316
    33rd Bombardment Squadron (medium), Patterson Field, 131, 313
    34th Bombardment Squadron (Heavy), 17th Bombardment Wing, 193, 318
    35th Medical Service Squadron, 320
    36th Transport Squadron, 314
    37th Transport Squadron, 314
    38th Transport Squadron, 314
    42nd Aero Squadron, redesignated Squadron I, 192
        Wilbur Wright Field, 59, 66, 80, 312
        Wilbur Wright Field, antecedent of 42nd Bombardment Squadron, 192
    42nd Bombardment Squadron, Wright-Patterson Air Force Base, 192, 193, 318
    43rd Aero Squadron, Wilbur Wright Field, 59, 312
    44th Aero Squadron, at Wilbur Wright Field, 59, 66, 80, 312
    44th Transport Squadron, 314
    45th Transport Squadron, 314
    47th Aero Squadron, at Wilbur Wright Field, 59, 66, 312
    49th Squadron, Langley Field, 84
    55th Depot Repair Squadron, 314
    55th Depot Supply Squadron, 314
    56th Depot Repair Squadron, 314
    56th Depot Supply Squadron, 314
    56th Fighter-Interceptor Squadron, formerly 97th FIS, 194, 317
    56th Tactical Training Squadron, 320
    57th Depot Repair Squadron, 314
    57th Depot Supply Squadron, 314
    66th Aviation Depot Squadron, 4043rd Strategic Wing, Wright-Patterson Air Force Base, 192, 317
    66th Munitions Maintenance Squadron, 317
    74th Aero Medical Squadron, 322
    74th Dental Squadron, 322
    74th Medical Operations Squadron, 322

74th Medical Support Squadron, 322
74th Surgical Operations Squadron, 322
85th Depot Repair Squadron, Patterson Field, 132, 314
86th Depot Repair Squadron, 315
87th Aerial Port Squadron, 320
    Cargo Deploying Function, 178, 320
881st Aero Squadron, 312
88th Civil Engineering Squadron, 322
88th Communications Squadron, 322
88th Depot Repair Squadron, Patterson Field, 132, 314
88th Equipment Maintenance Squadron, 322
88th Maintenance Squadron, 322
88th Mission Support Squadron, 321
88th Observation Squadron, 312
88th Operations Support Squadron, 322
88th Security Forces Squadron, Wright-Patterson Air Force Base, **186**, 320
88th Security Police Squadron, 320
88th Squadron, 312
88th Supply Squadron, 321
88th Transportation Squadron, 321
88th Weather Squadron, 175, 191, 321
89th Airlift Squadron, 322
89th Fighter Squadron, 320
89th Tactical Fighter Squadron, of 906th Tactical Fighter Group, 194, 320
94th Depot Supply Squadron, 315
97th Depot Supply Squadron, Patterson Field, 132, 314
97th Fighter-Interceptor Squadron, Wright-Patterson Air Force Base, 193-194, 316
    redesignated 56th Fighter-Interceptor Squadron, 194
98th Aviation Squadron, 314
103rd Airways and Air Communications Service Squadron, 315
137th Aero Squadron, Wilbur Wright Field, 106
149th Aero Squadron, Wilbur Wright Field, 59, 66, 312
151st Aero Squadron, Wilbur Wright Field, 59, 66, 312
159th Aero Squadron, training at Wilbur Wright Field, 66
162nd Aero Squadron, Wilbur Wright Field, 59, 66, 312
162nd Fighter Squadron, Ohio National Guard, 139, 241
163rd Aero Squadron, training at Wilbur Wright Field, 66, 312
166th Aero Squadron, Wilbur Wright Field, 59, 66, 312
172nd Aero Squadron, Wilbur Wright Field, 59, 66, 312
211th Aero Squadron, training at Wilbur Wright Field, 59, 66, 312
231st Aero Repair Squadron, Wilbur Wright Field, 68, 80, 312
246th [II] Aero Squadron, 312
246th Aero Squadron, Fairfield Aviation General Supply Depot, 79, 80, 312
    accommodations, **80**
255th Aero Squadron, 66, 312
256th Aero Squadron, 66, 312
257th Aero Squadron, 66, 312
258th Aero Squadron, 66, 312
259th Aero Squadron, 66, 312
260th Aero Squadron, 66, 312
265th Aero Squadron, 66, 312
287th Aero Squadron, 312
288th Aero Squadron, 312
315th Depot Repair Squadron, Patterson Field, 132, 314
338th Air Force Recruiting Squadron, 320, 323
342nd Aero Squadron, Wilbur Wright Field, 80, 312
342nd Communications Squadron, Depot, 316
345th Aviation Squadron, Patterson Field, 132, 314, 315
351st Aviation Squadron, 314
356th Airlift Squadron, 322
371st Aviation Squadron, 314
372nd Aviation Squadron, 314
401st Combat Logistics Support Squadron, Wright-Patterson Air Force Base, 180, 320
407th Service Squadron, Patterson Field, **132**, 314

412th Aircraft Maintenance Squadron Test Squadron Operating Location, 324
413th Flight Test Squadron Operating Location-HN, 324
435th Aviation Squadron, 314, 315
436th Aviation Squadron, 314, 315
437th Aviation Squadron, 315
445th Aeromedical Evacuation Squadron, 322
445th Aeromedical Staging Squadron, 322
445th Aerospace Medical Squadron, 322
445th Aircraft Maintenance Squadron, 322
445th Aircraft Squadron, 322
445th Civil Engineering Squadron, 322
445th Combat Logistics Support Squadron, 322
445th Logistics Readiness Squadron, 322
445th Maintenance Operations Squadron, 322
445th Maintenance Squadron, 322
445th Operations Support Squadron, 322
445th Security Forces Squadron, 322
463rd Aviation Squadron, 315
477th Base Headquarters & Air Base Squadron, 314
478th Air Base Squadron, Wright-Patterson Air Force Base, 222, 315
507th Aero Squadron, 312
555th Service Squadron, Springfield, Illinois, 132
612th Aero Squadron, Wilbur Wright Field, 80
615th Specialized Mission Squadron, 321
645th Civil Engineering Maintenance Squadron, 322
645th Civil Engineering Operations Squadron, 322
645th Materiel Squadron, 319
645th Mission Support Squadron, 321
645th Morale, Welfare, Recreation & Services Squadron, 321
645th Operations Support Squadron, 322
645th Security Police Squadron, 320
645th Supply Squadron, 321
645th Transportation Squadron, 321
645th Weather Squadron, 191, 321
669th Aero Squadron, Wilbur Wright Field, 80, 312
678th Aero Squadron, Wilbur Wright Field, 80, 312
788th Civil Engineering Squadron, 322
799th Aircraft Control and Warning Squadron, 317
807th Aero Squadron, 312
827th Aero Squadron, training at Wilbur Wright Field, 66, 312
836th Guard Squadron, 314
851st Aero Repair Squadron, Wilbur Wright Field, 67, 80, 312
874th Aero Repair Squadron, Wilbur Wright Field, 68, 312
901st Air Transport Squadron, 315
906th Civil Engineering Squadron, 321
906th Combat Support Squadron, 320
906th Consolidated Aircraft Maintenance Squadron, 320
922nd Air Refueling Squadron, Wright-Patterson Air Force Base, 192, 317
923rd Guard Squadron, 314
1062nd U.S. Air Force Communications Squadron, 316
1072nd Guard Squadron, 314
1126th Air Intelligence Service Squadron, 316
1289th Airways and Air Communications Service Squadron, 317
1350th Motion Picture Squadron, 317
1350th Photographic Service Squadron, 317
1401st Military Airlift Squadron, Detachment 2, Wright-Patterson Air Force Base, 163
1726th Air Transport Squadron, 315
1815th Test & Evaluation Squadron, 320
1828th Electronic Installation Squadron, 319
1914th Airways and Air Communications Service Squadron, 315
2046th Airways and Air Communications Service Squadron, 316
2046th Communications Squadron, 316
2700th Explosive Ordnance Disposal Squadron, 317
2702nd Explosive Ordnance Disposal Squadron, 317
2732nd Acquisition Logistics Operations Squadron, 204, 319
2750th Air Force Base, Headquarters Squadron, 140
2750th Air Police Squadron, 316

2750th Civil Engineering Squadron, 161, 319, 321
2750th Comptroller Squadron, 321
2750th Food Service Squadron, 316
2750th Logistics Squadron, 161, 319
2750th Materiel Squadron, 319
2750th Mission Support Squadron, 321
2750th Personnel Processing Squadron, 317
2750th Security Police Squadron, 319, 320
2750th Services Squadron, 319, 321
2750th Supply Squadron, 321
2750th Transportation Squadron, 321
2750th Women in the Air Force Squadron, 316
2762nd Logistics Squadron, 319
2762nd Maintenance Squadron, 319
2763rd Supply Squadron, 319
2764th Transportation Operations Squadron, 319
2790th Station Medical Squadron, 316
2863rd Ground Equipment Engineering and Installation Agency Squadron, 319
2956th Air Intelligence Squadron, 317
3000th Women in the Air Force Squadron, 316
3019th Base Services Squadron, 316
3020th Air Police Squadron, 316
3025th Management Engineering Squadron, 319
3030th Support Squadron, 319
3051st Air Base Support Squadron, 316
3060th Air Materiel Command Support Squadron, 316
3060th Support Squadron, 316
3061st Support Squadron, 317
3062nd Support Squadron, 317
3063rd Support Squadron, 317
3070th Computer Services Squadron, 319
3100th Specialized Mission Squadron, 321
3552nd U.S. Air Force Recruiting Squadron, 320
4043rd Airborne Missile Maintenance Squadron, 318
4043rd Armament and Electronics Maintenance Squadron, Wright-Patterson Air Force Base, 192, 317
4043rd Combat Defense Squadron, 317
4043rd Field Maintenance Squadron, Wright-Patterson Air Force Base, 192, 317
4043rd Organizational Maintenance Squadron, Wright-Patterson Air Force Base, 192, 317
4043rd Support Squadron, Wright-Patterson Air Force Base, 192, 317
4717th Ground Observer Squadron, 58th Air Division, 193, 317
4950th Armament & Electronics Maintenance Squadron, 320
4950th Field Maintenance Squadron, 320
4950th Organizational Maintenance Squadron, 320
4951st Test Squadron, 320
4952nd Test Squadron, 320
4953rd Test Squadron, 320
6500th Support Squadron, 317
6501st Support Squadron, 317
6592nd Management Engineering Squadron, 320
8401st Air Force Reserve Base Support Squadron, 318
8401st Materiel Squadron, 318
8525th Air Transport Squadron, 316
8548th Airways and Air Communications Service Squadron, 316
Cadet Squadron A, 59
Second Weather Squadron, 313
Squadron A, 231st Aero Squadron, 80, 312
Squadron B, 851st Aero Squadron, 80, 312
Squadron B, Aviation Armorers' School, Wilbur Wright Field, **68**
Squadron D, 4th Provisional Squadron, Armorers' School, 80
Squadron E, 5th Provisional Squadron, Armorers' School, 80
Squadron F, 2nd Provisional Squadron, Armorers' School, 80
Squadron I, formerly 42nd Aero Squadron, 80, 192, 312
Squadron K, 44th Aero Squadron, 80, 312
Squadron L, 246th Aero Squadron, 80, 312
Squadron M, 342nd Aero Squadron, 80, 312

Squadron N, 80
Squadron O, 80
Squadron P, 312
Squadron Q, 312
Squier, George O., **49**, 51, 54
SRIM press, 256
St Elizabeth's Hospital, Dayton, 20, 159
St. Anne's Hill Historic District, 237
St. Clair Street, and early settlement, Dayton, 12
Stamford, Connecticut, 242
Standard Aircraft Corporation, 59
Standard Oil Company, assistance with Around-the-World flight, 92
Standard Register, 268
STARBASE, see Science and Technology Academies Reinforcing Basic Aviation and Space Exploration
Start Your Engines art program, **258**
State Department, U.S., assistance with Round-the-World flight, 92
State of Ohio FIRST (For Inspiration and Recognition of Science and Technology), 267
State Route 128, 252
State Route 4, 233
Statue of Liberty, 244
Steele High School, **20**
    damage from flood of 1913, 26
    demolition, 233
    Katharine Wright as teacher, 5
    naming of, 15
Steele, James, 12
Steele, Mary, 15
Steele, Robert W., **14**, 15
Stephanides, Steven, 180
Stephens, R. L., **201**
Stevens Institute of Technology, 253
Stewart Air Force Base, New York, 193
Stewart, James T., 167, 220, **235**
    dedication of Building 557, 327
    at First Flight Ceremony on 75th Anniversary of powered flight, 170
Stinson Aeroplane Company, 23
Stinson Aircraft Corporation, 23
Stinson School of Aviation, 33
Stinson, Edward, **23**
Stinson, Jack, **23**
Stinson, Katherine, **23**
Stinson, Marjorie, **23**
Stitzel, David, 180
Stoddard Circle, 228
Stoddard Mansion, **25**, 36
Stoddard Manufacturing Company, **22**
Stoddard, Betty, **196**
Stoddard, Charles, 25
Stoddards, 228
Stout, Raymond K., and first automatic landing, 115
Strategic Air Command
    bomber alert facility, 232
    disestablished, 172
    Golf Course on West Ramp, 217
    joint tenancy agreement with AMC, 192
    nuclear-powered intercontinental Bomber, 218
    use LOGAIR system, 148
Strategic Arms Reduction Treaty, 171
Streett, St. Clair, 91
Stromme, J.L., **89**
Strongwind 99-1, Ohio Task Force One, exercise, 178
Stroop Agricultural Company, Dayton, 42
Stroud, Don, 287
Structural metals, 251
Structural Reaction Injection Molding (SRIM) press, 256
Structures Branch, Aircraft Laboratory, 260
Student Career Experience Program, 264
Student Temporary Employment Program, 262, 268

Sugar Camp, NCR, 43, 228
Summer Science and Engineering Enrichment Program, 268
Summit County, Ohio, 259
Sunrise Park, 233
Super magnets, 260
Supply Department, Fairfield Air Intermediate Depot, 89
Supply Division, Air Service, 77, 82, 88
Supply Division, Signal Corps, 58
Survivable Collective Protection System-Medical, 332, **333**
Svet, Frank
Swan, Leroy Amos, 106, 107
Sweet, George C., on Aeronautical Board, **49**
Switzer, J. M., mayor of Dayton, 75
Swofford, Ralph P., Jr., **144**
Sylvester, George H., General, **235**,
Symmes, John Cleve, 12
Systems Engineering Group, 318
Systems Research Laboratories, 332
T-2 Intelligence, Wright Field, 207
    in Building 12, 220; also see National Air and Space Intelligence
Center
Tactical Air Command, disestablished, 172
Taft, Bob, 258, 278
Taft, Hope, **266**
Taft, William Howard, 4, 24
Tagg, Brandyn, pictured anonymously, **253**
Take-off, newspaper, 169
Talbott, Harold E., Jr., 52
Talbott, Harold E., Sr., 52
Talbott, Nelson S., 326
Taliaferro Field, Texas, 60
Task Force 38, and USS *Dayton*, 42
Taylor, Charles "Chuck", **229**
Tech 21, 252
TECH CONNECT, 255
Tech Trek Mobile Research Laboratory, **267**
Tech Trek, 265, **267**
Technical Data Branch, Field Service Section, Materiel Division, **77**
Technical Section, Division of Military Aeronautics, 59, 74
Technology Council, 267
Technology Transfer Program (T²), 251
Technology transfer, 251, **252**, 254
Tecumseh, defeat at Battle of Thames, War of 1812, 13
Telegraph service, Dayton, 19
Telephone service, Dayton, 19
Tempelhoff Air Base, Germany, 244
Temperance Movement, Dayton, 20, 40
Tennessee Valley Authority, 28
Terpenning, James, **288**
Testing Squadron, Wilbur Wright Field, 74
Teterboro Field, Hasbrouck Heights, New Jersey, and Barling bomber,
70
Tetraethyl Lead Gasoline, developed by Kettering and Midgely, 32
Tewell, Dennis P., **158**, 215
Texas, 262
The Bishop's Boys: A Life of Wilbur and Orville Wright, 263
The Evening News, forerunner of *Dayton Daily News,* 3
The Journal Herald, newspaper, World War II, 42
The Mascot, newspaper, 169
The Wright Flyer, newspaper, 169
The Wright Idea, newspaper, 169
Third Frontier Project, 258
Thomas Dome exposure chambers, 221
Thomas Edison Award, **251**
Thomas Edison Program, 251, 255
Thomas, B. D., 59
Thomas, C. E., **88**, 95
Thomas, Charles, 98
Thomas, Edward, 180
Thomas, Eugene, 180

Thomas, Lowell, 91
    accounts of Round-the-World flight, 92
    attends First Flight Banquet, 170
    on Round-the-World flight pilot experience, 93
Thompson Trophy Race, 97
Thompson, Samuel, 12
Thrasher, C. O., **88**
Thunderbirds, U.S. Air Force, 167, 280, **281**
Thurling, Andrew, 244, **245**
Thurman, William E., 255
Time Flies, 280
Tinker Air Force Base, Oklahoma
    Detachment 1, 15th Weather Squadron, 190
    in LOGAIR system, 148
Tipp City, Ohio, 196, 250
Tobias, Connie, 245
Tokyo Raid, 45, 192
Tokyo, Japan, air raids on, 45, 192
Toledo, Ohio, 267
Topeka Sub-depot, Kansas, 135
Tops in Blue, at Festival of Flight, 171
Traffic congestion, 233
Traffic Section, Fairfield Air Depot, 124
Trailblazer Award, 263
Training Film Field Unit No. 2, Technical Data Branch, 77, 313
Training Film Production Laboratory, 313
Training
    aviation armorers, 67-68
    aviation mechanics, 65-67
Trans Western, 241
Transient Camp, Patterson Field, **110**, 111
Transportation Corps, Fairfield Air Depot, 124
Transportation, bicycles, 20
Travis Air Force Base, California, in LOGAIR system, 148
Treaty of Ghent, War of 1812, 13
Trepanier, Terry, 180
Triangle Park, North Field, potential for research and development
center, 33
TRICARE Heartland, Department of Defense Health Service Region 5,
323
Tritch, David, 180
Trotwood, Ohio, 161, 250
Troy, Ohio, 250, 284
    and Weaver Aircraft Company, 85
Tru 64 Unix operating system, 177
Truman, Harry S., 121, 187
Tudjman, Franjo, 185
Turbojet Engines
    research at Wright-Patterson Air Force Base, 145
    testing at Wright Field's Rabbit Patch, 123
Turner, Arthur, **93**, 94
Turner, Betty Stagg, 272
Tuskegee Airmen, 17
Twin Base Golf Course, Wright-Patterson Air Force Base, 217
Twin Base Rod and Gun Club, Wright-Patterson Air Force Base, 155
Twining, Nathan F., dedication of Building 641, 326-327
U. S Strategic Air Forces, Europe, 141
U.S. Air Force Academy, encampments on Wright-Patterson Air Force
Base, 149
U.S. Air Force Band of Flight, see Band of Flight
U.S. Air Force Centennial of Flight Office, see Centennial of Flight
Office
U.S. Air Force Central Audio-Visual Depository, 318
U.S. Air Force Central Motion Picture Film Depository, 318
U.S. Air Force Exhibit Group, 316
U.S. Air Force Hospital Wright-Patterson Air Force Base, see Wright-
Patterson Air Force Base, Medical Center
U.S. Air Force Marathon, see Air Force Marathon
U.S. Air Force Medical Center Wright-Patterson Air Force Base, see
Wright-Patterson Air Force Base, Medical Center

U.S. Air Force Orientation Group, see Orientation Group, U.S. Air Force
U.S. Air Force Radiological Health Laboratory, 318, 319
U.S. Air Force Recruiting Service, 318
U.S. Air Force Reserve Command, 248
U.S. Air Force Technical Intelligence School, 317
U.S. Air Force, see Air Force, United States
U.S. Airways Airbus 330, 245
U.S. Army Materiel Division Supply School, Fairfield Air Depot, 133
U.S. Army Nurse Corps, **277,**
U.S. Army, see Army, U.S.
U.S. Coast Guard, see Coast Guard, U.S.
U.S. Environmental Protection Agency, 166, 176
U.S. Military Academy, encampments on Wright-Patterson Air Force Base, 149
U.S. Munitions Standards Board, see Munitions Standards Board
U.S. Naval Reserve and U.S. Marine Corps Reserve, Naval and Marine Corps Reserve Center, 324
U.S. Navy, see Navy, U.S.
U.S. News & World Report, 243
U.S. Northern Command, creation of, 191
U.S. Postal Service, see Postal Service, U.S.
U.S. Weather Bureau, see Weather Bureau, U.S.
Umstead, Stanley, 200, **201**

Underground Railroad, 234
Unidentified Flying Objects, 207
Uniformed Services University of the Health Sciences, Bethesda, Maryland, 205
Unisys Company, Overland Park, Kansas, 260
United Aircraft Products Inc., 141, 310
United Brethren Building, later Centre City Building, **21**
United Brethren in Christ, Church of the, 263
United Brethren Publishing House, 266
United Color Press, 243
United Nations Command Truce Team, Korea, 141
United Nations, ban on Iraqi military flights, 181, 182
United States Air and Trade Show, 197
United States Air Force Band of Flight, see Band of Flight
United States Air Force Museum, 77, 249, **250**, 251, 263, 280
    aerial view 2003, **277**
    and Air Force Marathon, 188
    and Airborne Laser Laboratory, 201
    as tenant organization, 190, 324
    beginnings as Army's Aeronautical Museum, 205, 318
    celebration of 100th anniversary of powered flight, 190, 276, 279
    celebration of 75th Anniversary of Powered Flight, 170
    celebration of Air Force 50th Anniversary, 187
    Cold War Gallery, 210
    Eugene Kettering Gallery, 190, 210
    formation of Museum Foundation, 210
    Hall of Missiles, 210
    history of, 209-**210**
    Icarus, 274,
    IMAX Theater, 210
    Kettering Hall, 274
    loan of *New Orleans* aircraft, 95
    Modern Flight Gallery, 276
    Modern Flight Hangar, 210
    National Aviation Hall of Fame, 210
    on Aviation Trail, 284
    role in American Bicentennial, 167, **168**
    Space Gallery, 210
    Visitors Reception Center, 167
United States Army Corps of Engineers (MCLARE), 324
United States Consortium for Automotive Research, 256
United States Steel Corporation, 224
United States
    entry into World War I, 51
    pre-World War I aviation, 47-49

United Way, 250
Units
    2nd Combat Camera Unit, 316
    6th Photographic Technical Unit, 315
    15th Statistical Control Unit, 313, 314
    69th Army Air Forces Base Unit, 315
    72nd Army Air Forces Base Unit, 315
    342nd Finance Disbursing Unit, 316
    733rd Air Force Base Unit, 315
    733rd Army Air Forces Base Unit, 315
    2750th Air Force Base Unit, 316
    3100th U.S. Air Force Exhibit Unit, 316
    4000th Air Force Base Unit, Wright and Patterson fields, 140, 315, 316
        emblem, **146**
    4001st Army Air Forces Base Unit, 315
    4020th Air Force Base Unit, 315
    4020th Army Air Forces Base Unit, 315
    4021st Army Air Forces Base Unit, 315
    4100th Army Air Force Base Unit, Patterson Field, 126, 315
        merger with 4000th, 140
    4140th Air Force Base Unit, 316
    4140th Army Air Forces Base Unit (Research and Development Exhibition), 207; also see Orientation Group, U.S. Air Force
    4141st Army Air Forces Base Unit, 315
    4265th Army Air Forces Base Unit, Patterson Field, 315
        Separation Center, Patterson Field, 135
    4517th Army Air Forces Base Unit, 315
    5900th Army Air Forces Base Unit, 315
    5901st Army Air Forces Base Unit, 315 UNIVAC 1050-II Computer, 152
Universal Atlas Cement Division of United States Steel Corporation, 224
University of Dayton Research Center, 260-**261**
University of Dayton Research Institute, 243
    and National Composite Center, 256, 258
Upper Dayton View, 228
Urban renewal, 233, **236**
USS *Dayton*, 42
USS *Enterprise*, 153
USS *Hornet*, and Tokyo Raid, 45
Utah State University, 280
V Corps, Air Service, 74
Vallandigham, Clement, **16**, 18
Van Buren Township, Ohio, 228
Van Cleve Park, 36, **233**
Van Cleve, Benjamin, **12**, 15, 18
Van Cleve, John, 15, 45
Van DeGraff generator, **272**
Van Veen, Stuyvesant, 137
Vandalia, Ohio, 200, 231, 241, 250
Vandalia-Butler chambers of commerce, 282
Vandenberg, Hoyt S., 145
Vanloon, Hendrick, 188
Vaughn, George, 155
Vector Composites, Inc., 256
Vectren Dayton Air Show, 241, 245, 270, 276, 280, **282**
    2003 program cover, **281,**
Veridian Engineering, 332
Verville, Alfred, 40
Very Heavy Bomber Runway, Patterson Field, 118, 136
Veterans Administration, creation of, 11
Veterans Affairs Medical Center, Dayton
    cooperation with Wright-Patterson Medical Center, 11, 205
    on National Register of Historic Places, 11
Veterans Affairs Medical Centers, 11
Veterans Affairs, Department of, 11
Viccellio, Henry, 183
Victoria Theater, 237, **238**
Victory Theater, 237

Viemeister, Read, 274
Vietnam War
    return of prisoners of war, 154; also see Operation Homecoming
    role of Foreign Technology Division, 207
    role of Wright-Patterson Air Force Base, 152
Villa, Pancho, Punitive Expedition, 49
Visiting Officer Quarters, and Balkan Proximity Peace Talks, 183
Voinovich, George, 251, 256
Volandt, W., 91
Von Karman, Theodore, 36, 41
Waco aircraft factory, 284
Waco Historical Society, 85; also see Weaver Aircraft Company
Waco Learning Center, Troy, Ohio, 284
Waco Museum, 85; also see Weaver Aircraft Company
Waco, see Weaver Aircraft Company
Wade, Leigh, 93, 94
Wagner, Brian, 180
Walk, Dieter and Suzie, 265
Walker, Lois, and Arnold House Heritage Center, 222
Wallace-Kettering Neuroscience Institute at Kettering Memorial Hospital, 253, 254
Walton, Colonel James M., 255
Wapakoneta, Ohio, 263
War Department General Orders
    No. 20, 102
    No. 5
    No. 9, 103
War Department General Staff, 82
War Department Special Order 178, 76
War Department
    criteria for pilot training, 53
    established Air Corps Maintenance Command, 125
    establishes air intermediate depots, 86
    final approval for Round-the-World Flight, 93
    interest in aircraft, 2
    policy on air races, 97
    recognition of civilian employees, 122
    transfer of mechanics from Army to Signal Corps, 66
War of 1812, 13
War Relocation Authority, Cincinnati, 42
Ward, Edward, 326
Waring, Charles T., 81
    as commander of Wilbur Wright Field, 64
    commander of Wilbur Wright Air Service Depot, 80
    in charge of Wilbur Wright Field construction, 55
Warner Robins Air Depot, Georgia, 87
Warner Robins Air Logistics Center, Air Force Petroleum Office, Detachment 3, 324
Warren Dunes State Park, Michigan, 245
Warsaw Pact, 175
Washington County, Ohio, 259
Washington Court House, Ohio, 134
Washington Township, Ohio, 250
Waters, Jack, 85
Watson, Harold E., 206, 327
Watson's Whizzers, 206; also see Project Lusty; Harold E. Watson
Wattendorf, Frank, 41, 228
WAVES, see Women Appointed for Voluntary Emergency Service
Wayne County Sub-depot, Michigan, 135
Waynesville, Ohio, 250
Weather Bureau, U.S., 113, 198, 200
Weaver Aircraft Company (Waco), 85
Weaver, George E. "Buck", 85
Weaver, Walter R., 64, 66
Webcore Technologies, 259
Weeks, John W., 90
Welch, Clarence H., 81
Wellston Air Depot, Robins Field, Georgia, 87
Welsh, A. L., 24
West Carrolton, Ohio, 250

West Dayton, 33, 284
West Milton, Ohio, 161, 244
West Ramp, see Wright-Patterson Air Force Base, facilities
West, Jim, 334
Westerheide, Dr. John, 260, 261
Western Ohio Senior Executive Association, 262
Westover Air Force Base, Massachusetts, 148
Westvaco, 242
Whalen, Charles W., 170
Wheels of Dare, 271
Wherry-Spence Amendment to the National Housing Act, 211
White Consolidated Industries, 242
White House Millennium Council, 190
White, Dale, Sr., of Tuskegee Airmen, 17
White, Frederick A., 269
White, Janice, 288
Whiteman Air Force Base, Missouri, 244
Whitten, Mary, 15
Whitten-Brown, Arthur, 90
Widnall, Sheila E., 187
Wilberforce University, 262
Wilbourn, Arthur E., 62, 64
    1918 annual report, 58-60, 62, 63, 65
    as commander of Wilbur Wright Field, 55, 64
    description of problems at Wilbur Wright Field, 55
    management skills, 62
    transportation issues at Wilbur Wright Field, 58
    opinion of aircraft mechanics, 67
Wilbur and Orville Wright Commission, 168
Wilbur Wright Air Service Depot
    creation of, 74, 80, 140
    personnel strength, 74
    storage of materiel, 74
Wilbur Wright Field Officers' Golf Club, 217
Wilbur Wright Field, 262
    1917 flying season statistics, 59
    1922, 89
    42nd Aero Squadron, 192
    Air Service logistics hub, 33
    airplane testing, 69, 71
    and Barling bomber, 35, 70, 76
    and Building 88, Foulois House, 223
    Aviation Armorers' School, 67, 68, 74
        Squadron B, 68
    Aviation Mechanics' School, 65, 66, 67
    aviation records, 91
    aviation school, 55, 69, 80
        first season, 59-60
    becomes Area C, 139
    celebration of end of World War I, 75
    civilian work crew, 76
    commanders, 62, 64
    construction of, 54, 55, 58
    contributions to World War I, 71
    crash of Giovanni Pirelli, 149, 150
    creation of, 33, 47, 54, 55
    death of Frank Stuart Patterson, 35
    death of Pearson at 1924 Air Races, 100
    death of Skeel at 1924 Air Races, 100
    Deeds suggestion for, 31
    Department of Testing and Flying, 76
    description of, 55
    diagram of hangars, 55
    earliest Dayton military flying field, 165
    Engine Testing Rig, 63
    Engineering Department, 60, 62-63
        problems noted by Wilbourn, 62-63
    experimental testing programs, 86
    FFO airfield designator, 165
    first hospital building, 65